The Nalco Guide to Cooling Water Systems Failure Analysis

About Nalco, an Ecolab Company

Nalco, an Ecolab Company, is the global leader in water, hygiene, and energy technologies and services that protect people and vital resources. Ecolab delivers comprehensive solutions and on-site service to ensure safe food, maintain clean environments, optimize water and energy use, and improve operational efficiencies for customers in the food, healthcare, energy, hospitality, and industrial markets in more than 170 countries around the world.

The Nalco Guide to Cooling Water Systems Failure Analysis

Nalco, an Ecolab Company

Authored by

James J. Dillon

Paul B. Desch

Tammy S. Lai

Daniel J. Flynn Editor

Second Edition

McGraw Hill Education

New York Chicago San Francisco
Athens London Madrid
Mexico City Milan New Delhi
Singapore Sydney Toronto

1 2 3 4 5 6 7 8 9 0 CTP/CTP 1 2 0 9 8 7 6 5 4

ISBN 978-0-07-180347-2
MHID 0-07-180347-5

Sponsoring Editor
Robert Argentieri

Editorial Supervisor
Stephen M. Smith

Production Supervisor
Lynn M. Messina

Acquisitions Coordinator
Amy Stonebraker

Project Manager
Kritika Kaushik,
Cenveo® Publisher Services

Copy Editor
Kevin Campbell

Proofreader
Pia Oberoi

Art Director, Cover
Jeff Weeks

Composition
Cenveo Publisher Services

This book is printed on acid-free paper.

To our families, both professional and personal.

Harvey M. Herro and Robert D. Port, we are indebted to your laborious efforts to produce the first edition; it will forever be a classic work in the field. We truly appreciate the sacrifices that you made to make the guide a renowned success and for your guidance and inspiration.

Dawn, Shane, Kevin, and Tyler Dillon; Gretchen, Jenny, and Steven Desch; and Brian Ingram, you have our sincere gratitude for your understanding, patience, and support during the process. It provided an environment that allowed for completion of our work.

To our parents, we truly appreciate your devotion over the years, which helped us develop into the people we are today. Your love, support, and positive influence will forever be cherished.

Jim Dillon would like to extend special thanks to his father, Eugene F. Dillon, for inspiring him with his work ethic and guidance to pursue a career in engineering.

About the Authors

James J. Dillon obtained a B.S. degree in metallurgical engineering from the Illinois Institute of Technology. He joined Nalco, an Ecolab Company, in 1983 as a member of the metallurgy department, specializing in failure analysis both for industrial water and steam systems and for refinery, oil production, and various chemical process industry systems. He has authored numerous papers involving corrosion and failure analysis. Mr. Dillon is an active member of the National Association of Corrosion Engineers (NACE), for which he authored several papers, was chair for TEG-163X (Boiler Failure Analysis), and has been a member of several committees. He was a major contributor to the writing of the first editions of *The Nalco Guide to Boiler Failure Analysis* and *The Nalco Guide to Cooling Water Systems Failure Analysis.* He was a co-author of *The Nalco Guide to Boiler Failure Analysis,* Second Edition.

Paul B. Desch obtained a B.S. degree in metallurgical and materials engineering and a Ph.D. degree in metallurgical engineering from the Illinois Institute of Technology. Prior to joining Nalco, an Ecolab Company, he held positions at the Center for Materials Science and in the metallurgy group in the Materials Science and Technology Division of Los Alamos National Laboratory. Dr. Desch joined Nalco in 1997 as a member of the metallurgy group. He has authored or co-authored over 25 papers, including many on topics related to failure analysis. He has made presentations at professional societies such as the National Association of Corrosion Engineers (NACE) and the International Water Conference (IWC). He also has given training to a wide variety of industrial groups in the form of presentations that focus on failure and corrosion mechanisms, and has provided instruction for a course in failure analysis at the university level. He was a co-author of *The Nalco Guide to Boiler Failure Analysis,* Second Edition.

Tammy S. Lai obtained a B.S. in chemical engineering from Rice University, followed by a Ph.D. in materials science and engineering from Northwestern University. Dr. Lai joined Nalco, an Ecolab Company, in 2007 as a member of the metallurgy group. She has previous experience in the semiconductor industry and in teaching materials science at the university level. She was a co-author of *The Nalco Guide to Boiler Failure Analysis,* Second Edition.

Contents

Foreword

When cooling system failures occur, they are always at the most inopportune time. Not only are they an annoyance, but they also are an unplanned business expense and a safety concern. At Nalco, an Ecolab Company, we diligently work not only to extend the life of heat transfer equipment and minimize failures but also to understand why failure has occurred.

It can be difficult to determine the root cause of a failure because the factors that contributed to the failure are complex and often coexist. It takes an unbiased, investigative approach to determine the true root cause. By understanding the true nature of a failure, we are better positioned to prevent it from occurring in the future. Understanding the failure mechanisms and warning signs allow us to detect conditions favorable to failure and make recommendations to correct them, effectively operating a predictive failure process.

This second edition of *The Nalco Guide to Cooling Water Systems Failure Analysis* is a comprehensive, richly illustrated, authoritative field guide to cooling system failure mechanisms. This book details more than 100 different case histories of failures in cooling systems. The case histories combined with the in-depth discussions of the mechanisms involved will help you understand the aspects of why failures occur and how you can prevent them in your systems. We trust this book will guide you in your efforts to maintain the integrity of your equipment and extend its life while providing a continued safe and productive working environment.

Tim Mulhere
Executive Vice President and President, Global Water and Process Services
Nalco, an Ecolab Company

Preface

During our discussions about the changes between the first and second editions of this book, we centered on several areas of improvement for the benefit of the reader. We restructured the book into five separate sections: Cooling Water System Design and Operation, Waterside Corrosion, Cracking, Mechanical Damage, and Material and Design Issues. In each of these sections we grouped content where it made sense to do so, and expanded the discussions, including additional case histories as we looked deeper into each topic area. Our expanded discussions include significant new material. The 10 chapters covering the various materials most commonly found in cooling systems in Sec. 1, Cooling Water System Design and Operation, are completely new to this edition, as are the chapters on Localized Corrosion and Pitting, and Equipment Design and Operation.

In Sec. 1 where we discuss materials found in cooling systems, we cover the types of materials or various alloys, how they are prone to fail, and preventive measures that can be employed. In the remaining chapters, our content construction remains consistent with the first edition of this book. Each chapter includes the following topics, which we discuss in detail.

General description: Failure mechanisms are described. The causes and sources of attack are discussed.

Locations: Failure locations are pinpointed.

Critical factors: The most important or especially critical factors influencing failure are listed. Attack morphology and identification are stressed.

Identification: Visual inspection procedures are emphasized. The methods by which detailed metallographic procedures might elucidate failure mechanisms are also discussed.

Elimination: Necessary preventive and/or corrective steps to eliminate or reduce damage are listed and explained.

Cautions: Comparisons are drawn between failure mechanisms that produce damage with a similar appearance.

Related problems: Related problems and damages are described.

Each chapter finishes with a selection of case histories. These examples are chosen to illustrate both similarities and differences between actual and "classical" failures. These histories provide diverse perspectives on how failures actually occur.

We have made liberal use of graphics, tables, and photographs to illustrate the concepts and failures. The majority of illustrations and photographs used are new to this edition. The changes we have made to the content and presentation should make the book appealing across a broad section of industries and cooling systems.

While this book will help you in diagnosing failures, it is not a substitute for rigorous laboratory failure analysis. Many mechanisms are properly identified only by this type of in-depth evaluation. This book, however, will assist in the understanding of the mechanisms and locations of failures, allowing for steps to be taken in prevention.

DANIEL J. FLYNN
Editor

Acknowledgments

Undertaking the writing of the second edition of this book involved a great many individuals beyond the primary authors. When we started, we realized we would be requesting assistance from a good number of people. As the project progressed, the numbers grew as we cast our net further into the many disciplines and departments throughout Nalco, an Ecolab Company. We would like to take this opportunity to acknowledge the participation and efforts of the individuals involved with the second edition of *The Nalco Guide to Cooling Water Systems Failure Analysis*.

The authors of the first edition of this book, Harvey M. Herro and Robert D. Port, deserve special recognition. Their efforts, insight, and invaluable tutelage over the years have made this extension of their initial work possible. The authors of the second edition, James J. Dillon, Paul B. Desch, and Tammy S. Lai, worked diligently at updating existing materials, writing new case histories, and expanding the depth, breadth, and scope of the book. Together these authors have completed in excess of 13,000 failure analyses.

Our thanks to Anthony C. Biondo, Steve Hatch, and Larry Berger for managerial support throughout the process. We would also like to express our appreciation to Daniel Meier, Jasbir Gill, Craig Myers, Tzu-Yu Chen, Laura Rice, Barbara Moriarty, Andrew Cooper, Amit Gupta, F. Philip Yu, Peter TenEyck, Ben Brock, James Haff, Eric Rosenow, Darrell Wynkoop, Ron Kochik, Brad Stuart, Brian Jenkins, Barry Carroll, Jim Boak, Tony Mackovski, Heather Dubois, Richard Jacobs, David Herman, John Helge, Jim Halley, L. J. Aspinall, Brian Ornay, Don Johnson, John Pilsits, Kevin Boudreaux, Larry Hill, and Peter Jenner for ongoing in-depth discussions over the years that have helped further the understanding of the nature of failures. We are grateful to Steef Vrijhoeven, Dorothy Collins, Robert A. Gagliano, Michael S. Gagliano, Xiuping Jiang, Anthony Biondo, Jennifer Myczek, Daniel Meier, Elisa Luth, F. Philip Yu, and Peter TenEyck for their contributions to the case studies utilized throughout the book. Additional thanks are extended to Kevin Gehan, Daniel Meier, and Daniel Flynn for review of various chapters in the book and to Andrew Sorensen for legal review. We thank Greg Coy from G. L. Coy and Associates and Susan Molloy-Vesely for graphic illustration and assistance with figures.

We would like to acknowledge the contributions to this edition of the Global Customer Analytical Department, including the Deposit, Water, Microbiological, Coupon, and LIMS Support Groups. Specific mention should be made of Richard W. Cloud and Joan B. Lewis for scanning electron microscopy; John McGinn, Elizabeth Mueller, and Phil McKittrick for infrared spectroscopy; Carrie Keller-Schultz, Vic Keasler, and George Stearns for quantitative PCR and prokaryotic speciation analyses; Hayes O. Shackelford and Michael J. Danko for sample preparation and photographic documentation; and Karen Baumann for assisting with many needs.

SECTION 1

Cooling Water System Design and Operation

Introduction to Cooling Water Systems

Due to the fact that water is environmentally safe, abundant, easy to handle, and provides much better heat transfer than air, it is the most practical and commonly used medium for cooling of industrial processes. However, the effective use of water for cooling applications results in challenges in design and operation. Most components in a cooling water system are constructed from metallic materials, especially heat transfer equipment. These components can suffer failures due to many forms of corrosion, cracking, and other damage mechanisms. Nonmetallic components in the cooling water system may also experience degradation and failure. The varied types of damage are caused by differences in cooling water system design, temperature, flow, water chemistry, alloy composition, and operation.

Water chemistry is affected by dissolved gases, minerals, suspended solids, and biological matter. Water chemistries can be highly variable depending upon their source. For example, soft waters contain low concentrations of dissolved minerals, hard waters contain high dissolved mineral concentration, and brackish and seawaters contain very high concentrations of chloride. Natural water sources can vary seasonally and can be adversely affected by conditions such as drought and pollution. Pretreatment processes such as softening, clarification, demineralization, and reverse osmosis may be used to produce more suitable water chemistries for particular systems. Treatment chemicals are added to the cooling water to control corrosion, deposition of dissolved and suspended solids, and microbiological activity.

In addition to considering the effective transport and use of cooling water, design of the cooling system must also address contact with process fluids. An almost endless variety of process stream chemistries may be involved in cooling water systems. Facilities in the refinery and chemical process industries can employ hundreds of heat exchangers at a single plant that have many different process stream chemistries. The main purpose of this text is to discuss conditions on the cooling water side, which is formidable enough.

Types of Systems

Cooling water flow is either once-through or recirculating. Once-through cooling takes water from a plant supply, passes it through a cooling system, and sends it to a receiving body of water. Sources of water used in once-through systems commonly include river, pond, or lake water, which may contain significant concentrations of suspended solids. In addition, once-through cooling water may include brackish and seawater that contains high concentrations of aggressive chloride ions. Although dissolved and suspended solids do not concentrate in once-through systems, increasing temperatures in heat exchangers can result in deposition (scaling) if not controlled. Most deposition will result in decreased heat transfer, thus reducing heat exchanger efficiency. Moreover, deposition can promote corrosion on metal surfaces in some cases.

Recirculating cooling water systems are considered to be either closed or open. In a truly closed system, the cooling water is completely sealed off from the environment. The cooling water is partially exposed to the environment in an open system. The three basic types of cooling water systems—once-through, closed recirculating, and open recirculating—are shown schematically in Figs. I.1 to I.3.

Open recirculating systems employ cooling towers and spray ponds, which allow the dissipation of enormous heat loads. Typical cooling tower designs are shown in Figs. I.4 and I.5. In most designs,

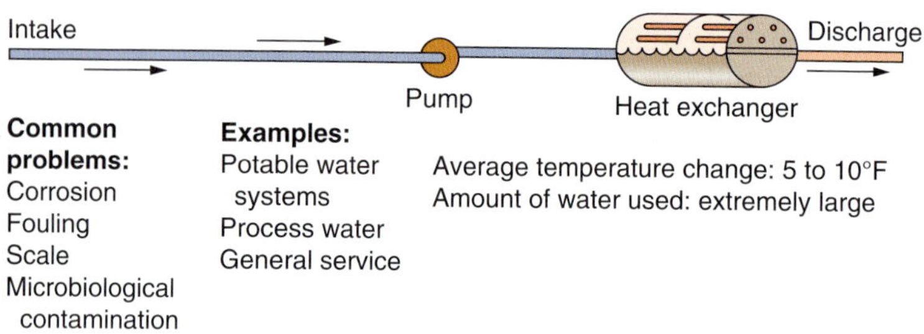

FIGURE I.1 Schematic of once-through cooling system.

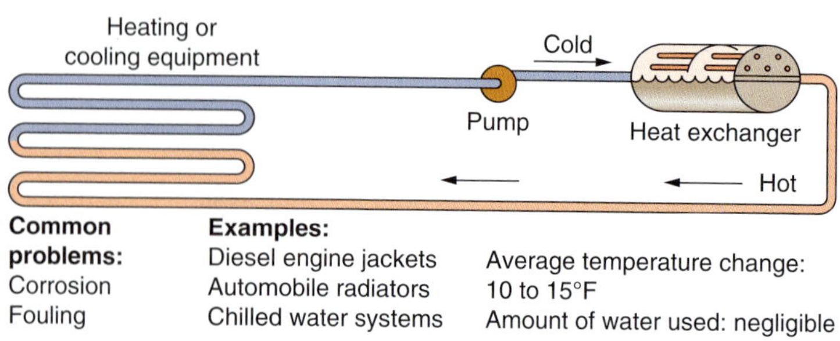

FIGURE I.2 Schematic of closed recirculating cooling system.

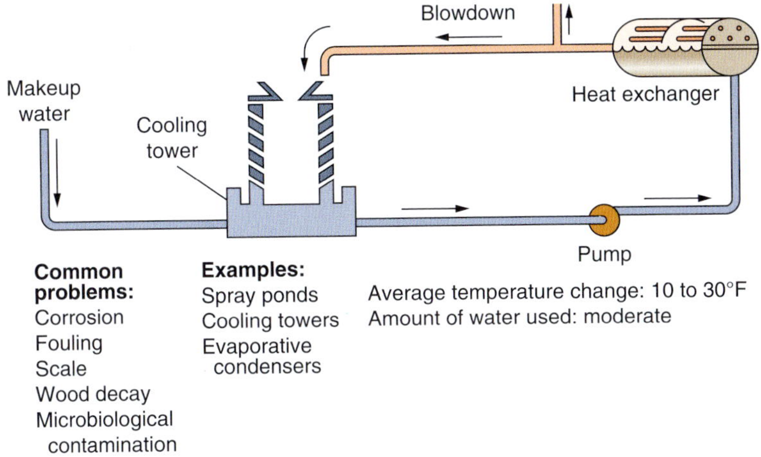

FIGURE I.3 Schematic of open recirculating cooling system.

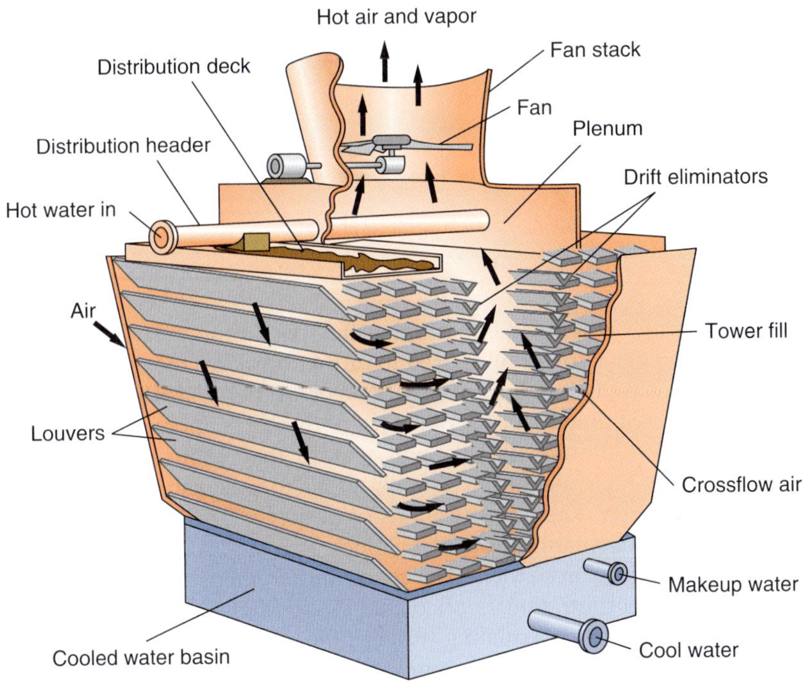

FIGURE I.4 Cooling tower with induced draft.

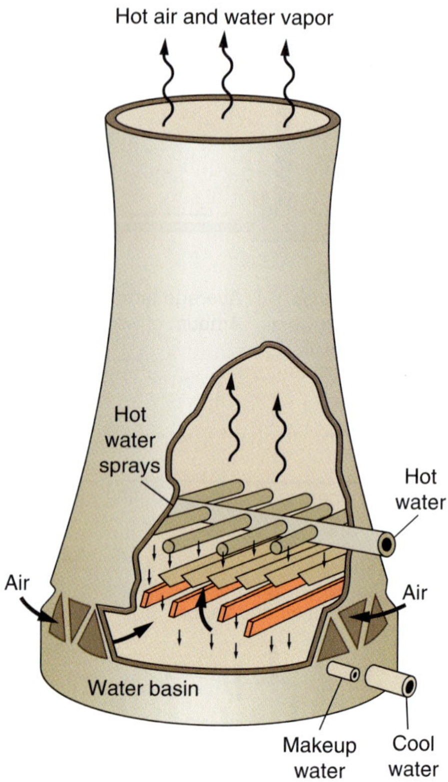

FIGURE I.5 Hyperbolic cooling tower with natural draft.

the cooling water is sprayed into droplets that pass through large volumes of flowing air. Tower fill is used to further break up the droplets as they fall by gravity into the tower basin. Open systems are designed to lose significant amounts of cooling water due to evaporation and drift into the environment. This will require that relatively large amounts of makeup water are continually added to replenish the losses due to evaporation. The concentrations of dissolved and suspended solids in open recirculating systems will increase over time due to the evaporation of water from them. The concentration ratio, also referred to as cycles of concentration, is calculated by dividing the concentration of any ion in the recirculating water of the cooling system by the concentration of that ion in the incoming makeup water. An increase in cycles of concentration is commonly known as cycling up. As a consequence, scaling, fouling, and associated corrosion are generally more significant in open systems than in true closed systems. Blowdown of the cycled-up or concentrated water is employed as an operational control to maintain appropriate water chemistry.

FIGURE I.6 Dark oxide and deposit lobes on a copper continuous caster mold from a steel-making operation. Since heat transfer is high, even small amounts of deposit are unacceptable.

In contrast to open recirculating systems, true closed systems should neither lose nor gain water during service. However, it is not uncommon for small amounts of water to leak from a closed cooling water system. It is necessary to replace the lost water with makeup water to maintain system volume. In closed systems it is common practice to use softened or demineralized water for filling the system and replacing water lost during operation. Since water losses are typically small, the total waterborne material entering the system is limited. Thus, deposited minerals accumulate at a much slower rate than in open systems, where large amounts of makeup water are added. Often, closed systems are used in critical cooling applications where even a small amount of deposit cannot be tolerated. Steel continuous caster molds that experience very high heat flux are a good example of this practice. Deposition and oxide layers on a copper caster mold are shown in Fig. I.6.

Complex cooling water systems may employ combinations of different types of closed, open, and once-through systems. Many heating, ventilation, and air conditioning (HVAC) systems use this approach. As an example, an open recirculating water loop joined to a closed loop used in an air conditioning system is diagrammed in Fig. I.7.

Cooling System Equipment

Cooling water system designs and equipment vary widely depending upon application. Included are cooling tower components, pumps, transfer piping, valves, and heat exchangers.

Cooling towers have many different designs, which depend on the system volume and cooling capacity required for the system.

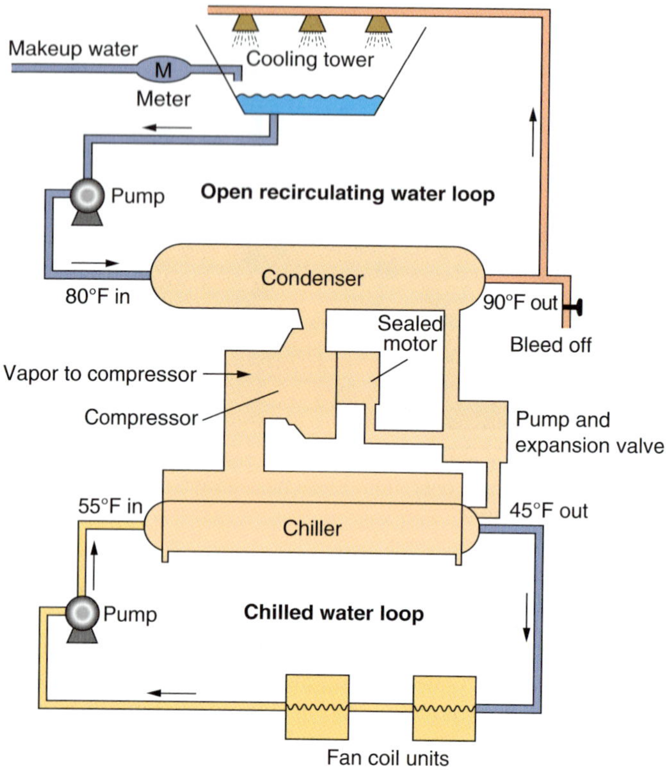

FIGURE I.7 Schematic of all components of a complete air conditioning system.

They are generally constructed with structural members that support the weight of the cooling tower components and the water that flows through the system. The components typically include the inlet water distribution system to disperse the water, tower fill to break up the water into droplets, mist eliminators to reduce the amount of entrained water lost to the environment, and a basin at the bottom to collect the water. Corrosion issues are most common in the basin of the tower. Areas that are wetted by mist can also be susceptible to attack. Screening and strainers are used to prevent large debris from being transported into the system. Pumps and transfer piping are used to distribute the cooling water throughout the system. In addition, chemical feed lines are required for dosing chemicals of the treatment program into the cooling water.

By far, the greatest variety of designs in cooling water systems involves heat exchangers. Heat exchangers have two common designs: shell-and-tube and plate-and-frame. Cooling is most often provided by heat transfer through metallic components in the exchangers.

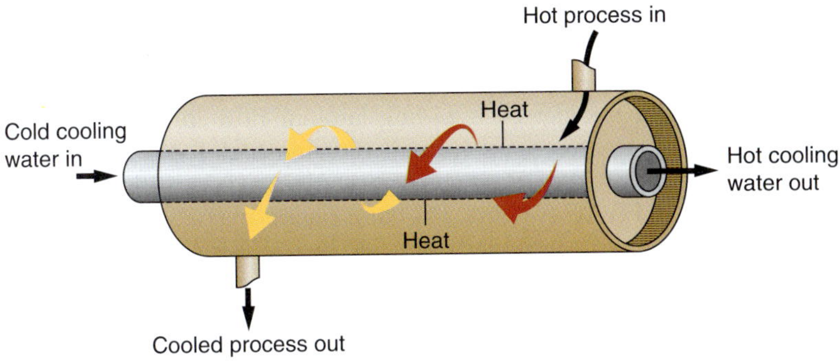

FIGURE I.8 Example of heat transfer through a single tube.

Shell-and-tube heat exchangers are very common. A schematic illustrating the heat transfer process through a single tube is shown in Fig. I.8. In most systems, cooling water flows through the tube, or tubeside, while the process fluids flow across the external surface or shellside. Under specific conditions, as when the process fluid is under high pressure and must be properly contained, it is necessary for the process fluid to flow through the tubeside and the cooling water to flow across the shellside. Another situation in which the process fluid may be on the tubeside occurs when its corrosive effects are more aggressive than those of the cooling water.

It is common for a bundle of several heat exchanger tubes to be installed in parallel using a shell-and-tube design; see Fig. I.9. Tubes

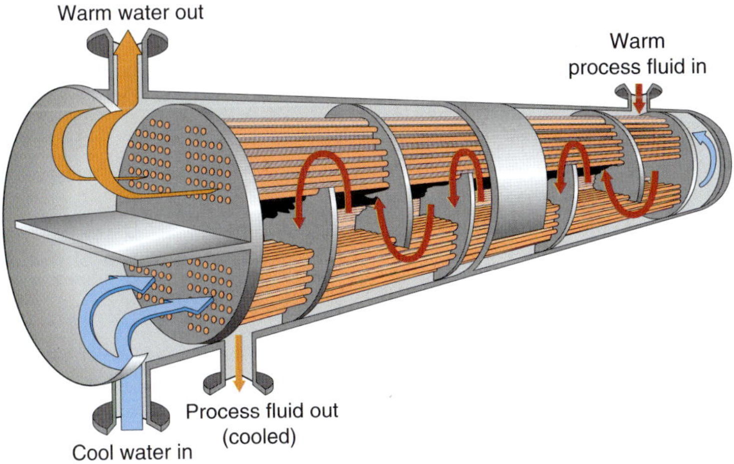

FIGURE I.9 Example of a shell-and-tube heat exchanger, with cooling water flowing through the tubes and process fluid on thc shell side.

are generally smooth on the internal and external surfaces, although some may have external fins or internal rifling that appreciably increases surface area for improved heat transfer. The internal rifling also improves heat transfer by avoiding laminar flow in the tubes. Tubes where these features are formed into the metal surface are known as enhanced tubes; these are commonly used in chiller and condenser units in HVAC systems. The components in shell-and-tube exchangers that are most frequently affected by corrosion and failure mechanisms are the tubes. Tube sheets, baffles, and water boxes may also experience damage.

Plate-and-frame heat exchangers consist of stacks of contacting corrugated metal sheets that are called plates, as shown in Figs. I.10 to I.12. The plates are squeezed together with gaskets between them to create watertight seals and to direct fluids from the corner ports across the sheet surface. Heat is transferred by passing cooling fluids and process fluids between alternating plates. Due to the tight flow passages and high efficiency of this design, the plates are particularly susceptible to crevice corrosion at locations such as plate-to-plate contact points and gaskets; therefore, plate alloys must be sufficiently resistant to crevice corrosion in the water and process environments. In addition, fouling of the surfaces needs to be prevented. The use of water with high levels of dissolved minerals, suspended solids, or microbiological activity, as in once-through

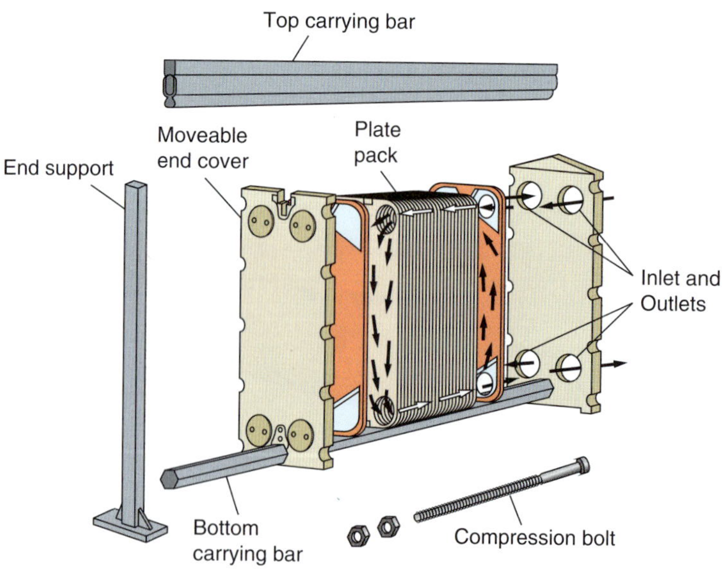

FIGURE I.10 Diagram showing construction elements of a plate-and-frame heat exchanger.

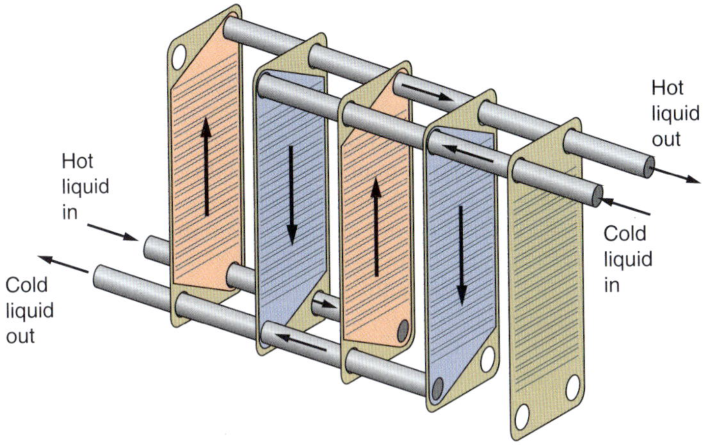

FIGURE I.11 Schematic illustrating the flow of hot and cold fluids through a plate-and-frame heat exchanger.

FIGURE I.12 Example of a corrugated plate from a plate-and-frame heat exchanger. Note the gasket along the edge.

cooling water systems, should generally be avoided. The use of such waters may require frequent cleaning of the plates to restore exchanger efficiency and avoid corrosion issues.

Some systems, such as those used in the continuous casting of steel and aluminum, supply cooling through direct contact of the cast metal with cooling water. Such systems are more challenging to treat since the water is more susceptible to contamination by materials used in the process.

Materials of Construction

Selection of materials for components throughout the cooling water system requires consideration of many parameters. One major parameter, if not often the most important, is cost. The overall cost of a component will include the material costs, fabrication and installation costs, and design costs. A profitable process requires economically sound decisions that do not reduce the reliability of equipment, so that lost production time can be prevented. The choice of materials may seem limitless, as virtually all commercially available alloys have been used in cooling water systems. However, past experience with materials in the intended service applications will provide insight into suitable choices for each set of conditions. Table I.1 lists

Alloy	Common Uses
Carbon steel, low-alloy steels	Transfer lines, heat exchanger shells, baffles, pump components, heat exchanger tubing, fan blades and shrouds, valves, screens, fasteners
Cast iron	Pump housings and impellers, valves, plumbing fixtures, large-diameter pipe
Galvanized steel	Cooling tower components, fan blades and shrouds, transfer pipes, plumbing fixtures
Bronzes	Bearings, pump impellers, special-purpose tubing, heat exchanger tubing, screens
Copper and brasses	Heat exchanger and surface condenser tubing, bearings, valve components, gaskets, brewing equipment
Cupronickel	Heat exchanger and surface condenser tubing
300 series stainless steel	Plate-and-frame exchangers, heat exchanger and surface condenser tubing, pump impellers and housings, heat exchanger shells, brewing equipment, drying equipment, oil refinery still tubes, pulp and paper equipment
Duplex stainless steel	Corrosion and stress-corrosion-cracking resistant tubing
400 series stainless steel	Special-purpose heat exchanger tubing, plate-and-frame heat exchangers
Aluminum	Heat exchanger tubing, transfer piping, plastic molds
Nickel	Special-purpose heat exchangers, caustic handling
Titanium	Fresh and seawater condenser tubing, severe corrosion condition equipment

Table I.1 Common Alloys Used in Cooling Water Systems

different metals and alloys and some of their applications in a cooling water system environment. Although metal alloys are by far the predominant materials of construction for most components, nonmetallic material usage has increased over the years for certain applications.

The first step in selecting materials is to identify candidates that satisfy criteria such as mechanical strength, corrosion resistance, and appropriate physical properties (such as thermal conductivity) for an application. After a list of candidate materials is compiled, the most cost-effective approach can be determined. Expected service life and maintenance during service must also be considered in the decision. Periodic monitoring, such as visual inspection and nondestructive testing, can be used to assure good condition of the tubes and unit during their service life. Corrosion coupons can be used to provide baseline information about corrosion rate and type (i.e., localized or general) for various metallurgies in a system, and to detect possible changes in the rate and type of corrosion.

Some important parameters that must be considered for exposure of a material to a cooling water environment include water quality, temperature, and operation. Water quality includes factors such as pH, dissolved and suspended solids, scaling tendency, chloride content, chlorination practices, and possible contaminants. Elevated temperatures will generally increase both the propensity for corrosion and deposition and corrosion rates. In addition, plastics can have considerably reduced strength at elevated temperatures, limiting their use. Operating practice for the system, such as flow rates through piping and tubing, can impact the successful use of a material. For instance, low flow and stagnant conditions will generally promote deposition and many forms of corrosion, while tubing and components subjected to excessive flow rates and/or turbulence can experience erosion-corrosion.

Materials used for heat exchangers and other cooling applications are either innately corrosion-resistant for the expected environment or have a strong tendency to passivate in a particular cooling water environment that employs proper chemical treatment. Materials in the latter category tend to be less costly and are often the preferred choice. By far the most commonly used material for transfer piping and heat exchanger tubing is carbon steel, primarily due to cost. Certainly, the chosen alloy must also possess the necessary material properties and corrosion resistance to the environment on the process side. Heat transfer is also important. Table I.2 compares the thermal conductivity of alloys typically used for heat exchanger tubing. Note that copper and some of its alloys have some of the highest thermal conductivities, making them preferred materials for efficiency in heat transfer. The high thermal conductivity of aluminum makes some aluminum alloys favored materials for use in water-cooled molds in the plastics industry. However, the alloy composition, water chemistry, and system

Alloy	Thermal Conductivity
Copper	225
Aluminum	131
Yellow brass	69
Admiralty brass	64
Aluminum brass	58
Nickel 200	33
Carbon Steel	30
90/10 Cupronickel	23
Low-alloy steel (1 Cr, ¼ Mo)	19
70/30 Cupronickel	17
410 Stainless steel	13
Monel 400	13
Titanium	12
304 Stainless steel	9
316 Stainless steel	9
Inconel 600	9
Hastelloy C	7

TABLE I.2 List of Thermal Conductivities of Some Alloys Used in Heat Exchanger Tubing at 70°F in kBtu/(hr · ft · °F)

design and operation are important factors that must be considered for successful use without corrosion issues.

Shell-and-tube heat exchangers can be fabricated from a variety of different alloys. Surface condensers typically use more corrosion-resistant alloy tubing. These alloys include austenitic stainless steels, which are widely used in freshwater service, and copper alloys, such as admiralty brasses, aluminum brasses, and cupronickels that may be used in freshwater and seawater. Titanium is commonly used in seawater service.

Corrosion-resistant alloy plates used in plate-and-frame heat exchangers are most commonly made from austenitic stainless steels. However, nickel alloys and titanium are also found.

As mentioned previously, one of the most important factors in choosing a replacement alloy for heat exchanger tubing is service experience. Causes of past corrosion and damage should be identified, and the service conditions should be reviewed. Mechanical, operational, and chemical factors must be considered. If possible, a few candidate alloy tubes can be installed in a test rack, preferably at the site, for a sufficient period of time to determine if they are suitable. After the test, candidate tubes should be carefully examined for signs of damage. Although old corrosion problems may be solved by an alloy substitution, new troubles may occur that are as bad as or worse than original problems.

Nonmetallic materials used in cooling water systems include polymeric material such as rigid plastics and elastomers. The fill used in cooling towers is often wood or plastic. Both materials can suffer degradation based on water chemistry, such as excessive chlorination. Microbiological activity, such as fungi, can promote appreciable attack on the wood. Elastomeric materials are used to create watertight seals and are used in valves, pipe couplings, mechanical seals for pumps, thread sealants, and gaskets for plate-and-frame heat exchangers. Chemically resistant polymer coatings are also used to protect metal components from corrosion by providing a barrier between the metal and the environment.

Chemical feed systems require the materials used for tanks, pumps, valves, and feedlines to be compatible with the chemical being delivered to the system. Many chemicals that act as corrosion inhibitors at dilute concentrations can be very aggressive in their neat (concentrated) forms. Stainless steel alloys typically provide sufficient corrosion resistance in most chemical feed systems. However, some formulations contain species that are aggressive to stainless steel and require the use of more corrosion-resistant alloys or chemically resistant nonmetallic materials for transport.

The most common alloys and nonmetallic materials used in cooling water systems are described in the following sections. Additional information concerning cooling water systems is provided in *The Nalco Water Handbook,* 3d ed. (ed. D. J. Flynn, McGraw-Hill, 2009).

CHAPTER 1

Carbon and Alloy Steel

Introduction

Steels are malleable, iron-based (ferrous) alloys that contain less than 2 wt% carbon. There are many different types and grades of steel, depending on alloy content. Steels are by far the most frequently used materials in cooling water systems due to their strength, ease of fabrication, and cost. This section will discuss the most commonly used steels, known as carbon steels and low-alloy steels. These steels will typically react with water in an oxygen-containing environment. Their susceptibility to corrosion must be addressed with respect to the application. Carbon steels are often the most economical choice for components in cooling water systems, provided that corrosion can be controlled so that they will have an acceptable service life. Approaches to limiting corrosion include chemical treatment to reduce corrosion rates in water-containing environments and preventing the metal from contacting the environment, such as by applying a protective coating.

Carbon steel is used for many components in cooling water systems. A major application is the use of piping for water intake into the system and distribution throughout the system. Heat exchangers in the system commonly employ carbon steel tubing unless better corrosion resistance or heat transfer is needed. Exchanger housing components that are often made of carbon steel include the shell, water box, and tubesheets. Due to the strength and low cost of carbon steel, it is used extensively for load-bearing structural members, such as support beams in cooling towers. The load-carrying capacity of steel components decreases as corrosion reduces their cross-sectional area. To help maintain structural integrity, corrosion-resistant coatings are applied to many steel components. Coatings are discussed in detail in Chap. 7.

Alloy Types

Carbon steels (also called plain carbon steels) contain carbon and manganese as their primary alloying elements. Manganese is a typical alloying element present in steel. Manganese improves some mechanical properties, such as strength and toughness.

Different classes of plain carbon steel alloys are characterized by the amount of carbon in the alloy. Low-carbon steels have less than 0.2 wt% carbon, medium-carbon steels contain between 0.2 and 0.5 wt% carbon, and high-carbon steels contain greater than 0.5 wt% carbon. The low-carbon steels are also called mild steels, with specified carbon contents between 0.06 and 0.18 wt%. Steel heat exchanger tubing is most often made of mild steel. Both carbon and manganese levels are higher in steel piping than in tubing. Piping is often a medium-carbon steel. The amount of carbon in the alloy strongly influences the weldability of the steel. Plain carbon steel alloys with less than 0.35 wt% carbon can be easily welded.

Low-alloy steels may contain a variety of different alloying elements used to improve certain properties, such as mechanical properties through heat treatment. To be considered a low-alloy steel, the overall alloying element content must be below 8 wt%. There are many types of low-alloy steels. Some alloys are commonly known in the United States by using their American Iron and Steel Institute (AISI) and Society of Automotive Engineers (SAE) designations. In the AISI/SAE designation, the first two numbers refer to an alloy type, as shown in Table 1.1. The last two or three numbers indicate the carbon content. Thus, a 1018 designation refers to a plain carbon steel with a nominal carbon content of 0.18 wt%. The Universal Numbering System (UNS) for steels has the letter G (for plain carbon and alloy steels) preceding the AISI/SAE designation.

The slightly increased alloying element additions to low-alloy steels do not generally result in marked improvements in corrosion resistance to freshwater environments compared to plain carbon steels. Chromium additions above 3 wt% can reduce corrosion in some cases. However, the dramatic increase in corrosion resistance offered by stainless steels requires chromium contents greater than 10.5 wt%.

Though the corrosion resistance of plain carbon and low-alloy steels is generally not vastly different, small additions of certain elements, such as chromium, copper, and silicon, can change the characteristics of the corrosion product layer that forms on the metal surface. Alloys that contain appreciable levels of chromium develop an oxide layer that is more resistant to some flow-related wastage mechanisms such as erosion-corrosion. The addition of small amounts of copper to steel can dramatically decrease the metal loss rate of steels with atmospheric exposure. Some chromium- and copper-containing low-alloy steels have been specifically developed for corrosion resistance with atmospheric exposure; these alloys are called weathering steels.

Although low-alloy steels may not provide additional corrosion resistance compared to carbon steel when exposed to freshwaters, they can provide increased corrosion resistance to various process streams. Thus, low-alloy steels have applications within the oil and

AISI/SAE Designation	UNS Designation	Typical Alloying Element Concentration (wt%)					Comment
		Manganese	Chromium	Molybdenum	Nickel	Silicon	
10XX	G10XX	0.25–1.65				0.00–0.30	Plain carbon
11XX	G11XX	0.50–1.65				0.00–0.30	Plain carbon: Resulfurized (0.08–0.33 wt% S)
13XX	G13XX	1.60–1.90				0.20–0.35	Plain carbon
40XX	G40XX	0.70–1.00		0.15–0.30		0.20–0.35	
41XX	G41XX	0.40–1.00	0.40–1.10	0.15–0.25		0.20–0.35	
43XX	G43XX	0.45–0.85	0.40–0.90	0.20–0.30	1.65–2.00	0.20–0.35	
44XX	G44XX	0.70–0.90		0.35–0.45		0.20–0.35	
46XX	G46XX	0.45–0.90		0.20–0.30	1.65–2.00	0.20–0.35	
47XX	G47XX	0.50–0.90	0.35–0.55	0.15–0.40	0.90–1.20	0.20–0.35	
48XX	G48XX	0.40–0.70		0.20–0.30	3.25–3.75	0.20–0.35	
50XX	G50XX	0.30–1.00	0.20–0.60			0.20–0.35	
51XX	G51XX	0.60–1.00	0.70–1.15			0.20–0.35	Carbon content 0.13–0.65 wt%
51XXX	G51XXX	0.25–0.45	0.90–1.15			0.20–0.35	Carbon content 0.95–1.10 wt%
52XXX	G52XXX	0.25–0.45	1.30–1.60			0.20–0.35	Carbon content 0.95–1.10 wt%
61XX	G61XX	0.50–0.90	0.50–1.10			0.20–0.40	Vanadium content 0.10–-0.15
81XX	G81XX	0.70–1.00	0.30–0.55	0.08–0.15	0.20–0.40	0.20–0.35	
86XX	G86XX	0.70–1.00	0.40–0.60	0.15–0.25	0.40–0.70	0.20–0.35	
87XX	G87XX	0.70–1.00	0.40–0.60	0.20–0.30	0.40–0.70	0.20–0.35	
88XX	G88XX	0.75–1.00	0.40–0.60	0.30–0.40	0.40–0.70	0.20–0.35	
92XX	G92XX	0.50–1.00	0.00–0.80			1.20–2.20	
93XX	G93XX	0.45–0.65	1.00–1.40	0.08–0.15	3.00–3.50	0.20–0.35	

TABLE 1.1 List of Carbon and Low-Alloy Steel Alloys by Their AISI/SAE and UNS Designations, and Typical Alloying Element Concentrations. The XXs in the designations are used in place of the carbon content, which can vary within the alloy class.

gas and refining industries. For example, increasing chromium content in a low-alloy steel will improve resistance to sulfidation attack in process environments that contain hydrogen sulfide.

Microstructure

Plain carbon steels are generally used in an annealed and normalized condition, where the metal is heated to certain elevated temperatures and slow cooled. This produces a microstructure in which the carbon in the alloy is in the form of pearlite colonies or iron carbides. These phases are distributed throughout a matrix of ferrite grains. Ferrite is essentially iron that has some trace elements, like manganese, in solution. Pearlite has a lamellar structure that consists of alternating layers of ferrite and iron carbide. Tubing and piping generally have this microstructure, as shown in Fig. 1.1.

Some other features often observed in steel alloys include pearlite banding, decarburized zones, and inclusions. The pearlite colonies can be aligned along the processing direction of the component due to microsegregation of elements in the steel, giving rise to the term *banded microstructure*. Narrow decarburized zones may be adjacent to the internal and external surfaces. These decarburized zones are normally present from the tube manufacturing process and do not represent an in-service condition. If pearlite banding and decarburized zones are present on tubing, piping, or plate steel, they can be used to help assess whether metal loss is primarily from internal or external surfaces.

Commercial alloys will typically have some inclusions in the microstructure. Inclusions are particles such as oxides, sulfides, and silicates that are considered to be impurities in the metal matrix. Manganese reacts with sulfur in the steel to preferentially form manganese sulfide inclusions. Elongated manganese sulfide inclusions, known as stringers, are typical in steel microstructures. Other sulfide compounds that would form will tend to promote cracking. In many cases, the locations of inclusions will be preferred sites to initiate corrosion. Most steels have restrictions on sulfur and phosphorus contents, as these impurity elements have a generally negative impact on corrosion resistance and mechanical properties. These elements can improve machinability, however, and higher contents are used in free machining grades of steel. Free machining grades of steel should generally not be used for water service.

Low-alloy steels will generally form similar microstructures when processed by annealing and normalization. However, the alloying elements are often added to these steels so that they can be processed by relatively fast cooling (quenching) and subsequent heat treatment (tempering) to produce desired mechanical properties. Detailed discussion of thermal processing of steel alloys is beyond the scope of this book. This type of processing is often used in fasteners such as high-strength bolts, which can be exposed to cooling water when used to connect large flanges on submerged piping.

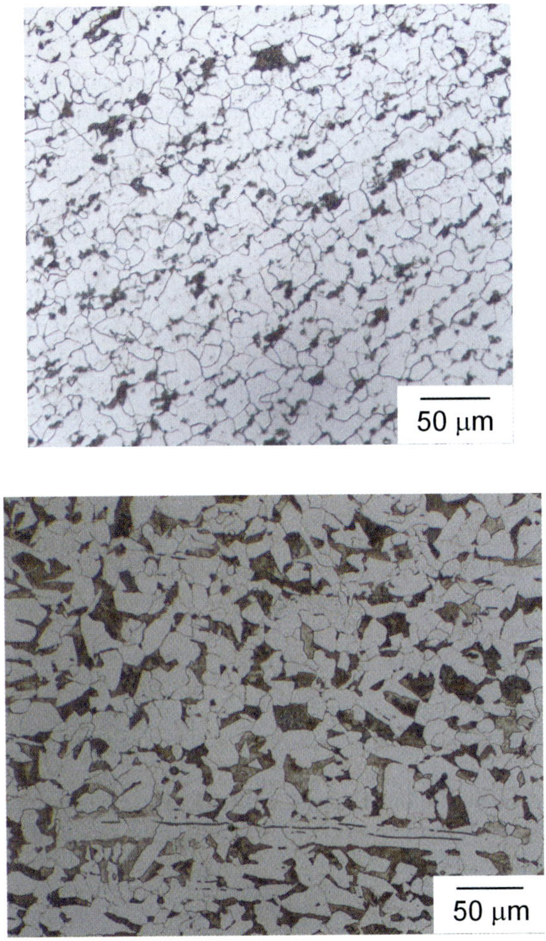

FIGURE 1.1 Typical microstructures of plain carbon steel tubing (top) and piping (bottom), consisting of pearlite colonies (dark etching constituent) in a matrix of ferrite grains. The metal was polished and etched with nital (a 3% nitric acid solution in methanol) to reveal the iron carbides and ferrite grain boundaries. Note the larger amount of pearlite in the pipe microstructure, which has higher carbon content. Also visible in the pipe microstructure are elongated inclusions (stringers).

Microstructural differences do not usually result in appreciable variation in corrosion rates of plain carbon and low-alloy steels. However, some exceptions exist, such as at welds and at plastically deformed areas in bends and rolled tube ends. These locations will have both increased internal energy and elevated residual stresses, making them susceptible to preferential or increased attack in a corrosive environment. This effect is known as stress-assisted corrosion.

Factors Affecting Corrosion

Iron, in its pure metallic state, is thermodynamically driven to revert to its oxide forms when exposed to aqueous environments that contain dissolved oxygen. This is known as oxygen corrosion. Mixtures of certain corrosion products formed by oxygen corrosion are often referred to as rust. Oxygen corrosion and the development of oxide-based corrosion products on a steel surface are discussed in detail in Chap. 12.

Many factors influence corrosion of plain carbon and low-alloy steels. The electrochemical description of the corrosion process can be beneficial to understanding the interactions between the environment and the metal surface that result in attack.

A schematic of an electrochemical corrosion cell on steel is shown in Fig. 1.2. In the electrochemical model, corrosion requires an anode, a cathode, a metallic path between the anode and cathode, and exposure of the anode and cathode to a common electrolyte solution. An oxidation reaction occurs at the anode, as the metal is converted to a metal ion and electrons. The oxidation reaction for iron in typical cooling water conditions is shown below, although other reactions or subsequent reactions are also possible:

$$Fe > Fe^{+2} + 2e^{-} \tag{1.1}$$

The metallic path allows electrons to be transported to the cathodic areas on the metal surface. At the cathode locations, the electrons can participate in many different reduction reactions. One of the

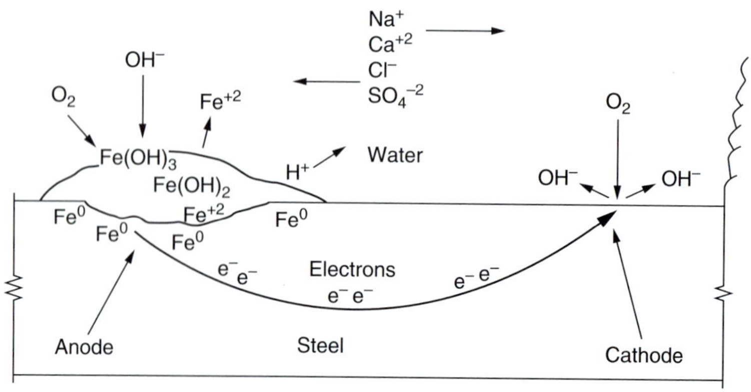

FIGURE 1.2 Basic electrochemical corrosion cell on a steel surface, showing the anode, cathode, the electron path from the anode to the cathode, and the electrolyte solution (cooling water).

most common cathodic reactions in cooling water systems is oxygen reduction, which is shown below.

$$\tfrac{1}{2}\,O_2 + H_2O + 2e^- > 2OH^- \tag{1.2}$$

It should be noted that other reduction reactions can occur at the cathode, such as hydrogen reduction in acidic solutions (which can be the predominant reaction as pH drops below 4) and reactions with oxidizing species in the environment, such as some biocides typically used in cooling water systems. The oxidation and reduction reactions result in positive charge buildup at the anode (due to the release of positively charged metal ions) and a negative charge at the cathode (due to hydroxide ion buildup). The electrolyte completes the electrochemical circuit, as positively charged species will be attracted toward cathodic regions and negatively charged species will migrate to the anodic sites. Secondary reactions between the released ions from the anode and cathode and species in the cooling water typically result in the formation of corrosion products, specifically iron oxide and iron oxide hydroxide compounds [see Eqs. (1.3) and (1.4)]. Deposits will also tend to form at cathodic regions.

$$Fe + H_2O + \tfrac{1}{2}\,O_2 \rightarrow Fe\,(OH)_2 \tag{1.3}$$

$$Fe\,(OH)_2 + \tfrac{1}{2}\,H_2O + \tfrac{1}{4}\,O_2 \rightarrow Fe\,(OH)_3 \tag{1.4}$$

The corrosion of steel generally depends on the electrochemical cell reactions. The slowest of the individual reactions will govern the overall corrosion rate. Thus, control of some of the many factors that influence these reactions may be implemented for water treatment strategies to minimize corrosion. Two of the main approaches include adjustments to water chemistry and the addition of chemicals known as corrosion inhibitors.

Water Chemistry

Characterizing the water source used in a cooling system and understanding the impact of several different parameters on corrosion are initial steps toward proper treatment. Some important water chemistry parameters that affect steel corrosion are discussed below, but the short list is by no means exhaustive. The following discussion highlights the effect of some individual parameters on corrosion. It is important to note that many factors can influence corrosion, so their interactions must also be considered. The following examples are used to highlight general concepts, but the exact relations may not be applicable to all conditions due to the diversity of water chemistry.

Dissolved Gases

Oxygen is the primary dissolved gas that affects steel corrosion in cooling water systems. The level of dissolved oxygen in the water depends on many factors, including the type of cooling system, temperature, and water quality. Maximum levels in open recirculating systems will generally be attained in the cooling tower basin, as the water will be saturated with dissolved oxygen. Dissolved oxygen concentrations in closed systems will vary, depending on the system design and operation. In a truly closed system, the concentration will decrease with time. As shown in Fig. 1.3, the corrosion of steel generally increases with the concentration of dissolved oxygen in the water. However, corrosion rates are time dependent and will decrease with time if protective iron oxide corrosion products form on the surfaces.

Other gases that can influence the corrosion of steel include carbon dioxide and hydrogen sulfide. These gases are most common if subsurface water from wells is used as a makeup water source. In some cases, elevated carbon dioxide concentrations can increase corrosion, if the dissolved CO_2 depresses pH or causes the formation of iron carbonate layers on the metal surface. However, the influence of dissolved carbon dioxide in cooling water systems is generally minimal, as the pH ranges in cooling water systems favor the formation of bicarbonate (HCO_3^-) and carbonate (CO_3^{-2}) ions rather than CO_2. Equilibrium reactions indicate no contribution of CO_2 in waters with pH greater than 8.5.

In addition to being present in some makeup water sources, hydrogen sulfide can be introduced as contamination from process leaks, especially in some refinery cooling water systems. These are often referred to as sour leaks. The introduction of H_2S into the system will affect corrosion in many ways, including interactions with oxidizing biocides. This may have the secondary effect of allowing

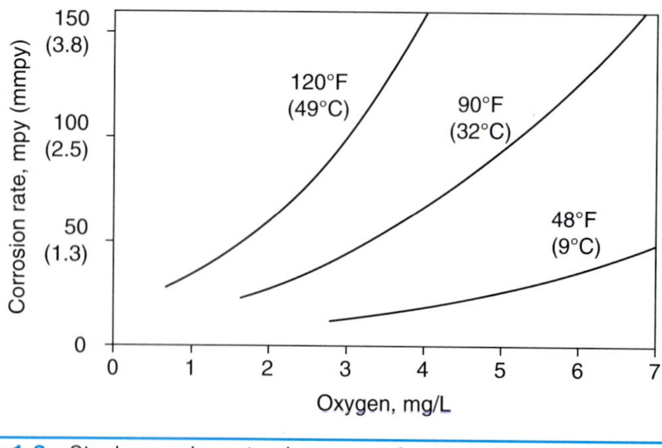

Figure 1.3 Steel corrosion rate shown as a function of dissolved oxygen. Note also the increasing corrosion rate with increasing temperature.

increased microbial growth on metal surfaces than can result in increased corrosion. In addition, the hydrogen sulfide ions can depolarize anodic sites on the metal surface, increasing localized attack at those locations. Even small amounts of hydrogen sulfide in the water can dramatically increase corrosion rates on steel.

pH and Alkalinity

For most cooling waters, the pH value is a critical parameter that must be maintained within a specific range. The operating pH range is generally chosen to optimize corrosion control and minimize excessive deposition by scaling. The range is dependent upon water chemistry and chemical treatment. The typical pH range is 6.5 to 9.5 in open recirculating systems. Over most of this range, the pH has minimal effect on the corrosion rate of steels in soft waters (Fig. 1.4). However, for mild steel exposed to freshwaters, corrosion rates can decrease if the system can operate at the higher pH values within the range. The alkalinity of the system is related to pH, and the increase in alkaline species in the water at higher pH values helps to limit corrosion of steel. There are pH- and alkalinity-related processes that can decrease cathodic reactions, such as the formation of protective

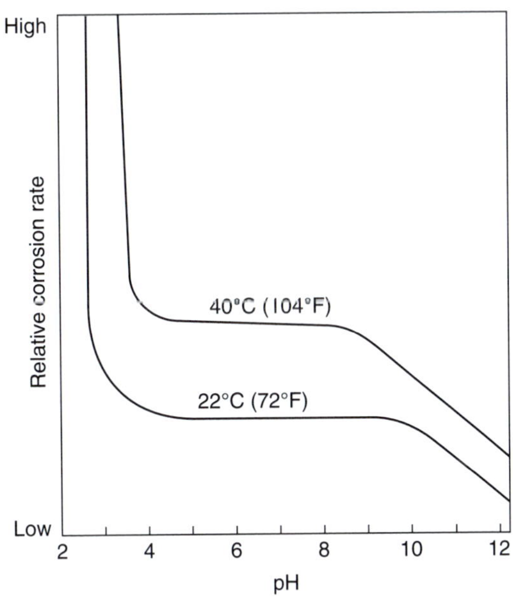

FIGURE 1.4 Steel corrosion rate as a function of pH in soft water, shown at two different temperatures. Note the decrease in corrosion rate at elevated pH ranges and the dramatic increase as the pH is reduced below 4. *(Reprinted with permission of ASM International. All rights reserved. www .asminternational.org.)*

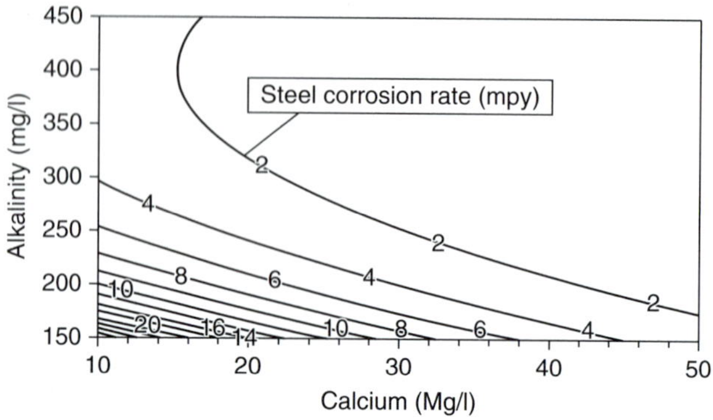

FIGURE 1.5 Isocorrosion curves for steel as a function of alkalinity and calcium concentration. Note the decrease in the corrosion rates as alkalinity and calcium concentration increase.

deposits and corrosion products. For instance, precipitation of calcium carbonate scale can provide a barrier that limits the availability of oxygen to the metal surface, and therefore oxygen reduction at cathodic sites. When carefully controlled as part of certain treatment programs, this can limit steel corrosion. The interaction between alkalinity and calcium concentration on the corrosion rate of mild steel is shown in Fig. 1.5.

The increased corrosion rate of mild steel when pH values decrease below 4 to 5 is also shown in Fig. 1.4. At these low pH values, hydrogen reduction becomes the primary reaction at the cathode, and corrosion rates can dramatically increase. These low pH values rarely occur in properly controlled cooling water systems. They are usually associated with an upset condition such as inadvertent acid feed or process leaks. Such low pH values can also be associated with acidic cleaning solutions, although these solutions should incorporate corrosion inhibitors to limit attack.

Aggressive Anions

Aggressive anions such as chloride and sulfate will generally adversely affect the corrosion of steel. The increase of chloride and sulfate content in cooling waters will increase the water's conductivity, and typically as a result, its corrosivity. In freshwaters, increases in mild steel corrosion rates have been correlated with the ratio of chloride and sulfate concentration to the bicarbonate and carbonate concentration. For a given bicarbonate level, the corrosion rate increases with increasing chloride content, as illustrated in Fig. 1.6. Chloride ions can impair the ability of protective oxide layers to form on the steel surface.

Elevated levels of chloride and sulfate in the water increase the likelihood that concentration cell corrosion will occur on steel surfaces. Concentration cells can develop beneath corrosion product layers such as tubercles, occluding porous deposit layers, and in

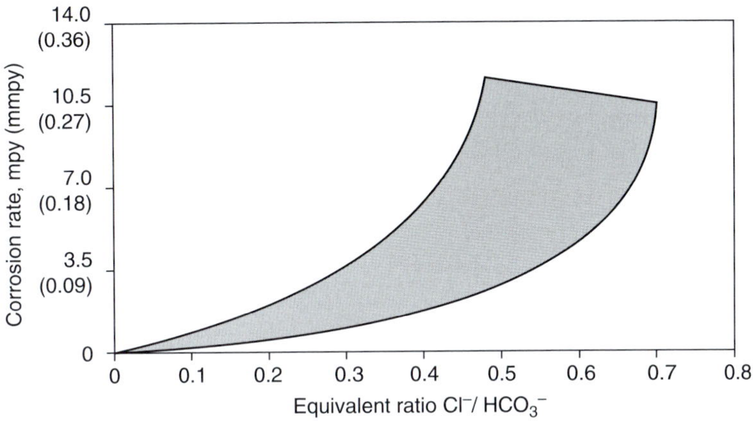

Figure 1.6 Illustration showing the increase in mild steel corrosion rate with increasing chloride/bicarbonate ratio for water having alkalinity in the 75–260 mg/L (as $CaCO_3$) range.[1] (*Courtesy of the Illinois State Water Survey.*)

crevices. Localized attack will typically occur at these locations with accelerated penetration rates compared to the surrounding areas.

Suspended Solids

Suspended solids can be introduced into the system from water sources or by removal of particulates in the air from the scrubbing action of cooling towers. In addition, salts precipitated in the system or corrosion products formed on metal surfaces can be dislodged and transported elsewhere. These solids can settle on metal surfaces downstream, allowing for underdeposit corrosion. The settling of particulate material is favored in areas of the system with low flow or periodic stagnant conditions. Suspended solids can be reduced through flocculation/clarification of source water and filtration in the system.

Oxidizing Biocides

Biological activity is controlled in cooling water systems by biocides. Sufficient levels of biocides must be maintained throughout the system to limit microbiological activity, thus avoiding fouling and corrosion issues. Oxidizing biocides, such as sodium hypochlorite and gaseous chlorine, are commonly used in once-through and open recirculating cooling water systems. However, increasing concentrations of oxidizing biocide tend to result in increasing corrosion rates on mild steel (Fig. 1.7). Oxidizing biocides provide a contributing reduction mechanism at the cathode to consume electrons, thus increasing corrosion rates. If high biocide levels are needed to help control microbiological activity, then slug feeding rather than a continuous feed is usually preferred. Additionally, nonoxidizing biocides can sometimes also be used to reduce microbiological activity and associated corrosivity on steel surfaces.

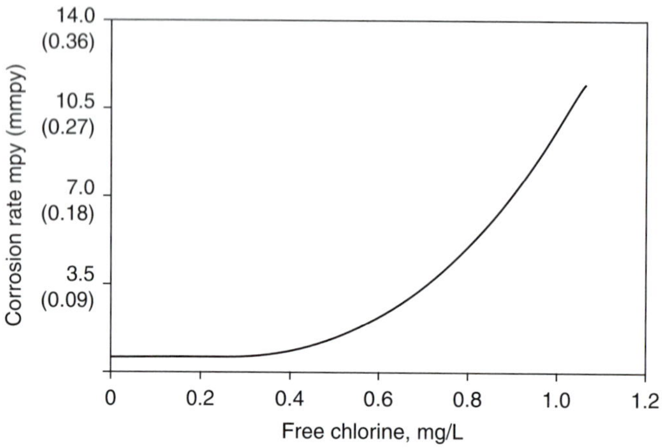

Figure 1.7 The effect of free chlorine on the corrosion rate of mild steel. Note that the corrosion rate increases significantly as free chlorine levels exceed 0.4 mg/L.[1] (*Courtesy of the Illinois State Water Survey.*)

Corrosion Inhibitors

Corrosion inhibitors are added to cooling water to control corrosion in the system. The corrosion inhibitors will interact with the surface to decrease attack, generally by reducing reactions at the anode and/or cathode sites on the surface. Thus, inhibitors are generally considered anodic or cathodic inhibitors. However, some chemicals can participate in both anodic and cathodic inhibition. Effective treatment programs often incorporate both anodic and cathodic inhibitors to reduce corrosion.

Anodic Corrosion Inhibitors

Anodic corrosion inhibitors decrease corrosion by blocking the reaction that occurs at anodic sites, specifically dissolution of the metal. The anodic inhibitors create a film that impedes the migration of iron ions from the metal surface to the electrolyte. The anodic inhibitors can be further classified as either oxidizing or nonoxidizing, which depends on the mechanism of film formation.

Oxidizing anodic inhibitors react to develop a tightly adherent layer of metal oxide. They are strong oxidizing species that do not require the presence of oxygen to be effective. The iron produced at the anode as Fe^{+2} is oxidized to form a passive oxide film that is protective. Examples of oxidizing anodic inhibitors include chromates and nitrites.

Chromates are highly effective oxidizing anodic corrosion inhibitors for steel and other ferrous alloys. They were used extensively in the distant past, but their current use is significantly restricted worldwide due to environmental concerns.

Nitrites are commonly used in closed cooling water systems. High dosages are required for effective treatment. Insufficient dosages of oxidizing anodic inhibitors, especially when used alone, will not provide adequate protection to all anodic sites, and severe pitting problems can result. Nitrites can be degraded to nitrate by some microorganisms, which will lower the actives in the treatment program over time. Because nitrites are readily oxidized by both microbial growth and reactions with oxidizing biocides, they are generally not used in open cooling water systems.

Nonoxidizing anodic inhibitors require the presence of some dissolved oxygen, or other oxidizing species, to provide protection. They react with the Fe^{+2} produced at the anode to form an insoluble compound that precipitates to increase the stability of the oxide film that covers the anodic site. Common nonoxidizing anodic inhibitors for steel include molybdates and inorganic phosphates (orthophosphate).

The influence of molybdates on corrosion inhibition depends on concentration levels and surface conditions. The presence of oxygen is needed to provide effective corrosion inhibition, although some inhibition is offered by molybdates at very high usage levels in oxygen-reduced environments. Under certain conditions, molybdates react with iron to form a complex that precipitates as a film at anodic sites. This film is insoluble and is absorbed into the oxide layer on the metal surface to help seal openings, making it more protective. However, the effectiveness of molybdates is reduced in the presence of chlorides, so the dosage level needs to increase as the chloride concentration in the water increases. To be cost effective, molybdates are used in conjunction with other inhibitors.

Orthophosphate reacts at the anode with iron ions to produce an iron phosphate complex. This reaction promotes the formation of a passive oxide film that reduces the migration of the iron ions from the surface of the anode. Like molybdate, the inhibition mechanism is sensitive to chloride concentration. A challenge to the proper use of orthophosphate is that excessive precipitation of calcium phosphate must be avoided. The deposition produced by excessive calcium phosphate precipitation will result in reduced heat transfer, will create a surface condition susceptible to underdeposit corrosion, and will possibly reduce the overall treatment dosage. Factors that increase deposition tendency include high calcium levels, high phosphate levels, increasing pH, and increasing temperature. Dispersant chemicals are generally required as part of the treatment program to maintain sufficient levels of orthophosphate in solution. In addition to chemical treatment additions, some orthophosphate may be introduced to the system with the makeup water, and by breakdown of polyphosphates or phosphonates (known as reversion). Orthophosphate is used in conjunction with polyphosphates in stabilized phosphate programs.

Cathodic Corrosion Inhibitors

Cathodic corrosion inhibitors diminish oxygen reduction rates by limiting the supply of oxygen to the cathode. They develop a barrier film that reduces oxygen transport and prevents electron transfer. Two common cathodic corrosion inhibitors are zinc and polyphosphates.

Polyphosphates require calcium, or another divalent ion, to be present in the water to form a barrier layer. If calcium is present in sufficient amounts, calcium polyphosphate can form. The locally high pH at the cathode from the formation of hydroxide promotes precipitation of the polyphosphate layer. Successful application of polyphosphate typically requires tight pH control. High-pH excursions can result in bulk precipitation. Low-pH excursions can strip the barrier film from the surface.

Zinc is a cathodic inhibitor that reacts to produce a zinc hydroxide barrier layer at the cathode. This barrier layer restricts oxygen transport to the metal surface. Zinc must be maintained in soluble form at sufficient levels in the cooling water to be an effective inhibitor. The amount of soluble zinc in the water is pH dependent, and it decreases with increasing pH. Dispersant polymers are used to stabilize the zinc in soluble form for use in higher pH ranges.

Some corrosion inhibitors, such as phosphonates, can act both as anodic and cathodic inhibitors. Phosphonates also can provide scale inhibition. There are many different phosphonates; some common molecules used in cooling water applications include 1-hydroxyethylidine-1,1-diphosphonic acid (HEDP), 2-phosphonobutane -1,2,4-tricarboxylic acid (PBTC), phosphinocarboxylic acid (PCA), and phosphino succinic oligomer (PSO). Blends of the phosphonates are used in all-organic programs.

Forms of Corrosion

Plain carbon and low-alloy steels can experience many corrosion and degradation mechanisms, including most of the mechanisms discussed in detail in this guide. Some of the more common mechanisms and some specific issues are highlighted below.

Oxygen Corrosion and Tuberculation

Perhaps the most prevalent cause of metal loss on carbon and low-alloy steels is oxygen corrosion. As mentioned previously, carbon and low-alloy steels will react with oxygenated water to cause general corrosion. This mechanism is called oxygen corrosion. Iron oxide corrosion products may form into mounds called tubercles. The service life of a component will depend on the rate and type of attack, such as whether corrosion is general or localized. Details on the oxygen corrosion mechanism are presented in Chap. 12.

Concentration Cell Corrosion

If porous corrosion products or permeable deposits form on metal surfaces, then underdeposit mechanisms related to differential oxygen and ion concentration cells can develop. This can result in localized attack that penetrates into the metal surface at accelerated rates. Plain carbon and low-alloy steels are also susceptible to accelerated attack at crevices. These mechanisms are discussed in detail in Chap. 11.

Microbiologically Influenced Corrosion

Microbiologically influenced corrosion can contribute to wastage on steels in many cooling water systems. Slime layers formed on the metal surface can promote corrosion indirectly by underdeposit corrosion mechanisms. Such layers may also create environments suitable for anaerobic organisms, such as sulfate-reducing bacteria and acid-producing bacteria. These microorganisms are capable of directly causing significant wastage on steel. Some microbiological activity can reduce concentrations of specific chemical inhibitors, particularly nitrite in closed systems. Diagnosis that microbiologically influenced corrosion contributed to metal loss at corroding sites requires evidence of specific microorganisms, and the particular features associated with their attack. The presence of potentially corrosive microorganisms in the system alone is not proof that microbiologically influenced corrosion occurred. Details and examples of attack related to microorganisms are presented in Chap. 13.

Acid Corrosion

Steels can be corroded at high rates when solutions become acidic. This is usually related to an upset condition in the cooling water system. In addition, a low-pH excursion will also solubilize many of the deposits and corrosion products on the surfaces. If dissolved solids precipitate as the system returns to normal operation, then underdeposit corrosion can occur. Features associated with acid corrosion of steels are presented in Chap. 14.

Alkaline corrosion is generally not an issue for steels in cooling water service.

Intergranular Corrosion

Intergranular corrosion refers to preferential attack along the grain boundaries in the microstructure. Some environments can cause intergranular corrosion of carbon and low-alloy steels. Solutions with concentrated nitrates can produce intergranular attack in some situations, such as storage of ammonium nitrate. Intergranular attack on steel at typical cooling water chemistries is very rare.

Preferential Weld Corrosion

Welds used to join carbon and low-alloy steels can be susceptible to preferential corrosion in aggressive environments. Welds are often more susceptible to corrosion due to high residual stresses, microstructural differences, and compositional variation. In addition, sometimes defects are present at welds, such as incomplete penetration, and these can act as crevices. Discussion of weld-related attack and welding issues is presented in Chap. 26.

Galvanic Corrosion

If steel is coupled directly to a metal that is more noble in the environment, such as a copper alloy, then galvanic corrosion can result. For example, galvanic attack can occur when steel piping is connected to brass fittings in a distribution line. The potential difference between the metals provides a driving force for preferential corrosion of the steel. Corrosion will be most severe at the interface between the two alloys. Careful consideration of the design of a system can prevent dissimilar metal couplings and galvanic corrosion of steel. In addition, appreciable levels of dissolved copper in the cooling water can stimulate corrosion by the galvanic mechanism if they form metallic deposits directly on the steel surfaces. Examples of galvanic corrosion are presented in Chap. 16.

Conversely, the concept of galvanic corrosion can also be used to protect steel components in cooling water systems. Sacrificial anodes made from relatively active metals, such as zinc alloys, are often installed in the water boxes on heat exchangers to reduce the corrosion of the surrounding steel.

Surface Finish

The condition of the metal surface can influence corrosion of steel, especially the initial stages. In general, rougher metal surfaces will tend to be more susceptible to corrosion. Surfaces mechanically cleaned by bead or sand blasting typically will experience higher corrosion rates than finely ground or typical mill-finished surfaces due to increased surface area, higher residual stresses from deformation, and possibly embedded contaminants.

A mill scale layer that covers a steel surface may promote corrosion. Mill scale is the thermally formed iron oxide layer that develops on the metal surface due to reaction with air during elevated temperature processing such as hot rolling. This layer is cathodic to the underlying steel, and it will contain many cracks. Consequently, pitting can occur with water exposure if mill scale covers a large percentage of the total surface area. Mill scale can be removed from the steel surface by pickling in inhibited solutions of either sulfuric or hydrochloric acid.

Cleaning and Passivation

Proper cleaning and passivation of metal surfaces for critical applications will effectively minimize rapid attack on the steel when it is initially placed into cooling water service. New components may be covered by mill scale, oils, and greases. Components that were in service can be covered by layers of deposits and corrosion products. In each case, removal of the material by an effective cleaning program will prepare the steel surface for passivation. The cleaning and passivation solutions are generally formulated with surfactants, dispersants, and sequestrants, along with corrosion inhibitors. Corrosion inhibitors limit metal loss and prevent flash rusting by initializing the formation of protective films on the metal surface. Following the passivation treatment, elevated levels of the corrosion inhibitors in the treatment program are maintained in the system for an extended period. The chemical inhibitor levels can be reduced to maintenance levels after a protective layer is established.

Reference

1. *Corrosion by Domestic Waters*, T. E. Larson, ISWS-75-BUL59.

CHAPTER 2

Cast Iron Alloys

Introduction

Common cast irons are iron-based alloys that have a carbon content typically in excess of 2.0 wt% and more than 0.4 wt% silicon. The higher carbon content results in a lower melting point compared to plain carbon steel. In addition, excellent fluidity and low shrinkage when solidifying make these alloys particularly suitable for forming components by casting. The ease of production and relatively low raw material costs make cast iron alloys one of the most economical choices for metallic components.

Some typical applications for cast iron alloys in cooling water systems include their use as pump and valve housings and pump impellers. They are also often used for the end caps and cover plates on heat exchangers. Piping can be made from cast iron alloys, especially larger diameter pipes such as those used for water distribution. In addition, pipe fittings such as elbows, tees, and end caps are often made from cast iron as are water-cooled engine blocks and pistons.

The most common cast iron alloys are inherently brittle, limiting their use in high-pressure systems or in applications that can experience high tensile stress levels or impact (shock) loading. Although there are many uses for cast iron, its mechanical properties and corrosion susceptibility in some applications typically preclude more extensive use.

Types

The types of cast iron can be divided into categories based on their alloy content and microstructure. The types of cast iron that are commonly used for cooling water applications are essentially unalloyed. These include the gray, ductile, and white cast iron. Alloy cast irons are typically referred to by their primary alloying addition, such as high-silicon, high-nickel, and high-chromium. The types of cast irons are discussed below, and Table 2.1 summarizes some of the differences between them by their microstructural constituents and typical alloying element ranges. Specifications for unalloyed cast iron are

Cast Iron Type	Matrix	Main Carbon Form	Total Carbon (wt%)	Silicon (wt%)	Nickel (wt%)	Chromium (wt%)	Manganese (wt%)	Magnesium (wt%)
Gray	Ferrite / Pearlite	Graphite Flakes	2.5–3.8	1.2–2.6			0.5–1.0	
Ductile	Ferrite / Pearlite	Graphite Nodules	3.2–4.1	1.8–2.8			0.8 max	0.04–0.10*
White	Pearlite	Carbides	2.2–3.6	0.4–1.6			0.3–0.7	
High-Nickel	Austenite	Graphite Flakes or Nodules	1.8–3.0	1.0–2.8	14.0–32.0	1.8–5.5	0.4–1.5	
High-Silicon	Ferrite	Graphite Flakes	0.8–1.0	14.0–17.0			0.4–1.0	
High-Chromium	Ferrite	Carbides	1.2–4.0	0.5–3.0	5.0 max	12.0–35.0	0.3–1.5	

*Magnesium is a typical elemental addition to ductile cast iron, although other elements can also be used.

Table 2.1 Typical Composition Ranges for Selected Elements in Some Common Types of Cast Iron. These values are not all-inclusive; they are meant to highlight differences between the major constituents for the different alloy types.

typically associated with mechanical properties rather than alloy constituents. For instance, many of the specifications have classes or grades that are associated with minimum tensile strength levels. However, chemical requirements are an important specification for alloyed cast irons.

Gray Cast Iron

Based on worldwide casting production, gray cast iron is by far the most common cast alloy.[1] The main characteristic of gray cast iron is that the uncombined carbon in the metal is in the form of graphite flakes. It is called "gray" due to the gray fracture appearance when broken, as the color is related to the graphite flakes on the surface. A typical microstructure of gray cast iron is shown in Fig. 2.1; it consists of graphite flakes dispersed throughout the matrix. The matrix of gray cast iron alloys can be pearlitic, ferritic, or a combination of the two. The form and distribution of the flakes is important, as these parameters can significantly affect mechanical properties. The graphite flakes can form in rosette groupings or show interdendritic segregation rather than being randomly dispersed.

The size, morphology, and distribution of the graphite flakes also influence the mechanical properties of the alloy. Applied stresses will locally concentrate at the sharp tips of the graphite flakes, which promote mechanical cracking. This results in low toughness for gray cast iron. Compacted cast iron contains alloy additions that round the tips of the graphite flakes, which increases the strength and toughness of the material.

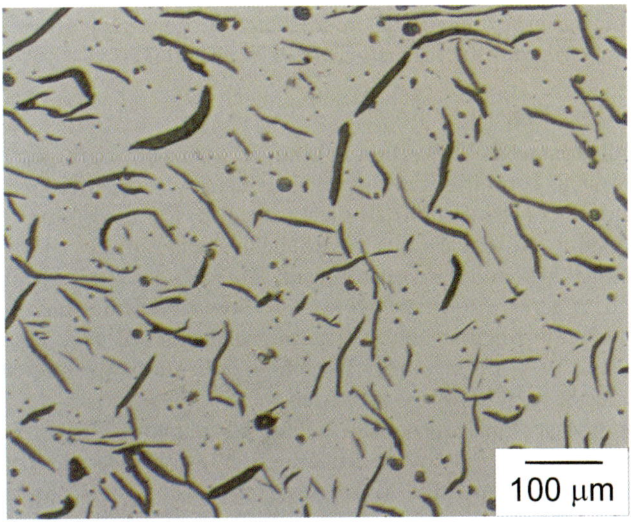

FIGURE 2.1 Microstructure of gray cast iron showing randomly oriented dispersed graphite flakes.

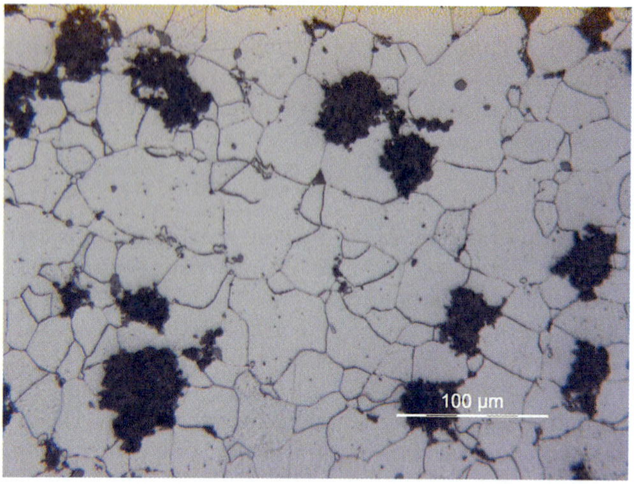

FIGURE 2.2 Graphite nodules dispersed in a matrix of ferrite grains, a typical microstructure of ductile cast iron.

Ductile (Nodular) Cast Iron

The graphite in ductile cast irons is spheroidal in shape, otherwise known as a nodular form (Fig. 2.2). Additions to the melt, such as magnesium or cerium, act as inoculants to form the graphite nodules. The spheroidal shape of the graphite provides improved mechanical properties compared to gray cast iron. In addition, the spheroidal form of the graphite in the microstructure generally offers better resistance to graphitic corrosion than graphite flakes.

White Cast Iron

For white cast irons, the carbon in the alloy forms into carbides, which are known as combined carbon. These alloys have slightly lower carbon and silicon contents than gray cast iron. Rapid cooling during solidification causes carbides to form instead of graphite. This is often referred to as a chilled microstructure. The carbides are typically distributed in a pearlite matrix (Fig. 2.3). Some graphite may also be present in the microstructure. White cast irons are very hard and brittle, and they are primarily used for applications where abrasion resistance or high hardness is needed.

White cast irons can be heat treated to transform the carbides to graphite. During a lengthy heat treatment in a controlled atmosphere, the carbides are converted to irregularly shaped clusters of graphite. This morphology significantly reduces the brittleness of the component. A cast iron processed in this way is called malleable iron, and the properties are generally similar to those of ductile cast iron.

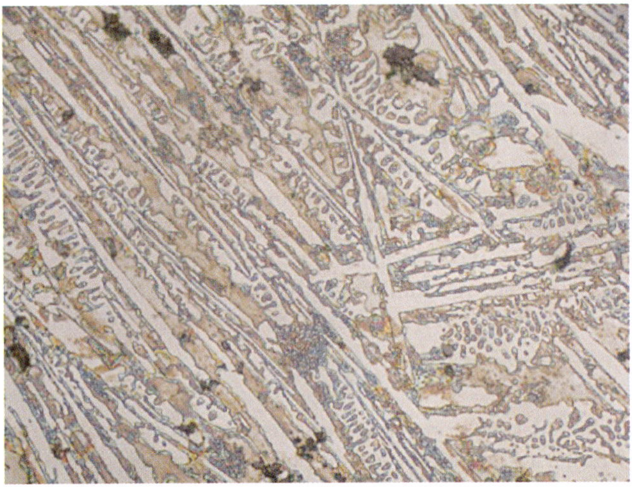

FIGURE 2.3 Typical microstructure of white cast iron, consisting primarily of carbides (light phase) in a dendritic matrix of pearlite (gray phase).

Alloyed Cast Irons

The more common, unalloyed cast iron alloys are considered to have similar corrosion resistance to unalloyed cast steels in many cases. Much like steel, appreciable alloying additions can provide much improved corrosion resistance for different environments. However, with more alloying elements and more complicated processing, the alloyed corrosion-resistant cast irons are appreciably more costly.

One class of the corrosion-resistant cast iron alloys is high-nickel cast iron. Sufficient nickel and other austenite stabilizing elements result in the formation of an austenitic matrix rather than a ferritic one. These alloys have carbon in the form of graphite flakes or nodules that are dispersed in an austenitic matrix. Some high-nickel cast iron alloys are known by the trade name Ni-Resist (Inco). These alloys typically show good chemical resistance to caustic and other alkaline environments, especially with nickel contents above 18 wt%. The use of high-nickel cast irons in cooling water service is very limited.

High-silicon cast irons have silicon as the primary alloying element. The corrosion resistance increases dramatically when the silicon concentration exceeds 14.2 wt%. The improved corrosion resistance has been attributed to the development of an adherent and protective silicon oxide-based film on the surface. The carbon in high-silicon cast iron can have a spheroidal graphite or graphite flake form, which influences both mechanical properties and corrosion resistance. These alloys have shown good corrosion resistance in many acidic environments such as sulfuric, nitric, and organic acids. Two commercial high-silicon cast iron alloys are known as Duriron and Durichlor. The Durichlor alloy has increased alloying additions for improved resistance to hydrochloric acid. However, these corrosion-resistant alloys can be attacked by hydrofluoric acid. The high-silicon cast irons are not commonly used in cooling water systems.

High-chromium cast irons have chromium concentrations ranging from 12 to 30 percent that result in a microstructure of chromium-iron carbides in a ferrite matrix of iron and chromium. Corrosion resistance is related to the development of a protective oxide film on the surface of the exposed matrix. Alloys having more than 20 wt% chromium are very resistant to oxidizing acids. Corrosion can occur in environments with oxygen-depleted or reducing conditions.

Corrosion Issues

The unalloyed cast iron alloys are susceptible to corrosion mechanisms similar to those that attack plain carbon steel and other low-alloy steels when exposed to cooling water systems. In addition, the microstructures of some cast iron alloys make them susceptible to graphitic corrosion. Some specific issues associated with failures of cast iron alloys are highlighted next.

Graphitic Corrosion

Graphitic corrosion occurs due to a microstructural galvanic interaction. Though graphitic corrosion is sometimes also referred to as selective leaching similar to dealloying, graphitic corrosion does not involve the preferential removal of one element from a solid solution of elements. The driving force for corrosion is a large potential difference between the graphite phase in the cast iron microstructure and the surrounding iron-based matrix. The iron-based matrix is anodic to the graphite phase, and iron can experience preferential corrosion if exposed to suitable conditions. Attack of the iron matrix penetrates along the graphite/matrix interface (Fig. 2.4). Graphitic corrosion is promoted in high conductivity, mildly acidic, or soft waters. General or pitting attack can overwhelm graphitic corrosion if the water is very aggressive.

Although graphite is present in multiple types of cast iron, attack is generally most severe in gray cast iron alloys. This is because the overall severity of the attack depends on the form and distribution of the graphite in the microstructure. The flake form of graphite, when well dispersed throughout the microstructure, provides a considerably higher cathode-to-anode surface area ratio than does the nodular form of graphite in ductile cast irons. Although ductile (and malleable) cast iron alloys can still experience graphitic corrosion, the attack is usually superficial.

Graphitic corrosion results in the formation of a layer of iron-oxide-based corrosion products containing dispersed graphite particles. This layer is mechanically weak and cannot support appreciable

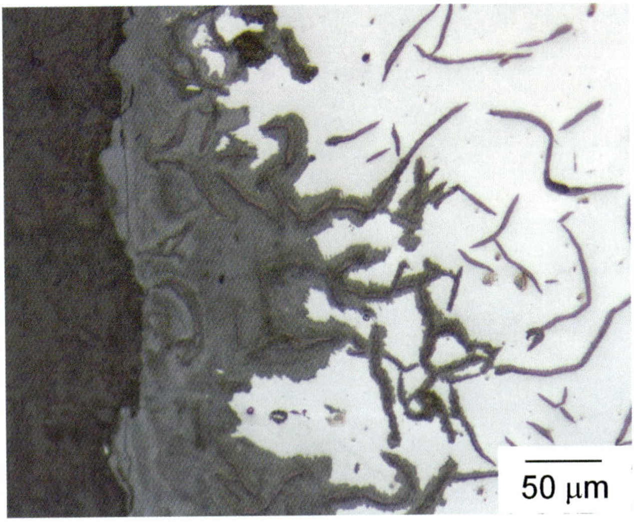

FIGURE 2.4 Example of graphitic corrosion of gray cast iron.

stresses. There may be no observable change in the dimensional appearance of a component that has experienced appreciable graphitic corrosion. This situation can be a major concern in pressurized systems, as the remaining, thinner intact metal is generally weaker, inherently brittle, and can experience sudden overload-type failures. An example of a failed gray cast iron end cap on a dead leg of a piping system is shown in Fig. 2.5. The cap experienced some graphitic corrosion on the internal surface that weakened the component slightly. A pressure surge during operation resulted in elevated stress

FIGURE 2.5 Brittle fracture of a gray cast iron end cap on a dead leg of a piping system. Failure occurred when a pressure surge in the line was sufficient to cause a fracture in a thin portion of the cap that was further weakened by graphitic corrosion.

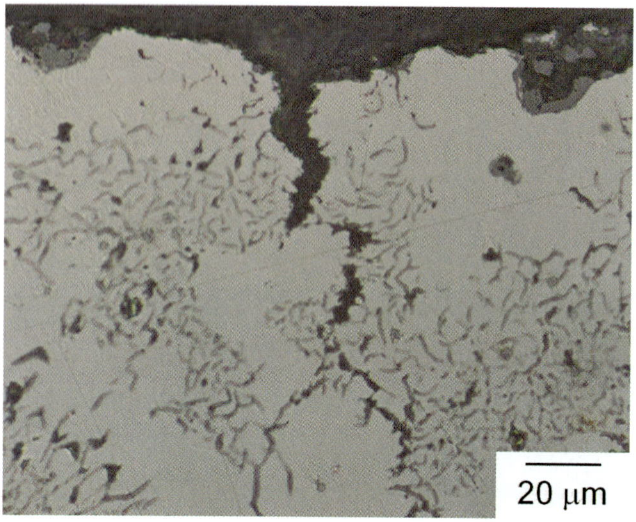

FIGURE 2.6 Brittle fracture path along regions containing fine graphite flakes in the microstructure of a gray cast iron end cap.

levels that caused failure along regions containing graphite flakes in the microstructure (Fig. 2.6).

Galvanic Corrosion

In addition to graphitic corrosion of cast iron due to microgalvanic interactions between phases in the microstructure, attack due to galvanic corrosion on a macro scale can also occur. A special situation arises when a new cast iron component is attached to a cast iron component that has been in service for an extended period of time and has developed a layer of corrosion products containing graphite. An example of this scenario occurs when a new impeller is placed into a cast iron pump housing that has been in service. This can create a galvanic couple between the impeller and the housing. The newly installed impeller with a clean metal surface is anodic to the graphite-laden corrosion product layer present on the housing. The newly installed impeller can experience accelerated attack.

Erosion-Corrosion

As mentioned previously, the corrosion products that form on cast iron components, especially due to graphitic corrosion, are typically soft and weakly adherent to metal surfaces. If the cast iron component is subjected to high water velocity and/or turbulent conditions, the corrosion product layer can be dislodged from the metal surface. Removal of the overlying layer results in the re-formation of the corrosion products on the exposed metal, promoting accelerated wastage in affected areas. Pump impellers are components that can be susceptible to wastage from erosion-corrosion. Alloyed cast irons generally offer increased resistance to erosion-corrosion.

Microbiologically Influenced Corrosion

Cast iron alloys can be subject to microbiologically influenced corrosion (MIC) under certain conditions specific to the microorganism. For instance, sulfate-reducing bacteria (SRB) can contribute to attack in oxygen-depleted environments that develop beneath deposits or in stagnant areas of a distribution line. Some references indicate that attack related to SRB on cast iron can be similar in appearance to a graphitically corroded structure,[2] and that comparison of the sulfur content of the corrosion products to the metal can indicate that SRB activity may be the primary cause of corrosion. It is assumed that, if the ratio of sulfur in the corrosion products to that in the metal is substantially more than a similar ratio for other elements in the corrosion products and metal, then the sulfur that accumulated is probably due to SRB activity. However, it is important to distinguish between sulfur as sulfide and sulfur as sulfate in the corrosion products, as aggressive anions such as sulfate may also accumulate in wasted areas to further promote graphitic corrosion. Sulfates, as well as chlorides, can locally increase beneath deposit and corrosion product layers due to the development of ion concentration cells. Careful diagnosis of MIC requires consistency of all factors necessary for corrosion to be active, which should include confirmation of the presence of microorganisms responsible for the wastage at attack sites.

Microbiologically influenced corrosion related to sulfate-reducing bacteria is known to be a primary mechanism that can affect buried pipelines.[3] However, other corrosion mechanisms related to soil parameters, such as resistivity, moisture drainage, degree of aeration, pH, and chloride concentrations must also be considered in determining the cause of corrosion. Corrosion in soils can be mitigated by application of protective coatings, the proper use of cathodic protection, and the use of backfill that allows for good drainage from the metal surfaces.

Intergranular Corrosion

Intergranular corrosion is rare in cast iron alloys. Ammonium nitrates have been linked to intergranular attack of unalloyed cast iron,[4] a condition similar to intergranular corrosion of carbon steels by nitrate solutions.

Stress Corrosion Cracking

Corrodents known to cause stress corrosion cracking of steel, such as concentrated nitrates and hot concentrated caustic, are also possible stress corrosion cracking agents for unalloyed cast iron. However, design stress levels are generally low for cast iron components due to

FIGURE 2.7 Nodular cast iron pipe coupling that likely fractured due to stresses caused by water freezing in the line.

the inherent brittleness for some types. In addition, the severe environments necessary to cause such cracking are very uncommon in cooling water systems. Therefore, failures of cast iron due to stress corrosion cracking are extremely rare.

Brittle fractures that occur in cast iron components are more typically associated with overload conditions. One situation that can cause unsuspected mechanical overloading is freezing of the water in the component. Freezing of water into ice is accompanied by expansion, which can cause elevated stress levels to develop in the component. An example of a nodular cast iron pipe coupling that was fractured by stresses that developed due to water freezing in the line is shown in Fig. 2.7.

Surface Finish

Cast iron alloys will generally have a surface finish related to the mold material used to cast the component. For instance, sand castings will have a fine, pebbly surface contour. Surface defects or open porosity from the casting process can provide sites that are more susceptible to corrosive attack. Gray cast iron and nodular cast iron can be easily machined, which results in a smooth surface finish with very shallow, parallel grooves.

Corrosion Inhibitors

Corrosion inhibitors used to protect carbon and low-alloy steels will generally reduce the corrosion of unalloyed cast iron alloys also. However, aggressive water chemistries that promote attack by graphitic corrosion, such as high concentrations of chlorides and sulfates, can significantly limit the effectiveness of corrosion inhibitors.

References

1. P. A. Schweitzer (ed.), *Corrosion Engineering Handbook*, Marcel Dekker, Inc., New York, 1996, p. 258.
2. L. L. Shreir, R. A. Jarman, and G. T. Burstein (eds.), *Corrosion Metal Environment Reactions,* 3d ed., Butterworth Heinemann, Oxford, 1994, p. 3:105.
3. S. D. Cramer and B. S. Covino, Jr. (eds.), *ASM Handbook,* Volume 13B Corrosion: Materials, Materials Park, Ohio, 2005, p. 46.
4. S. D. Cramer and B. S. Covino, Jr. (eds.), *ASM Handbook,* Volume 13B Corrosion: Materials, Materials Park, Ohio, 2005, p. 46.

CHAPTER 3

Stainless Steel Alloys

Introduction

Stainless steel is typically defined as a steel alloy with chromium content in excess of 10.5 percent. This chromium concentration generally imparts high corrosion resistance to these steels in mildly corrosive environments such as moist air or freshwater. As such, dull or dark-colored corrosion products do not form and stain the surfaces. This is the basis for the term *stainless steel.* Stainless steels are very familiar for their use in flatware, cutlery, and cookware since they remain clean under consistent exposure to environments that would be highly corrosive to carbon and low-alloy steels and cast iron. For similar reasons stainless steels are widely used in industrial water systems.

Stainless steels are typically and primarily chosen for their superior corrosion resistance in water service. Under proper conditions, stainless steels will provide extended service life and will allow thinner tube or pipe walls to be used than for carbon and low-alloy steels. Another benefit to the use of stainless steel is the substantial reduction in corrosion product formation that would be experienced on less corrosion resistant carbon and low-alloy steels. Fewer corrosion products will allow greater heat transfer and fluid flow. In addition, reduced corrosion product formation will control or prevent operational problems such as blockage of narrow orifices by spalled corrosion products that are carried downstream, such as for fine spray nozzles. Stainless steels may also be more suitable for use due to superior mechanical properties such as strength, wear, and erosion resistance where copper and aluminum alloys would be unsuitable. Common uses for stainless steels in cooling water systems include heat exchanger tubing, surface condenser tubing, service water system piping, and steel continuous caster secondary cooling system piping.

As described in Chap. 1, the corrosion product layers that form on carbon and alloy steels provide limited corrosion protection. Conversely, stainless steels with chromium contents that exceed 10.5 percent produce a thin, very stable, very protective, passive chromium oxide layer under oxidizing conditions. Oxidizing conditions are produced when surfaces are exposed to oxygen in the air and dissolved oxygen in water. The passive film may be damaged by chemical attack, mechanical abrasion, or the presence of microstructural irregularities, such as nonmetallic inclusions at the surface. Chemical attack may be caused by exposure to aggressive chemical environments. The passive film will spontaneously re-form when exposed to oxidizing conditions.

Due to the passive nature of the chromium oxide film, which causes the metal to polarize readily, stainless steels tend to experience localized corrosion. Localized corrosion refers to metal loss that interrupts a mostly unattacked surface at scattered sites. Localized corrosion begins when the protective chromium oxide film is removed at discrete sites. These sites are not exposed to oxidizing conditions that are sufficient to repair the passive film. In addition, the difference in corrosion resistance between the large passive surface and the small localized corrosion sites is driven by galvanic effects. Galvanic corrosion and galvanic effects will be described in more detail in Chap. 16, "Galvanic Corrosion." Also, occluded sites will promote diffusion and concentration of aggressive anions, such as chloride in water solution, which will tend to remove protective oxide films.

Localized corrosion produces progressively deeper metal loss at the sites of corrosion without causing significant overall weight loss to a component. A severe form of localized corrosion is called pitting, where the depth of metal loss penetrates more deeply than the opening at the surface. Some pitting may penetrate significant distances below the surface in lateral directions along microstructural features; this is referred to as "tunnel pitting." In some cases, localized corrosion will penetrate through a stainless steel component more rapidly than a carbon steel component will corrode through due to general corrosion.

Resistance of stainless steels to localized corrosion and pitting depends upon several factors, including temperature, pH, concentration of aggressive anions, surface conditions, and alloy composition. Addition of certain alloying elements to stainless steels enhances corrosion resistance. For instance, at the minimum chromium concentration of 10.5 percent that classifies an alloy as stainless steel, corrosion resistance provided by the passive film is rather weak. Much greater protection is provided by higher chromium contents. Molybdenum additions improve resistance to pitting and crevice corrosion in the presence of reducing acids and halogen ions such as chlorides and bromides. Nickel additions increase resistance to corrosion in reducing environments and to mineral acids.

Perhaps the most significant contributor to corrosion of stainless steels in cooling water systems is dissolved chloride ions. Chloride ions may remove the protective chromium oxide layer to allow localized corrosion and pitting to occur. A relative indicator of resistance to pitting by chloride may be expressed by the well-known pitting resistance equivalent number (PREN).[1] This number is based upon a calculation involving the concentration of three elements, chromium, molybdenum, and nitrogen. The higher the PREN, the greater the resistance to chloride pitting will be. The PREN is expressed by the following calculation:

$$\text{PREN} = \%\ \text{Chromium} + (3.3 \times \%\ \text{Molybdenum}) + (16 \times \%\ \text{Nitrogen}) \tag{3.1}$$

% = weight percent

Types of Stainless Steel

Stainless steels are classified into one of five categories according to their crystalline structures. These categories are austenitic, ferritic, duplex, martensitic, and precipitation hardening stainless steels. The different types of stainless steel provide different benefits. Some alloys are more suitable for certain applications and environments than others. A list of the more commonly used grades of stainless steel is shown in Table 3.1. The five different types of stainless steels, including their benefits and limitations, are described in detail next.

Austenitic Stainless Steels

Austenitic stainless steels, which are the most commonly used stainless steels in cooling water systems, must contain more than 16 percent chromium and over 6 percent nickel by weight. The most common austenitic stainless steels have an American Iron and Steel Institute (AISI) designation in the 300s and are described as 300 series stainless steels. The 304 and 304L grades of stainless steel account for most of the stainless steel that is used in the United States.

Typically, the more commonly used 304 and 316 grades contain at least 18 percent chromium and 8 percent nickel to form a microstructure that is primarily austenitic (Fig. 3.1) with very little or no residual ferrite. However, in some cases, residual ferrite may form in the mostly austenitic microstructure (Fig. 3.2). Galvanic corrosion may occur along the interfaces of the ferrite and austenite (Fig. 3.3). Galvanic corrosion is described in Chap. 16. The residual ferrite decreases the overall corrosion resistance of the steel, and it should be minimized or prevented. Nickel is the primary alloying element that promotes the formation of the austenitic microstructure. In some steels, residual ferrite may form when nickel contents are near the lower threshold for the alloy specification and/or when improper manufacturing processes are used.

Austenite has a face-centered cubic crystalline structure. This crystal structure imparts favorable mechanical properties to the alloy, such as high ductility and excellent formability that are not

UNS Alloy Number	AISI Alloy Or Other Number	C (max)	Mn (max)	P (max)	S (max)	Si (max)	Cr	Ni	Mo (max)	Other Elements	PRE Number (max)
Austenitic Stainless Steels											
S30100	301	0.15	2.00	0.05	0.03	1.00	16.00–18.00	6.00–8.00			18.0
S30200	302	0.15	2.00	0.05	0.03	1.00	17.00–19.00	8.00–10.00			19.0
S30300	303	0.15	2.00	0.20	0.15 min	1.00	17.00–19.00	8.00–10.00	0.60		21.0
S30400	304	0.08	2.00	0.05	0.03	1.00	18.00–20.00	8.00–10.00			20.0
S30403	304L	0.03	2.00	0.05	0.03	1.00	18.00–20.00	8.00–10.00			20.0
S30500	305	0.12	2.00	0.05	0.03	1.00	17.00–19.00	10.00–13.00			19.0
S30800	308	0.08	2.00	0.05	0.03	1.00	19.00–21.00	10.00–12.00			21.0
S30900	309	0.20	2.00	0.05	0.03	1.00	22.00–24.00	12.00–15.00			24.0
S30940	309S	0.08	2.00	0.05	0.03	1.00	22.00–24.00	12.00–15.00			24.0
S31000	310	0.25	2.00	0.05	0.03	1.50	24.00–26.00	19.00–22.00			26.0
S31008	310S	0.08	2.00	0.05	0.03	1.50	24.00–26.00	19.00–22.00			26.0
S31600	316	0.08	2.00	0.05	0.03	1.00	16.00–18.00	10.00–14.00	2.00–3.00		28.0
S31603	316L	0.03	2.00	0.05	0.03	1.00	16.00–18.00	10.00–14.00	2.00–3.00		28.0
S31700	317	0.08	2.00	0.05	0.03	1.00	18.00–20.00	11.00–15.00	3.00–4.00		33.0
S31703	317L	0.03	2.00	0.05	0.03	1.00	18.00–20.00	11.00–15.00	3.00–4.00		33.0
S31800	318	0.08	2.00	0.05	0.03	1.00	17.00–19.00	13.00–15.00	2.00–3.00	Cb 10XC min	29.0
S32100	321	0.08	2.00	0.05	0.03	1.00	17.00–19.00	9.00–12.00		Ti 5XC min	19.0
S33000	330	0.20	2.00	0.03	0.03	1.50	14.50–16.50	34.00–37.00			16.5

S34700	347	0.08	2.00	0.05	0.03	1.00	17.00–19.00	9.00–13.00		Cb-Ta 10XC min	19.0
S31254	254 SMO	0.02	2.00	0.03	1.00	1.00	19.5–20.5	17.5–18.5	6.0–6.5	Cu 0.5–1.0, N 0.18–0.22	45.0
N08367	Al-6XN	0.03	2.00	0.04	0.03	1.00	20.0–22.0	23.5–25.5	6.0–7.0	Cu 0.75, N 0.18–0.25	49.0
N08904	904L	0.02	2.00	0.04	0.03	1.00	19.0–23.0	23.0–28.0	4.0–5.0	Cu 1.0–2.0	39.5
N08020	Alloy 20	0.07	2.00	0.04	0.03	1.00	19.0–21.0	32.0–38.0	2.0–3.0	Cu 3.0–4.0, Cb + Ta Cb + Ta 8XC-1.0	31.0
Ferritic Stainless Steels											
S40500	405	0.08	1.00	0.04	0.03	1.00	11.50–14.50			Al 0.10–0.30	14.5
S40900	409	0.08	1.00	0.05	0.05	1.00	10.50–11.75	0.50		Ti 6XC - 0.75	11.75
S43000	430	0.12	1.00	0.04	0.03	1.00	14.00–18.00				18.0
S43400	434	0.12	1.00	0.04	0.03	1.00	14.00–18.00		0.75–1.25		22.0
S43500	435	0.12	1.00	0.04	0.03	1.00	14.00–18.00			Cb + Ta 0.40–0.60	18.0
S43600	436	0.12	1.00	0.04	0.03	1.00	14.00–18.00		0.75–1.25	Cb + Ta 0.40–0.60	22.0
S44200	442	0.20	1.00	0.04	0.03	1.00	18.00–23.00				23.0
S44300	443	0.20	1.00	0.04	0.03	1.00	18.00–23.00			Cu 0.90–1.25	23.0
S44600	446	0.20	1.50	0.04	0.03	1.00	23.00–27.00			N 0.25 max	31.0
S44635	Monit	0.03	1.00	0.04	0.03	0.75	24.50–26.00	3.5–4.5	3.5–4.5	N 0.035, Nb + Ti 0.2 + 4(C + N)-0.8	41.0

Table 3.1 Commonly Used Grades of Stainless Steel[9]

UNS Alloy Number	AISI Alloy Or Other Number	C (max)	Mn (max)	P (max)	S (max)	Si (max)	Cr	Ni	Mo (max)	Other Elements	PRE Number (max)
Ferritic Stainless Steels											
S44660	Sea-Cure	0.03	1.00	0.04	0.03	1.00	25.00–27.00	1.5–3.5	2.5–3.5	N 0.035, Nb + Ti 0.2 + 4(C + N)-0.8	39.0
S44735	AL-29-4C	0.03	1.00	0.04	0.03		28.00–30.00	1.00	3.6–4.2	N 0.045, Nb + Ti 6(C + N)	44.5
S44800	AL-29-4-2	0.01	1.00	0.40	0.03		28.00–30.00	2.0–2.5	3.5–4.2		44.0
S44627	E-Brite	0.01	0.40	0.02	0.02	0.40	25.00–27.00	0.50	0.75–1.5	Nb 0.05–0.2, Cu 0.2, N 0.015	32.0
Martensitic Stainless Steels											
S40300	403	0.15	1.00	0.04	0.03	0.50	11.50–13.00				13.0
S41000	410	0.15	1.00	0.05	0.03	1.00	11.50–13.50				13.5
S41400	414	0.15	1.00	0.05	0.03	1.00	11.50–13.50	1.25–2.50			13.5
S41600	416	0.15	1.25	0.06	0.15 min	1.00	12.00–14.00		0.60		16.0
S42000	420	0.15 min	1.00	0.04	0.03	1.00	12.00–14.00				14.0
S43100	431	0.20	1.00	0.04	0.03	1.00	15.00–17.00	1.25–2.50			17.0

Duplex Stainless Steels											
S32900	329	0.08	2.00	0.04	0.02	1.00	23.0–28.00	2.5–5.0	1.0–2.0		28.0
S32304	2304	0.03	2.00	0.04	0.02	1.00	21.5–24.50	3.0–5.5	0.05–0.6	N 0.05–0.2	30.0
S32205	2205	0.03	2.00	0.04	0.02	1.00	22.0–23.00	4.5–6.5	3.0–3.5	N 0.14–0.2	38.0
S32950	7-Mo PLUS	0.03	2.00	0.04	0.02	1.00	26.0–29.00	3.5–5.2	1.0–2.5	N 0.15–0.35	43.0
S32550	Ferralium 255	0.04	2.00	0.04	0.02	1.00	24.0–27.00	4.5–6.5	2.9–3.9	Cu 1.5–2.5, N 0.1–0.25	44.0
S32750	2507	0.03	2.00	0.04	0.02	1.00	24.0–26.00	6.0–8.0	3.0–5.0	Cu 0.5, N 0.24–0.32	47.5
Precipitation Hardening Stainless Steels											
S17400	17-4PH	0.04	0.40	0.04	0.03	0.50	16.50	4.25		Cb 0.25, Cu 3.6	16.5
S17700	17-7PH	0.07	0.70	0.04	0.03	0.40	17.00	7.00		Al 1.15	17.0
S35500	AM-350	0.10	0.75	0.04	0.03	0.35	16.50	4.25	2.75	N 0.1	27.0
Cast Austenitic Stainless Steels											
	CF8	0.08	1.50	0.04	0.04	2.00	18.00–21.00	8.00–11.00			21.0
	CF8M	0.08	1.50	0.04	0.04	2.00	18.00–21.00	9.00–12.00	2.0–3.0		31.0
	CN7M	0.07	1.50	0.04	0.04	1.50	18.00–22.00	21.00–31.00		Cu 3.0	22.0
	CB30	0.30	1.00	0.04	0.04	1.50	18.00–22.00				22.0

TABLE 3.1 Commonly Used Grades of Stainless Steel[2] *(Continued)*

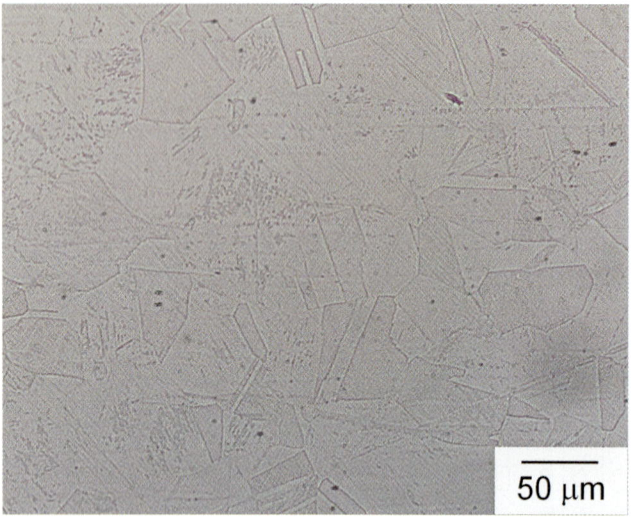

Figure 3.1 Microstructure of austenitic stainless steel consisting of twinned grains of austenite. Etchant: electrolytic oxalic acid

Figure 3.2 Austenitic stainless steel containing stringers of residual ferrite phase. Etchant: electrolytic oxalic acid

available with the other types of stainless steels. Due to this crystal structure, the austenitic alloys are generally not ferromagnetic, so they will not be attracted to a magnet. The nickel in austenitic stainless steels also imparts improved resistance to reducing environments. Higher nickel content in austenitic stainless steels increases

FIGURE 3.3 Corrosion penetrated preferentially along residual ferrite stringers. Etchant: electrolytic oxalic acid

resistance to corrosion in media that are only slightly oxidizing or not oxidizing at all.

Molybdenum improves resistance to corrosion in the presence of reducing acids or halogen ions (chlorides, bromides, iodides). The 316 grade of stainless steel is essentially the 304 grade with substitution of about 2 to 3 percent molybdenum for iron in the alloy. Consequently, the 316 grade has substantially greater chloride pitting resistance than the 304 grade. Other alloys containing higher molybdenum concentrations are also used, which have even greater pitting resistance.

The austenitic stainless steels are considered to have better corrosion resistance than the lower chromium content ferritic stainless steels, as well as the martensitic and precipitation hardening stainless steels. However, the more commonly used 304 and 316 grades of austenitic stainless steel tend to be less resistant to corrosion than the duplex stainless steels.

Alloy selection for cooling water service may be based upon guidance provided by the PREN number for the alloy as described in the introduction to this chapter. The more common 304 and 316 stainless steels are unsuitable in high-chloride and seawater service. As such, the use of austenitic stainless steels with higher chromium and molybdenum concentrations is required. These alloys are in the classification of superaustenitic stainless steels. Superaustenitic grades contain higher concentrations of alloying elements that improve corrosion resistance, including chromium, molybdenum, nickel, nitrogen, and copper. Chromium concentration in austenitic stainless steels may be as high as about 30 percent. Higher chromium concentrations in such alloys cause difficulty in fabrication and welding. Consequently, control of chloride pitting in austenitic stainless steels is achieved through suitable molybdenum additions, as high as 7 percent. Nitrogen may also be added to some stainless steel alloys as a substitute for carbon, since nitrogen increases pitting resistance. Superaustenitic grades that may be suitable for high-chloride content waters or seawater include Al6XN and 254SMO.

It may be necessary to specify low carbon concentration in austenitic stainless steels, particularly when the component is to be welded or will be exposed to high temperatures. An alloy with the L designation, such as 304L or 316L, will contain less than 0.03 percent carbon. Low carbon content is desirable to control sensitization within the weld-heat-affected zones during welding. Sensitization refers to the formation of chromium carbides along grain boundaries in stainless steels. When these carbides form, they deplete the surrounding metal of chromium, thus increasing susceptibility to corrosion. The chromium carbides form when the steel is heated to high temperatures such as during heat treatment, welding, or high-temperature service. Sensitization may also be controlled with additions of stabilizing elements such as niobium and titanium. A more detailed discussion of the sensitization of stainless steels appears in Chap. 26. If the steel will not be welded or heated to high temperatures such as 1000 to 1600°F (538–871°C), there will be no need to specify an L grade or stabilized grade, which is more costly.

The austenitic stainless steels are particularly susceptible to chloride stress corrosion cracking (SCC) in cooling water service under specific conditions. In particular, the more commonly used 304 and 316 grades are susceptible to chloride SCC at metal temperatures exceeding about 120 to 130°F (49–54°C), even at relatively low chloride concentration in the water. More detailed information will be presented in Chap. 22. The use of substantially higher concentrations of nickel (typically >30 percent) will reduce or eliminate chloride stress corrosion cracking in austenitic stainless steels, depending upon the chloride concentration as shown in Fig. 3.4. Though austenitic stainless steels containing higher concentrations of certain critical elements may be beneficial, they are less commonly used, will be more expensive, and typically have limited availability especially in large quantities.

Though such alloys are not necessary for cooling water service, certain austenitic stainless steel alloys contain other alloy additions that promote resistance to specific environments. For instance Alloy 20, also known as Carpenter 20, which contains a significant concentration of copper (2–4 percent), provides excellent resistance to corrosion by sulfuric acid.

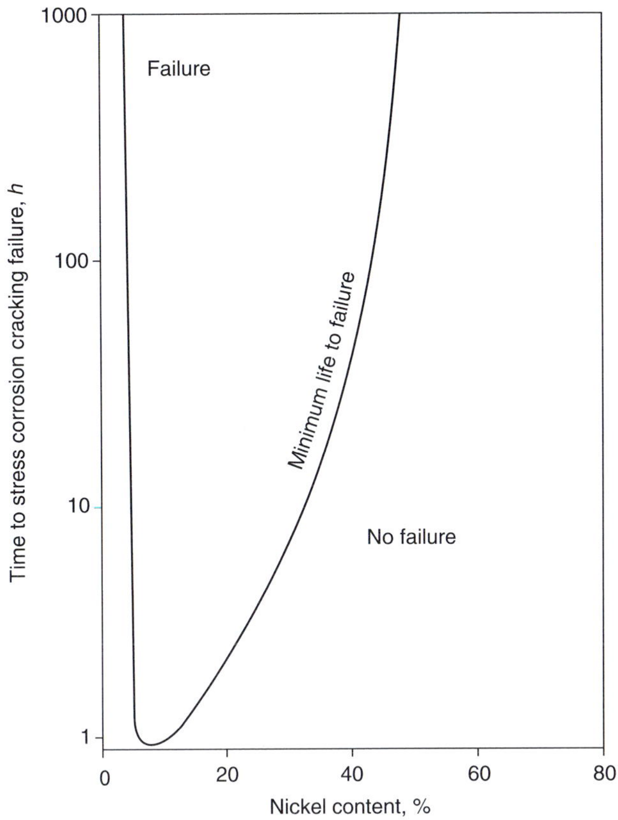

Figure 3.4 Graph illustrating resistance of austenitic stainless steels to chloride stress corrosion cracking as a function of nickel concentration.[3] (*Courtesy of McGraw-Hill.*)

The more commonly used austenitic stainless steels in cooling water systems are listed in Table 3.1.

Another important consideration when specifying austenitic stainless steel is the method of processing. Most austenitic stainless steel components used in cooling water service are in the wrought form, such as tube, pipe, sheet, and plate. These forms are produced by various rolling, drawing, and other forming operations. However, some components in cooling water systems may be cast to their shape. Though the composition (primarily the chromium and nickel concentration) may be similar, a cast microstructure will be significantly different than a wrought microstructure. For instance, 316 wrought stainless steel will have chromium, nickel, and molybdenum concentration similar to that of CF8M cast stainless steel, but the microstructures of the wrought and cast alloys will be substantially different. Due to the higher concentrations of silicon (1.5–2 percent)

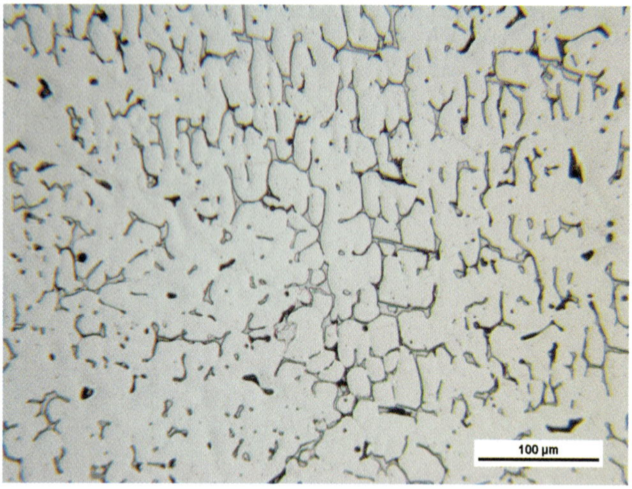

FIGURE 3.5 Microstructure of cast stainless steel consisting of ferrite phase in an austenite matrix. Etchant: electrolytic oxalic acid

needed to promote fluidity, cast stainless steels will tend to contain significant amounts of ferrite phase, Fig. 3.5, while a wrought material will not, Figs. 3.1 and 3.2. In addition, many of the cast stainless steels will contain high carbon concentrations (>0.08 percent), which will tend to form carbides within the ferrite phase. These carbides will decrease corrosion resistance significantly. Due to the second phase and carbides in a cast microstructure, it will be more susceptible to corrosion than the equivalent wrought grade. The corrosion will penetrate preferentially along the ferrite phase. As such, cast stainless should not be considered to have equivalent corrosion resistance to wrought stainless steel. A cast austenitic stainless steel will be slightly attracted to a magnet due to the ferrite content.

Ferritic Stainless Steels

These steels contain between 10.5 and 30 percent chromium, but little (0–4.5 percent) nickel or other alloying elements, except for molybdenum. At this composition, these steels have a microstructure consisting of ferrite (Fig. 3.6). The most common ferritic stainless steels have an AISI designation in the 400s and are described as 400 series stainless steels. Some of the commonly used ferritic stainless steels are listed in Table 3.1. Some martensitic grades, which will be described later, also have 400 series designations. Ferrite has a body-centered cubic crystal structure. Due to this crystal structure, the ferritic alloys are ferromagnetic, so they will be attracted to a magnet. Ferritic stainless steels are not as widely used as the austenitic grades in cooling water systems. This is largely due to the

FIGURE 3.6 Microstructure of ferritic stainless steel consisting of grains of ferrite containing dispersed carbides. Etchant: electrolytic oxalic acid

less desirable mechanical properties of ferritic stainless steels, particularly low ductility, as well as lower strength, which would make it less desirable for high-pressure applications. Ferritic stainless steels tend to be formed into thin components due to challenges in manufacturing processes.

As a group, the low-chromium ferritic stainless steels are less corrosion resistant than the austenitic, duplex, and precipitation hardening grades. Unlike austenitic grades, ferritic stainless steels are very resistant, or even immune to chloride stress corrosion cracking. Ferritic stainless steels that contain higher concentrations of chromium (25–30 percent), and molybdenum (up to 4.5 percent) will have greater resistance to pitting by aggressive anions such as chloride. These alloys also have lower carbon concentration (<0.03 percent), which is substituted with nitrogen. They also contain additions of stabilizing elements, such as niobium and titanium to control sensitization. Some of these alloys have exhibited good resistance to pitting in high-chloride, brackish, and seawater service. Typically it is recommended that the alloy have a minimum PREN of 32 for seawater service. Therefore the alloys used for such service typically contain high concentrations of chromium and molybdenum. Ferritic stainless alloys used in seawater service include SEA-CURE, AL-29-4C, and AL-29-4-2.

Duplex Stainless Steels

These steels contain chromium concentrations ranging from 18 to 29 percent, 0 to 5 percent molybdenum, and 4 to 8 percent nickel and/or

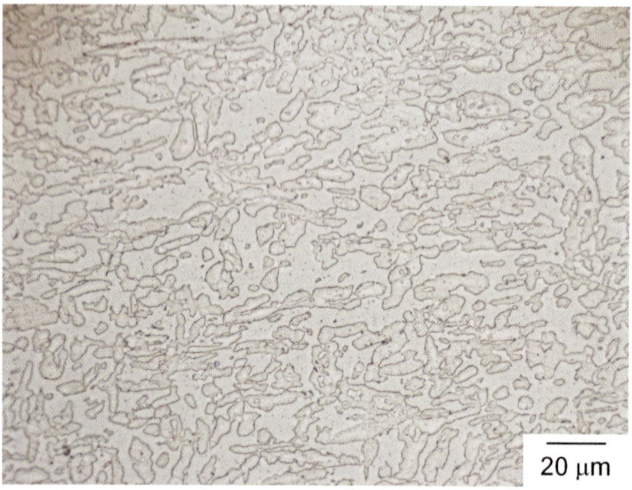

Figure 3.7 Microstructure of duplex stainless steel consisting of ferrite and austenite. Etchant: electrolytic oxalic acid

other austenite stabilizers such as nitrogen and manganese, to promote the formation of a balanced microstructure of ferrite and austenite at room temperature (Fig. 3.7). The volume fraction of each phase is usually near 50 percent. The higher nickel content duplex alloys tend to contain higher volume fractions of austenite, since nickel is an austenite stabilizer. The two-phase alloy possesses favorable characteristics of both phases. Due to the presence of ferrite phase, the duplex alloys are ferromagnetic, so they will be attracted to a magnet. The high concentrations of chromium and molybdenum provide excellent corrosion resistance, while the austenite phase provides increased ductility. The strength of the alloy also exceeds that of either phase alone. Super duplex stainless steels, such as grade 2507, contain higher concentrations of chromium and molybdenum (at the upper limits of the ranges listed above). The general corrosion resistance of duplex stainless steels is considered to be superior to all other types of stainless steel, including the 18 percent chromium, 8 percent nickel austenitic steels. While the austenite phase is susceptible to chloride stress corrosion cracking under certain conditions, the ferrite phase is not. Therefore, resistance of duplex stainless to chloride stress corrosion cracking is much greater than the austenitic grades. If proper welding procedures are followed, welding should not significantly reduce the corrosion resistance of duplex stainless steels.

Martensitic Stainless Steels

These steels may be used in applications where both mechanical wear and corrosion must be controlled. These steels typically contain 11.5 to 17 percent chromium and higher carbon contents (0.15–0.25 percent) than the other types of stainless steel. The amounts of other alloying elements such as nickel and molybdenum are low (0.0–2.5 percent). The higher carbon concentration allows the alloy to be heat treated to produce a microstructure of martensite. Martensite has a body-centered tetragonal crystal structure. Due to this crystal structure, the martensitic alloys are ferromagnetic, so they will be attracted to a magnet. Heat treatment to produce a martensitic microstructure involves heating the component to high austenitizing temperatures, followed by rapid cooling, referred to as quenching. Quenching produces a hard and brittle structure that is susceptible to cracking under shock loads, so a subsequent tempering heat treatment is commonly employed. The martensitic and tempered martensite microstructures provide high hardness and strength (Fig. 3.8). High hardness values impart excellent resistance to mechanical wear. These properties allow the alloy to be used for applications such as steam turbines, shafts, and corrosion-resistant hardfacing, such as for continuous caster rolls.

Due to the low chromium and high carbon content of these alloys, their corrosion resistance is limited. The higher concentration of carbon that is needed to impart strength and hardness also makes the martensitic stainless steels more susceptible to sensitization at high temperatures, including during welding. Sensitization will substantially reduce the corrosion resistance of the steel. Sensitization is described in more detail in Chap. 26. Therefore, martensitic stainless steels are the least corrosion resistant of all stainless steel types. Since these steels do not contain austenite, they are considered to be resistant to chloride stress corrosion cracking.

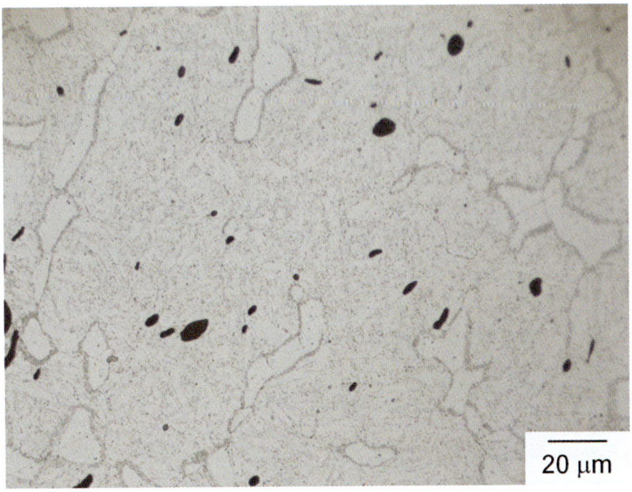

Figure 3.8 Microstructure of martensitic stainless steel consisting of tempered martensite. Etchant: electrolytic oxalic acid

Precipitation Hardening Stainless Steels

These steels contain 12 to 18 percent chromium and 3 to 8.5 percent nickel. Some contain small amounts of molybdenum (up to 2.75 percent). These alloys may also contain copper (1.25–3.6 percent) or aluminum (0.9–1.35 percent), which form intermetallic precipitates following proper high-temperature heat treatment or aging. The precipitates form throughout the alloy to cause it to harden. Aging means that a hardened structure forms gradually over time at lower temperatures due to diffusion of the precipitate within the alloy. Most precipitation hardening steels have a duplex microstructure of ferrite and austenite (Fig. 3.9), but some have a martensitic structure similar to that shown in Fig. 3.8. The precipitates that promote hardening are not resolvable during metallographic examination. Due to the ferrite or martensite phases being present, the precipitation hardening alloys are ferromagnetic, so they will be attracted to a magnet. Similar to the martensitic stainless steels, the precipitation hardening steels have excellent strength and hardness. These steels are not commonly used in cooling water systems. However, the mechanical properties of these alloys make them useful for shafts, turbines, and hardfacing. The precipitation hardening alloys are not suitable for high-temperature applications, since their mechanical properties will be compromised. The precipitation hardening alloys are considered to have corrosion resistance which approaches that of the austenitic stainless steels, and they are superior to ferritic and martensitic stainless steels. Their corrosion resistance is inferior to that of standard duplex stainless steels. Due to the heat treatment or aging process for hardening these steels,

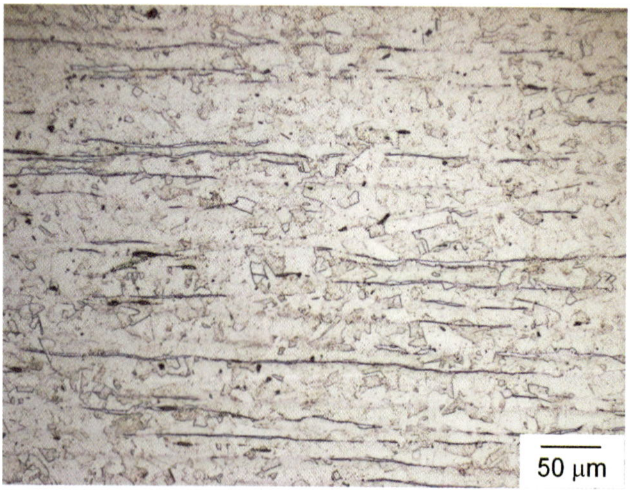

FIGURE 3.9 Microstructure of precipitation hardening stainless steel consisting of islands of ferrite in an austenite matrix. Etchant: electrolytic oxalic acid

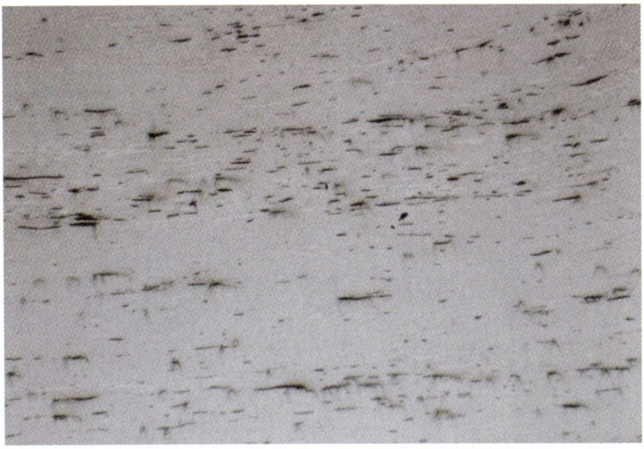

FIGURE 3.10 High population of manganese sulfide inclusions in free-machining grade of stainless steel. Etchant: electrolytic oxalic acid

any heat treatment or welding will negatively affect material properties and will also promote sensitization and associated corrosion. Welding should be conducted prior to the final age hardening heat treatment.

Free-Machining Grades

Alloys referred to as "free-machining" grades contain higher concentrations of certain alloying elements, such as sulfur and selenium, in order to promote the formation of chips during machining. Increased sulfur content (0.15–0.25 percent) causes a larger population of manganese sulfide inclusions to form within the matrix (Fig. 3.10). The inclusions have no structural integrity. Therefore, during machining processes, steel chips will tend to fracture away from the matrix along the inclusions. The sites of the inclusions also will serve as preferential locations for corrosion to initiate and propagate (Fig. 3.11). Therefore, free-machining grades of stainless steel are susceptible to corrosion in cooling water service.

Forms of Corrosion

General Corrosion

Under certain rare, severe upset conditions, stainless steels can experience general corrosion. This may occur if the passive film is uniformly stripped from the surface and cannot re-form. Such corrosion may occur when the surface is exposed to reducing environments, particularly by certain strong mineral acids such as hydrochloric acid at high concentration, or sulfuric acid at certain concentrations.

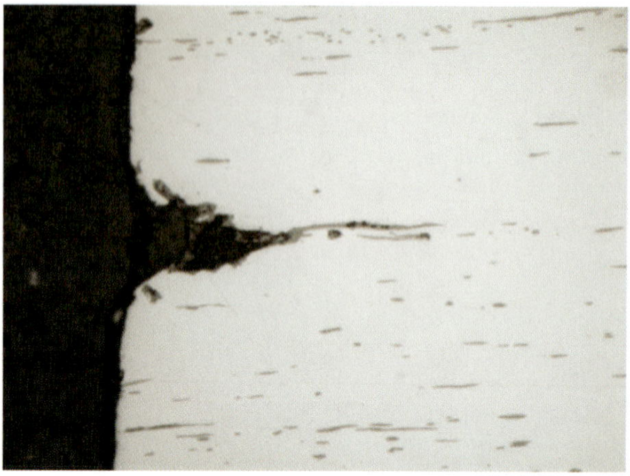

FIGURE 3.11 Corrosion penetrating preferentially along inclusion stringers. Etchant: electrolytic oxalic acid

Pitting

Perhaps the most insidious form of corrosion that occurs on stainless steels is pitting. Pitting of stainless steels is influenced by the concentration of aggressive anions in the water. Concentrated chloride ions commonly cause pitting on stainless steels. Pitting is characterized by localized corrosion where the width at the surface is narrower than the depth of penetration. Pits may penetrate through a tube or pipe wall within a very short period of time under certain circumstances (Fig. 3.12).

FIGURE 3.12 Stainless steel beaker which perforated within 64 hours [4106 mpy (104 mmpy)] due to exposure to water that contained a chlorine tablet.

Table 3.1 lists the PREN for various grades of stainless steel, which provides information concerning relative chloride pitting resistance of the alloys. Another important factor that influences pitting is the metal temperature. A guide to the chloride pitting resistance of the various stainless steels is the critical pitting temperature (CPT). The graphs shown in Fig. 3.13 illustrate the critical pitting temperatures of a select group of stainless steels as a function of varying metal temperatures and chloride concentrations.[4] These curves were generated by electrochemical testing at neutral pH. Another important factor that influences pitting is the pH of the water. Graphs which highlight the influence of pH, chloride, and temperature on pitting of 316 stainless steel are shown in Fig. 3.14.

In addition to chloride concentration in the water, another contributing factor is the oxidizing potential of the solution. This is typically measured by the free chlorine level. Since chlorine, commonly used for biocontrol in cooling water systems, produces aggressive chloride compounds, the concentration should be controlled properly to prevent pitting of stainless steels. Free chlorine concentrations above 0.5 ppm generally cause the pitting potential of stainless steels to rise rapidly. However, some references indicate that corrosion may be controlled on 304 stainless steel up to 3 ppm free chlorine in potable water service when deposition is under control.[5]

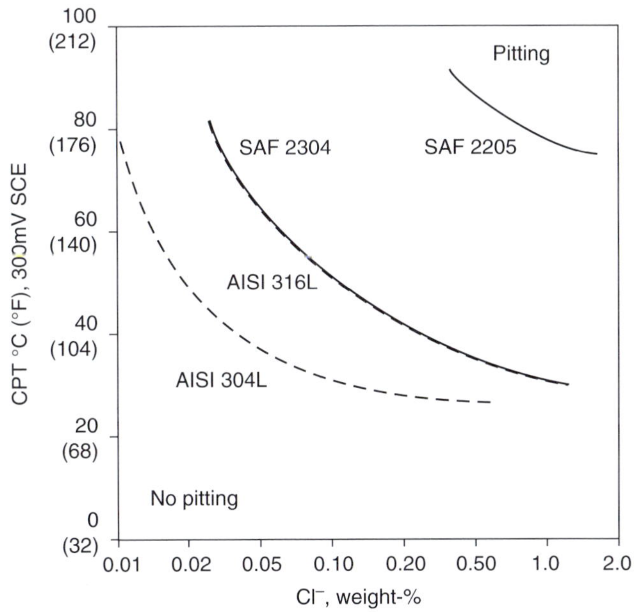

FIGURE 3.13 Graph illustrating resistance of various stainless steels to chloride pitting as a function of temperature and chloride concentration. (*Courtesy of AB Sandvik Materials Technology*.)

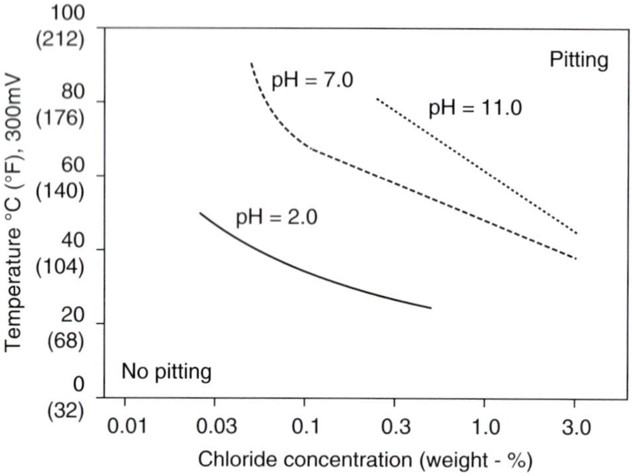

FIGURE 3.14 Graph illustrating resistance of 316 stainless steels to chloride pitting as a function of temperature, pH, and chloride concentration.[6] (*Courtesy of Alpha Laval.*)

Crevice Corrosion

Crevice corrosion is a distinct and severe form of pitting. A crevice may form between tightly contacting metal or metal to nonmetal surfaces. Occlusive deposit layers will also produce conditions similar to mechanical crevices. The resultant corrosion is commonly referred to as underdeposit corrosion. A tight crevice tends to restrict the diffusion of oxygen to the stainless steel surface, thus preventing repassivation. Aggressive anions, such as chloride, will diffuse into the crevice region autocatalytically. This will produce environments that are substantially higher in anion concentration than in a bulk water solution and may decrease pH markedly. Such environments are very aggressive to stainless steels. Crevice corrosion tends to develop over time, depending upon the anion concentration. This is referred to as the "incubation period". In some cases the incubation period may last several months. Once crevice corrosion begins, it will continue as long as the crevice is unsealed and aggressive anions are available. Critical crevice corrosion temperatures (CCT) of various stainless steel alloys have also been measured.[7] In general, the CCT tends to be significantly lower than the critical pitting temperature (CPT) described earlier. This highlights the need to design and operate systems to avoid crevices and deposition. Crevice corrosion resistance of stainless steels varies as a function of pH, as illustrated by Fig. 3.15. Also, as with pitting, crevice corrosion resistance increases as a function of increasing chromium and molybdenum concentration in the steel. A more detailed discussion of crevice and underdeposit corrosion of stainless steels appears in Chap. 11.

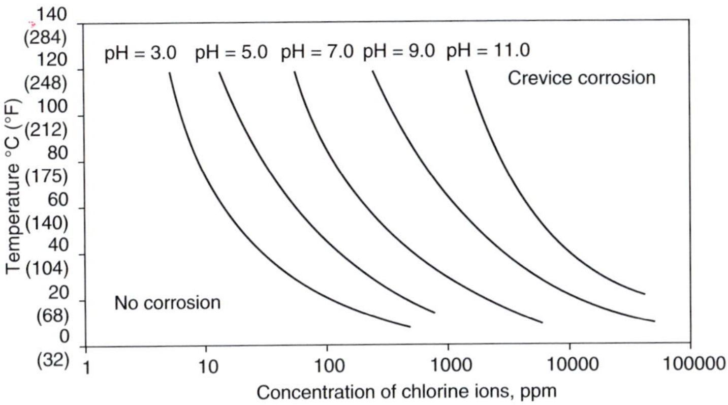

FIGURE 3.15 Graph illustrating resistance of 316 stainless steels to crevice corrosion as a function of temperature, pH, and chloride concentration.[8] (*Courtesy of Alpha Laval.*)

Intergranular Corrosion

Intergranular corrosion refers to preferential corrosion along grain boundaries in the metal. Such corrosion in stainless steels occurs primarily due to "sensitization" as discussed earlier. All grades of stainless steel may be affected, but low-carbon and stabilized grades are less susceptible. More detailed discussion of intergranular corrosion and sensitization appears in Chap. 18 and Chap. 26, respectively.

Galvanic Corrosion

Stainless steels in their passive state are more corrosion resistant than most industrial alloys that are used in cooling water systems. Therefore, they will only experience galvanic corrosion if they are in direct metal-to-metal contact with a much more corrosion-resistant alloy. The environment must also be highly corrosive to the stainless steel to strip the passive film. Stainless steel alloys may promote corrosion of less noble metals and alloys when in direct metal-to-metal contact. A more detailed discussion concerning galvanic corrosion appears in Chap. 16.

Microbiologically Influenced Corrosion

Microbiologically influenced corrosion (MIC) may occur on stainless steels due to exposure to certain bacteria in water. Certain bacteria such as slime-formers produce occlusive layers on the surface that will promote underdeposit corrosion. Some bacteria will cause corrosion directly, such as sulfate-reducing bacteria and possibly certain acid-producing bacteria. Manganese-oxidizing bacteria produce manganese oxide deposits that can promote corrosion of exposed stainless steel. A more detailed discussion concerning MIC of stainless steel appears in Chap. 13.

High-Temperature Corrosion

High-temperature corrosion, particularly thermal oxidation, will occur on components in certain systems. High temperatures are usually not experienced in cooling water systems, since cooling water is typically used to reduce metal temperatures in order to control or prevent high-temperature corrosion. However, stainless steels are used in high-temperature applications where cooling water is used since they are resistant to thermal oxidation at temperatures that are substantially higher than carbon and low-alloy steels. For example, martensitic stainless steels are used in applications such as cladding on the continuous caster rolls used in steel production. These rolls contact and convey the cast strand of produced high-temperature steel. The caster rolls are also exposed to cooling water that is sprayed onto the strand to supply secondary cooling. Martensitic stainless steel has much greater resistance to high-temperature thermal oxidation and aqueous corrosion than the carbon and low-alloy steels.

Stress Corrosion Cracking

Stress corrosion cracking occurs in metals that are exposed to specific corrodents for the particular alloy, while being subjected to applied and/or residual stresses. Residual stresses are supplied by original metal-forming operations and welding. The commonly used AISI 304 and 316 austenitic stainless steels are susceptible to stress corrosion cracking when exposed to water containing chlorides, even at low concentrations. Cracking tends to occur when metal temperatures exceed about 120 to 130°F (49–54°C) and when dissolved oxygen is present. Figure 3.16 illustrates the dramatic contribution of metal temperature to chloride stress corrosion cracking of austenitic stainless steels. The other families of stainless steels are either immune or highly resistant to chloride stress corrosion cracking. A more detailed discussion concerning stress corrosion cracking of stainless steels appears in Chap. 22.

Erosion

Stainless steels have superior erosion resistance. Austenitic, ferritic, and duplex stainless steels are resistant to erosion up to flow rates as high as 30 ft/s (9.14 m/s). Erosion at high flow rates typically occurs when the water contains significant amounts of entrained abrasive particles such as silt. Martensitic and precipitation hardening stainless steels are significantly more erosion resistant than these steels since they are used for abrasive wear applications.

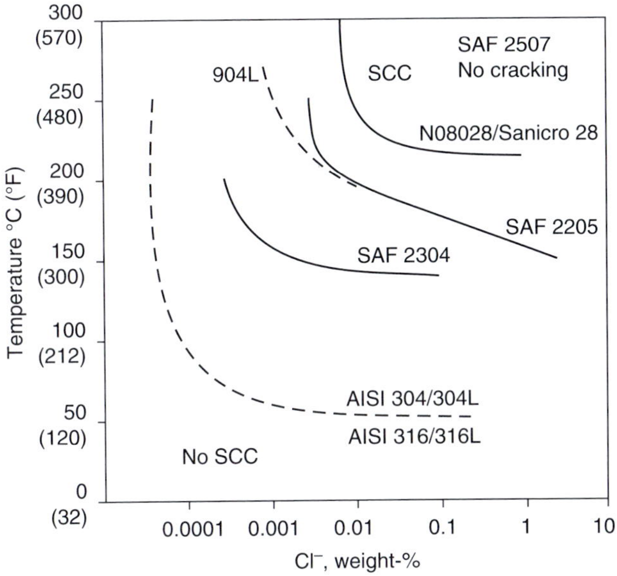

FIGURE 3.16 Graph illustrating resistance of various stainless steels to chloride stress corrosion cracking as a function of temperature and chloride concentration. (*Courtesy of AB Sandvik Materials Technology*.)

Operational and Design Considerations

Several guidelines for stainless steel components used in cooling water systems are recommended. Flow rates greater than 3 ft/s (0.914 m/s) should be maintained. Stagnant flow conditions of less than 1 ft/s (0.305 m/s) and deadlegs must be avoided. This will prevent settling of deposits that can promote underdeposit corrosion. Occlusive deposits should be removed, and clean surfaces should be maintained to prevent underdeposit corrosion. In addition, stagnant flow may produce anaerobic conditions that may promote corrosion, since passivation of stainless steels is provided by dissolved oxygen. Anaerobic conditions may also promote the growth of sulfate-reducing bacteria that can cause corrosion. Crevices should be avoided if possible. If crevices are necessary, they should be sealed properly. Metal-to-metal contact of stainless steels with carbon and low-alloy steels and other less noble metals should be avoided. This will prevent galvanic corrosion of the less noble metal.

Welding Stainless Steels

In addition to the control of carbide formation to prevent sensitization, as described previously, other criteria are important to consider during welding of stainless steels. Welding processes should be used that do not decrease the resistance of the alloy to corrosion. Proper base metal and filler metal composition should be used, as well as proper welding practice. When welding stainless steel to carbon steel, the difference in composition between the two steels will dilute the stainless steel at the weld interface and may cause cracking of the weld. Such cracking may be avoided by using proper stainless steel filler metal and proper welding techniques.

Beads for circumferential welds on the more common austenitic stainless steel grades such as 304L and 316L will contain small amounts of residual ferrite (Fig. 3.17). The amount of residual ferrite is usually about 5 to 10 percent in order to prevent cracking in the weld. The residual ferrite will not reduce corrosion resistance significantly in mildly corrosive environments. However, preferential corrosion of the ferrite may occur under highly corrosive conditions, such as through exposure to acidic, high-chloride environments.

Weld spatter, arc strikes, and slag produced during welding should be mechanically wire brushed or lightly ground away from surfaces since they will tend to initiate corrosion. In addition, thermally formed oxides, commonly known as heat tint, may be produced during welding (Fig. 3.18). Heat tint should be controlled adequately through proper use of inert shielding gases such as argon and helium. However, if such measures are not taken, or upset conditions occur, an excessive heat tint layer may form that must be removed. A heat tint layer should at most have a slight gold or straw color. A black or blue color is excessive and must be removed. Thermally formed oxides on stainless steels are not protective

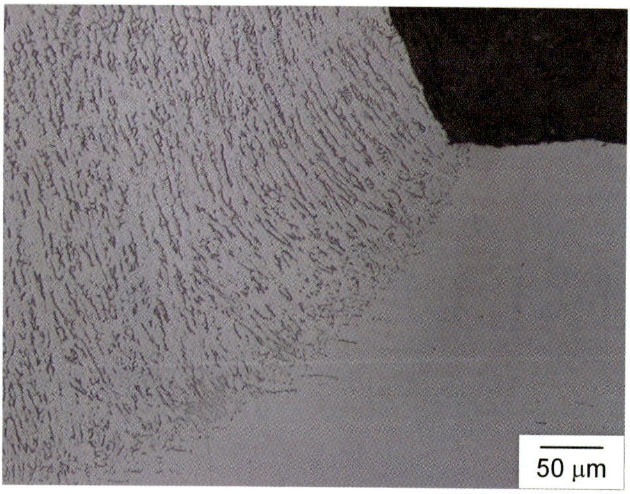

FIGURE 3.17 Austenitic stainless steel weld containing residual ferrite. Etchant: electrolytic oxalic acid

Figure 3.18 Heat tint oxides formed along weld heat–affected zones. Pitting occurred only within heat-tinted areas.

for surfaces in cooling water service, and they should be removed to allow the normal passive film to form on the surface. In addition, when heat tint layers form, chromium is removed from the underlying steel to a shallow depth. This chromium-depleted steel is susceptible to corrosion and may initiate pitting. Heat tint oxides should be removed using appropriate chemical cleaning processes such as pickling.

Pickling, Passivation, and Cleaning

In many cases it is recommended that stainless steel components be mechanically cleaned, followed by appropriate chemical cleaning and passivation prior to use. Organic materials should be removed with solvents and alkaline cleaners prior to welding and heat treatment to prevent carburization and carbon smut formation, both of which may promote corrosion. Where excessive heat tint oxides form at welds, it would be prudent to consider pickling treatments. Alkaline cleaners and solvents should be used to remove organic contaminants prior to a pickling or passivation treatment.

Pickling treatments involve the use of specific mixtures of suitable oxidizing acids at proper temperatures and exposure times. The pickling process will remove heat tint oxide layers and degraded metal that penetrates to shallow depths beneath them, as well as residual embedded iron from machining and forming tools. Pickling typically employs nitric acid–based solutions that contain added ingredients such as hydrofluoric acid (HF) to increase cleaning efficiency and reduce cleaning times. Proper mixtures of oxidizing nitric acid and reducing hydrofluoric acid are used in order to prevent corrosion of the stainless steel by the HF. Hydrofluoric acid alone will cause pitting or even general corrosion on stainless steel. The pickling treatment may employ a fill and soak technique. However, if areas to be pickled are isolated and accessible, a safer technique may be to locally spray surfaces with pickling acids or apply a pickling paste. Since pickling acids may cause corrosion, the safest and least intrusive method is recommended. Following pickling, the unit should be properly rinsed to remove all residual pickling acid. A detailed pickling procedure is outlined in the *ASM Metals Handbook*.[9]

In some high-purity water systems, pickling is recommended in order to remove thin films of red iron oxide that may spall and contaminate the water. The red oxide layer formation is commonly described as "rouging." This phenomenon may be caused by the oxidation of residual embedded flakes of nonstainless tooling materials that were used to form the component. Rouging may also be produced by the oxidation of surface metal that experienced reduction of chromium concentration during processing. Proper pickling treatment will remove both the embedded iron and the degraded surface metal. Once the pickling is completed and residual pickling acids are rinsed away, the surface will have a proper, passivated surface. Pickling will produce surfaces that are covered with a uniform chromium oxide layer that will provide improved overall corrosion resistance and will prevent the establishment of preferential corrosion in welded and other affected areas.

Passivation will spontaneously occur on a bare stainless steel surface due to exposure to the air or oxygenated water. It is not necessary to increase the oxidation reaction by exposing stainless steels to more highly oxidizing agents such as hydrogen peroxide since oxygen in the air or dissolved in water will react at the same rate. It has been found, however, that a passivation treatment tends to produce a more corrosion-resistant surface than an untreated surface. In new installations where heat tint layers are not present, there may be an advantage to using a milder passivation treatment rather than a pickling treatment. Passivation uses only nitric acid, which will remove small amounts of steel contaminants that may be present on the surface, and will produce a more robust passive layer.

Following initial pickling and passivation treatments, most systems will not require subsequent passivation treatments. If chronic corrosion problems occur on stainless steel components, additional passivation treatments will not be as effective for controlling corrosion as altering the corrosive environment. It should be mentioned that it may not be possible to adequately pickle and passivate surfaces that experience deep, undercut pits. Pickling and passivating agents may be unable to properly flow into occluded, undercut pits in order to be suitably effective. For the same reason, occluded pits will tend to be susceptible to concentration cell corrosion, which will be described in more detail in Chap. 11. The effectiveness of a pickling or passivation treatment may be determined with chemical spot tests. ASTM A380 describes the testing which uses ferroxyl or copper sulfate solutions. These spot tests are also useful for determining if surfaces of manufactured components are contaminated with iron that would require pickling.

Following service, periodic cleaning may be necessary to control possible underdeposit corrosion mechanisms. In systems that contain stainless steel components, it is vitally important to use the proper type of acid for cleaning and/or passivation. In some cases, the intent is not to passivate, but simply to clean deposits from stainless steel components. Since nitric acid is more difficult to handle and will emit noxious fumes, it is common for other acids to be used for cleaning. It is vitally important to choose an acid that will not cause significant damage during the exposure. Weaker acids, such as citric or sulfamic acids, will be suitable for removing iron oxides and water solids such as calcium carbonate, but they will not be significantly corrosive to stainless steels. Citric acid has been described as an oxidizing acid that can be used to passivate stainless steel surfaces. However, citric acid is significantly less effective at passivating stainless steels than nitric acid. Reducing acids such as hydrochloric acid must never be used alone to clean stainless steels. Essentially, due to its inherent corrosivity to stainless steels, there is no safe concentration of hydrochloric acid that can be recommended for exposure to stainless steels.

Surface Finish

Surface finish on stainless steel components may influence corrosion to some degree. A very smooth surface is less susceptible to deposition and associated underdeposit corrosion, and it is easier to clean. In most applications in cooling water systems, such as piping and tubing, a normal cold-rolled 2B or 2D mill finish is smooth and bright, and is adequate enough to provide proper resistance to deposition and corrosion. Such finishes are also more cost effective. Polished surfaces, produced either mechanically or through electropolishing, are used in some chemical feed and high-purity water systems to control corrosion. An electropolished surface will not contain surface folds, laps, and other imperfections. A mechanically polished surface may contain folds or laps that can produce crevice corrosion. Scratches, gouges, and nicks on a stainless steel surface will also fold metal on the surface to promote crevice corrosion.

Corrosion Inhibitors

Unlike carbon and low-alloy steels, which require the use of corrosion inhibitors for control of oxygen corrosion, stainless steels will passivate when exposed to oxygenated waters. Therefore, under normal conditions during service, when surfaces are submerged in cooling water containing moderate concentrations of aggressive anions, it is unnecessary for corrosion inhibitors to be used to control corrosion. If the concentration of aggressive anions is excessive and causes corrosion, there is no inhibitor available that can be used in recirculating cooling water to control the corrosion. Under such circumstances, a more resistant alloy must be used.

Stainless steels may provide excellent corrosion resistance during service, but following system drainage, residual cooling water will evaporate. This may produce water solutions that have much higher concentrations of aggressive anions than in the bulk water, particularly chloride. Higher anion concentration may cause pitting to occur on stainless steels either immediately or shortly after system drainage. Chloride pitting of stainless steels during idle periods following system drainage may be controlled by using three to five parts of sodium nitrate ($NaNO_3$) to one part of chloride that is present in the water. Adjustment of pH may be provided by appropriate additions of a suitable alkalinity supplement. Though this treatment may be useful for layup practice, it is not suitable or practical for use in recirculating cooling water during service.

Though the more common austenitic stainless steels, such as the 304 and 316 grades, have superior corrosion resistance, they may be unsuitable for certain applications where service conditions are highly corrosive, such as seawater and brackish water service, or where metal temperatures are high enough to support stress corrosion cracking. Certain grades of stainless steel have shown promising results in such service, such as some high-chromium ferritic stainless steels, high-nickel austenitic stainless steels, or duplex stainless steels. However, if experience with such alloys reveals them to be unsuitable, other alloys such as nickel alloys and titanium alloys may be considered. These alloys are also discussed in Chap. 6.

References

1. Curtis H. Kovach (ed.), *High-Performance Stainless Steels,* NIDI, p. 5.
2. Compiled from *ASM Metals Handbook,* 9th ed., Volume 13, pp. 548–549 and NIDI, *High-Performance Stainless Steels,* ed. Curtis H. Kovach.
3. D. Peckner and I. M. Bernstein, *Handbook of Stainless Steels,* McGraw-Hill, 1977, pp. 12–33.
4. Sandvik product literature, *Sandvik Duplex Stainless Steels–fighting corrosion worldwide,* undated.
5. R. E. Avery, S. Lamb, C. A. Powell, and A. H. Tuthill, *Stainless Steel for Potable Water Treatment Plants,* Nickel Development Institute, Technical Literature 10087, undated.
6. Alfa Laval product literature, undated.
7. Sandvik product literature, *Sandvik Duplex Stainless Steels–fighting corrosion worldwide,* undated
8. Alfa Laval product literature, undated.
9. *ASM Metals Handbook,* 8th ed., Volume 2, Heat Treating, Cleaning and Finishing, ASM, 1964, pp. 599–606.

CHAPTER 4

Copper Alloys

Introduction

Copper was the first metal utilized in any considerable quantity by humans. According to some estimates, copper metal was used as early as 6000 years ago. Thousands of years later, copper and its alloys remain important materials in human society. Copper and copper alloys, commonly referred to as yellow metals, have long been used in a variety of environments, particularly seawater. They offer good corrosion resistance in many environments, especially when compared to plain carbon steel. Copper-based metals also have excellent formability and machinability. In industrial cooling water systems, copper and its alloys are most commonly used in heat exchanger and condenser tubing, where both corrosion resistance and high thermal conductivity are important considerations. Refer to Table I.2 in Section 1, "Cooling Water System Design and Operation," for additional information. Pump components and valves are also commonly fabricated from copper-based alloys. Table I.1 in Introduction to Cooling Water Systems lists various different alloys and their uses.

Common Copper Alloy Types

A few classes of copper-based alloys commonly found in cooling water systems are discussed next. Major alloying elements for some wrought and cast copper alloys commonly used in water-based environments are given in Tables 4.1 and 4.2.

Some copper alloys contain lead, which improves machinability due to the easy formation of chips. Lead is also added to some alloys for reduction of friction in applications such as bearings. Concern about lead for environmental and health reasons has led to increasing use of lead-free copper alloys.

Copper

Copper metal is used often used where good corrosion resistance and high thermal conductivity are needed, but it is generally more susceptible to erosion-corrosion damage than copper-based alloys due to

UNS Number	Alloy Type	Copper	Tin	Aluminum	Nickel, incl. Cobalt	Lead (max)	Iron	Zinc	Manganese	Silicon	Arsenic	Antimony	Phosphorus
C10200	Oxygen-free copper	99.95 min	–	–	–	–	–	–	–	–	–	–	–
C10800	Oxygen-free low phosphorus copper	99.95 min	–	–	–	–	–	–	–	–	–	–	0.005–0.012
C23000	Red brass	84.0–86.0	–	–	–	0.05	0.05 max	Balance	–	–	–	–	–
C26000	Cartridge brass	68.5–71.5	–	–	–	0.07	0.05 max	Balance	–	–	–	–	–
C26800	Yellow brass	64.0–68.5	–	–	–	0.15	0.05 max	Balance	–	–	–	–	–
C28000	Muntz metal	59.0–63.0	–	–	–	0.30	0.07 max	Balance	–	–	–	–	–
C44300	Inhibited admiralty brass	70.0–73.0	0.9–1.2	–	–	0.07	0.06 max	Balance	–	–	0.02–0.06	–	–
C44400	Inhibited admiralty brass	70.0–73.0	0.9–1.2	–	–	0.07	0.06 max	Balance	–	–	–	0.02–0.10	–
C44500	Inhibited admiralty brass	70.0–73.0	0.9–1.2	–	–	0.07	0.06 max	Balance	–	–	–	–	0.02–0.10
C46400	Naval brass	62.0–65.0	0.50–1.0	–	–	0.20	0.10 max	Balance	–	–	–	–	–

C60800	Aluminum bronze (5%)	Balance	–	5.0–6.5	–	0.10	0.10 max	–	–	–	0.02–0.35	–	–
C65100	Silicon bronze, low	Balance	–	–	–	0.05	0.8	1.5	0.7	0.8–2.0	–	–	–
C65500	Silicon bronze, high	Balance	–	–	0.6	0.05	0.8	1.5	0.50–1.3	2.8–3.8	–	–	–
C68700	Aluminum brass, arsenical	76.0–79.0	–	1.8–2.5	–	0.07	0.06 max	Balance	–	–	0.02–0.06	–	–
C70600	Cupronickel, 90-10	Balance	–	–	9.0–11.0	0.05	1.0–1.8	1.0 max	1.0 max	–	–	–	–
C71000	Cupronickel, 80-20	Balance	–	–	19.0–23.0	0.05	0.50–1.0	1.0 max	1.0 max	–	–	–	–
C71500	Cupronickel, 70-30	Balance	–	–	29.0–33.0	0.05	0.40–1.0	1.0 max	1.0 max	–	–	–	–

Table 4.1 Major Alloying Elements for Some Wrought Copper Alloys Commonly Used in Water-Based Environments (wt.%)

UNS Number	Alloy Type	Copper	Tin	Lead	Zinc	Iron	Nickel, incl. Cobalt	Aluminum	Manganese	Silicon
C83600	Leaded red brass	84.0–86.0	4.0–6.0	4.0–6.0	4.0–6.0	–	1.0	–	–	–
C84400	Leaded semired brass	78.0–82.0	2.3–3.5	6.0–8.0	7.0–10.0	–	1.0	–	–	–
C85400	Leaded yellow brass	65.0–70.0	0.50–1.5	1.5–3.8	24.0–32.0	–	–	-	-	–
C85700	Leaded yellow brass	58.0–64.0	0.50–1.5	0.8–1.5	32.0–40.0	–	–	–	–	–
C87500	Silicon bronze	79.0 min.	–	0.50	12.0–16.0	–	–	–	–	3.0–5.0
C90700	Tin bronze	88.0–90.0	10.0–12.0	0.50	0.50	–	0.50	–	–	–
C95400	Aluminum bronze	83.0 min.	–	–	–	3.0–5.0	1.5	10.0–11.5	0.50	–
C95500	Aluminum bronze	78.0 min	–	–	–	3.0–5.0	3.0–5.5	10.0–11.5	3.5	–
C95700	Manganese-aluminum bronze	71.0 min	–	0.03	–	2.0–4.0	1.5–3.0	7.0–8.0	11.0–14.0	–
C96200	Cupronickel (90-10)	Balance	–	0.01	–	1.0–1.8	9.0–11.0		1.5	0.50
C96400	Cupronickel (70-30)	Balance	–	0.01	–	0.25–1.5	28.0–32.0	–	1.5	0.50

TABLE 4.2 Major Alloying Elements for Some Cast Copper Alloys Commonly Used in Water-Based Environments (wt.%)

its relatively low hardness and mechanical strength. Copper's formability and high thermal conductivity make it a common material for enhanced tubing, in which external surface fins and/or internal rifling serve to enhance heat transfer.

Copper is a relatively noble metal, which translates to good corrosion resistance. In water-based environments, a thin, adherent film of corrosion products forms on the metal surface and provides corrosion protection. Copper corrosion is complex, and the chemical species that result from corrosion processes strongly depend upon water chemistry. Cuprous oxide (Cu_2O) is the primary species normally found in the protective corrosion product film in freshwater. Cupric oxides are generally more soluble and less stable than cuprous oxides and thus are not strongly associated with the formation of a protective film. Basic copper carbonate is another common species found in the corrosion product film. The film may also contain various other species, depending upon the composition of the environment. For example, the presence of chloride, sulfate, and sulfide in the water would likely result in the formation of copper chloride, copper sulfate, and copper sulfide, respectively.

Soft water tends to be corrosive to copper, especially when it contains a sufficient concentration of carbon dioxide. In hard water, calcium-based compounds tend to form on the surface, increasing resistance to corrosion. Low-pH conditions generally promote metal loss (Fig. 4.1). Over the range of typical cooling water pH values, pH itself is not

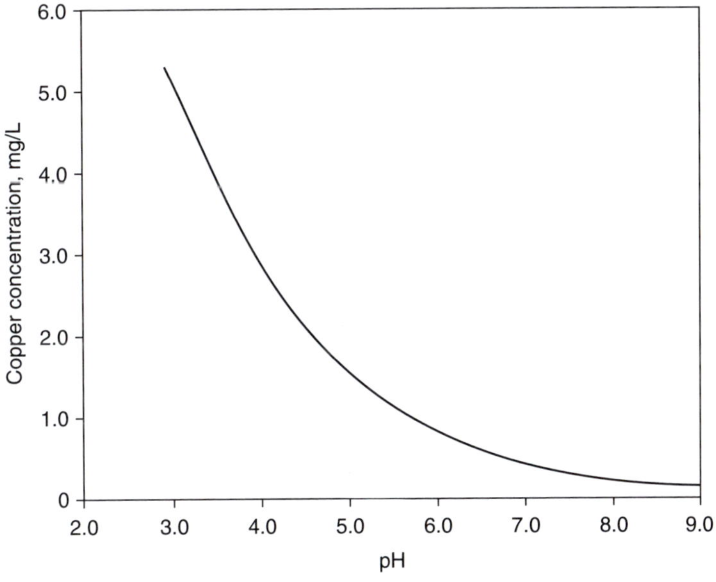

FIGURE 4.1 Effect of pH on the release of copper into solution.[1] (*Reprinted with permission of AWWA. All rights reserved.*)

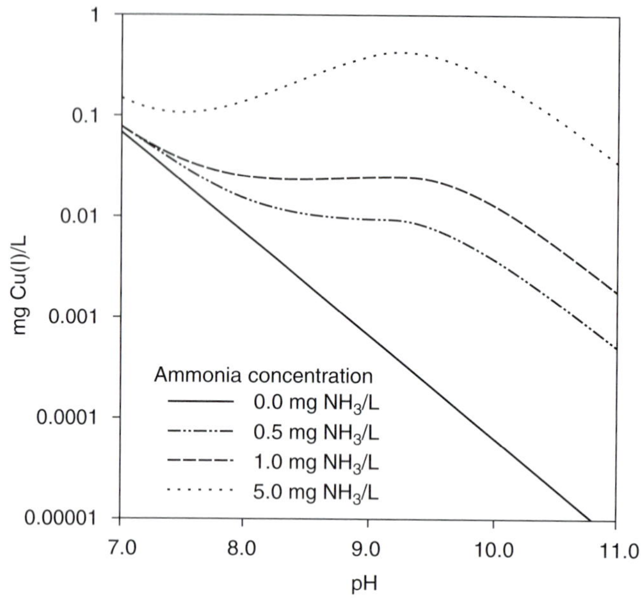

FIGURE 4.2 Calculated effect of ammonia complexation on CuOH solubility at a temperature of 25°C and a fixed ionic strength of I = 0.01 mol/L. *(Reprinted with permission of AWWA. All rights reserved.)*

expected to have a significant effect on corrosion; therefore, other species and factors tend to have greater effects. Ammonia-based and chlorine-containing species may be found in industrial waters and are specifically known to be aggressive toward copper. Figures 4.2 and 4.3 illustrate the calculated increase in CuOH solubility that occurs with increasing concentrations of ammonia and chloride, respectively. Cu_2O would be 60 to 100 times less soluble than CuOH, although the trends and relative positions of the curves depicted in Figs. 4.2 and 4.3 would not be expected to change.[2]

Copper is two to five times more resistant to corrosion in seawater than mild steel when constantly immersed.[3] However, its resistance to flow-related attack in seawater is low. Therefore, copper is generally only used in seawater when the water is stagnant or flow velocities are low (less than approximately 3 ft/s [0.941 m/s]—see Table 23.1 in Chap. 23). Other copper alloys, such as aluminum brasses and cupronickels, are significantly more resistant to flow-related attack in flowing seawater.

Brasses

Brasses have copper and zinc as their major constituents, and are the most widely used class of copper alloys. Brasses offer enhanced mechanical strength and resistance to erosion-corrosion damage

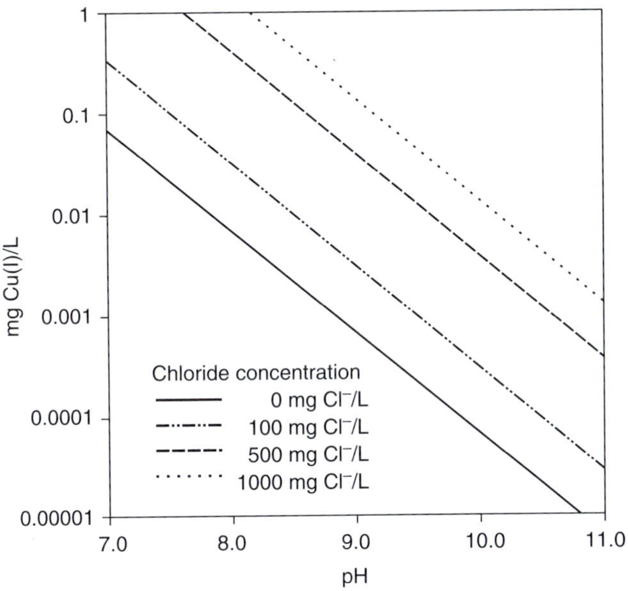

FIGURE 4.3 Calculated effect of chloride complexation on CuOH solubility at a temperature of 25°C and a fixed ionic strength of I = 0.02 mol/L. (*Reprinted with permission of AWWA. All rights reserved.*)

compared to plain copper. Strength generally increases with increasing zinc content; however, the likelihood of dealloying also increases. With a zinc content of up to 32 percent, brass generally has a single-phase microstructure consisting only of a copper-zinc solid solution known as alpha phase (Fig. 4.4). Above 32 percent zinc, zinc-rich beta phase can form, which increases the potential for dezincification (Fig. 4.5). See the "Common Corrosion Mechanisms" section later in this chapter and Chap. 17 for more in-depth information on dealloying. Various elements, such as tin and aluminum, can be added to brasses in order to further enhance performance. Some brasses contain lead for improved machinability.

Tin is added to some brasses to decrease susceptibility to dezincification. Admiralty brass is widely using in cooling water systems, particularly in tubing, and contains approximately 28 percent zinc and 1 percent tin. Antimony, arsenic, or phosphorus may also be added to admiralty or naval (37 percent zinc, 1 percent tin) brasses in low concentrations (up to 0.10 percent) to control dezincification; such alloys are known as inhibited brasses.

The most commonly used wrought aluminum brass contains approximately 20.5 percent zinc and 2 percent aluminum, as well as a small concentration of arsenic for inhibition of dezincification. The aluminum content of aluminum brass provides increased resistance to erosion-corrosion damage in salt water relative to other brasses.

FIGURE 4.4 Single-phase microstructure of brass condenser tubing, consisting of twinned grains of alpha phase copper-zinc solid solution. Etchant: potassium dichromate

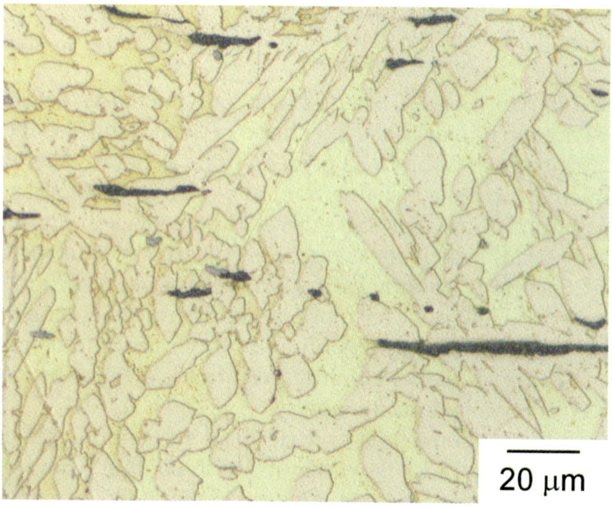

FIGURE 4.5 Duplex microstructure of a brass containing more than 32 percent zinc, consisting of alpha phase solid solution (pinkish yellow) in a matrix of zinc-rich beta phase solid solution (darker yellow).

Like pure copper, soft waters tend to be more corrosive to brasses than hard water. Very pure water, such as condensate, is not intrinsically very corrosive to brass, though dissolved gases such as carbon dioxide or ammonia, along with oxygen, significantly increase the corrosivity of such waters.

Cupronickels

The most commonly used cupronickels in industrial applications include 70-30, 80-20, and 90-10 Cu-Ni ratios. Since nickel is completely soluble in copper, cupronickels have a single-phase microstructure. Copper-nickel alloys offer excellent resistance to corrosion, erosion-corrosion, and stress corrosion cracking relative to plain copper and brasses. The addition of a small amount of iron to the alloy results in the formation of a surface film that has been shown to enhance resistance to erosion-corrosion as well as attack in stagnant and polluted waters. The addition of chromium also provides excellent resistance to erosion-corrosion. Dealloying (denickelification) can occur, but it is relatively rare. Cupronickels are most often used in condenser and heat exchanger tubing in steam systems due to their excellent resistance to oxidation in steam and humid air. Though cupronickels have excellent properties that would make them good candidates in a variety of applications, their high cost relative to other copper alloys limits their use to situations in which the conditions justify the expense.

Bronzes

Traditionally, bronzes are copper alloys in which copper and tin are the major alloying elements. The corrosion resistance of tin bronzes in water and steam environments varies widely depending on the presence of corrosive species, as previously discussed.

Other bronzes, such as aluminum bronzes, which primarily consist of copper and aluminum, are also important engineering materials. Aluminum bronze has a single-phase microstructure up to an aluminum content of 9.6 percent. Corrosion resistance generally increases with increasing aluminum content up to that concentration. The aluminum content results in the formation of a very protective surface film, which makes aluminum bronzes more corrosion-resistant than both tin bronzes and brasses in seawater. The surface film also makes aluminum bronzes good candidates for applications that require high resistance to erosion-corrosion and cavitation relative to other copper alloys.

Silicon bronzes are comparable in corrosion resistance to pure copper and may be used where high strength is necessary.

Common Corrosion Mechanisms

Despite the numerous desirable characteristics of copper alloys, they are vulnerable to failure by a number of mechanisms, including those listed next, all of which will be discussed in greater detail in other chapters within this volume.

General Corrosion

General corrosion of copper and copper-based alloys in cooling water service is rare due to their tendencies toward localized corrosion under typical cooling water conditions. General corrosion usually occurs under acidic conditions, at pH values much lower than those of most cooling waters. Corrosion rate tends to increase with decreasing pH (Fig. 4.1). Chemical species that are specifically aggressive toward copper and its alloys include oxidizing acids, ammonia and ammonia compounds, and some sulfur compounds.

Concentration Cell Corrosion

The formation of pits or depressions in copper and copper-based alloys is a common problem and can result in pinhole failures. Though localized metal loss can occur for a number of different reasons, pits and depressions often form because of the development of differential oxygen concentration cells and ion concentration cells in locations such as at waterlines or beneath deposit layers. Aggressive anions such as sulfate or chloride can concentrate to levels many times those in the bulk water if deposit material, including biofilms, covers a metal surface. Pits and depressions that form because of concentration cells are often covered with corrosion products of various colors, depending on the compounds. Such corrosion products may be sparkling and crystalline in appearance. Examples of common corrosion product colors include red from cuprous oxide, bluish green from copper chloride, and brilliant blue from copper sulfate (Fig. 4.6). Of course, many other compounds may be formed as well. Copper-based compounds are often blue-green. Low flow and stagnant conditions should be avoided, and metal surfaces should be kept clean in

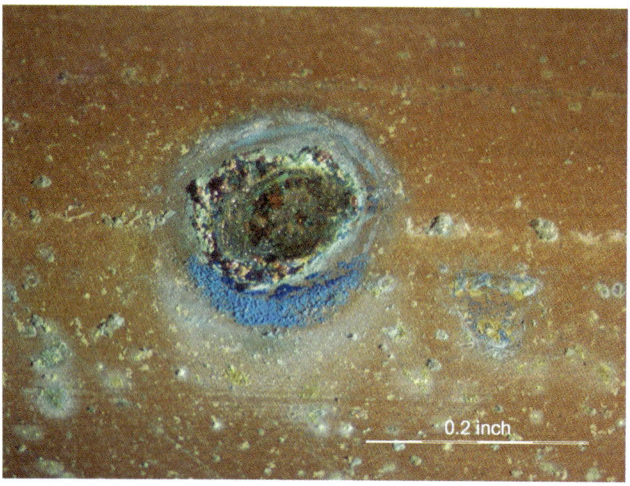

Figure 4.6 Corrosion product–lined pit that formed at a waterline in a copper tube from an adsorption chiller. The corrosion products lining the pit are blue-green and blue; SEM-EDS analysis detected elevated concentrations of chlorine and sulfur in the corrosion products.

order to reduce the likelihood of concentration cell formation. See Chap. 11 for more information about concentration cell corrosion.

Dealloying

Dealloying of copper alloys is a phenomenon in which a major alloying element is selectively removed, leaving behind mechanically weak metallic copper corrosion products. It most commonly occurs in brasses when zinc is removed from the primarily copper-zinc alloy, thus the term dezincification. Brasses with zinc contents of 15 percent and above are susceptible to dezincification, while those with zinc contents below 15 percent are essentially immune to dezincification in most cases. Brasses containing 32 percent or more zinc are particularly prone to dezincification due to their typical microstructure. Unlike the single-phase microstructure shown in Fig. 4.4, such brasses usually have a duplex (two-phase) microstructure consisting of alpha phase solid solution, which is a copper-rich phase, and beta phase solid solution, which contains an elevated zinc level relative to the alpha phase (Fig. 4.5). Dezincification occurs preferentially in the beta phase due to its higher zinc content and resulting lower corrosion resistance relative to the alpha phase (Fig. 4.7). These two-phase (referred to as alpha-beta or just beta) brasses are often selected for their low cost (relative to brasses with higher copper contents) and high hot ductility, which is desirable in applications involving casting or hot working.

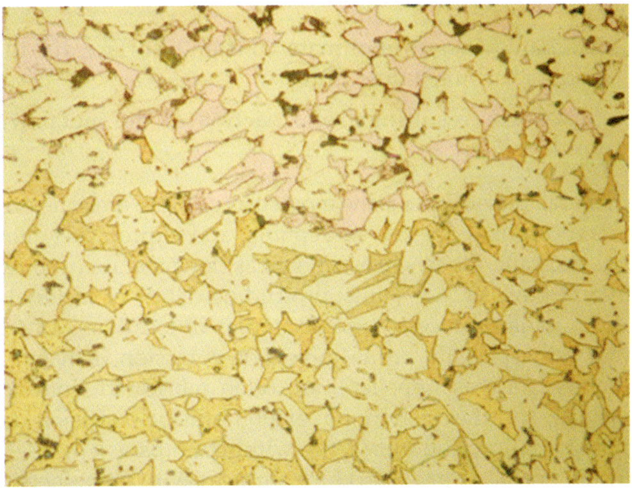

FIGURE 4.7 Micrograph showing dealloying along the beta phase in a brass with a duplex microstructure. The porous reddish copper corrosion products penetrate preferentially along the beta phase, the darker yellow constituent in the microstructure.

Dezincification may be controlled through the use of inhibited brasses, which contain small concentrations of antimony, arsenic, or phosphorus. It should be noted that inhibiting elements are only of benefit in single-phase (alpha only) brasses, as the inhibiting elements do not inhibit dezincification in the beta phase. Special dezincification-resistant (DZR) brasses have zinc contents over 32 percent, but are heat treated to contain only alpha phase at room temperature, and they contain a dezincification-inhibiting element such as arsenic.

In cast components such as pump impellers, compositional segregation that occurs during solidification (known as coring) can result in dealloying despite an overall bulk alloy composition that would not normally be considered vulnerable to dealloying. For instance, an alloy may contain less than 15 percent zinc total, but dezincification can occur preferentially in some phases relative to others due to compositional differences between them.

Dealloying can also occur in copper-based alloys other than brasses, such as cupronickels and bronzes. Aluminum bronzes with multiphase microstructures, that is, those with high aluminum contents, are particularly vulnerable to dealuminification. Aluminum-rich phases in such alloys promote dealloying in the same manner in which the duplex microstructure of some brasses promotes dezincification.

A more in-depth discussion of dealloying can be found in Chap. 17.

Erosion-Corrosion

Copper-based alloys are much more susceptible to erosion-corrosion than ferrous alloys due to their relatively soft nature. Erosion-corrosion occurs due to a combination of:

1. Mechanical removal of material from a surface due to rapid and/or turbulent fluid flow
2. Metal loss resulting from reaction of the metal with its environment

In general, pure copper is the most susceptible to erosion-corrosion, brasses and bronzes are less susceptible, and cupronickel alloys are the most resistant. Elbows, tees, and areas downstream from features that disrupt flow, such as joints, protruding debris or residual metal burrs from joining processes, are examples of locations of turbulent flow that are especially prone to erosion-corrosion damage. Pump components, such as impellers, are often fabricated from copper-based alloys, and they commonly exhibit erosion-corrosion damage due to the high flow and turbulence within pumps.

Erosion-corrosion may produce surfaces with smoothened contours, sand dune–like patterns, and/or horseshoe-shaped depressions. Bare surfaces indicate recent attack. If design or operational changes (e.g., avoiding sharp bends, decreasing flow velocity) are not feasible, it may be necessary to replace affected components with more erosion-corrosion-resistant alloys. For instance, admiralty brass components experiencing erosion-corrosion may be replaced with cupronickel, which is considered to be very resistant to damage by erosion-corrosion. In heat exchangers, erosion-corrosion damage often occurs near tube inlets due to turbulence, and may be controlled through the use of ferrules. Ferrules are inserts made from materials that are more erosion resistant than the tube material, such as a cupronickel or stainless steel. Erosion-corrosion is discussed in greater depth in Chap. 23.

Stress Corrosion Cracking

Stress corrosion cracking (SCC) of copper and its alloys is relatively uncommon in actual cooling water environments; however, it sometimes occurs on the condensate or air sides of condensers or air coolers, where small amounts of condensing moisture can contain relatively high concentrations of corrodents. The most common corrodents associated with stress corrosion cracking of copper and its alloys are ammonia, amines, sulfur dioxide, nitrates, and nitrites. The susceptibility to stress corrosion cracking due to ammonia varies significantly between the different copper alloy types. In general, pure copper and cupronickel alloys are very resistant to SCC in ammonia-containing environments. In brasses, which are much more vulnerable to SCC, susceptibility to cracking in an ammonia atmosphere increases with increasing zinc content. More information on stress corrosion cracking can be found in Chap. 22.

Intergranular Corrosion

Intergranular attack can result in entire grains of metal detaching from affected areas; it is often associated with high heat flux and the presence of deposits on the metal surface. In some cases of intergranular attack, metallic copper corrosion products have been observed along the grain boundaries, suggesting localized dealloying as a mechanism of attack; however, metallic copper corrosion products are not always observed in cases of intergranular corrosion. Intergranular corrosion in discussed in greater detail in Chap. 18.

Formicary (Ant Nest) Corrosion

Formicary corrosion, also known as ant nest corrosion, primarily affects enhanced copper tubing used in HVAC applications. Such attack is typically characterized by oxide-filled, cavernous pits and interconnecting tunnels resembling those in an ant nest. Formicary corrosion requires the simultaneous presence of moisture, oxygen, and a specific corrodent species, usually an organic acid. Organic acids may form by the degradation or hydrolysis of residual oils and/or organic solvents used in tube manufacturing and cleaning processes. Carboxylic acids, such as formic and acetic acids, are known to aggressively attack copper and cause significant corrosion. Formicary corrosion is discussed further in Chap. 14.

Galvanic Corrosion

Due to the relatively noble nature of copper and copper-based alloys compared to alloys such as nonstainless ferrous metals (e.g., carbon steel and low-alloy steels), care must be taken when copper and copper alloys are electrically coupled to more active metals. In the case of a shell-and-tube heat exchanger, for instance, a carbon steel tube sheet in electrical contact with copper tubes may experience accelerated corrosion due to the formation of a galvanic couple. Measures that can help protect against such galvanic corrosion include sacrificial anodes and protective tubesheet coatings.

Dissolved copper in water can also promote galvanic-type attack of less-noble alloys such as carbon steel and aluminum. The difference in solution potential provides a driving force for copper to be reduced and deposited as metal, promoting corrosion of the less-noble metal. Galvanic corrosion is discussed in greater detail in Chap. 16.

Solder Flux Corrosion

Fluxes used to prepare metal surfaces for joining by soldering and brazing may cause corrosion if they are not properly removed from metal surfaces after joining. Fluxes are inherently corrosive, as one of their main purposes is to remove oxides from metal surfaces in order to ensure good bonding. Surfaces should be adequately cleaned before components are placed into service.

Microbiologically Influenced Corrosion

Copper and its alloys are known to have a certain level of toxicity to organisms and have thus been valued for their ability to deter the attachment of marine organisms to surfaces in seawater applications. Copper and copper alloys also tend to be more resistant to microbiologically influenced corrosion (MIC) than other alloys such as carbon steels, though they are not immune to it. MIC is discussed in detail in Chap. 13.

Corrosion Inhibition

Though copper and copper-based alloys are much more corrosion resistant in typical cooling water environments than plain carbon steel, it is common practice to utilize chemical corrosion inhibitors to prevent metal loss. Some commonly used treatment chemicals, such as phosphonates used for scale inhibition, can promote yellow metal corrosion and are thus often formulated with yellow metal corrosion inhibitors. The most common types of copper alloy corrosion inhibitors are azoles and thiazoles, which are organic molecules that adsorb onto the metal surface to form a very thin protective film. Such corrosion inhibitors also form complexes with dissolved copper, helping to prevent it from depositing onto other metal surfaces and possibly causing galvanic-type attack.

Some commonly used azole-based inhibitors are benzotriazole (BZT) and tolyltriazole (TT or TTA). Halogens, such as chlorine used for control of microbiological activity, can degrade some azoles. Azoles that are more resistant to oxidizing agents, such as chlorotolyltriazole, may be used in systems that are heavily chlorinated. Counterintuitively, adding chlorine to the molecule minimizes the reaction of chlorine in the water with nitrogen atoms in the azole ring portion of the molecule. Mercaptobenzothiazole (MBT), one of the earliest azoles in wide use, is a very effective inhibitor, a very fast film former, and effective at high temperatures. However, MBT is sensitive to rapid degradation by microorganisms and rapid oxidation by halogens and dissolved oxygen; therefore, it is now mainly used in closed loops and once-through cooling systems.

References

1. "Corrosion in Potable Water Systems," USEPA, 1982.
2. John F. Ferguson, Otto von Franque, and Michael R. Schock, "Corrosion of Copper in Potable Water Systems," Foundation, American Water Works Association Research, in *Internal corrosion of Water Distribution Systems*, 2d ed., Denver: DVGW-Technologiezentrum Wasser, 1996. pp. 231–268.
3. C. A. C. Sequeira, "Copper and Copper Alloys," in *Uhlig's Corrosion Handbook*, 3d ed., ed. R. Winston Revie, Hoboken: John Wiley and Sons, Inc., 2011,pp. 757–785.

CHAPTER 5

Aluminum Alloys

Introduction

Due to its low density, aluminum has the advantage of being lighter in weight per unit of volume than other conductive materials used in cooling water systems. Aluminum is ductile, which allows components to be formed or machined easily into a variety of shapes. Aluminum alloys are also easily joined by welding, brazing, soldering, and riveting. Aluminum is also nontoxic, which may be useful for certain applications. This, and its low cost, make it a desirable material of construction for a variety of uses.

Aluminum is not as commonly used in cooling water systems as steel and copper alloys. Aluminum is typically used in cooling water systems when specific applications require high thermal and electrical conductivity. Aluminum will be chosen over other conductive alloys because of some of its other properties. One of the main reasons for its use is that in air, a stable protective oxide layer forms on an aluminum surface, which will impart a bright, silvery metallic finish (Fig. 5.1). When the surface of an aluminum component is scratched to bare metal, the aluminum oxide layer spontaneously reforms. In addition, aluminum doesn't tend to tarnish easily. For all of these reasons, aluminum is used extensively for water-cooled components used for the production of silicon wafers and for plastic injection molding.

For certain specific cooling water service conditions, the more corrosion-resistant aluminum alloys will be suitable for long-term service. However, under most conditions experienced in cooling water systems, aluminum is much less corrosion and erosion-resistant and less mechanically durable than other, more commonly used alloys, such as steel and copper alloys.

Aluminum is one of the most active metals in the galvanic series, so it is necessary for the protective oxide layer to be maintained in order to control corrosion. Therefore, conditions on the waterside must be maintained properly to ensure that the oxide layer is not chemically or mechanically removed to expose the susceptible underlying metal.

FIGURE 5.1 Shiny aluminum surface exposed to atmospheric conditions and for molding of two-liter plastic bottles.

The oxide layer that forms on the waterside is typically thicker than those that form in the air. The crystalline form of the oxide layer that is typically produced at normal cooling water temperatures is aluminum trihydroxide [$Al(OH)_3$]. Certain alloying elements such as manganese and magnesium will tend to strengthen the protective properties of the oxide, while others such as copper tend to weaken the protective properties.[1]

Types of Aluminum

Solid aluminum has a face-centered cubic crystal structure, and is not ferromagnetic. Pure aluminum has low strength. However, when aluminum is alloyed with certain elements, such as copper, magnesium, silicon, zinc, or manganese and proper forming operations and heat treatments are applied, mechanical properties such as tensile strength and hardness may be improved substantially. This allows aluminum alloys to be used for applications that require high strength.

However, alloying, forming operations, and heat treatment may significantly affect corrosion resistance. The degree to which the alloy composition affects corrosion resistance depends upon the particular alloying element as well as its concentration. For example, aluminum alloys that contain elements that are more active than aluminum in the galvanic series such as magnesium and zinc may be more susceptible to corrosion than pure aluminum. In addition, the presence of intermetallic precipitates in the microstructure, especially the heat-treatable alloys, may influence the corrosion resistance of an alloy.

The extent of deformation of a component, and the resultant shapes of the grains in the microstructure, may also affect corrosion resistance.

Aluminum components used in cooling water systems may be in the wrought or cast form. Tubing tends to be produced in the wrought form. Wrought aluminum alloys are designated by the Aluminum Association[2] with a four-digit numbering series. The thousands digit in the four-digit wrought designation depends upon the primary alloying element. The remaining digits distinguish differences between alloys within the series. The wrought aluminum alloys that are most commonly used in cooling water systems are listed in Table 5.1.[3] Details about the specific series of alloys are presented in the following section.

1000 Series

The 1000 designation indicates essentially unalloyed aluminum, with aluminum concentration of 99 percent or higher. These alloys have excellent corrosion resistance and high thermal and electrical conductivity. Low concentrations of alloying elements increase formability, but also reduce mechanical strength significantly. Moderate increases in strength may be obtained by strain hardening (cold working). However, this alloy will not be suitable where high strength or hardness is required. The 1100 alloy has been used for components in certain high-purity cooling water systems in the electronics industry where high heat transfer and corrosion resistance was desired, but applied stresses were not significant.

2000 Series

Copper is the principal alloying element in this series, typically ranging from 3.5 to 6.0 wt%. When heat-treated, the mechanical properties are similar to and sometimes exceed those of mild steel. These alloys are commonly used for aircraft components, particularly alloy 2024. High copper content alloys are generally more susceptible to corrosion than other alloy types. Copper is much more noble than aluminum. When corrosion occurs on copper-rich aluminum alloys, the copper from the alloy dissolves into the water and plates onto the surface of the aluminum to cause subsequent galvanic corrosion. Such corrosion occurs along grain boundaries in the microstructure. In addition, when the grains in the microstructure are heavily deformed during processing, they are susceptible to a phenomenon referred to as "exfoliation corrosion." Consequently, the high copper content aluminum alloys are poorly suited to cooling water service, so their use should be avoided in such systems.

3000 Series

Manganese is the primary alloying element in this series, typically between 0.4 and 1.5 wt%. These alloys are not heat treatable. They have superior corrosion resistance and formability. For this reason,

AA Alloy Number	Si	Fe	Cu	Mn	Mg	Zn	Ni	Ti	Cr	Other Elements (Each)	Other Elements (Total)	Al
Wrought												
1100	Si + Fe ≤1	Si + Fe ≤1	0.20	0.05		0.10				0.05	0.15	Balance
1099											0.01	99.99 min.
1060	0.25	0.35	0.05	0.03	0.03	0.05		0.03				Balance
2011	0.04	0.07	5.0–6.0	0.03	1.00	0.30				0.05	0.15	Balance
2024	0.50	0.50	3.8–4.9	0.3–0.9	1.2–1.8	0.25			0.10	0.05	0.15	Balance
3003	0.60	0.70	0.20	1.0–1.5		0.10				0.05	0.15	Balance
3004	0.30	0.70	0.25	1.0–1.5	0.8–1.3	0.25				0.05	0.15	Balance
4032	11–13.5	1.00	0.5–1.3		0.8–1.3	0.25	0.5–1.3		0.10	0.05	0.15	Balance
4043	4.5–6.0	0.80	0.30	0.05	0.05	0.10		0.02		0.05	0.15	Balance
5052	Si + Fe ≤ 0.45	Si + Fe ≤ 0.45	0.10	0.10	2.2–2.8	0.20			0.15–0.35	0.05	0.15	Balance
5056	0.30	0.40	0.10	0.05–0.2	4.5–5.6	0.10			0.05–0.2	0.05	0.15	Balance
5086	0.40	0.50	0.10	0.2–0.7	3.5–4.5	0.25		0.15	0.05–0.25	0.05	0.15	Balance
6061	0.4–0.8	0.70	0.15–0.4	0.15	0.8–1.2	0.25		0.15	0.15–0.35	0.05	0.15	Balance
6063	0.2–0.6	0.35	0.10	0.10	0.45–0.9	0.10		0.10	0.10	0.05	0.15	Balance

6151	0.6–1.2	1.00	0.35	0.20	0.45–0.8	0.25		0.15	0.15–0.35	0.05	0.15	Balance
7075	0.50	0.70	1.2–2.0	0.30	2.1–2.9	5.1–6.1		0.20	0.18–0.4	0.05	0.15	Balance
7079	0.30	0.40	0.4–0.8	0.1–0.3	2.9–3.7	3.8–4.8		0.10	0.1–0.25	0.05	0.15	Balance
7178	0.50	0.70	1.6–2.4	0 30	2.4–3.1	6.3–7.3		0.20	0.18–0.4	0.05	0.15	Balance
						Cast						
208	3.00	<1.2	4.00		<0.1	<1.0						Balance
319	6.00	<1.0	3.50		<0.1	<1.0						Balance
356	7.00	<0.6	<0.25		0.35	<0.35						Balance
444	7.00	<0.2	<0.1		<0.05	<0.1						Balance
515	0.75	<1.3	<0.2	0.50	3.25							Balance
710	<0.15	<0.5	0.50		0.70	6.50						Balance

TABLE 5.1 Components of Cast Aluminum Alloys

the 3004 alloy is commonly used for manufacturing beverage cans. However, these alloys should not be used where high strength and hardness are required. Consequently, these alloys are not commonly used in cooling water service.

4000 Series

Silicon is the major alloying element in this series, typically ranging between 1.0 and 13.5 wt%. Silicon lowers the melting range. The silicon additions result in the formation of silicon particles in the microstructure, which are cathodic to the aluminum matrix. However, these alloys have fairly good corrosion resistance, since the silicon particles in the matrix are highly polarized, which results in low corrosion current density. Due to brittleness, the 4000 series alloys are not commonly used in cooling water service. However, cladding of these alloys has been applied to less resistant aluminum alloys to provide better corrosion resistance in some applications.

5000 Series

Magnesium is the major alloying element in this series, typically ranging from 0.8 to 5.8 wt%, which is one of the most effective and widely used alloying elements for aluminum. When it is used as the major alloying element or combined with manganese, the result is a moderate-to high-strength alloy. Alloys in this series are readily weldable and have excellent resistance to corrosion, even in marine applications. Most alloys in this series are suitable for use in cooling water systems. However, alloys with magnesium concentration exceeding about 3 percent are significantly more susceptible to corrosion, since an aluminum magnesium precipitate forms along grain boundaries, which promotes intergranular corrosion and exfoliation corrosion.

6000 Series

Alloys in this series contain 0.4 to 1.3 wt% magnesium, and 0.2 to 1.5 wt% silicon in a variety of combinations. These alloys can be heat treated to improve mechanical properties. The magnesium-silicon alloys have good formability and corrosion resistance with high strength. Small concentrations of copper (0.3 percent) are added to some 6000 series alloys, which increases strength, but does not significantly reduce corrosion resistance. A major alloy in this series is 6061, which is perhaps one of the most versatile of the heat-treatable alloys. Alloys in this series are more commonly used in cooling water systems than alloys in the other series listed.

7000 Series

Zinc is the primary alloying element in this series at concentrations ranging from 3.5 to 7.6 wt%. The high zinc concentration substantially reduces corrosion and stress corrosion cracking resistance.

When the zinc is combined with smaller percentages of magnesium and, in some cases, copper, it results in heat-treatable alloys of very high strength. Certain 7000 series alloys, such as 7075, contain high concentrations of copper (typically 1.2–2.0 wt%) to provide high strength for applications such as aircraft components. Similar to the 2000 series alloys, the 7000 series alloys containing high copper concentration are susceptible to corrosion, particularly when they are heavily deformed during processing. Heavy deformation will promote "exfoliation corrosion." Alloys that are susceptible to exfoliation are not suitable for use in cooling water systems.

Although alloy composition is a main factor that affects corrosion resistance, the processing of the alloy can also strongly influence corrosion resistance.

Temper Designations

The alloy number designation is followed by either a letter or a combined letter and number designation which defines the temper of the alloy. Temper designations indicate the type of heat treatment and/or mechanical processing that is applied to the alloy. The specific letter designations are as follows:[4]

T – Designates that the alloy was heat treated to produce specific mechanical properties. The numbers following the letter T indicate specific degrees of heat treatment, which varies in time at temperature.

H - Denotes that mechanical properties were imparted by forming operations such as cold rolling. Numbers following the letter designation correspond to specific amounts of cold work, which may be combined with heat treatment.

O – Designates an alloy that was heat treated by annealing. Such heat treatments typically produce alloys that have lower mechanical strength. Annealed alloys are also more easily formed.

W – Designates an alloy that is solution heat treated. Certain alloying elements will increase strength when the component is solution heat treated.

F – Designates that the alloy has the properties produced by the fabrication of the component, and is usually associated with cast alloys.

Cast Alloys

Cast alloys use a three-digit designation where the hundreds digit denotes the primary alloying element.[5] However, the number designating the primary alloying element in the cast alloy designation doesn't always match the number in the wrought alloy designation. For example, the primary alloying element in a 300 series cast alloy is silicon, while the primary alloying element for 3000 series wrought aluminum is manganese. Letter and number designations following the main number designation indicate processes that were used to impart mechanical properties to the casting. The effect of alloying elements upon the corrosion resistance of cast alloys is similar to that for the wrought alloys discussed earlier. Some cast alloy compositions are listed in Table 5.1.

Forms of Corrosion

High and Low pH Corrosion

In untreated water, the protective aluminum oxide layer covering an aluminum surface is quite stable within its passive range of pH. Aluminum alloys are amphoteric, indicating that the oxide layer is susceptible to destruction at both high (>8.5) and low (<4) pH. The oxide layer is soluble in many different acids and bases. Acids tend to dissolve the oxide and underlying metal to put aluminum ions (Al^{3+}) into water solution, while bases put aluminate ions (AlO_2^-) into solution. When the pH is outside of the passive range, both high and low pH will tend to produce general or uniform corrosion. The type of acid or base is significant. For instance, at similar pH, a sodium hydroxide solution will be more corrosive to aluminum than an ammonia solution. A reducing hydrochloric acid solution will be more corrosive to aluminum than an oxidizing nitric or chromic acid solution.

Aluminum is very resistant to corrosion in neutral, high-purity water. As such, it is commonly used to line condensate storage tanks at power plants. Many small, closed cooling water systems are also constructed with aluminum components that employ deionized water with great success.

Pitting

When aluminum is within its passive range of pH, it is rare for surfaces to experience uniform or general corrosion. Metal loss tends to occur in the form of pitting or localized corrosion. This corrosion occurs at scattered discontinuities in the protective oxide layer or where aggressive anions concentrate to form locally acidic environments that dissolve the protective oxide layer. Since it is one of the most active metals listed on the galvanic series, the exposed aluminum is very susceptible to corrosion if the protective oxide layer does not re-form. Chloride is the most significant contributor to localized corrosion of aluminum in cooling water systems. Chloride ions most commonly concentrate beneath occlusive deposit and corrosion product layers, which locally produces a hydrochloric acid solution due to hydrolysis. When aqueous corrosion occurs, voluminous, porous,

FIGURE 5.2 Porous, white aluminum oxide corrosion products formed due to aqueous corrosion.

white corrosion product mounds and layers, which primarily consist of hydrated aluminum oxide, form on the surface (Fig. 5.2).

Aluminum may be used successfully in seawater service provided that a resistant alloy is used. Therefore, copper bearing alloys should be avoided in seawater applications. When corrosion occurs in seawater, it will be in the form of pitting, and it will be more significant within crevices and beneath occlusive deposits where chloride may concentrate. The corrosion rate tends to increase in seawater service as a function of decreasing temperature, pH, and flow velocity, as well as increasing dissolved oxygen and noble metal ion concentration.

Galvanic Corrosion

Aluminum is one of the most active metals in the galvanic series. It is very susceptible to galvanic corrosion when directly coupled with much more noble metals such as copper, lead, nickel, and tin. Galvanic corrosion may also occur when these noble metal ions are dissolved in water and plate onto the surfaces of aluminum components. Such deposition and corrosion may occur even at low concentrations of noble metal ions in water solution. For instance, galvanic corrosion has been known to occur on aluminum exposed to waters containing dissolved copper ions at concentrations as low as 20 parts per billion in neutral to acidic water.[6] Certain aluminum alloys are more susceptible to corrosion due to their alloy composition. This is due to microgalvanic interactions between the matrix and the second phase particles that form within the microstructure. Alloys containing high concentrations of copper, zinc, or magnesium may be particularly susceptible. In addition, when alloys containing high copper concentrations experience corrosion, the copper in the alloy is released into water and plates onto the aluminum surface, causing galvanic corrosion.

Stainless steels are much more noble than aluminum alloys in the galvanic series. However, since stainless steels easily polarize in mild environments, such as in mildly corrosive cooling water service, aluminum alloys can be in direct contact without experiencing significant galvanic corrosion. Susceptibility will be reduced when the exposed surface area of the aluminum is over five times that of the stainless steel. More in-depth discussions concerning galvanic corrosion are presented in Chap. 16.

Intergranular Corrosion

Intergranular corrosion may occur when the grain boundaries in the microstructure become anodic, without significant corrosion of the grains themselves. Such corrosion is promoted by composition differences at the grain boundaries, such as the presence of anodic precipitates. Cathodic precipitates at grain boundaries may also promote intergranular corrosion.

Exfoliation Corrosion

The alloy composition and the method of forming an aluminum component may promote an intergranular corrosion mechanism called exfoliation corrosion. One contributing factor is heavy deformation (cold work) of the component during forming operations that produces thin, elongated grains. Corrosion penetrates preferentially along the grain boundaries. Depending upon the alloy composition, precipitates that form along the grain boundaries may promote substantial corrosion as well. In particular, alloys containing high concentrations of copper are particularly susceptible to exfoliation corrosion. Heavily cold worked, high copper content alloys, such as 2024 and 7075, have high strength and are commonly used for aircraft components. As corrosion occurs, the copper in the alloy is released into water solution and plates onto the surface along the grain boundaries in the metal, causing galvanic corrosion. Corrosion penetrates preferentially along the grain boundaries, and it may follow along a network of interconnecting grain boundaries to substantial depths into a component. A schematic drawing which shows exfoliation corrosion on aluminum alloy skins that are riveted to aircraft fuselages is shown in Fig. 5.3. Corrosion occurs when protective coatings fail and allow water to seep between the rivet and aluminum skin.

In some applications where durable aluminum components are required, alloys that are normally deployed in aircraft applications have been used. A 7075 aluminum two-liter plastic bottle mold that

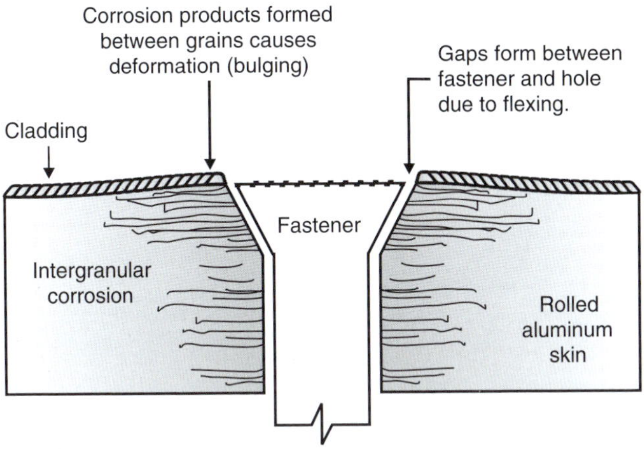

FIGURE 5.3 Schematic drawing depicting exfoliation corrosion of aircraft skin at riveted joint.[7]

experienced exfoliation corrosion is shown in Fig. 5.4. During the corrosion process, aluminum oxide corrosion products formed within the gaps between the grains. Since the corrosion products were much more voluminous than the aluminum, the grains were forced apart (Figs. 5.5 and 5.6). This led to bulging and breakaway, or exfoliation of the aluminum grains. Water flow can accelerate the exfoliation of the grains. Figure 5.7 shows exfoliated aluminum grains that were retrieved from a bag filter. These grains of uncorroded aluminum will be flushed downstream and may plug narrow orifices. In components

FIGURE 5.4 Aluminum alloy mold used for production of two-liter plastic bottles.

FIGURE 5.5 Narrow, elongated aluminum grains forced apart by aluminum oxide corrosion products.

FIGURE 5.6 Microstructure showing intergranular corrosion and deformation of narrow, elongated aluminum grains.

such as plastic injection molds, such flow obstruction cannot be tolerated, since sufficient water flow is required in order to provide proper cooling to produce defect-free products. Eventually, such corrosion will penetrate the wall of the mold to cause water leaks into the mold cavity (Figs. 5.8 and 5.9). Aluminum alloys that are susceptible to exfoliation corrosion are not well suited for use in cooling water systems.

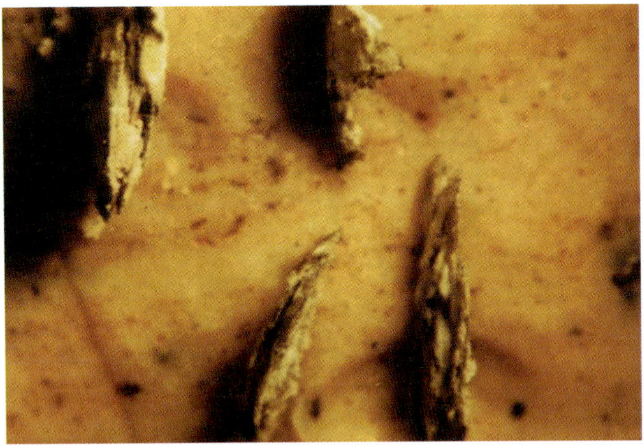

FIGURE 5.7 Exfoliated aluminum grains that were removed from a bag filter.

FIGURE 5.8 Exfoliation corrosion penetrated through wall of mold.

Stress Corrosion Cracking

Stress corrosion cracking of aluminum almost exclusively penetrates intergranularly. Such cracking results from the combination of intergranular corrosion and applied and/or residual stresses. Residual stresses are supplied by forming of the component. More heavily cold worked aluminum will contain larger residual stresses that are more susceptible to stress corrosion cracking. Unlike stress corrosion cracking of other alloys, such as steels and copper alloys, stress corrosion cracking of aluminum alloys does not require exposure to specific corrodents. As such, the difference between stress corrosion

FIGURE 5.9 Metallographic cross-section showing exfoliation corrosion penetrating from cooling water channel to molding surface.

cracking and intergranular corrosion of aluminum alloys is not particularly distinct. Stress corrosion cracking tends to occur mainly in the 2000 and 7000 series aluminum alloys, and the 5000 series alloys that contain greater than 3 percent magnesium.[8]

Crevice and Underdeposit Corrosion

The mechanisms of crevice and underdeposit corrosion for aluminum alloys are similar to those for other alloys, particularly due to the formation of ion concentration cells involving chloride. Such corrosion is promoted by low or stagnant flow conditions. Aluminum is more susceptible to such corrosion than copper and steel alloys. The crevice and underdeposit corrosion mechanisms are described in more detail in Chap. 11.

Microbiologically Influenced Corrosion

The mechanisms of microbiologically influenced corrosion (MIC) for aluminum alloys are similar to those for other alloys. Aluminum is more susceptible to such corrosion than copper and steel alloys. The microbiologically influenced corrosion mechanisms are described in more detail in Chap. 13.

Erosion and Erosion-Corrosion

Aluminum is a soft metal that is highly susceptible to erosion due to impingement and/or turbulent flow of water across surfaces. Erosion may be accelerated by excessive suspended solids and/or entrained gas bubbles. The aluminum oxide layer provides significant resistance to erosion. However, when the oxide layer is removed mechanically or due to corrosion, rapid metal loss will occur under erosive conditions. Erosion is described in more detail in Chap. 23.

High-Temperature Corrosion and Effects

Above a temperature of about 150°C (302°F), aluminum alloys will experience accelerated intergranular corrosion.[9] In addition, mechanical properties of strain-hardened alloys begin to decrease at about 150°C (302°F), and above about 100°C (212°F) for age-hardenable alloys. The recommended upper temperature limit for aluminum is about 150°C (302°F).

Cleaning

Removal of deposits and corrosion products from aluminum surfaces through the use of chemical cleaning methods can be very difficult. Alkaline and acidic cleaning agents that will be effective for cleaning steel and copper alloys that are used in cooling water systems will be highly corrosive to aluminum alloys. Organic solvents and alkaline cleaners are required for the removal of organic materials, but they are not effective for removing aluminum oxide corrosion products. Multistep cleaning may be required to remove different materials.

Alkaline cleaning agents such as caustic, sodium carbonate, and alkaline phosphates, which produce solutions with a pH exceeding 9, will produce high corrosion rates and will require the use of suitable inhibitors during the cleaning process, typically silicates. The amount of silicate must be controlled to low concentrations, since otherwise excessive amounts of aluminum silicate may be produced. When thin silicate layers form on surfaces, they will inhibit corrosion successfully, but at higher concentrations, more substantial amounts of aluminum silicate may be produced that will cause fouling.

Removal of deposits such as calcium carbonate may be done with mildly acidic cleaners, including organic acids such as sulfamic acid or citric acid along with small amounts of acid phosphates. Removal of aluminum oxide corrosion products will require the use of acidic cleaners. It is common to use a mixture of certain acids that will minimize corrosion, such as chromic and sulfuric acids. Though the intent is to minimize corrosion, acid cleaning will etch the metal surface to some degree.

Surface Finish and Anodizing

Differences in surface finish on aluminum components do not significantly influence the formation of the protective oxide layer. Therefore, polishing a mill-finished surface of an aluminum component will provide no significant added benefit.

Some aluminum alloys are anodized, which refers to the formation of a thicker, more protective aluminum oxide layer than is produced naturally. The oxide layer that forms during anodizing is produced by an electrolytic oxidation process. The aluminum component is submerged in a chemical solution that typically consists of acids such as sulfuric, chromic, or organic acids. A DC current is applied, with the aluminum component serving as the anode in the circuit to produce the thicker oxide layer. Anodizing layers may be dyed to a variety of different colors. Figure 5.10 shows the surface of an aluminum alloy component that is covered with a lustrous black anodizing layer.

In some applications a thicker and denser oxide layer may be formed on aluminum components; this is referred to as a hard coating. Such layers are more erosion resistant than a simple anodized coating, and they are produced by a different anodizing process. In general, anodizing is considered to be more useful in atmospheric service or to produce a decorative finish. In cooling water systems, however, anodizing will not provide significant additional protection. The oxide layer that is produced by anodizing will be dissolved by the same corrosion processes that affect oxide layers that cover unanodized aluminum surfaces. In addition, where there are small

FIGURE 5.10 Shiny black anodized layer covering aluminum component.

breaks in anodized coatings to expose bare aluminum, the large cathode-to-anode surface area ratio may produce deep pits in the underlying aluminum. Figures 5.11 and 5.12 show exfoliation corrosion that occurred at breaks in an anodizing layer.

Figure 5.11 Metallographic cross-section showing exfoliation corrosion that occurred where anodizing layer broke away from the surface and cracked.

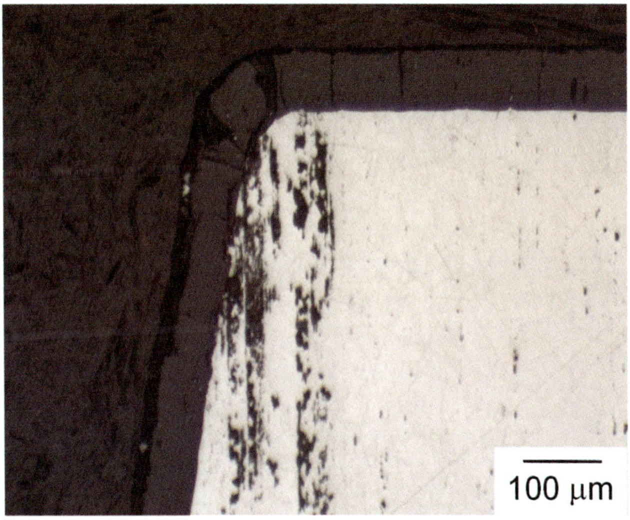

Figure 5.12 Higher magnification view of corrosion that occurred where the anodizing layer cracked.

Corrosion Inhibitors

The primary corrosion inhibitors used for aluminum alloys are silicates. The use of silicates will allow the use of aluminum at a slightly higher pH than the recommended threshold of 8.5 in untreated water. Molybdate has also been found to provide good corrosion inhibition. It is essential to control the corrosion of copper alloys in mixed alloy systems. This will control the amount of dissolved copper in the water that can plate onto the aluminum to cause galvanic corrosion.

Exfoliation corrosion and intergranular corrosion of high copper content aluminum alloys cannot be reliably controlled with corrosion inhibitors. Water containing proper concentrations of corrosion inhibitors cannot penetrate into the tight crevices where exfoliation and other intergranular corrosion occur. The galvanic effects between the plated copper and the aluminum may overwhelm the ability of an inhibitor to control the corrosion. This is another reason why aluminum alloys that are susceptible to exfoliation corrosion should be avoided in cooling water systems.

References

1. Christian Vargel, *Corrosion of Aluminum*, Elsevier, 2004, p. 105.
2. Christian Varel, *Corrosion of Aluminum*, Elsevier, 2004, p. 25.
3. *ASM Metals Handbook*, 8th ed., Volume 1, p. 917.
4. Christian Varel, *Corrosion of Aluminum*, Elsevier, 2004, p. 46.
5. *Aluminum Properties and Physical Metallurgy*, ASM, pp. 322–325.
6. *ASM Metals Handbook*, 9th ed., Volume 13, p. 589.
7. J. R. Davis, *Corrosion of Aluminum Alloys*, ASM, 1999, p. 69.
8. *Aluminum Properties and Physical Metallurgy*, ASM, p. 264.
9. Christian Varel, *Corrosion of Aluminum*, Elsevier, 2004, pp. 314–318.

CHAPTER 6

Corrosion-Resistant Alloys

Introduction

When carbon steels, stainless steels, or copper alloys do not provide adequate resistance to corrosion and cracking due to exposure to aggressive conditions on the cooling water and/or process sides, it may be necessary to use corrosion-resistant alloys (CRAs). Corrosion-resistant alloys commonly contain primary alloying elements such as nickel, titanium, and cobalt. Certain nickel- and titanium-based corrosion-resistant alloys tend to be very resistant to corrosion and stress corrosion cracking by aggressive cooling waters, and even brackish water or seawater. The resistance is primarily due to the presence of very protective oxide layers on their surfaces. These alloys also have excellent mechanical properties that are suitable for heat exchanger tubing.

The common types of corrosion-resistant alloys used in cooling water systems are described here along with applications. Proper screening or corrosion testing should be conducted to ensure compatibility with the materials of construction on the process side. Note that corrosion-resistant cobalt alloys that are appropriate for high-temperature applications are not appropriate for cooling water service. However, cobalt is used in certain components of cooling water systems, such as cemented carbide seal rings and hardfacing, where abrasive wear resistance is required.

Types of Corrosion-Resistant Alloys

Nickel Alloys

As discussed in Chap. 3, austenitic stainless steels, such as the commonly used AISI 304 and 316 grades, are susceptible to pitting and stress corrosion cracking above certain metal temperatures and chloride concentrations in cooling water. Under such conditions, more resistant stainless steels may be substituted. However, under severe conditions involving chloride, even the more corrosion and cracking-resistant stainless steels will be insufficient. Therefore, suitable alternate alloys are required to provide acceptable corrosion and cracking resistance. For applications where copper alloys also do not have sufficient chloride corrosion resistance or mechanical properties, nickel alloys should be considered.

Pure nickel, such as alloy 200, has excellent resistance to corrosion in freshwater and nonoxidizing acidic solutions, as well as caustic-based alkaline solutions. However, pure nickel has limited mechanical properties. For instance, pure nickel has lower strength than stainless steel alloys. Nickel alloys may contain higher concentrations of alloying elements than iron-based alloys. This has allowed nickel alloys to be developed with improved mechanical properties and corrosion resistance for environments where pure nickel is unsuitable.

Nickel-based alloys that are alloyed primarily with copper, such as monel (Alloy 400), are very resistant to chloride corrosion and stress corrosion cracking. Monel is suitable for brackish and seawater applications in flowing systems, but stagnant conditions should be avoided. Monel is less erosion resistant and has lower mechanical properties than the alloys that contain varying amounts of alloying elements such as chromium, molybdenum, tungsten, iron, copper, and cobalt. Some of the common nickel-based alloys used in cooling water service that contain these alloying elements include Hastelloy, Inconel, and Incoloy alloys. Unlike pure nickel and monel, these alloys are very resistant to corrosion by nonoxidizing acids, so they significantly resist pitting by chlorides in water solution, and they are immune to chloride stress corrosion cracking. Resistance to pitting by chloride may also be predicted by pitting resistance equivalent number (PREN) calculations as described in Chap. 3. Iron and cobalt alloy additions are typically used in nickel alloys for high-temperature applications, and they are not commonly used in water systems. Chromium in nickel alloys also promotes resistance to corrosion by oxidizing acids. Some of the commonly used corrosion-resistant nickel alloys are listed in Table 6.1. Though nickel alloys are very resistant to corrosion and cracking by chloride, they are not commonly used for surface condenser tubing in brackish and seawater service.

The choice of a nickel alloy in a cooling water system depends significantly upon the conditions on the process side. Nickel and nickel alloys are very resistant to many chemical environments, particularly caustic-based aqueous environments as well as mineral acids and highly aggressive chloride environments. Certain Hastelloy alloys are used for the manufacture of hydrochloric acid. Some Hastelloy alloys have good resistance to strong oxidizers such as chlorine and chlorine dioxide solutions. However, nickel alloys tend to be susceptible to corrosion by aggressive sulfur compounds, such as hydrogen sulfide, as well as ammonia solutions, and they are generally not used in environments containing such agents.

Titanium Alloys

Titanium alloys have a high strength-to-weight ratio. These alloys are highly corrosion and wear resistant due to the presence of a tenacious, protective titanium dioxide layer that covers the surfaces. Consequently, titanium alloys are resistant to erosion at very high flow velocities [>100 ft/s (30.5 m/s)] and by abrasive suspended solids. If the resistant titanium dioxide layer is damaged, it will quickly be repaired when exposed to oxidizing conditions such as oxygenated cooling water. Titanium tubing is typically produced with thin walls due to its strength and corrosion resistance. Additional considerations for the use of thin-walled tubing are that titanium has a lower heat transfer coefficient relative to ferrous and copper alloys, and it is relatively expensive. Due to the high cost of titanium alloys, there are few industrial applications.

For cooling water service, unalloyed titanium is used, since it is more easily formed into tubing. Unalloyed titanium is also more corrosion resistant than alloyed titanium. The common titanium alloys that are used in cooling water service are listed in Table 6.1. In the unalloyed condition, small concentrations of aluminum, nickel, molybdenum, tin, vanadium, and other elements may be used to impart higher strength and improved mechanical properties. Mechanical properties of unalloyed titanium are sufficient for cooling water and process chemical exposure. Titanium alloys are commonly used for surface condenser tubing where the cooling water source is brackish water or seawater. Titanium is the preferred choice for seawater condenser tubing, and it typically provides long-term, trouble-free service. Titanium alloys may be used in heat exchangers to cool aggressive process fluids. Titanium is resistant to aqueous environments containing oxidizing species such as hypochlorite and chlorine dioxide. However, titanium is susceptible to corrosion when exposed to acidic, reducing conditions, particularly at high temperatures and high acid concentrations. Such conditions may be produced on the cooling water side due to chloride concentration within crevices or beneath occlusive deposit layers. However, titanium is much more resistant to corrosion under such conditions than stainless steels or even nickel alloys.

Embrittlement and associated cracking damage may result from hydrogen-induced cracking due to absorption of monatomic hydrogen. In order for significant hydrogen absorption to occur, it is necessary for the titanium material to experience severe corrosion at low (<3) or high (>12) pH, to be coupled to a less noble metal such as aluminum or zinc, or for excessive cathodic currents to be applied by an impressed current system. Excessive cathodic currents may produce hydrides at ambient temperatures. Such a combination of factors may occur in a heat exchanger containing titanium tubes that are in contact with tubesheets and baffles that are made of less noble alloys. Hydride formation may also occur at temperatures greater

UNS Alloy Number	Alloy Name	C (max)	Ni (max)	Cr (max)	Co (max)	Mo (max)	Cu (max)	Fe (max)	Al (max)	Ti (max)	W (max)	Other Elements	PRE Number (max)
Nickel Alloys													
N02200	Nickel 200		99.60					0.20				Mn 0.2	0
N04400	Monel		67.00				31.50	1.20					0
N10276	Hastelloy C-276		59.00	16.00		16.00		5.00			4.00		69
N06022	Hastelloy C-22		59.00	22.00		13.00		3.00			3.00		65
N06200	Hastelloy C-2000		59.00	23.00		16.00	1.60						76
N06455	Hastelloy C-4		68.00	16.00		16.00							69
N06030	Hastelloy G-30		44.00	30.00	4.00	5.00	2.00	15.00			2.50		46.5
N10665	Hastelloy B-2		72.00			28.00							92
N10675	Hastelloy B-3		68.50	1.50		28.50		1.50					95.5
N06600	Inconel 600		76.00	15.50				8.00					15.5
N08825	Incoloy 825		44.00	21.00		3.00	2.20	30.00		1.00			31

Titanium Alloys													
R50250	Ti Grade 1	0.10						0.20		99.20		N 0.05, O 0.25, H 0.015	
R50400	Ti Grade 2	0.10						0.20		99.20		N 0.05, O 0.25, H 0.015	
R52400	Ti Grade 7	0.10						0.30		99.00		N 0.02, O 0.25, H 0.015, Pd 0.12–0.25	
R53400	Ti Grade 12	0.08	0.60–0.90			0.20–0.40		0.30				N 0.03, O 0.25, H 0.015	
Cemented Carbide Alloys													
	DC7				7.00						93 as carbide		
	D100				5.00						85.3 as carbide	TaC 5, TiC 4, Cr_3C_2 0.7	
	DN6		6.00								94 as carbide		
	DS1				7.50						71.5 as carbide	TaC 4, TiC 17	
Hardfacing Alloys													
	Stellite 1	2.45	3.00	31.00	46.05			2.50			13.00	Si 1, Mn 1	
	Stellite 6	0.90–1.40		27.00–32.00	Balance						4.00–6.00		
	Stellite 12	1.40–1.85	3.00	29.50	Balance			2.50			8.50	Si 1.5, Mn 1.5	

TABLE 6.1 Common Corrosion-Resistant Alloys Used in Cooling Water Systems

than 80°C (176°F) due to exposure to high concentrations of monatomic hydrogen that exceed the solubility in the metal. Titanium alloys may experience stress corrosion cracking when exposed to specific corrodents such as alcohols and glycols. In theory, such cracking is also caused by hydrogen absorption.

Titanium is a very stiff material. Therefore, titanium alloy tubing is susceptible to fatigue cracking if it is not adequately supported. For this reason, when condenser tubing is retrofitted from copper alloys or stainless steels to titanium, it is necessary to install additional tube supports and baffles.

Cemented Carbide Components and Hardfacing

Seal rings for cooling water pumps are made of alloys that contain hard, brittle tungsten carbide that must be bound or cemented with suitable ductile alloying elements such as cobalt and nickel. Tungsten carbide is very resistant to abrasive wear under normal sliding contact with the mating soft seal ring, which is commonly made of graphite. Tungsten carbide is also very corrosion resistant. However, the binder material in tungsten carbide may experience corrosion in aggressive cooling waters such as brackish or seawater, or in applications where cooling water evaporates to allow aggressive agents such as chlorides to concentrate. One example, shown in Fig. 6.1, is a roll that was used to form steel into a round rod shape. Corrosion of the cobalt binder occurred due to exposure to high chloride content cooling water that concentrated due to evaporation at high temperatures. When the cobalt binder in tungsten carbide components experiences excessive corrosion, substitution of nickel binder will tend to improve corrosion resistance. Some commonly used cemented carbide alloys are listed in Table 6.1.

FIGURE 6.1 Chloride pitting of cobalt binder on surface of roll used to form steel rods.

Abrasive wear on some components is controlled by the application of hardfacing material, which contains a combination of various hard alloying elements. Some commonly used hardfacing materials are cobalt based and contain high concentrations of chromium and significant amounts of tungsten. Some commonly used hardfacing alloys are listed in Table 6.1. Hardfacing materials are typically also very corrosion resistant. Some hardfacing materials are applied using a plasma spray technique, which produces a porous structure that will allow water to contact the underlying base metal. When the base metal is much less resistant to corrosion than the hardfacing, corrosion of the base metal will be accelerated by galvanic corrosion effects. The hardfacing will also tend to become undermined and peel away from the surface. Some hardfacing may also be applied using welding techniques, which will produce a uniform, dense layer that will provide better protection than a porous plasma sprayed layer.

Forms of Corrosion

Galvanic Corrosion

Corrosion-resistant alloys should not be in direct contact with much less noble materials (lower in the galvanic series), since the CRAs will tend to promote preferential galvanic corrosion of such metals. This can be an issue for heat exchangers with titanium tubing. Recommended tubesheet materials for titanium tubes include titanium alloys, resistant stainless steels, and nickel-based alloys. The titanium tubes may experience embrittlement due to hydride formation when hydrogen is evolved due to overprotection of the less resistant tubesheet metals by impressed current cathodic protection systems. More in-depth discussions concerning galvanic corrosion are presented in Chap. 16.

Crevice and Underdeposit Corrosion

The mechanisms of crevice and underdeposit corrosion for nickel and titanium alloys are similar to those for other alloys where occluded areas develop differential oxygen concentration cells, and subsequently anion concentration cells, particularly involving chloride. Such corrosion is promoted by low or stagnant flow conditions. However, corrosion-resistant alloys are substantially more resistant to such corrosion than steel and copper alloys. The crevice and underdeposit corrosion mechanisms are described in more detail in Chap. 11.

Microbiologically Influenced Corrosion

In some cases, microbiologically influenced corrosion (MIC) processes can affect certain corrosion-resistant alloys, particularly nickel-based alloys. However, corrosion-resistant alloys are much less susceptible to such corrosion than steel and stainless steel alloys. Titanium is highly resistant, and possibly immune to MIC based upon field experience and testing.[1] The microbiologically influenced corrosion mechanisms are described in more detail in Chap. 13.

Cleaning and Passivation

Cleaning and passivation treatments prior to service are not as important for corrosion-resistant alloys as for steel and copper alloys. Corrosion-resistant alloys are generally very resistant to underdeposit corrosion. However, if localized, aggressive conditions develop beneath occlusive deposit layers, or heat transfer losses are excessive, then appropriate mechanical or chemical cleaning would be recommended on a periodic basis. Corrosion-resistant alloys are very resistant to corrosion by cleaning agents that are used to remove normal waterside deposits. Care must be exercised when chemically cleaning to minimize galvanic corrosion of less noble metals that contact corrosion-resistant alloys.

Surface Finish

Differences in surface finish on CRA components do not significantly influence the formation of the protective oxide layer.

Corrosion Inhibitors

Corrosion inhibitors are not required for control of corrosion on the cooling water sides of nickel-, titanium-, and cobalt-based corrosion-resistant alloys.

Reference

1. Brenda Little, *International Materials Review,* 1991; Little, B., Wagner, P., and Mansfield, F., (1991) *Microbiologically Influenced corrosion of metals and alloys,* International Materials Review.

CHAPTER 7
Coatings

Introduction

In simple terms, a coating is an applied layer of material that overlies the surface of another material, known as the substrate. Coatings are often an economical choice for corrosion control, especially for exposure to the atmosphere or very aggressive chemical solutions. To provide corrosion control, barrier coatings essentially isolate the substrate from the environment. Certain metallic coatings will corrode preferentially to the substrate. Protective coatings are often used to preserve the strength and integrity of structural components that are susceptible to corrosion in the environment. Some examples include steel I-beams in wet or moist environments or pipes that are buried in soil, which are often externally coated.

Coatings have many applications in cooling water and industrial systems, including surfaces that are exposed to water continuously or intermittently during operation, or are subjected to high-humidity conditions. Chemically resistant coatings are often required for applications where metals would otherwise be exposed to a very aggressive environment, such as acidic solutions or vapors. Piping or tubing can be lined with protective coating layers. Other examples of components that may be protected from aqueous exposure by coatings include heat exchanger tubesheets and end cap surfaces, valves, pump parts, tanks, and cooling tower basins.

In addition to corrosion resistance, coatings can also be used to provide resistance to erosive environments. Some examples include chromium-plated ball valves and hardfacing layers to cover areas susceptible to mechanical degradation. Anodized aluminum is also considered to be a coated material, although technically the protective layer is an oxide formed from the aluminum itself, rather than a material that is applied to the aluminum surface.

The success of a protective coating during service requires it to remain intact and adherent to the substrate. Coatings generally have an expected service life, but premature failures can result from poor choice of a coating type for the environment, improper surface preparation of the substrate, incorrect or insufficient application of the coating, or unexpected chemical or mechanical degradation during service.

This section reviews some coating types, discusses factors that influence their effectiveness, and highlights some examples of applications in cooling water systems. The purpose is to provide a general understanding of coatings rather than to present detailed information about coating selection. Coating specialists should be consulted regarding the selection of appropriate coating types for a specific application. Testing may be conducted to ensure that a chosen coating is suitable for use in a particular environment.

Types

The wide variety of coatings that exist can generally be classified into metallic coatings, nonmetallic inorganic coatings, and organic coatings.

Metallic Coatings

Metallic coatings are typically applied to the surface of a component as a finishing step in the manufacturing process. Techniques to apply metallic coatings include electroplating, electroless plating, hot dipping, metal spraying, and cladding.

Electroplating is one of the more common methods of producing metallic coating layers on a component. Electroplating is essentially an electrochemical cell, with the component to be plated as the cathode. The component is submerged in a plating bath. An anode is also installed in the bath that releases metal ions into solution. When the component (cathode) and anode are coupled, a voltage is applied. The metal ions in the solution are reduced onto the surface of the component, forming a thin, uniform, adherent metallic coating under proper conditions. Multiple layers of different metals can be plated onto a surface.

Some metallic coatings can be produced by electroless plating, which utilizes chemical reduction in a special solution rather than applied voltage to form a metallic layer. One of the most common electroless coatings is electroless nickel. Electroless nickel coatings will typically consist primarily of nickel with small amounts of phosphorus or boron that are supplied by the reducing agents in the bath. Metallic layers formed from an electroless process generally have a very uniform thickness that is independent of the geometry of the part.

Another common method used to develop a metallic coating layer is hot dipping. During the hot dipping process, a metal substrate is submerged into molten metal. The hot dipping process is mainly used for producing layers of low-melting-point metals such as zinc, aluminum, and tin. Hot dipping to produce a zinc-based coating layer on steel, known as hot dip galvanizing, is by far the most common metallic coating technique for components in cooling water systems.

There will be a potential difference between the metallic coating and a metal substrate if they have sufficiently different compositions. Due to this, the overlying metallic layers can be considered anodic or cathodic coatings. For corrosion protection, a cathodic coating must behave as a barrier layer. It must not contain pores or other flaws that result in exposure of the substrate to liquid water. When a break in the coating is present, the potential difference between the dissimilar metals accelerates corrosion of the underlying anodic substrate (Fig. 7.1). Since the surface is almost completely covered by the cathodic coating, there is a large cathode-to-anode surface area ratio. This will intensify galvanic corrosion of the substrate at the openings in the coating. Due to the fact that most commercial coatings will contain discontinuities, the use of components with cathodic coatings for applications with significant exposure to water should be avoided. An example of corrosion of a steel tube electroplated with nickel and chromium cathodic coatings is shown in Fig. 7.2. The tube was taken from a test heat exchanger.

Anodic coatings can provide protection in two main ways. As long as the coating layer remains intact, it will act as a barrier to isolate the substrate from the environment. When the underlying substrate is exposed to water, either at openings in the coating layer or as the layer is consumed by the environment, additional protection will be offered by the coating due to the potential difference between the coating and substrate. The anodic coating will corrode sacrificially to the cathodic substrate (Fig. 7.3). Probably the most common example of a sacrificial coating is galvanized steel. Galvanized steel consists of a carbon steel

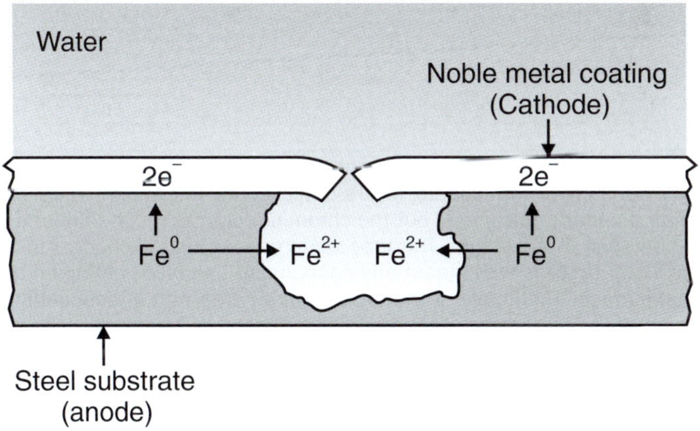

FIGURE 7.1 Schematic illustrating the effect on the corrosion process of the potential difference between a metallic coating layer and an underlying metal substrate when the coating layer is more noble than the substrate. The substrate is anodic to the overlying layer, resulting in preferential attack on the substrate. (*Reprinted with permission of ASM International. All rights reserved. www.asminternational.org.*)

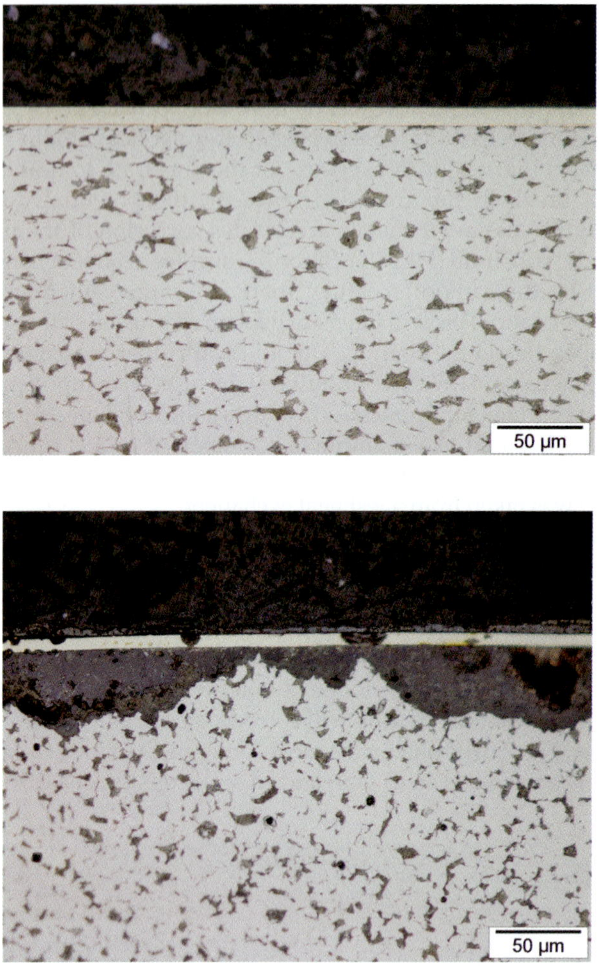

Figure 7.2 Micrographs of a steel heat exchanger tube that was plated on the external surface with coating layers of nickel and chromium. The nickel layer has a uniform thickness, but the chromium layer is much thinner and cannot be seen in the image. The tube was in a test heat exchanger in a cooling water system, with steam and condensate used for heating on the external surface. The image at the top shows an area with intact coatings. Corrosion of the underlying steel occurred at locations where the protective coating layers had openings. (Etchant: Picral.)

substrate that is covered by a zinc layer formed either by electroplating or hot dipping. The zinc layer is anodic to the steel, and it will corrode rather that the steel in most environments. Figure 7.4 shows attack of the zinc layer on a galvanized steel sheet, with no attack occurring on the exposed underlying steel. Generally, the underlying steel will not appreciably corrode until a substantial percentage of the surface area covered by the zinc layer is consumed.

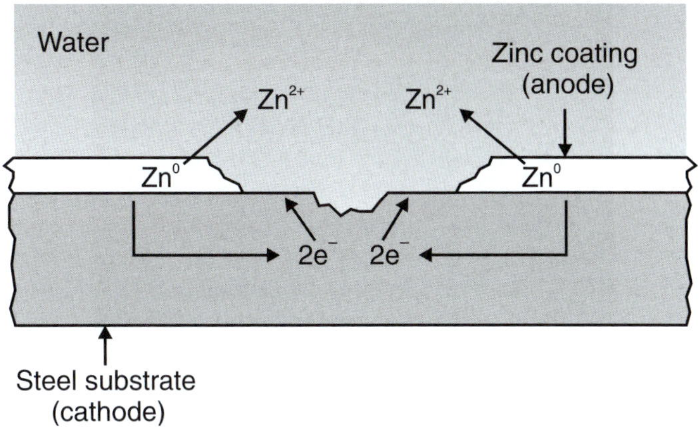

Figure 7.3 Schematic illustrating the protective galvanic effect due to the potential difference between a metallic coating layer and an underlying metal substrate when the coating is anodic to the substrate. Corrosion of the underlying substrate is minimal due to the sacrificial coating. (*Reprinted with permission of ASM International. All rights reserved. www.asminternational.org.*)

The wide use of galvanized steel in cooling water systems merits further discussion. Some applications for galvanized steel include sheets for cooling tower basins, decks, and structures, piping for water distribution, and tubing in heat exchanger equipment. In some

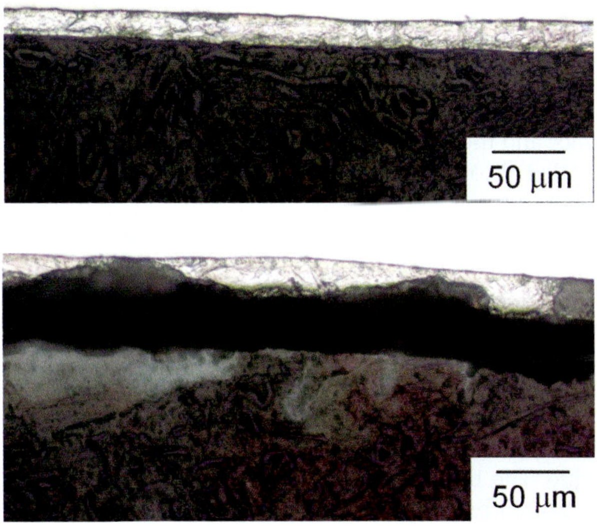

Figure 7.4 Micrographs of a galvanized steel sheet that was continuously exposed to moisture. The sheet is covered by a zinc-based coating, shown in the image at the top. The image on the right shows a location where the zinc coating was consumed in places, but no attack on the underlying steel (which appears white in the image) is apparent. (Unetched.)

FIGURE 7.5 Spangled appearance on the surface of galvanized sheet steel.

instances, a galvanized component can be easily identified by the appearance of spangles on the coated surface (Fig. 7.5). The spangled appearance is due to the development of dendritic structures that form during solidification of the molten zinc. In addition, reactions can occur between the molten zinc and the steel substrate that result in the formation of iron-zinc intermetallic layers. Both the spangled appearance and the development of iron-zinc intermetallic layers can be controlled through small alloying additions to the molten zinc bath for hot dipping.

The thickness of the zinc coating on galvanized steels strongly influences its service life in many environments, especially with atmospheric exposure. As the coating layer becomes thicker, longer periods of time are required before the underlying steel shows evidence of corrosion in the form of visible rust on the surface (Fig. 7.6). The amount of zinc coating on galvanized steel is typically represented by a weight per area [oz/ft^2 (g/m^2)] measurement rather than a thickness. Specifications for galvanized coatings typically require a minimum value. The thickness of the zinc-based coating on steel can be easily measured by magnetic thickness gages. The thickness value is then converted to the weight measurement, as a 0.001 in. (0.025 mm) thick zinc coating layer roughly corresponds to 0.59 oz/ft^2 (180 g/m^2).

The resistance of the galvanized layer to aqueous corrosion depends on many factors related to water quality. The corrosion rate

Figure 7.6 Orange-brown iron oxide rust spots surrounded by white zinc-based corrosion products on the surface of a galvanized sheet. Small depressions in the steel substrate are located beneath the orange-brown spots.

of zinc can be relatively low if protective corrosion products or deposits are formed on the metal surface. Corrosion rates are generally low in hard waters. One of the most important parameters influencing the corrosion of zinc is pH, as both high- and low-pH conditions can result in excessive corrosion. White corrosion products, which can contain hydroxides, oxides, and carbonates, can develop on corroding surfaces. The porous, friable, nonprotective corrosion products are known as white rust (Fig. 7.6). The formation of these nonprotective corrosion product layers should be avoided, as corrosion can be rapid. Guidelines from cooling tower manufacturers have been developed to limit the formation of white rust.[1] First, the system should be prepassivated with appropriate levels of inorganic phosphate and hardness for an extended period of time with pH levels below 8.0. During operation, some cooling water parameters known to promote white rust formation include:

- High cooling water pH values
- High alkalinity levels [levels typically greater than 300 mg/L (as $CaCO_3$) can promote attack]

- Soft water conditions [calcium hardness less than 50 mg/L (as $CaCO_3$) should be avoided]
- High levels of chlorides or sulfates (levels below 250 mg/L are preferred)
- High concentrations of free halogens, such as chlorine or bromine (values in excess of 0.5 mg/L) should be avoided.
- Appreciable levels of copper (copper can plate onto the surface, creating a galvanic couple that results in attack of the anodic zinc layer)

Temperature has two important effects on the corrosion of galvanized steel. First, the corrosion rate of zinc increases substantially over the temperature range from about 120 to 160°F (49–71°C) (Fig. 7.7). This makes galvanized piping a poor choice in many hot water systems. In addition, the potential difference between zinc and iron varies with temperature and water chemistry. Polarity or galvanic reversal can occur at elevated temperatures, making the steel anodic to the zinc layer. This typically arises at temperatures above 140°F (60°C), but

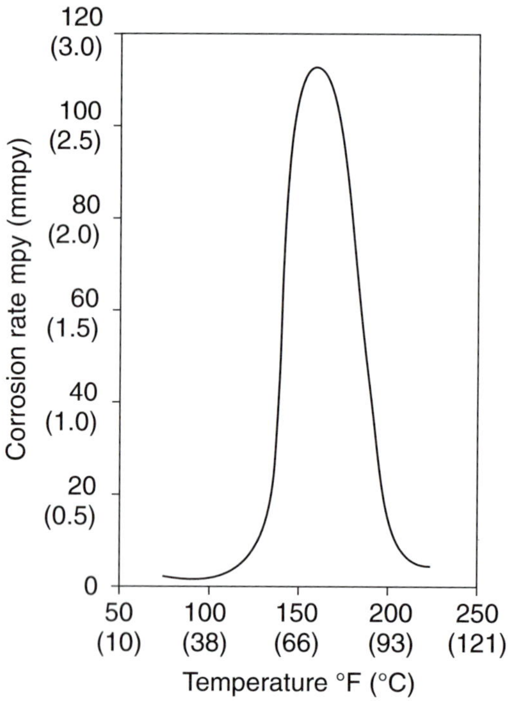

FIGURE 7.7 Effect of temperature on the corrosion rate of zinc in distilled water.[2] (*International Lead Zinc Research Org., Inc., with permission.*)

polarity reversal has been shown to develop at lower temperatures for certain water chemistry additions.

The corrosion of galvanized steel with atmospheric exposure can vary considerably, depending on species present in the environment. Studies have shown considerable differences between rural, marine, and industrial environments, with corrosion being most severe in the industrial environments. The increased attack on the zinc coating in industrial environments is primarily associated with sulfur dioxide as a contaminant in the air. Similarly, the presence of chloride in seacoast environments generally increases the corrosion rate when compared to rural environments.

Alloys of aluminum and zinc have been developed to provide hot dip coatings that have increased corrosion resistance in certain atmospheric and aqueous environments. These coatings consist of approximately 55 wt% aluminum and 45 wt% zinc and have a two-phase microstructure with aluminum-rich dendrites and a zinc-rich interdendritic phase. Like zinc-based coatings, they provide both barrier and sacrificial protection to the underlying steel. The increased corrosion resistance compared to the zinc-based coatings is due to the aluminum-rich phase. The aluminum-zinc coatings are often better than zinc-based coatings for applications that can experience wet/dry cycling.

Nonmetallic Inorganic Coatings

Nonmetallic inorganic coatings include metal oxides and other compounds. This category includes what are known as conversion coatings, in which the coating layers typically develop during immersion in an appropriate solution. Some of the more common nonmetallic inorganic coatings include anodized layers and phosphate coatings. These coatings are often used as a precursor for the application of painted coatings, as they can improve paint adhesion. Cement- or mortar-lined piping is another example of a nonmetallic inorganic coating.

The anodizing process (also known as anodic oxidation) produces a layer of oxide that is thicker than the oxide layer that naturally forms on the metal surface. The process is primarily used on aluminum and its alloys, although titanium and magnesium can also be commercially anodized. A detailed discussion of anodized aluminum is presented in Chap. 5.

Phosphate coatings are produced by precipitation of phosphate ions with a divalent metal cation. The coating process typically involves exposure of a metal substrate to a dilute phosphoric acid–containing solution. The substrates are often steel, but zinc and aluminum can also be phosphate coated. The solutions will generally contain iron, zinc, or manganese. Some of the substrate metal will dissolve when exposed to the acidic solution. The reaction between the metal substrate and the elements in the solution determine the type of phosphate coating that forms. The coatings do not offer significant improvement in corrosion resistance, but generally provide a suitable finish for further application of a paint coating. In cooling water systems, phosphate coatings are sometimes applied to steel coupons that are used for corrosion monitoring to provide a pretreated condition to the surface. The pretreated condition is used to simulate a passivated steel surface after a pretreatment program is applied to a system. Such coupons may also be used after exposure to a cooling water environment that is properly treated with phosphate-containing corrosion inhibitors. The goal of such pretreatment is to decrease the amount of corrosion that occurs immediately after installation of the coupon in a system, relative to untreated coupons.

Cement-mortar is used as a coating to line piping for some applications. Typical applications include cast iron cement-lined pipe in water distribution systems and sewage lines (Fig. 7.8). The alkaline environment that develops around the cement lining helps protect the steel or cast iron, and the lining also acts as a barrier to the environment.

Organic Coatings

There are many types of organic coatings that can be used for corrosion protection in cooling water systems. The different coating types are often categorized by the type of binder resin of the coating layer, as the binder is generally the primary constituent in the coating that provides the corrosion resistance. Some categories include oxygen-reactive

FIGURE 7.8 Cement-lined cast iron pipe.

resins (which include alkyds and epoxy esters), thermoplastic resins (including vinyls, chlorinated rubber, acrylic, and bituminous coatings like coal tar and asphalt), and cross-linked thermosetting resins (including epoxy, phenolic, and urethane). Table 7.1 provides a list of some coating materials and their uses. The coatings will typically also contain additives such as pigments and filler particles.

Proper selection of a coating to provide a corrosion-resistant barrier to the underlying metal depends on many factors such as environmental resistance, the substrate and its condition, expected service life, and consequences of failure. Environmental conditions that can influence resistance in water systems include chemical species that degrade the coating, temperature, and mechanical abrasion.

Even if an appropriate coating is chosen for an environment, its success depends on proper application. Successful application starts with surface preparation of the substrate, which will depend on the type of coating to be applied. In general, surface cleanliness and roughness will promote adhesion. Blasting the surface with abrasive grit is beneficial to remove surface corrosion products, mill scale, and contaminants. It also can increase surface roughness, which improves bonding of the coating. Thorough cleaning to remove grit after blasting is also important to achieve good bonding.

Coatings are often applied in multiple layers, with a primer layer covering the metal surface and a topcoat exposed to the environment. The primer layer should provide a strong bond to the metal substrate. The topcoat should be chemically resistant in the environment, provide an overlying seal to the underlying coating layers, and have toughness and wear resistance. Intermediate layers can be used to increase the overall coating thickness, provide good adhesion between the primer and topcoat layers, and increase the impermeability of the coating. The use of multilayer coatings helps to significantly reduce the likelihood of pinholes or holidays in a coating layer that allow exposure of the metal substrate to the environment.

Many coatings are viscous liquids. They can be applied by brushing, rolling, and spraying. Troweling is often used for thick coatings. Application skill is critical to provide a good, uniform coating without significant defects. Sharp corners, threads, edges, crevices, and projections are troublesome areas where coating layers will generally be thinner. Coatings should be periodically inspected and repaired as needed to increase their longevity.

The corrosion protection afforded by many types of coatings is primarily due to the formation of barrier layers on the metal surface that reduce the permeation of water and/or oxygen from the environment (Fig. 7.9). Openings in the coating layers, either due to defects during application or to defects that develop during service, allow the environment to contact the underlying metal. In the case of steel exposed to cooling water environments, this can result in oxygen corrosion or other aqueous corrosion mechanisms (Fig. 7.10).

Coating Type	Advantages	Limitations	Applications
Phenolic	Excellent solvent, acid resistance	Poor resistance to strong oxidants, alkalis	Most widely used lining material
	Nontoxic	Poor flexibility	Hot water immersion, food service
Epoxy (amine cured)	Best chemical resistance of epoxy coatings	Less flexible than other epoxy coatings	Tank linings
	Excellent adhesion to steel and concrete	Strong solvents may lift coatings	Maintenance coatings
Epoxy (polyamide cured)	Superior epoxy for water resistance	Poor acid resistance	Coatings for water service
	Excellent adhesion to steel and concrete	Fair alkali resistance	Maintenance coatings
Urethane	Can be formulated to enhance resistance	Only fair solvent resistance	Exterior structures
	Noted for glossy appearance, aesthetic appeal	High humidity can cause bubbling	
Vinyls	Excellent resistance to strong acids	Initial adhesion can be poor	Potable water tanks, sanitary equipment
	Insoluble in oils, greases, and alcohols	More likely to have pinhole defects	Fresh and saltwater immersion
Chlorinated rubber	Excellent water resistance	Can dissolve in strong solvents	Lining for concrete
	Good abrasion resistance	Degraded at slightly elevated temperatures	Fresh and saltwater immersion
Coal tar	Best water resistance	Softens at high temperature, hardens at low temperature	Water tanks
	Relatively inexpensive	Can degrade and crack with ultraviolet exposure	Exterior coating for buried pipelines
Alkyds	Low cost, ease of application	Not chemically resistant	Nonaggressive atmospheric exposure
	Can provide coverage to poor surfaces	Not for water immersion service	Primer coating on steels

Table 7.1 List of Some Coating Types Along with Their Advantages, Limitations, and Some Applications[3]

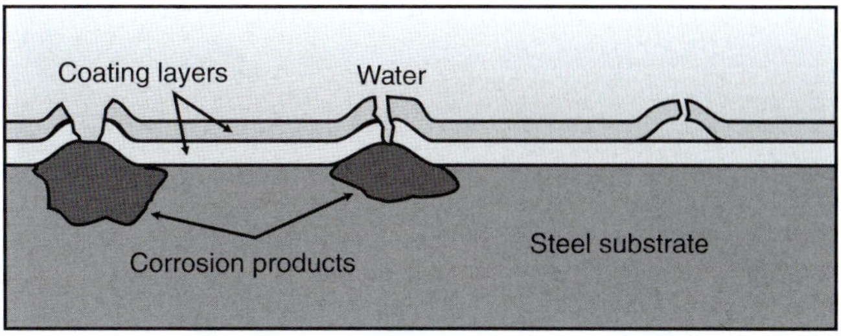

FIGURE 7.9 Schematic of multilayer organic coatings providing a protective barrier layer on a steel substrate. Attack can occur at openings in the coating, and the voluminous iron oxide corrosion products result in further coating degradation.

FIGURE 7.10 Micrograph showing attack on the underlying steel at a location where an organic coating had deteriorated. The bracket highlights the coating layer on the tube surface. (Unetched.)

Additives can be included in coating formulations to further limit corrosion. In these coatings, the additives provide mechanisms to help reduce attack in addition to the physical barrier that the coating provides. Some additives can provide chemical corrosion inhibition on the metal surface by forming a protective film. Coatings can also offer cathodic protection if they contain additives that are anodic to the steel, such as zinc particles.

Field- and factory-applied organic coatings have been used extensively in cooling water systems to retard attack on a wide variety of components. These components include cooling tower basins and structural members, exchanger surfaces such as water boxes and tubesheets (Fig. 7.11), and liners for piping and tubing. A list of common pipe linings and their advantages is provided in Table 7.2.

Following coating failure, the formation of voluminous corrosion products from the metal substrate will cause the coating to further degrade. The coating is pushed away from the metal surface, cracks

FIGURE 7.11 Cross section through a tubesheet with rolled tubes, showing a thick epoxy-based coating covering the surfaces. Note the minor porosity in the coating, which is not considered detrimental in this case due to the appreciable thickness of the coating. The image also highlights that the coating is much thinner at the tube end.

form, and the coating becomes undermined. Figure 7.12 shows damage that occurred beneath a coating on heat exchanger tubing. Locations where the coating becomes detached from the metal surface can act as crevices, promoting the formation of concentration cells and possibly the activity of anaerobic microorganisms.

An example of an organic coating layer that was used on the internal surface of the shell in a heat exchanger with shellside cooling is shown in Fig. 7.13. The exchanger was exposed to once-through cooling water that could not be treated with corrosion inhibitors due to discharge restrictions, so a protective organic coating was applied to the shell to prevent corrosion. However, defects in coating application and/or degradation in service allowed for openings in the barrier coating and subsequent attack.

A failed organic coating on a carbon steel heat exchanger tubesheet is shown in Fig. 7.14. Analysis of the coating indicated it contained multiple layers, which is generally beneficial for corrosion resistance. However, the layers were very thin, and the outermost layer contained a very large volume fraction of inert filler particles containing silicon, aluminum, and magnesium. This morphology likely contributed to the short service life of the coating. The coating became brittle and developed cracks, allowing the underlying steel to be corroded.

Coatings are often applied for corrosion protection to structural components having atmospheric exposure, as well as for aesthetic

Material	Use	Advantages	Disadvantages
Hot-applied coal tar enamel	Lining for steel pipes	- Long service life, >50 years - Good erosion resistance to silt and sand -Resistance to biological attachment	- Need to reapply to welded areas - Extreme heat may cause cracking -Extreme cold may cause brittleness - May cause an increase in trace organics in water
Epoxy	Lining for steel or ductile iron pipes	- Smooth surface results in reduced pumping costs - Formulated from components approved by the Food and Drug Administration (FDA)	- Relatively expensive - Less resistance to abrasion than coal tar enamel - Service life <15 years
Cement-mortar	Standard lining for ductile iron pipes, sometimes used in steel or cast iron pipes	- Relatively inexpensive - Easy to apply - Calcium hydroxide release may protect uncoated metal in pipe joints	- Rigidity of lining may lead to cracking or sloughing - Thickness of coating reduces cross-sectional area of pipe and reduces carrying capacity
Polyethylene	Lining used in ductile iron and steel pipes	- Long service life, >50 years-Good erosion resistance to abrasives - Good resistance to bacterial corrosion - Smooth surface results in reduced pumping costs	- Relatively expensive - Potential health concern

TABLE 7.2 Advantages and Disadvantages of Some Materials Used to Line Pipes for Water Service[4]

appearance. Selection of an appropriate coating material must be considered with respect to the aggressiveness of the environment. Figure 7.15 shows carbon steel support beams that were exposed to a warm, moist environment around a paper machine. The beams were coated with two different materials (one epoxy-based and one polyester or alkyd-based). The epoxy coating, which was also significantly thicker, provided a protective barrier while the other coating degraded and allowed for the underlying steel to rust. Note that the attack of the steel was predominantly along the corners of the beam, where the coating was likely thinner.

Lastly, organic coatings are often used to protect buried pipelines. Bituminous coatings are especially suited for this application, as the

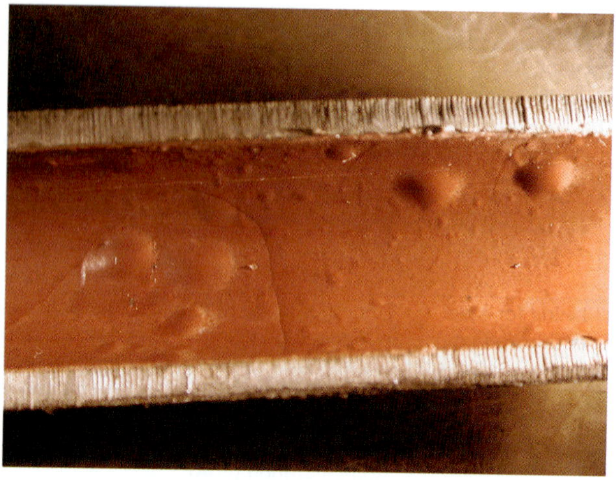

FIGURE 7.12 Coated carbon steel tube from a heat exchanger that developed blisters with cracks over its service life. Openings in the coating layer allowed for oxygen corrosion to occur, and the corrosion products resulted in large blisters in the coating.

FIGURE 7.13 Partially cleaned surface of a heat exchanger shell that was covered by an organic coating. Corrosion occurred at many openings in the coating, producing areas of deep metal loss that perforated the exchanger shell in places. Note the blistering and cracking of the coating along the edges of the metal loss area. The metal has a smooth contour at locations where the coating remained intact on the surface.

appearance of the piping is not a concern. Such coatings are also much less expensive. Cathodic protection can be used with coated buried pipelines to limit attack that occurs at openings in the coating where the underlying metal is exposed.

FIGURE 7.14 Corrosion occurring on the surface of a tubesheet at an opening in the organic coating layers. Corrosion occurred within a few years of service. Analysis of the coating revealed that it was thin and contained appreciable amounts of filler particles.

FIGURE 7.15 Coated steel beams in the press section of a paper machine. During service, the beams are continuously exposed to high humidity with elevated temperatures. Although the beams had approximately the same service life, one beam experienced breakdown of the coating and subsequent corrosion of the underlying steel. Infrared analysis of the coatings indicated a polyester or alkyd-type coating was on the attacked beam, while an epoxy-based coating covered the unattacked beam.

References

1. *White Rust: An Industry Update and Guide Paper*, Association of Water Technologies (AWT) , 2002.
2. C. J. Slunder and W. K Boyd, *Zinc: Its Corrosion Resistance*, 2d ed., International Lead Zinc Research Organization, New York, 1983, p. 118.
3. Table compiled from: *Corrosion Engineering Handbook*, ed. P. E. Schweitzer, Marcel Dekker, 1996, p. 588, and *Metals Handbook*, 9th ed., Volume 13, Corrosion, ASM International, 1987, pp. 401–402.
4. *Technologies for Upgrading Existing or Designing New Drinking Water Treatment Facilities, USEPA*, 1984, p. 140.

CHAPTER 8

Nonmetallic Materials

Introduction

The term *nonmetallic* encompasses an extremely wide variety of materials that includes polymers, ceramics, and glasses. Nonmetallic materials also include many composites such as concrete and fiber-reinforced plastic (FRP), which are blends of different materials. Wood is an example of a naturally occurring composite. The wide variety in types of materials results in equally wide variations in properties. The properties of the nonmetallic materials make them the preferred choice for many applications in cooling water systems. This section provides a brief overview of some different types of nonmetallic materials, discusses their applications in cooling water systems, and highlights some forms of degradation associated with the different materials. In-depth discussion on the specific types of materials, their structure, properties, and degradation mechanisms is covered in other texts.[1]

Types of Nonmetallic Materials

Polymers

The term *polymer* indicates a material having a structure based on the repetitive assembly of a simple chemical unit, known as a *mer*. Polymers are generally organic compounds that can be natural or synthetic. An example of a naturally occurring polymer is cellulose fiber, while polyethylene is one of the most common synthetic polymers. A polymer chain built from repeating mers of ethylene (C_2H_4) is shown in Fig. 8.1. The number of mer units in a polymer chain can be deduced from the molecular weight. For most polymeric materials, the lengths of the polymer chains are not the same but rather fall into a distribution. Therefore, the average molecular weight is a parameter typically used to characterize a polymer, as it often influences the properties.

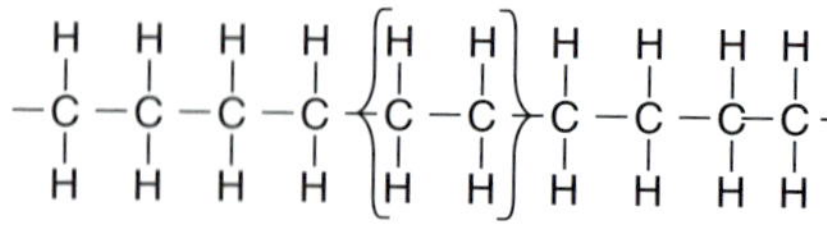

FIGURE 8.1 Example of a basic polymer structure, shown for polyethylene. The portion of the chain in the brackets highlights the ethylene mer.

In addition to molecular weight, the properties of polymeric materials depend on many factors that include the structure of their chemical building blocks (mers), the structure of the polymer chains, and reactions between the chains. Reactions between the chains usually result from processing techniques, such as cross-linking. Furthermore, polymers are often compounded with additives to impart certain properties. Some additives include antioxidants for protection against atmospheric oxidation, colorants, plasticizers, and reinforcing fibers to increase the strength and toughness of the polymer. Fillers may also be added to reduce cost or improve specific properties. The term *plastic* refers to a synthetic polymeric material that is processed into a load-bearing shape.

Polymeric materials are generally classified as thermoplastics, thermosets, and elastomers. The thermoplastic materials typically consist of long polymer chain molecules that are bonded to each other by weak secondary bonds. This structure allows most thermoplastic polymers to be easily formed repeatedly by heating and cooling. On the other hand, thermosets are generally cross-linked, resulting in stronger bonds between the polymer chain molecules. Thermosets will not melt upon reheating after processing, but will degrade instead. They are generally stronger and more brittle than thermoplastics. Polymer coatings used for corrosion protection of metals are generally thermoset polymers. Thermosetting polymer resins are also typically used to make the most common elastomers. An elastomer is a polymeric material which can be substantially stretched (or compressed) by applied stresses and will return to its original shape when the stress is removed.

Specific polymeric materials are identified by a chemical name or often an associated abbreviation. Many materials are known by a customary name instead of their chemical name, such as polyesters, vinyls, and acrylics. In addition, trade names are commonly used to refer to some materials. A list of some polymeric materials, their abbreviations, and types is provided in Table 8.1. The list is by no means inclusive, but it includes many common polymers found in cooling water systems. Discussion on the differences in the structure, processing, and properties of the different polymer types can be found in other texts specific to those materials.[2]

Plastic materials have many attractive properties that make them suitable choices, and often the necessary materials, for components in

Chemical Name	Abbreviation	Type
Acrylonitrile-butadiene-styrene	ABS	Thermoplastic
Chlorinated polyvinyl chloride	CPVC	Thermoplastic
Ethylenetetrafluoroethylene	ETFE	Thermoplastic
High-density polyethylene	HDPE	Thermoplastic
Low-density polyethylene	LDPE	Thermoplastic
Polyamide (Nylon*)	PA	Thermoplastic
Polycarbonate	PC	Thermoplastic
Polyethylene	PE	Thermoplastic
Polyetheretherketone	PEEK	Thermoplastic
Polyethersulfone	PES	Thermoplastic
Polymethylmethacrylate	PMMA	Thermoplastic
Polypropylene	PP	Thermoplastic
Polytetrafluoroethylene (Teflon*)	PTFE	Thermoplastic
Polyurethane	PUR	Thermoplastic
Polyvinyl chloride	PVC	Thermoplastic
Polyvinylidene fluoride (Kynar†)	PVDF	Thermoplastic
Epoxy esters	EP	Thermoset
Furans		Thermoset
Phenolics		Thermoset
Polyesters		Thermoset
Vinyl esters		Thermoset
Chloroprene rubber (Neoprene*)	CR	Elastomer
Ethylene-propylene rubber	EPM	Elastomer
Fluoroelastomers	FKM	Elastomer
Isoprene rubber	IR	Elastomer
Natural rubber	NR	Elastomer
Nitrile rubber (Buna N)	NBR	Elastomer
Polyurethane	AU	Elastomer
Perfluoroelastomers	FFKM	Elastomer

*Trademark of DuPont; †Trademark of Elf Atochem; All trademarks are the property of their respective owners.

TABLE 8.1 List of Common Polymers and Their Abbreviations

cooling water systems. In addition, components made from plastics often offer considerable cost savings relative to other materials, due to their relatively low material cost and ease of fabrication. Some applications include fill in cooling towers, water distribution piping, and valve and pump components. Rigid piping systems are generally constructed of thermoplastic materials such as polyvinyl chloride (PVC), chlorinated polyvinyl chloride (CPVC), polypropylene (PP), polyethylene (PE), or reinforced thermoset resins. Design of plastic piping systems requires consideration of allowable pressures, which depend on pipe dimensions, expected temperatures, and joint types. In addition, the design must account for thermal expansion and contraction and provide proper support. Flexible tubing can be produced by compounding the thermoplastic materials with plasticizers. Steel components, such as structural members and tubesheets that are exposed to cooling water, process fluids, or otherwise aggressive environments, are often coated with polymer-based chemically resistant materials for corrosion protection (see Chap. 7, "Coatings"). Some common coating materials include epoxy, vinyl ester, and phenolic-based resins. In addition, the chemical resistance of some plastics makes them preferred materials for chemical feed systems.

Failures of plastic materials in service are often associated with response to applied stresses, chemical degradation, or possibly a combination of both. Plastic materials have significantly lower strength than the common metals used for structural applications. Thus, they may fail due to mechanical overload if they are not properly designed for the expected stress level. An example of an overload failure on a PVC ball valve is shown in Fig. 8.2. The strength of most plastics will decrease considerably with increasing temperature. In addition, many plastics exhibit ductile-to-brittle transitions, with

FIGURE 8.2 Mechanical failure in a PVC ball valve at the location where the valve handle couples to the ball to rotate it. Scale formation in the valve required extra force to be applied to the valve handle, resulting in fracture along the thin edge of the molded plastic ball.

brittle-type failure promoted at lower temperatures and higher strain rates, such as those developed during impact loading. Other mechanical failure mechanisms of plastic components include creep, fatigue, and wear. Creep is the slow, permanent deformation of a material when subjected to a constant stress over an extended period of time. Fatigue failures are caused by the application of cyclic tensile stresses on a component. Most plastics are relatively soft, making them susceptible to wear damage when subjected to a harder material rubbing across the surface.

Exposure to an aggressive environment can result in degradation of plastic materials by a wide variety of mechanisms. Comprehensive discussion of the mechanisms that can occur in polymeric materials requires in-depth knowledge of polymer structures and chemical interactions, which is well beyond the scope of this guide. Attack on polymeric materials can be related to a number of specific water chemistry or environmental factors, including acidic conditions, alkaline conditions, oxidative species, solvents, and certain salts in solution. It is important to note that chemical resistance is very material-dependent. Chemical degradation is typically prevented by proper material selection for the environment or by avoiding conditions that are known to be detrimental to the materials in the system. Some potentially deleterious conditions that can occur in cooling water systems include high- or low-pH excursions, excessive levels of oxidants, or contamination by process leaks. Degradation of some types of polypropylene used for components in cooling towers, such as cooling tower fill and water distribution grids, has occurred with exposure to certain water chemistries (Fig. 8.3). By-products from

FIGURE 8.3 Degraded polypropylene water distribution grid in a cooling tower. (*Courtesy: Nalco Europe BV–S. Vrijhoeven.*)

microbiological activity can also cause degradation in some cases. Cleaning solutions typically contain higher concentrations of chemical species that can be aggressive to nonmetallic materials (as well as metals), so their application must be carefully considered prior to use. The use of nonmetallic materials for storage containers and distribution lines for treatment chemicals generally requires exposure tests to determine if the material is compatible with the chemical.

Stresses on components often will accelerate chemical attack on polymers. Stresses can develop from applied loads on a component, or residual stresses may arise by fabrication or installation. Molded plastic components will often develop residual stresses on cooling. A mechanism that involves the interaction of stress on polymeric materials with exposure to specific chemicals is environmental stress cracking (ESC). ESC typically results in a brittle fracture. Cracking initiates on the surface exposed to the specific chemical. Susceptibility to ESC is generally promoted by a higher concentration of the specific chemical, elevated stress levels, increased temperature, and longer exposure times. An example of ESC in a chemical feedline is shown in Fig. 8.4.

Some features associated with chemical degradation may include embrittlement, surface cracking, blistering, swelling, shrinkage, and the formation of voids or holes. Although these characteristics are typically associated with chemical degradation, determining the

Figure 8.4 Environmental stress cracking of a PVC pipe used as a chemical feedline. The cracks originate on the internal surface due to exposure to a solution containing an ESC agent. The longitudinal orientation of the cracks indicates circumferentially oriented stresses were primarily responsible for cracking, likely from a combination of pressurization of the line and residual stresses associated with pipe manufacture.

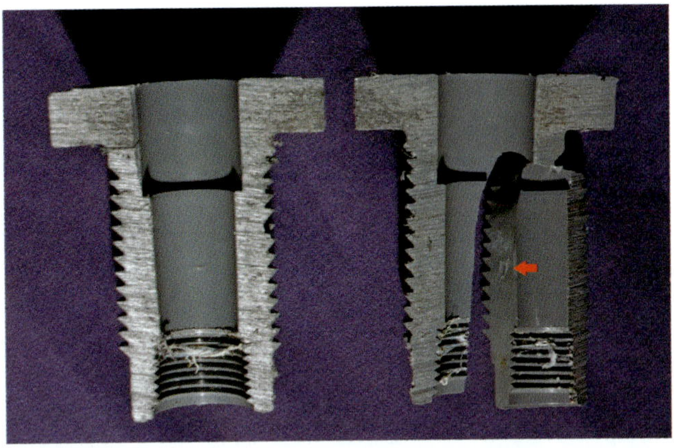

FIGURE 8.5 PVC bulkhead fitting that was longitudinally split to reveal the internal surface. The fitting was located between a tank and a metal valve. Failure occurred when the fitting was subjected to appreciable bending stresses during operation of the valve. The fracture was centered on a defect caused by incomplete fusion along the weldline of the part during molding (arrow).

cause requires careful analysis of the affected material and consideration of possible interactions with the environment. It is also critical to check for improper processing or defects from fabrication that could have contributed to the failure, as shown in Fig. 8.5.

Elastomers

Although mentioned previously, the properties of elastomers rate special attention. Elastomers are natural or synthetic materials that can be appreciably stretched and are resilient. For example, natural rubber is a polymer of methyl butadiene. The polymer chains are cross-linked during the curing process (such as vulcanization). For elastomers, the amount of cross-linking is carefully controlled so that the polymeric chains can stretch appreciably when stress is applied and spring back to their original shape when the stresses are removed. To be called an elastomer, a material is generally required to stretch about 100 percent or more without breaking and then return to its original shape after the stress is eliminated.

The elastic properties of these materials make them especially useful in cooling water applications where a seal is required to prevent leakage of water from the system. Compression of the elastomer allows it to deform and make tight contact with rigid surfaces. For example, elastomers can be used in valves as seals or diaphragms as shown in Fig. 8.6. When sufficiently compressed, O-rings and gaskets create seals between water-containing components, such as the plates in a plate-and-frame heat exchanger (Fig.I.12). Elastomeric materials are also used for gaskets and couplings to join pipes.

FIGURE 8.6 Elastomeric gasket in a diaphragm valve. Note that the water-contacting surface of the gasket is covered by brown and white deposits. The deposit layers on the thin gasket resulted in some degradation of the elastomer and prevented the valve from operating correctly.

FIGURE 8.7 Deposition and degradation on the internal surface of a reinforced chloroprene coupling used to join stainless steel piping in a closed cooling water system.

Figure 8.7 shows a degraded elastomeric coupling used to join piping in a cooling water system.

Much like other polymers, elastomers can fail due to mechanical or chemical interactions with the environment. Mechanical failures, such as over-compression and extrusion, are often related to improper design of the seal. Chemical and thermal degradation can also cause failures.

Signs of chemical degradation may include blisters, cracks, voids, or discoloration on the elastomer surface. Chemical absorption will cause swelling and loss of strength. The degree of chemical attack will generally increase with increasing temperature and concentration of aggressive species. Failures of elastomers in cooling water systems are generally associated with leaks when the elastomer fails to provide an adequate seal. This can result in contamination of cooling water with process fluids and vice versa.

Elastomeric seals often fail prematurely in applications because of improper design or poor material choice. The type of elastomer should be suited for the environment. Ethylene-propylene rubber (EPM or EPDM) is a common elastomer used in water systems, as it is considered to have excellent resistance to water and oxidizing chemicals such as ozone and oxidizing biocides. It also performs well at elevated temperatures, as it is considered to have excellent resistance to steam. However, EPDM has generally poor resistance to degradation in environments containing oil. Nitrile rubber (NBR) can provide improved chemical resistance to oil-containing environments. Under certain cooling water conditions, particularly high pH, some fluoroelastomers (FKM) can experience some degradation related to deposition.

Composites

Composites consist of mixtures of different materials, generally processed in a way to provide optimum properties. One common example is fiber-reinforced plastic, also known as FRP. Most FRPs used in cooling water systems are comprised of glass fibers in a thermosetting resin. The resin provides chemical corrosion resistance and a matrix to transfer applied stresses to the fibers which increase the strength of the composite. Some typical resins for FRP include epoxy, vinyl-esters, phenolic, and polyester. Chemically resistant tanks and piping can be manufactured from FRPs using resins suitable for the service environments. Fiber reinforced polyester composites are used for structural materials in cooling towers.

Concrete is a composite material that is comprised of aggregate material such as stones in a cement matrix that acts as a binder. Concrete has applications in the fabrication of many types of cooling towers. For instance, it is typically used for the basin in large, field-erected cooling towers. Concrete basins should be watertight, durable, and resistant to degradation. Sulfate attack can occur in some concrete mixtures that are not chemically resistant. This can be an issue in cooling towers that use sulfuric acid for pH control and operate with high cycles of concentration. Soft water conditions can cause leaching of calcium from the concrete. Proper processing of concrete during fabrication and the use of waterproof barrier systems can minimize degradation. Cracking of the concrete should also be avoided, especially for concrete that is reinforced with steel bars.

A naturally occurring composite material used for fill and structural members in cooling towers is wood. Wood contains cellulose and hemicellulose, lignin, and other material. The cellulose and hemicellulose provide the basic structure to the wood, and lignin is the matrix that binds the cells together.

The cooling tower environment can result in degradation of wood by chemical and biological attack. Degradation of the wood can cause sagging fill and slats and fallen fill racks. Out-of-position fill can reduce the efficiency of the tower. Severe attack can cause loss of structural integrity, which can result in the collapse of entire sections of tower structures if not identified and remediated. Some typical forms of attack are highlighted next.

Chemical attack on the surface of lumber is generally observed as wood delignification, or selective removal of lignin from the wood. Loss of lignin will expose stringy fibers on the wood surface. An example of delignification attack is shown in Fig. 8.8. Water chemistry conditions that can cause this type of attack include excessive chlorination, excessive alkalinity, and low pH. Appreciable degradation requires exposure to these conditions for extended periods of time. Attack will be most severe on wood that is submerged or frequently wetted. Detection of this chemical attack by periodic inspection can allow water chemistry to be corrected. It is virtually nonexistent within the pH range of 6.0 to 8.0.

Other forms of chemical attack include iron rot and salt crystallization. Iron rot can occur in areas contacting steel components, such as fasteners. The corrosion reaction of steel results in conditions that promote the degradation of the wood structure. Wood that is weakened by iron rot has a charred, darkened appearance. Surface attack

FIGURE 8.8 Chemical attack on a wood surface caused by delignification. Note the exposed fiber strands on the wood surface.

due to iron rot can also predispose the wood to deep rot. The use of steel fasteners should be avoided. Salt crystallization results from repeated periods of wetting and drying. Mineral salts concentrate in the wood during drying periods and crystallize. The crystals penetrate into the wood structure and weaken it, eventually resulting in rupture of material from the surface. Attack due to salt crystallization is characterized by the formation of long, coarse fibers in the affected wood surfaces. Although chemical attack is generally a thin surface form of degradation, it can make the wood more susceptible to biological attack.

Biological attack of wood is generally caused by fungal decay. Surface attack, also known as soft rot, is selective attack of the cellulose in very damp wood by fungi, leaving behind the lignin matrix. This attack is characterized by a dark appearance, and the wood surface becomes soft and spongy. A checked morphology from fissures that develop on the surface forms in decayed areas if the wood is allowed to dry (Fig. 8.9). For soft rot, the active growth of the wood-destroying fungi is confined to the surface layers and remains superficial unless the wood is allowed to dry out frequently. Deep penetration of fissures will allow fungi to penetrate into the wood to cause more internal damage. Internal attack, also known as deep rot, occurs below the surface of the wood. Fungal decay of the wood can attack both the cellulose and the lignin (referred to as white rot) or primarily the cellulose (called brown rot, as lignin remains). Deep rot

FIGURE 8.9 Checked pattern of decay related to fungal activity on the surface of wood. This form of biological attack of wood is known as soft rot, as the surface of the wood becomes soft. The checked appearance develops when the wood dries.

FIGURE 8.10 Spongy appearance in a large internal area of a wood slat from a cooling tower. This attack is called deep rot, and it is caused by fungal activity.

results in spongy internal cavities (Fig. 8.10). It can spread undetected throughout the wood and cause severe structural damage.

The woods chosen for cooling water applications should be types that are naturally resistant to fungal decay, such as redwood and Douglas fir. In addition, cooling tower lumber is often pressure treated with chemicals to help limit fungal decay. Once a tower is placed in service, periodic post-treatment with fungicides should be implemented as a preventive measure as part of the chemical treatment program. Preventing biological attack on the surface of the wood is important, as fungal treatments cannot effectively penetrate the wood to prevent further growth of internal fungal masses. Deterioration of lumber is irreversible, and severely attacked wood will require replacement.

References

1. For additional information refer to Myer Ezrin, *Plastics Failure Guide—Cause and Prevention*, Hanser Publishers, 1996; J. Moalli, *Plastics Failure—Analysis and Prevention*, William Andrew Publishing/Plastics Design Library, 2001; J. Scheirs, *Compositional and Failure Analysis of Polymers—A Practical Approach*, John Wiley and Sons, New York, 2000.
2. For additional information refer to P. A. Schweitzer, *Mechanical and Corrosion-Resistant Properties of Plastics and Elastomers*, Marcel Dekker, New York, 2000; *Characterization and Failure Analysis of Plastics*, ASM International, Materials Park, Ohio, 2003.

CHAPTER 9

Brazed and Soldered Joints

Introduction

Copper alloy piping and tubing is commonly joined by soldering for domestic plumbing applications. Both soldering and brazing techniques are used to join piping and other components in some industrial cooling water systems. Both soldering and brazing techniques employ suitable filler metals that are heated above their melting points and are applied to the narrow gaps within socket joints. The filler metal flows into the joint by capillary action and bonds to the base metals when it cools and solidifies.

Certainly the base metal surfaces to be joined must be cleaned properly to a bare and bright condition prior to the joining process. Oxides and other deposits or residues will prevent the molten filler metal from wetting the base metal surfaces properly to disrupt bonding. Soldering and some brazing processes employ an initial application of an aggressive fluxing chemical solution. These fluxing agents are applied within the joint to remove the oxide layers from the base metal surfaces and to prevent oxidation during the joining process. This will allow the molten solder or brazing alloy to bond to the metal. Common fluxing agents for soldering copper alloys contain chloride and bromide compounds, or mixtures with phosphoric acid. Fluxing agents used for brazing may include mixtures of boric acid, borates, fluorides, and fluoroborates. The fluxing agent that is used will depend upon the base metals and filler metals that are involved. Certain copper-based brazing filler metals contain fluxing agents such as phosphorus compounds that are mixed into the brazing metal. When using these filler metals, a separate fluxing agent is generally not needed.

Soldering employs filler metals with melting points that are lower than 800°F (427°C). These alloys typically consist primarily of mixtures of lead and tin, with tin contents ranging from 2 to 100 percent.

General-purpose solder that contains about 40 to 50 percent tin is commonly used for plumbing applications. Lower concentrations of tin (10–25 percent) are used for applications where metal temperatures are higher during service, such as for automobile radiators to prevent softening or melting of the solder material. Small concentrations of silver may be used to increase joint strength. Common solder metal compositions that are used to join copper alloy components are listed in Table 9.1. Corrosion resistance of lead-tin solder decreases as a function of increasing lead concentration. Solder metal with lead contents exceeding about 70 percent are most susceptible to corrosion in cooling water service. The lead portion of the solder metal is particularly susceptible to corrosion by soft and demineralized waters. In waters that contain suitable amounts of hardness, protective, insoluble lead carbonate and lead oxide layers will form. Dissolution of lead can be controlled by maintaining pH within the range of 5 to 9.

Brazing filler metal alloys melt at temperatures greater than 800°F (427°C). Brazing may be used to join copper alloy components that operate at higher temperatures or require higher joint strength than those that are produced using the solder alloys. Certain alloys and dissimilar alloy combinations used in cooling water systems that may be joined by brazing include carbon and low-alloy steels, cast irons, stainless steels, copper, and copper alloys. The filler metals that can be used for brazing depend upon the base metal combination. The filler metals that are typically used for copper and copper alloys are self-fluxing copper phosphorus and copper phosphorus zinc alloys, as well as copper-zinc and silver alloys. The filler metals that are typically used for carbon and low-alloy steel alloys are copper-zinc and silver alloys. The filler metals that are typically used for stainless steel alloys are copper, nickel, and silver alloys. The self-fluxing phosphorus-containing alloys are not used for steels since brittle iron phosphides and intermetallics may form at the joint interface. Cast iron components, which are difficult or impossible to weld, are joined by brazing using silver alloys. Common brazing filler metal compositions and the metal combinations that may be joined by them are listed in Table 9.1.

Forms of Corrosion

Galvanic Corrosion

The susceptibility of solder and brazing filler metals to galvanic corrosion depends upon their position on the galvanic series relative to the base metals that are joined. The potential difference between the coupled alloys in the environment will indicate if galvanic corrosion is likely to occur, and on which metal. For instance, lead- and tin-based solder materials are less noble, and thus anodic to the copper

	Ag	Cu	Zn	Cd	Ni	Sn	Fe	Mn	Si	Pb	Sb	Cr	Au	B	P	Other	Metals to Be Joined
Common Solder Filler Metals																	
Antimonial Tin						95.00					5.00						Copper alloys
Tin-silver	5.00					95.00											Copper alloys
Soft solder						70.00				30.00							Copper alloys
Soft solder						63.00				37.00							Copper alloys
Soft solder						50.00				50.00							Copper alloys
AWS Alloy	**Torch Brazing Filler Metals**																
BAg-1	45.00	15.00	16.00	24.00													Carbon and low-alloy steels, cast iron, stainless steel, copper alloys
BAg-1a	50.00	15.50	16.50	18.00													Carbon and low-alloy steels, stainless steel, copper alloys
BAg-2	35.00	26.00	21.00	18.00													Carbon and low-alloy steels, copper alloys
BAg-2a	30.00	27.00	23.00	20.00													Carbon and low-alloy steels
BAg-3	50.00	15.50	15.50	16.00	3.00												Carbon and low-alloy steels, stainless steel, copper alloys
BAg-4	40.00	30.00	28.00		2.00												Carbon and low-alloy steels

TABLE 9.1 Types of Soldering and Brazing Alloys[1] *(Continued)*

	Ag	Cu	Zn	Cd	Ni	Sn	Fe	Mn	Si	Pb	Sb	Cr	Au	B	P	Other	Metals to Be Joined
BAg-5	45.00	30.00	25.00														Carbon and low-alloy steels, copper alloys
BAg-6	50.00	34.00	16.00														Carbon and low-alloy steels
BAg-7	56.00	22.00	17.00			5.00											Carbon and low-alloy steels, stainless steel
BAg-13	54.00	40.0	5.00		1												Carbon and low-alloy steels, stainless steel
BAg-18	60.00	30.0				10.00											Carbon and low-alloy steels, stainless steel
BAg-19	92.50	7.5															Carbon and low-alloy steels, stainless steel, copper alloys
RBCuZn-A		59.00	40.00			0.60											Carbon and low-alloy steels, copper alloys
RBCuZn-B		48.00	41.00		10.00				0.15						0.25		Carbon and low-alloy steels
RBCuZn-D		48.00	42.00		10.00												Copper alloys
RCuZn-B		58.00	38.00		0.50	0.95	0.70	0.25	0.08								Carbon and low-alloy steels
RCuZn-C		58.00	39.00			0.95	0.70	0.25	0.08								Carbon and low-alloy steels
BNi-1					73.25		4.50			4.00		14.00		3.50		Carbon 0.75%	Stainless steel

BNi-3					90.9		1.50		4.50					3.10			Stainless steel
BNi-7					77.0							13.00			10.00		Stainless steel
BCu-1		99.0 min															Stainless steel
BCu-2		86.5 min															Stainless steel
BAu-4					18.5								81.50				Stainless steel, copper alloys
BCuP-1		95.00													5.00		Copper alloys
BCuP-2		92.75													7.25		Copper alloys
BCuP-4	6.00	86.75													7.25		Copper alloys
BCuP-5	15.00	80.00													5.00		Copper alloys
BAlSi-2									7.50								Aluminum alloys
BAlSi-3		4.00							10.00								Aluminum alloys
BAlSi-4									12.00								Aluminum alloys
BAlSi-5									10.00								Aluminum alloys

TABLE 9.1 Types of Soldering and Brazing Alloys[1] (*Continued*)

alloy base metals, and are susceptible to corrode preferentially. In addition, the cathode-to-anode surface area ratio is very high at any brazed or soldered joint. Consequently, galvanic corrosion of the solder material would be expected. However, galvanic corrosion is strongly dependent upon water chemistry. Galvanic corrosion will occur more readily in higher-conductivity waters and waters with high chloride concentrations. In mildly corrosive waters, a suitably resistant lead carbonate or oxide corrosion product layer forms on the surface of the exposed solder, and corrosion will be controlled. Brazing alloys are more noble than the solder alloys, and are therefore more resistant to galvanic corrosion. Brazing metal that is used to join copper alloys is very resistant to corrosion in cooling water service, particularly the silver-based alloys. The small cathode-to-anode surface area ratio significantly limits corrosion of the copper alloy base metal. A more in-depth discussion concerning galvanic corrosion is presented in Chap. 16.

Crevice Corrosion

Soldered and brazed joints may contain significant amounts of porosity, or they may not be completely sealed with filler metal (Case History 9.1). These defects allow water to infiltrate into tight crevices between the metal components, which can promote crevice corrosion. The mechanisms of crevice corrosion for soldering and brazing alloys are similar to those for other alloys, particularly due to the formation of differential oxygen concentration cells and anion concentration cells. Such corrosion is promoted by low or stagnant flow conditions. The concentration cell corrosion mechanisms are described in more detail in Chap. 11.

Corrosion by Residual Flux

Proper fluxing, soldering, and brazing practice will reduce or prevent the leakage of flux residues from joints. However, in some cases flux residues leak out of joints and may drip and collect at nearby locations. Consequently, proper soldering and brazing practice also includes a subsequent flushing step to ensure removal of potential residual corrosive fluxing material from the areas surrounding the joints. Aggressive flux remnants such as chlorides and fluorides are particularly corrosive. If such residual material is not removed prior to service, corrosion will occur due to hydration of the flux remnants by condensed moisture that forms in an empty pipe, or due to exposure to water when the piping is filled (Case History 9.2). Corrosion by aggressive flux residue will continue for a period during subsequent service as long as the corrosive agents remain. Additional aggressive anions in the cooling water will tend to diffuse into crevices or beneath corrosion products that contain aggressive anions from flux residue. Eventually, the corrosive flux residue may rinse away. However, even after flux residue rinses away, corrosion may continue at the site at a slower rate due to concentration cell corrosion.

Microbiologically Influenced Corrosion

The mechanisms of microbiologically influenced corrosion (MIC) for certain soldering and brazing alloys are similar to those for other alloys. Copper- and silver-containing brazing alloys will be very resistant to bacteria growth and MIC. The microbiologically influenced corrosion mechanisms are described in more detail in Chap. 13.

Corrosion Inhibitors

Corrosion inhibitors are less effective within soldered and brazed joints that are incompletely fused, are incompletely filled with filler metal, or have excessive porosity. In order for corrosion inhibitors to produce proper protective films, it is necessary for the cooling water to flow at proper rates across the surfaces. Proper corrosion inhibitor film production cannot be assured due to the uncontrolled flow of cooling water into the defective joint. In addition, aggressive anions that diffuse and concentrate into crevices will tend to remove any protective films that do form.

Case History 9.1

Industry:	Hospital Building
Specimen location:	Domestic hot water piping
Specimen orientation:	Horizontal
Environment:	Internal: Hot domestic water, low hardness
Time in service:	30 years
Sample specifications:	2½ in. (6.4 cm) outer diameter copper piping joined to copper bend coupling using lead-tin solder

The received section of copper domestic hot water piping experienced a perforation near a solder joint after 30 years of service. The section was longitudinally sawcut to reveal the presence of distinct mounds of black and green material only within localized areas closely adjacent to the solder joints at the locations of the perforations. Metallographic cross sections through the solder joints revealed that they were only partially filled with solder at the locations near the perforations (Figs. 9.1 and 9.2). A crevice was present at the edge of the joint that faced the waterside due to incomplete penetration of filler metal into the joint. The solder material at the crevice mouth experienced shallow metal loss (Figs. 9.3 and 9.4). Where the mound of deposits and corrosion products formed on the internal surface shown in Fig. 9.2, remnants of corroded solder material were present (Fig. 9.5). Beneath the corroded solder remnants, dense copper oxide corrosion products formed that filled a deep depression (Fig. 9.6).

Solder deposits and residual flux covered the internal surface when the pipe was originally soldered together. Fluxing agents are typically very corrosive, and they should have been removed following the soldering process. It is probable that flux was entrapped beneath the solder deposits to initiate the corrosion. Over time, the flux rinsed away, but the corrosion products and deposits that remained continued to cause underdeposit corrosion over a very

FIGURE 9.1 Mounds of residual solder and corrosion products on internal surface of copper piping adjacent to soldered joints.

FIGURE 9.2 Close-up of residual solder deposits and corrosion products covering surface at location where deep metal loss and a perforation occurred.

long period of time until the perforations occurred. Underdeposit corrosion resulted from the formation of differential oxygen concentration cells and ion concentration cells. No metal loss occurred away from the solder deposits at any location examined. Corrosion would not have occurred in the absence of the solder deposits. When proper soldering techniques are used, solder

FIGURE 9.3 Partially corroded solder at mouth of crevice. Joint is devoid of solder material for a short distance to opening of joint. (Unetched.)

FIGURE 9.4 Higher-magnification view of corroded solder shown in Fig. 9.3.

deposition should not occur on the internal surface. Shallow corrosion of the solder within the open crevice was also caused by concentration cell corrosion. Such corrosion will be controlled by ensuring that the solder joints are completely sealed through proper soldering practice.

FIGURE 9.5 Corroded solder layer overlying copper oxide corrosion products covering the corroded internal surface.

FIGURE 9.6 Deep, burrowing depression filled with copper oxide on internal surface near perforation. (Unetched.)

Case History 9.2

Industry:	Automotive
Specimen location:	Robotic welder transformer
Specimen orientation:	Horizontal
Environment:	Internal: Closed-loop cooling water
Time in service:	2 years
Sample specifications:	Copper piping joined to copper coupling using silver-based brazing filler metal

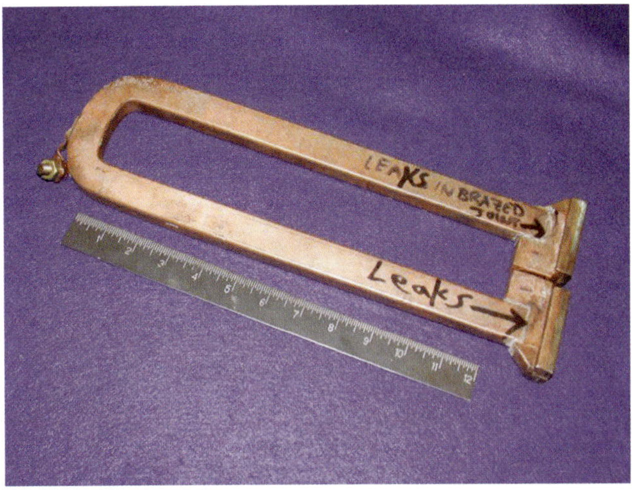

FIGURE 9.7 Robotic welder transformer cooling coil.

FIGURE 9.8 Close-up of brazed joint that experienced water leakage.

A water-cooled welder transformer component was received for metallographic analysis (Fig. 9.7). The component experienced water leakage in service at the brazed joint at the locations marked by arrows (Fig. 9.8). The water leak caused the transformer to fail by short-circuiting. Figure 9.9 is a close-up of the sawcut water channel. Figure 9.10 shows a portion of a brazed joint that was broken open to reveal numerous pores. Some pores contained light-colored, acidic deposits.

A metallographic cross section through the joint is shown in Fig. 9.11. Energy dispersive X-ray analysis (EDS) indicated that the composition of the bulk brazing metal consisted of about 6.4 percent silver and 4.8 percent phosphorus, with

FIGURE 9.9 Internal surface at leaked brazed joint.

FIGURE 9.10 Portion of brazed joint that was mechanically fractured to reveal numerous pores of various sizes.

the balance consisting of copper. The gap between the components at the external surface, where the brazing metal was applied, was found to be about 0.004 in. (0.1016 mm). Within this portion of the joint, the brazing is mostly dense, except for a single large pore. However, where the joint turned 90° on the way to the internal surface, the gap widened to about 0.019 in. (0.4826 mm). Within this portion of the joint, the brazing metal was found to be quite porous, particularly at the internal surface of the water channel, where the brazing metal did not penetrate completely through the gap.

FIGURE 9.11 Metallographic cross section of joint. Note the narrow gap within the joint at the external surface where the brazing metal was supplied, and the much wider gap at the internal surface. The brazing metal incompletely fills the wider gap, and it contains numerous pores. (Unetched)

Figure 9.12 is a close-up of the many pores in the filler metal, denoted by E. The microstructure of the brazing filler metal consisted of three separate components that had different compositions. One component was copper-rich material denoted by an A. Some of the copper-rich material was present within eutectic zones. Another component, which appeared gray in color, denoted by B, consisted of copper phosphorus compounds. This gray material was also present within the eutectic zones shown at both A and C. EDS indicated that the phosphorus composition of this material ranged from 12.5 to 13.3 percent, with the balance consisting of copper. The third component, which was located within the eutectic zones, consisted of light-colored copper-silver compounds, denoted by C, containing about 61 to 65 percent silver, with the balance consisting of copper. Wastage and associated corrosion product formation penetrated to shallow depths from the internal surface. The dark gray corrosion products, denoted by D, formed where corrosion occurred both in the eutectic portion and within large nodules in the copper-rich material only. EDS indicated that this material consisted mostly of copper as copper oxide, with a small amount of sulfur (3.6 percent). No corrosion occurred in the copper-phosphorus or copper-silver components.

At the interfaces of the brazing and base metal, corrosion and corrosion product formation occurred in the copper base metal and within the copper-rich components of the brazing in narrow strips that were located directly adjacent to the copper-phosphorus and copper-silver components (Fig. 9.13).

The brazed joint shown contained nonuniform gaps between the joined components. In order to ensure proper joint quality for the BCuP brazing filler metal alloy that was used, it is recommended that the gap between the components should range from about 0.001 to 0.005 in. (0.0254–0.1270 mm). If the gap is wider than 0.005 in. (0.1270 mm), porosity is likely to form in the brazing metal. In fact, porosity was much more significant within the wider portions of the joint, as should be expected. Joint strength decreases as a function of increasing porosity.

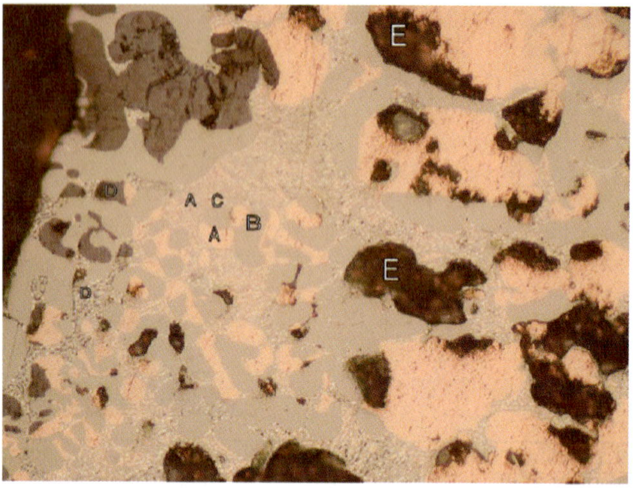

FIGURE 9.12 High-magnification view of brazing. Internal surface is at left of photo. The microstructure consists of copper-rich material (A, both within the eutectic zones and wider zones), copper-phosphorus material (B, both within the eutectic zones and wider zones), and silver-rich material within the eutectic zones (C). Corrosion occurred within portions of the copper-rich material at the exposed surface (D), where copper oxide corrosion products and voids formed. Internally, several scattered voids formed of various sizes, some of which are quite large (E). (Unetched.)

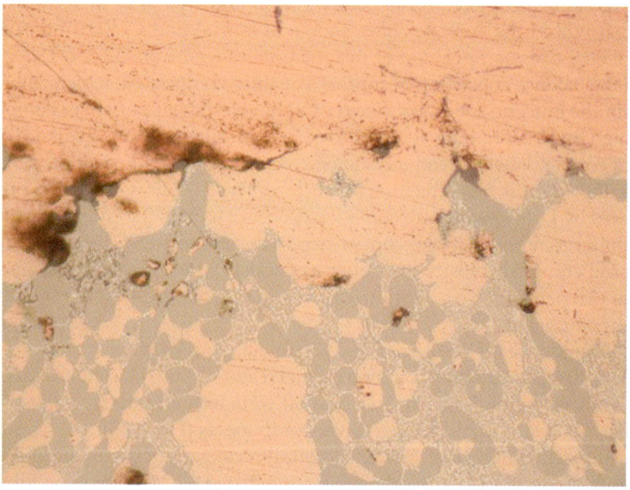

FIGURE 9.13 Corrosion of copper-rich components and corrosion product formation at braze/ base metal interface. (Unetched.)

In addition, pores provide easier pathways for water to wick though the joint. Potentially, corrosion may occur more readily inside a porous joint due to the formation of concentration cells that will tend to form inside tight crevices. In addition, due to restricted water flow into and out of the crevices, the corrosion inhibitors in the cooling water cannot be relied upon to form a proper film that will control corrosion. This is due to the fact that there will not be a controlled amount of inhibitor supplied to the crevices as within the bulk water that flows through the water channels.

Even though the brazed joints contained an unacceptable amount of porosity, no significant corrosion occurred within the brazing metal. Corrosion penetrated to fairly shallow depths where water was able to wick in along pores and gaps to contact the brazing and brazing/base metal interfaces. Where corrosion occurred, only the copper base metal and copper-rich brazing metal components were affected. The copper-phosphorus and copper-silver components experienced no corrosion since they were both more noble than the copper components. As such, they promoted corrosion of the copper-rich components due to galvanic effects.

When a portion of the joint was broken open, a large number of pores were present. Some of the pores contained entrapped phosphorus-rich deposits that were acidic in a distilled water solution. It is probable that these deposits consisted of phosphates that formed phosphoric acid when dissolved in water. The brazing filler metal contained phosphide that was used in part to flux the copper surface during the brazing process. The fluxing process removes copper oxide, which allows the brazing metal to wet a bare copper surface in order to form a proper bond. During fluxing, the phosphide reacts with the oxide to form phosphate. Under ideal conditions, the phosphate will flow out of the joint. However, when pores form, some phosphate may remain. If water is able to reach the phosphate deposits, aggressive phosphoric acid solutions may form that could cause copper corrosion.

It was strongly recommended that the joints which were brazed should be welded, or an alternate design should be considered that removes the joint from a location where a leak would cause catastrophic failure of the transformer. Due to the many problems that can occur within brazed joints, it is recommended that all joints should be nondestructively tested prior to installation to determine if they contain excessive porosity or other defects. Components that contain poorly brazed joints should never be installed. Alternate brazing alloys should be chosen with care. For instance, silver-based brazing filler metal alloys are believed to produce weaker joint strength than copper-based brazing filler metal alloys. In addition, silver is significantly more noble than copper. Therefore, interfacial corrosion of copper may be significant. Corrosion will tend to be much more pronounced when the brazed joint contains substantial porosity, poor penetration, and other brazing defects.

Reference

1. *Heat Treating, Cleaning and Finishing*. ASM Metals Handbook, 8th ed., Vol. 2, 1964.

CHAPTER 10

Corrosion Monitoring

Introduction

Monitoring certain parameters of the cooling water system provides vital information that is used to evaluate performance of the treatment program. In addition, these parameters can also be used to optimize system control. Areas of concern regarding performance include deposition, corrosion, microbiological activity, and fouling. This section reviews methods typically used to monitor for corrosion, as corrosion is generally the primary cause of failures of components in the cooling water system. Thus, corrosion monitoring provides an important measure of system performance, especially the effectiveness of the treatment program. Trends based on corrosion monitoring results can be correlated to other parameters, and changes in water chemistry can be implemented to minimize corrosion.

Many techniques are available for corrosion monitoring, and new approaches for measurements are continuously being developed. Some of the proven methods for monitoring corrosion in cooling water systems are presented in Table 10.1. This table compares the measurements and applications of the methods. In addition, it is important to assess the condition of equipment exposed to cooling water periodically when it is not in service. This section also includes a brief discussion of some inspection methods that can be used to evaluate the extent of corrosion or degradation that has occurred on equipment.

Corrosion Monitoring Techniques

Corrosion Coupons

One of the most common and time-tested ways to measure corrosion rates of metals in a cooling water system is by using corrosion coupons. Corrosion coupons are carefully prepared specimens of metal that are exposed to the cooling water. Proper monitoring using corrosion coupons requires consideration of many factors that can influence results, some of which are discussed in this chapter. The coupon

Method	Measures	Applications
Corrosion coupons (small, metal specimens)	Average corrosion rate is determined by mass change of coupon over a period of typically 30–90 days. Can be used for pit depth measurements.	Best for uniform corrosion under steady-state conditions. Can show localized attack by pitting and under-deposit corrosion. Coupons are available for most alloys.
Linear polarization resistance (LPR) (Examples: Nalco NCM100 and Corrater*)	Instantaneous corrosion rate is measured by electrochemical polarization between two electrodes. Tendency for pitting (pitting index) can be measured.	Water must have sufficient conductivity to provide suitable measurement of corrosion rate. Suitable for common engineering alloys.
Electrical resistance (ER) (Example: Corrosometer*)	Corrosion rate is determined by electrical resistance change from thickness loss of a corroding metal element, such as a wire. Uniform corrosion rate is assumed in the calculation.	Works in liquid or vapor environments for most alloys. Does not require conductivity in the solution, gives general corrosion rate even if localized corrosion is occurring.
Analytical monitoring	Concentration of dissolved or suspended species in water, such as corrosion products, pH, and conductivity. Does not measure corrosion rate directly.	Trends of dissolved or suspended corrosion products may be proportional to changes in corrosion rate. Precipitation of these corrosion products limits technique's effectiveness.

* Corrator® and Corrosometer® are trademarks of Rohrback Casasco Systems.

Table 10.1 Comparison of Different Corrosion Monitoring Techniques, What They Measure, and Their Application. (*Courtesy of McGraw-Hill, The Nalco Water Handbook, 3d ed., New York, 2009.*)

method for corrosion monitoring requires following best practices consistently to achieve reproducible, reliable results. Standard practices have been developed to address and minimize the effect of some parameters on the measurements.[1]

Coupon Selection

A wide variety of corrosion coupon types are available. The coupon forms include disk, rod, and bar styles that can be provided in many dimensions. The form chosen often depends on how they are to be

FIGURE 10.1 Bar-style carbon steel corrosion coupons, highlighting the appearance of an unexposed coupon, a coupon after exposure in a cooling water system, and the same coupon after cleaning to remove corrosion products and deposited material. Examples of coupons with general and localized corrosion are also shown.

inserted into the system. Generally, the coupon form will include a drilled or cut hole for attachment to a holder, or possibly a threaded end. Fabrication techniques that introduce appreciable plastic deformation into the metal, such as drilling and shearing, should be avoided as they can promote localized attack on the coupon. Bar-style coupons are a typical form used in cooling water systems. An example of a bar-style coupon is shown in Fig. 10.1.

Many different alloys are available for metal coupons. Low-carbon steel is the most common alloy chosen for monitoring, as the majority of tubing and piping used in cooling water systems consists of carbon steel. For other metals, it is important to choose the alloy or alloys that represent the metallurgies in the system. As an example, the corrosion resistance of aluminum alloys significantly depends on alloy composition. For the yellow metals, there can also be appreciable differences between copper, brass, and cupronickel alloys.

In some cases, the effect of stress on corrosion or cracking may be of interest. Coupons can be prepared so that they have areas with appreciable residual or applied stress levels that are introduced in a controlled manner. Some stressed coupon types include welded coupons, U-bend coupons, and teardrop coupons. Stressed coupons can be used in some cases to check for susceptibility to corrosion and cracking mechanisms that require sufficient tensile stresses, like stress corrosion cracking.

Surface Preparation

The surface finish of a coupon can appreciably affect the measured corrosion rates. Coupons can be supplied with many different surface finishes, including mill, ground, and bead- or sandblasted finishes. Bead- or sandblasted coupon finishes are rougher and are more susceptible to surface contamination. This makes ground surface finishes generally preferred for consistency. Special surface finishes, such as double disk ground, use abrasive discs that produce a fine surface finish that leaves minimal residue. Although the surface finish is believed to predominately affect the initial corrosion of the coupon, the formation of corrosion products during the initial exposure can often impact long-term corrosion rates.

In addition to the mechanical influence of surface finish, some coupons may be chemically treated prior to exposure. For cooling water systems, pretreated low-carbon steel coupons are sometimes used. The pretreated coupons are generally exposed to solutions that produce a very thin film, such as a phosphate coating, on the steel surface. This approach is used to produce a condition on the coupon that is similar to a steel surface after a passivation treatment of the cooling water system. The pretreated coupon may experience less corrosion over a short exposure period when compared to an untreated coupon. However, over a long exposure period, corrosion on treated and untreated coupons will be about the same. The use of pretreated or untreated coupons will depend on the cooling system being monitored. Both can be used in some situations.

Coupon Installation

Coupons should not be handled with bare hands, as oils and salts from the skin can contaminate the coupon surface and affect corrosion. Coupons should be appropriately installed into a sidestream corrosion test rack for consistent exposure to conditions experienced in the cooling water system. Suspending coupons in the cooling tower basin or other similar locations in the system must be avoided because there will invariably be flow pattern differences around the coupon surface. The position of the rack in the system should be carefully chosen to provide the most representative conditions and information. The rack is often placed on a return line to the cooling tower. The cooling water at this location will be at higher temperatures, which can promote increased corrosion. In addition, concentrations of treatment chemicals such as corrosion inhibitors can be slightly reduced compared to the supply to the cooling water system. A typical corrosion test rack is illustrated in Fig. 10.2. The rack should be designed to allow proper flow around coupons and avoid coupon contact with the rack surfaces. Periodically cleaning the internal surfaces of the rack will remove deposits and microbiological fouling that can influence coupon results.

The inlet end of the corrosion test rack should be fitted with a valve to control the cooling water flow through the rack, as flow velocity is an important parameter that can affect corrosion coupon results. Flow velocities above 3 ft/s (1 m/s) should generally be used,

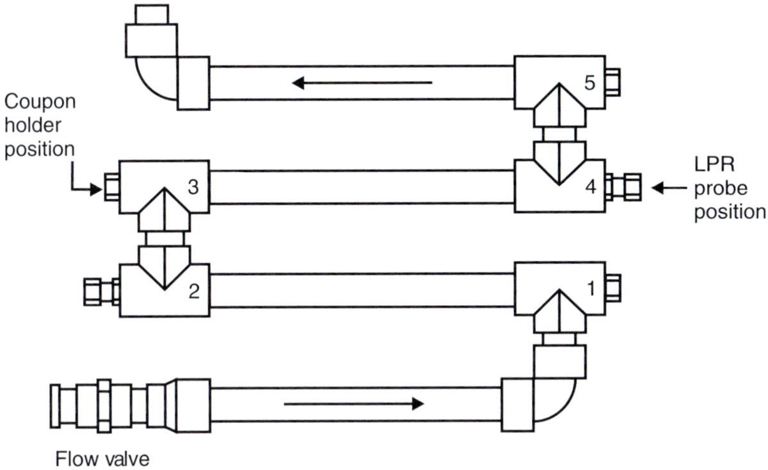

FIGURE 10.2 Schematic of a corrosion test rack, showing appropriate locations for installation of a coupon holder (locations 1, 3, and 5) and a linear polarization resistance (LPR) probe (locations 2 and 4). (*Courtesy of McGraw-Hill,* The Nalco Water Handbook, *3d ed., New York, 2009.*)

and the flow rate should be consistent throughout the exposure period. Low flow or stagnant conditions must be avoided. The flow velocity should not exceed the maximum recommended fluid velocities to avoid erosion of copper alloys. Strainers should be used in systems with significant levels of suspended solids. The outlet end of the rack should be open to the atmosphere to avoid backpressure.

The coupons must be firmly attached to a holder for insertion into the rack. In addition to securing the coupon in the corrosion test rack, the holder will keep the coupon electrically isolated from other metals in the system. Holders are made from insulating materials that are chemically resistant to the cooling water environment and have sufficient mechanical strength to properly support the coupon. Appropriate plastic materials are typically used for the attachment hardware.

The coupon holder should be installed into the rack so that the water flows across the coupon holder and then the coupon. Proper positions for corrosion coupon placement in the rack are shown in Fig. 10.2. The plane of bar-style coupons should be oriented perpendicular to the ground (Fig. 10.3). This orientation limits the influence of material settling on the coupon surface, which can promote underdeposit corrosion mechanisms.

Lastly, multiple metal alloys are typically used for corrosion monitoring by coupons. The location of the different alloys in the corrosion test rack is important. The different metals should be placed in positions according to the galvanic series, with the most active metal on the upstream end of the rack. As an example, aluminum alloys

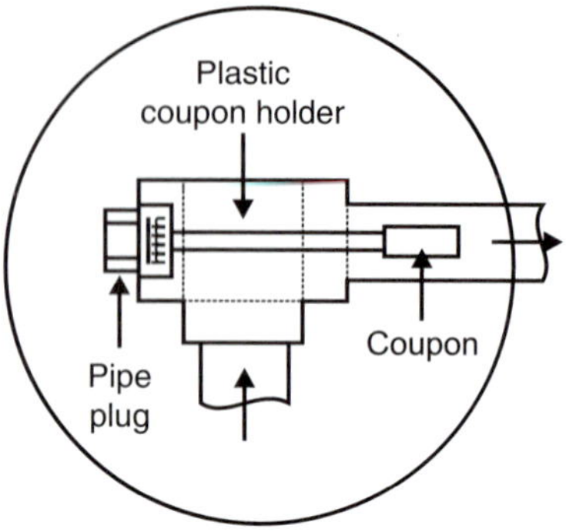

FIGURE 10.3 Schematic showing the preferred position and orientation of a bar style corrosion coupon in a corrosion test rack. The water flow is from the bottom of the rack. (*Courtesy of McGraw-Hill*, The Nalco Water Handbook, *3d ed., New York, 2009.*)

would be placed before carbon steel, and copper alloys would be placed after carbon steel. The release of copper can possibly cause galvanic corrosion of a more active metal that is placed downstream in the rack, hence the importance of placing more active metals upstream.

Exposure Period

The exposure period is an important parameter for corrosion monitoring by metal coupons, as an insufficient exposure period can produce misleading results. For example, oxide layers can form quickly on a steel coupon surface when it is initially exposed to oxygenated water. Once the oxide forms, corrosion slows because the accumulated oxide retards diffusion. The corresponding corrosion rates are initially high and then decrease rapidly (Fig. 10.4). Eventually, a nearly steady-state corrosion rate is achieved.

Hence, a minimum exposure period must be satisfied to obtain consistent corrosion rate data for coupons exposed in cooling water systems. The appropriate minimum exposure period generally depends on the metal type and the environment. According to standards for corrosion testing in field applications,[2] the following relations can be used to determine an appropriate minimum time period for exposure:

$$\text{MET (in hours)} = 2000/\text{ECR (in mpy)} \tag{10.1}$$

$$\text{MET (in hours)} = 50/\text{ECR (in mmpy)} \tag{10.2}$$

MET = minimum exposure time
ECR = expected corrosion rate

Typical exposure periods are between 30 and 90 days, although longer periods may be required in some cases. Alloys that form passive

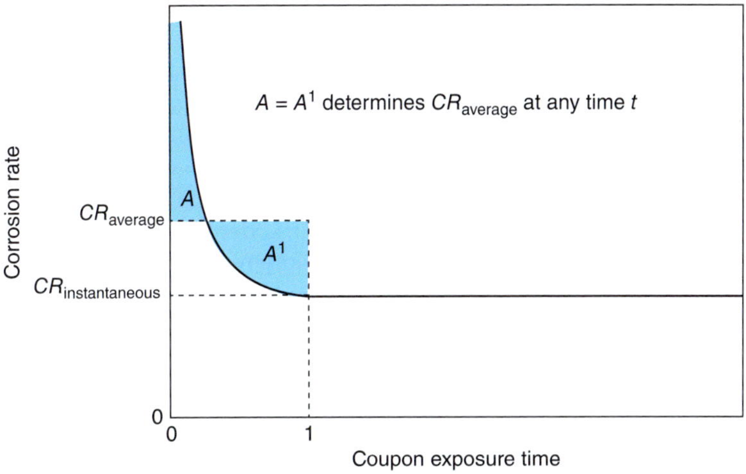

FIGURE 10.4 Schematic of carbon steel corrosion rate versus exposure time of a coupon in oxygenated cooling water. The average corrosion rate decreases with time until extended exposure periods result in a steady-state corrosion rate. Note that the average corrosion rate ($CR_{average}$) is always larger than the instantaneous corrosion rate ($CR_{instantaneous}$) and that this difference is largest at small exposure times.

oxide films for corrosion protection, such as stainless steel alloys, require long exposure periods (typically at least 90 days) in the cooling water system for a representative test. In some cases, multiple coupons of the same metallurgy will be exposed for different periods, an approach known as planned interval testing. The purpose of this testing is to provide both short-term measurements for trending and more representative long-term corrosion rates.

Evaluation

After an appropriate exposure period, coupons are removed for analysis. Analysis of the exposed coupons typically provides information on corrosion rate and the type of corrosion occurring on the coupon. The type of corrosion can be described as general or localized. General corrosion is characterized by uniform metal loss. The attack extends over most of the surface area (Fig. 10.1). Localized corrosion refers to metal loss that is confined to distinct regions or isolated pits or depressions. Groups of these regions may extend over much of the surface area, but they do not compose more than half of the total area. Pitting is a form of localized corrosion where the depth of penetration is greater than the width of the opening on the surface. Localized corrosion occurring along the coupon edges or at the location of the attachment hole should also be noted. Corrosion located in the area surrounding the attachment hole may indicate susceptibility to concentration cell corrosion, such as crevice corrosion or underdeposit corrosion.

Improper (i.e., electrically conductive) attachment hardware may also promote attack due to galvanic effects.

Corrosion rates from coupons are calculated from weight loss measurements. This approach assumes that uniform or general metal loss occurred on the coupon surface. However, if the coupon surface experiences localized metal loss, the corrosion rate calculated by weight loss can significantly underestimate the actual depth of penetration. For localized corrosion, the depth of deepest penetration for the exposure time is often a more representative measure of corrosion.

To calculate a corrosion rate by weight loss, the initial coupon weight, surface area, alloy density, and exposure period must be known. The material covering the surfaces of the coupon, including deposits and corrosion products, is removed using appropriate cleaning procedures. The cleaning procedure must completely remove the material on the surface without causing significant metal loss. Appropriate cleaning procedures will be different for different metals.[3] After suitable cleaning of the coupon surface, the coupon is weighed. The difference between the initial and final weight is the weight loss. The corrosion rate can be calculated from the following relation:

$$\mathrm{CR} = \frac{(K \times W)}{(A \times T \times D)} \tag{10.3}$$

where CR = the corrosion rate by weight loss

K = a constant that contains conversion factors for the units

W = corrosion weight loss

A = original surface area of the coupon

T = time of exposure

D = coupon alloy density

In addition to providing quantitative corrosion rate measurements, the examination of corrosion coupons can provide insight into the corrosion mechanism and factors that contributed to the corrosion. Observations of the deposit and corrosion product morphology and features of the metal loss are important. Analysis of the material on the surface can provide valuable information. If microbiologically influenced corrosion is suspected, then sampling material from the coupon surface for microbiological analysis (such as with a sterile swab) is recommended when the coupon is initially removed from the system.

It is important to note that the corrosion rate measured by corrosion coupons is not the actual corrosion rate occurring on components in the system. Coupons are not subjected to heat transfer, which prevents them from being completely representative of corrosion on heat exchanger tubing. The effect of heat transfer on corrosion can be investigated by using test heat exchangers with an applied heat flux.

Examination of the tube surfaces after an exposure period in a test heat exchanger can provide information on the type and extent of attack, but measurement of a corrosion rate is difficult. In addition to the effect of heat transfer, exposure conditions of coupons will generally be significantly different from the conditions elsewhere in the cooling water system. For instance, when coupons are installed their surfaces are bare, while system components may be covered with corrosion product and deposit layers. If significant corrosion is occurring in an exchanger, but the corrosion coupons indicate good corrosion control for the periods over the service life, then it is likely that mechanical or operational issues are the primary factors promoting attack rather than the water chemistry.

Corrosion Rate Guidelines

Corrosion rate measurements from corrosion coupons represent the average corrosion rate over an extended exposure period. The calculated corrosion rates using coupons depend on many factors. The measured corrosion rates and type of corrosion do not represent the actual conditions of system components. Rather, the results should be evaluated for changes in trends over successive exposure periods. The results provide information about the relative effectiveness of a water treatment program during particular exposure periods. General guidelines concerning the classification of corrosion rates have been proposed by multiple sources.[4,5,6] Guidelines will generally depend on multiple factors, with the type of system and the metallurgy being primary. Table 10.2 provides some generally accepted ranges for corrosion rates in an open recirculating cooling water system. Corrosion rate guidelines for closed cooling water systems are lower than for open systems. The ranges for the guidelines stated for closed systems from the different references[7,8,9] are typically between 50 and 90 percent less than the guidelines for mild steels in open systems.

Corrosion Rate, mpy (μm/y)		Description of Results
Mild Steel Coupons	Copper Alloys	
< 1 (< 25)	< 0.1 (< 2.5)	Excellent (negligible corrosion)
1–3 (25–76)	0.1–0.25 (2.5–6.4)	Very good (minimal corrosion)
3–5 (76–127)	0.25–0.35 (6.4–8.9)	Good (moderate corrosion)
5–8 (127–203)	0.35–0.5 (8.9–12.7)	Poor (moderately high corrosion)
8–10 (203–254)	0.5–1.0 (12.7–25)	Bad (high corrosion)
> 10 (> 254)	> 1.0 (> 25)	Unacceptable (severe corrosion)

TABLE 10.2 Generally Accepted Corrosion Rate Guidelines for Mild Steel and Copper in an Open Cooling Water System (*Courtesy of McGraw-Hill,* The Nalco Water Handbook, *3d ed., New York, 2009.*)

The guidelines presented here are for general corrosion only. Localized corrosion requires separate evaluation. Trends of multiple, consecutive exposure periods when coupons experience significant localized corrosion may indicate a tendency toward pitting in the cooling water system. Special attention should be given to any changes in water chemistry or operating conditions that occurred in the time periods when the type of corrosion changed from general to localized. Incidental occurrences of localized corrosion may indicate that upset conditions occurred in the water system or possibly within the corrosion rack during the exposure period.

Linear Polarization Resistance

The linear polarization resistance (LPR) method of monitoring corrosion is an electrochemical technique. The technique uses a probe that has two or more metallic electrodes that are exposed to the cooling water. A small voltage is applied across the electrodes, and the resulting current is measured. The resistance to current flow between the electrodes is the sum of the polarization resistance values at each electrode and the resistance of the solution between them. Because the solution resistance is significant, the LPR technique can be inaccurate in low-conductivity waters. Some models of LPR devices, including the Nalco NCM100, can compensate to allow measurement in lower conductivity solutions. However, in most cooling water applications, the solution resistance is negligible.

The measured current can be converted to a corrosion rate using appropriate relations that account for electrode metallurgy, surface area, anodic reactions, and other parameters. The calculated corrosion rate assumes uniform corrosion. A high current (low polarization resistance) will indicate a high corrosion rate. The formation of inhibitor films on the electrode surfaces or other conditions that limit corrosion increase the polarization resistance and decrease the measured current. Two measurements with reversed potentials across the electrodes are made, and these measurements are averaged. This compensates for slight differences between the electrodes. A large difference between these measurements may indicate a tendency for localized corrosion.

Much like corrosion coupons, good corrosion monitoring results require best practices to be followed. The LPR corrosion probe should be installed in the system so that the electrodes have similar water flow across their surfaces. To achieve this, the probe should be inserted into a corrosion test rack so that the electrodes are pointed into the flow of water (Fig. 10.5). When inserted perpendicular to the internal surface of a pipe or tube, the electrodes should be oriented so that the water flows through the space between the electrodes to prevent fouling. Bridging of conductive deposits or corrosion products between the electrodes can affect the measured corrosion rate. Fouled electrodes should be cleaned periodically to avoid accumulations of material that

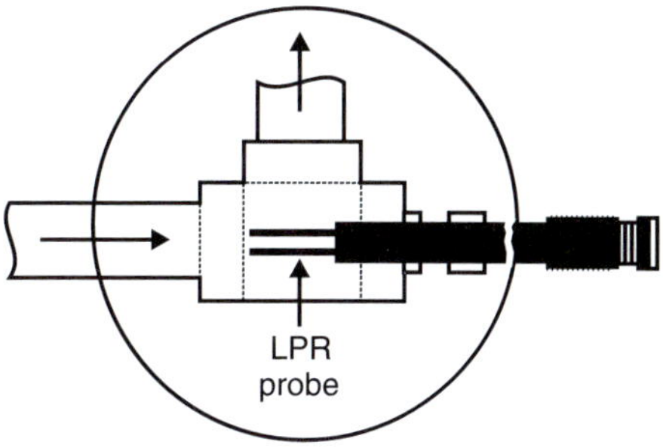

FIGURE 10.5 Schematic showing the preferred position and orientation of an LPR probe in a corrosion test rack. (*Courtesy of McGraw-Hill,* The Nalco Water Handbook, *3d ed., New York, 2009.*)

can produce inaccurate readings. When new or freshly cleaned probes are initially installed into the system, the surfaces will need to be conditioned for an aging period before readings are representative. The probe electrodes should be replaced if they show evidence of significant pitting or if their diameters are appreciably reduced.

A typical scan of two potentials across the electrodes takes less than a minute. Thus, corrosion rates are provided in real time, which is a main benefit of the technique. Corrosion rates can be continuously trended, which allows fast detection of upset conditions in the system that affect corrosion. Figure 10.6 illustrates the dramatic increase of the

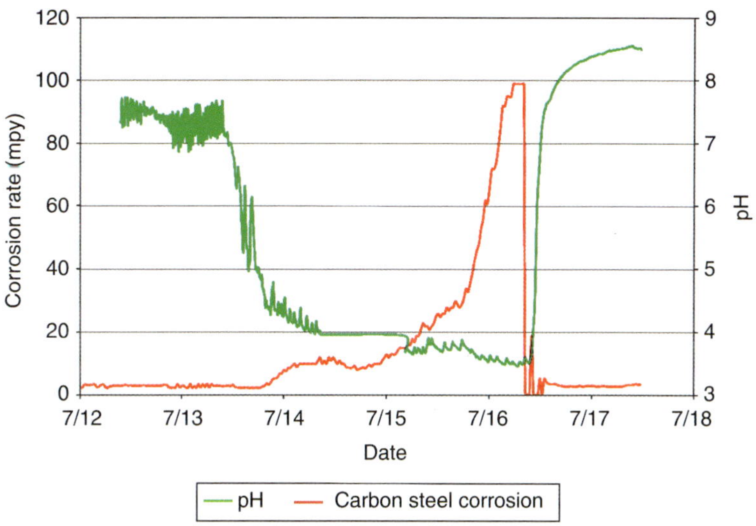

FIGURE 10.6 Graph showing the trend of pH and mild steel corrosion rate monitored by an LPR probe (NCM100). The corrosion rate increases dramatically during a severe low-pH upset, and returns to normal values when pH control is restored.

mild steel corrosion rate in a system that experienced a low-pH upset. With data from an LPR probe, corrective actions (such as increased corrosion inhibitor feed) can be implemented quickly to minimize the damage associated with an upset (Fig. 10.7).

Electrical Resistance

Another online technique to monitor corrosion uses electrical resistance (ER) probes. These probes are constructed of a metal element that is typically in the form of a wire, strip, or tube. As the element corrodes, its cross-sectional area is reduced, which results in an increase in the electrical resistance. To determine the resistance of the element, a current is passed through it and the voltage across it is measured. The amount of metal loss can be determined by the resistance measurements. The metal loss occurring over a time period represents a corrosion rate (Fig. 10.8).

One of the main benefits of the electrical resistance method is that it is applicable to corrosion monitoring in almost any environment, including cooling water, low-conductivity solutions, or in vapors. However, there are some considerable limitations for the method. First, ER probes do not provide accurate corrosion measurements if the metallic element experiences localized corrosion. In addition, corrosion rate measurements are not instantaneously determined. The time period for obtaining a corrosion rate depends on the probe metal, element shape, and amount of metal loss. In a cooling water system, a time period of a week may be required to produce a significant change in resistance that is needed to determine an accurate corrosion rate. For these reasons, the linear polarization resistance method is the preferred online corrosion monitoring technique for most recirculating cooling water systems.

Water Chemistry Analysis

Periodic analysis of water chemistry can provide a qualitative indication of corrosion in a cooling water system. This method analyzes the water for the concentration of metal species, such as iron and copper. The concentration of these species may be related to the corrosion of steel and copper alloy components in the cooling water system. Guidelines have been established that relate iron concentrations in a recirculating water system to corrosion. According to criteria based on total iron concentrations, 0.1 to 0.2 mg/L is considered normal, 0.5 to 1.0 mg/L is excessive, and more than 1.0 mg/L can be alarming.[10]

Both total and dissolved ion concentrations can be important. The concentration of dissolved ions (represented as the value after filtering the water sample) is especially a concern related to dissolved copper in the water. Concentrations over 0.1 mg/L of copper can induce galvanic corrosion issues if deposited in metallic form on metal surfaces.[11]. For dissolved iron, it tends to precipitate as oxides or salts,

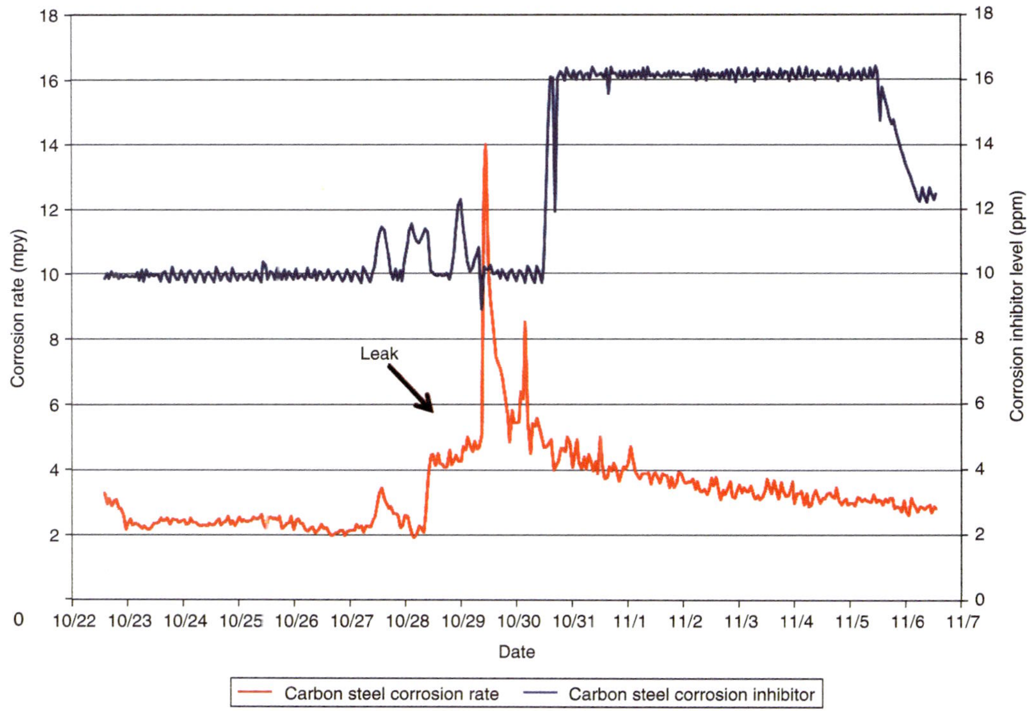

FIGURE 10.7 Corrosion rate for mild steel measured using an LPR probe (NCM100), showing a sharp increase due to contamination of the cooling water with a hydrocarbon process leak containing hydrogen sulfide (arrow) and a gradual decrease in response to an increased corrosion inhibitor level.

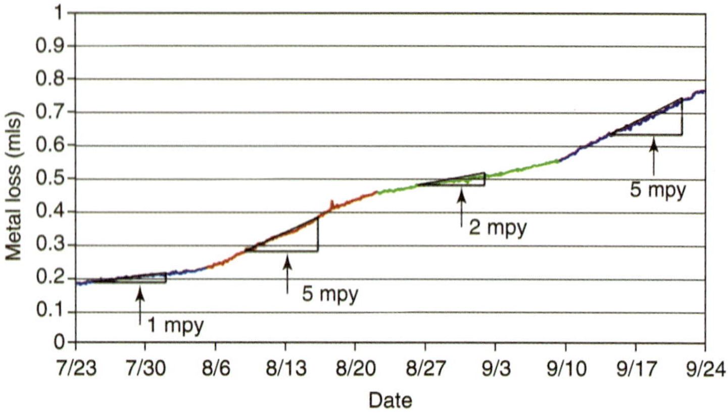

FIGURE 10.8 Graph of metal loss versus time obtained from a carbon steel electric resistance probe exposed to an environment around a paper machine. The changes in slopes correspond to periods of time having different average corrosion rates.

which can cause fouling and deposition problems. Water with high concentrations of metal ions may also influence the effectiveness of the corrosion inhibitors. If the inhibitor is attracted to the suspended material in the water, the amount available to protect the tube surfaces may be insufficient.

However, it is important to note that the dissolved and suspended metal ion concentration in the water can be affected by a variety of factors. Some factors include variations in the makeup water, the tendency for the metal ions released from corrosion reactions to oxidize and deposit in the system, and inconsistent sampling techniques.

Equipment Inspection Methods

Periodic inspection of components in cooling water systems should be routinely conducted on some equipment when it is not in service, such as critical heat exchangers. Visual examination of surfaces can provide valuable information regarding deposition, corrosion, and fouling in the cooling water system. Inspection of surfaces that are obstructed from direct viewing can be accomplished using boroscopes and fiberscopes. However, visual examination typically only provides qualitative assessment of the components. Corrosion monitoring to provide detail on the extent of metal loss on a component generally requires more involved inspection methods. Some nondestructive testing (NDT) methods capable of providing quantitative measurements of metal loss include ultrasonic testing (UT), eddy current testing, and radiographic inspection. These inspection methods can also indicate defects and cracking in metals, if the testing is properly conducted. Surface defects and cracks can also be highlighted using liquid penetrant inspection or magnetic particle inspection. Most of the inspection methods require considerable skill and experience to provide reliable results. Some of the more common inspection methods used in cooling water systems and their uses are briefly highlighted next.[12]

Ultrasonic Inspection

This technique requires an ultrasonic transducer to be placed in contact with the surface of the component to be examined. The transducer produces acoustic waves that are transmitted to the surface and propagate through the component. The acoustic waves are reflected and scattered within the test object. Reflection of the waves from internal surfaces can be used to provide thickness measurements. A common use of ultrasonic testing is to measure the wall thickness of piping. Scattering of the waves from internal features can be used to indicate defects in the metal.

A specialized ultrasonic technique for examination of tubing uses an internal rotary inspection system (IRIS) probe. The probe consists of an ultrasonic transducer that is centered in the tube and has a rotating element. As the probe is pulled through the tube, a scan of wall thickness along the circumference and length of the tube is generated. The method can also detect surface defects on the inner diameter or outer diameter. The IRIS method is generally used for inspection of carbon steel tubing.

Eddy Current Inspection

Eddy current inspection is an electromagnetic-induction technique. The technique employs an eddy current probe that contains an electric coil through which an alternating current flows. When placed near the surface of a component, this current induces eddy current flow in the component as a result of electromagnetic induction. The magnitude and phase of the induced current flow are sensed by a secondary coil or by change of inductance in the primary coil probe. A localized change in induced current flow indicates the presence of a discontinuity in the component. The extent of the change in the induced current flow correlates to the size of the discontinuity, which can represent localized metal loss on a surface or a defect in the metal. Specific probe geometries are used for examination of tubing that allows the probe to be scanned along the length of a tube. This technique can provide a thorough assessment regarding the extent and location of metal loss in the tubes of a heat exchanger. Eddy current testing is more useful for materials that are not ferromagnetic, such as stainless steels and copper alloys. It is not as accurate for inspection of carbon steel.

Liquid Penetrant Inspection

Liquid penetrant inspection is a useful technique for identifying surface cracking or defects that are not observable by visual inspection.

The surface of the component being inspected must be clean and must have a relatively smooth surface finish. A penetrant fluid is applied to the surface. The penetrating fluid can be either a visible color or fluorescent. The fluid remains on the surface for an extended period so it can penetrate into openings on the surface by capillary action. The excess penetrant fluid is then removed from the surface with care taken not to remove the fluid that penetrated the openings. A developer coating is applied to the surface to highlight the locations where the penetrant remains in openings on the surface. In the case of a fluorescent fluid, a black light is then used to make the penetrant fluoresce visibly. This technique is very useful for examining tight crack patterns, such as stress corrosion cracking on stainless steels (Fig. 10.9).

Metallographic Analysis

If tubes, piping, or other damaged components can be removed from the cooling water system, then a metallographic analysis can be conducted to provide details on the failure or corrosion. Metallographic analysis incorporates detailed visual and microscopic examination to determine the cause of failure or corrosion, the extent of the metal loss on the section, and factors that influenced the attack. Metallographic analysis will often include characterization of the composition of deposit and corrosion product layers on the surfaces. When coupled with observations of the sample, a thorough consideration of mechanical factors, system operation, and water chemistry will provide details on the root cause of failure or corrosion of the component.

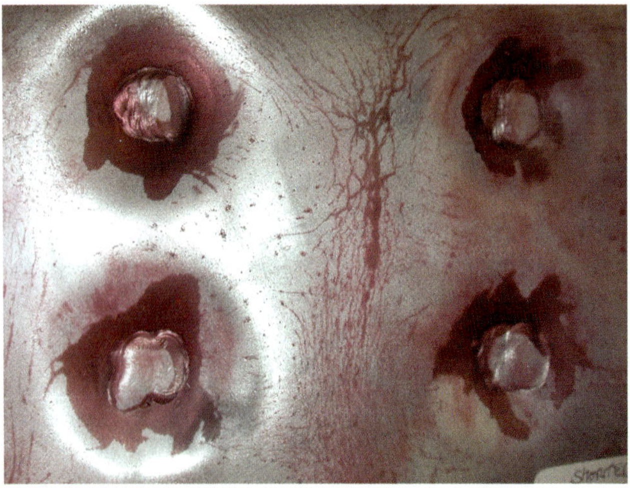

Figure 10.9 Surface of a stainless steel reactor, showing the use of dye penetrant inspection to reveal an extensive pattern of tight cracks on the surface.

References

1. ASTM D2688-11, Standard Test Method for Corrosivity of Water in the Absence of Heat Transfer.
2. ASTM G4 – 01 (2008), Standard Guide for Conducting Corrosion Tests in Field Applications.
3. Cleaning procedures are presented in detail in the ASTM G1 – 03(2011), Standard Practice for Preparing, Cleaning, and Evaluating Corrosion Test Specimens.
4. B. P. Boffardi, "Water Treatments," in *Corrosion Tests and Standards—Application and Interpretation,* ed. R. Baboian, American Society for Testing and Materials, Philadelphia, PA, 1995, p. 699.
5. C. Frayne, *Cooling Water Treatment Principles and Practice,* Chemical Publishing Co., Inc., New York, 1999, p. 96.
6. P. Puckorius, "Cooling Water System Corrosion Guidelines, Part 1," *Process Cooling,* July 2003.
7. Boffardi, "Water Treatments," p. 699.
8. Frayne, *Cooling Water Treatment Principles and Practice,* p. 96.
9. Puckorius, "*Cooling Water System Corrosion Guidelines.*"
10. J. W. McCoy, *The Chemical Treatment of Cooling Water,* 2d ed., Chemical Publishing Company, New York, 1983, p. 29.
11. Frayne, *Cooling Water Treatment Principles and Practice,* p. 401.
12. Detailed information on the techniques and their applications can be found in *ASM Handbook* Volume 17: Nondestructive Evaluation and Quality Control, ASM International, 1989.

SECTION 2

Waterside Corrosion

CHAPTER 11

Crevice and Underdeposit Corrosion (Concentration Cells)

Introduction

Concentration cell corrosion occurs when the environment near the metal surface differs from region to region. These differences create distinct localized anodes. The anodic surface areas are surrounded by cathodic regions that generally occupy much larger surface areas. Shielded and occluded areas are particularly susceptible to corrosion, as they often act as anodes (Fig. 11.1). Differences in concentration of dissolved anions such as chloride and sulfate, and/or dissolved gases, particularly oxygen, eventually develop between shielded and surrounding regions. Shielded areas may be produced by design or by changes in surface condition. Crevices may be produced by design at locations of metal-to-metal contact, or where metals contact dense nonmetallic materials. Crevices may form when protective coatings that cover metal surfaces are damaged and become detached. Pits that form on surfaces may also develop concentration differences. Occluded regions may also form beneath accumulations of deposits, corrosion products, and biological material. Any of these conditions may promote concentration cell corrosion.

Virtually all metallurgies may experience concentration cell corrosion if the local environment becomes sufficiently aggressive. Alloys that strongly passivate in oxygenated waters, such as stainless steels, are more susceptible to localized ion concentration cell corrosion than alloys with weaker passivation tendencies. Under conditions that develop at ion concentration cells, the protective oxide layer that covers the metal surface may be removed, and it will not be replaced. Chloride, sulfate, thiosulfate, or other potentially corrosive ions may concentrate to produce conditions that weaken protective oxide layers,

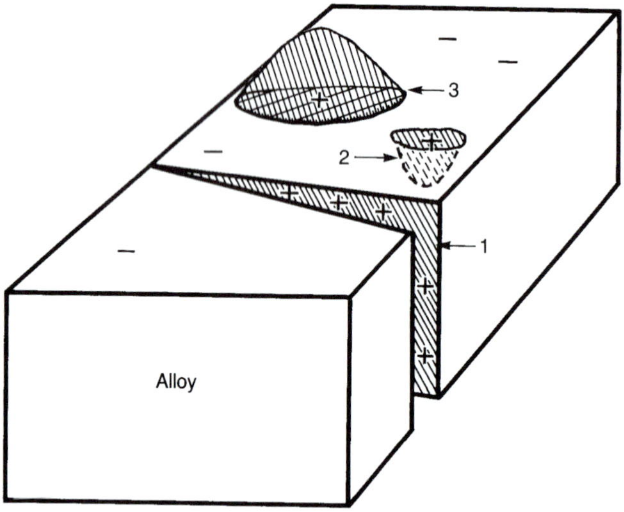

FIGURE 11.1 Region of local action cells in an alloy immersed in an electrolyte. (1) Crevice, (2) pit, (3) deposit. Shaded regions are anodic (+) to surrounding metal (−).

making the metal surface susceptible to corrosion. Corrosion is focused on localized areas where aggressive anions concentrate, sometimes producing deep pits. Even very corrosion-resistant titanium may be attacked under suitably severe conditions. Weakly passivating metals are also affected. However, attack is likely to be more general and pitting less pronounced. Where crevices are produced at contact points of dissimilar metals, accelerated metal loss may occur on the more active metal.

For each situation, solute concentration differences arise primarily by diffusion-related processes. Concentration gradients supply the energy that drives diffusion. As a consequence, stagnant conditions may promote corrosion, since concentration gradients near affected areas are reduced by flow. High concentrations of dissolved species that are corrosive to the metal increase corrosion. Elevated temperature usually stimulates corrosion by increasing both diffusion and reaction rates.

Exceptions to these rules of thumb abound, however. For example, diffusion is faster at elevated temperatures, which would tend to increase ion concentration cell activity. However, at high temperatures, dissolved oxygen concentration may be lower, which would decrease oxygen concentration cell activity. Also, at high and widely varying temperatures, increased turbulence and thermal stresses may dislodge deposits to decrease concentration cell activity on surfaces.

Finally, pitting may be viewed as a special form of concentration cell corrosion as localized corrosive conditions may develop within pits. Most alloys that are susceptible to crevice corrosion also experience pitting. However, many metals may pit under conditions that are not produced by concentration cell corrosion. Further, although sharing many common features with concentration cell corrosion, pitting is sufficiently different to warrant a separate categorization.

The two major forms of concentration cell corrosion are crevice corrosion and underdeposit corrosion. Both forms of corrosion are common in cooling systems. Many corrosion-related problems in cooling water systems are caused by these forms of wastage. These two mechanisms are described in detail next.

Crevice Corrosion

General Description

For corrosion to occur, the crevice must be filled with water. The water must also contact surfaces adjacent to the crevice. The crevice typically is a few thousandths of an inch wide. Corrosion may extend up to several feet into a crevice mouth in high-conductivity waters.

The crevice geometry and normally occurring corrosion combine to produce accelerated metal loss in the shielded region. The corrosion mechanism is an autocatalytic process. Initially, corrosion in oxygenated water of near-neutral pH occurs by Reactions (11.1) and (11.2), where *M* denotes the metal involved:

$$\text{Anode: } M \rightarrow M^{+n} + \text{ne}^- \tag{11.1}$$

$$\text{Cathode: } O_2 + 2H_2O + 4e^- \rightarrow 4OH^- \tag{11.2}$$

These reactions are shown schematically within a crevice and in areas surrounding the crevice mouth in Fig. 11.2. Many other reactions may occur at anodes and cathodes, but Reactions (11.1) and (11.2) predominate on carbon steel, for example, in near-neutral pH, oxygenated water.

Eventually oxygen becomes depleted in the crevice. Replenishment of oxygen by convective transport cannot occur because the tight crevice geometry limits the flow of oxygenated water into the crevice. Although diffusion of oxygen in water near room temperature is relatively rapid, the cross-sectional area of the crevice mouth relative to the internal crevice area is small. A typical crevice might have an internal area that is several orders of magnitude greater than the crevice mouth area. Hence, the total amount of oxygen transported into the crevice cannot sufficiently replace the oxygen that is consumed.

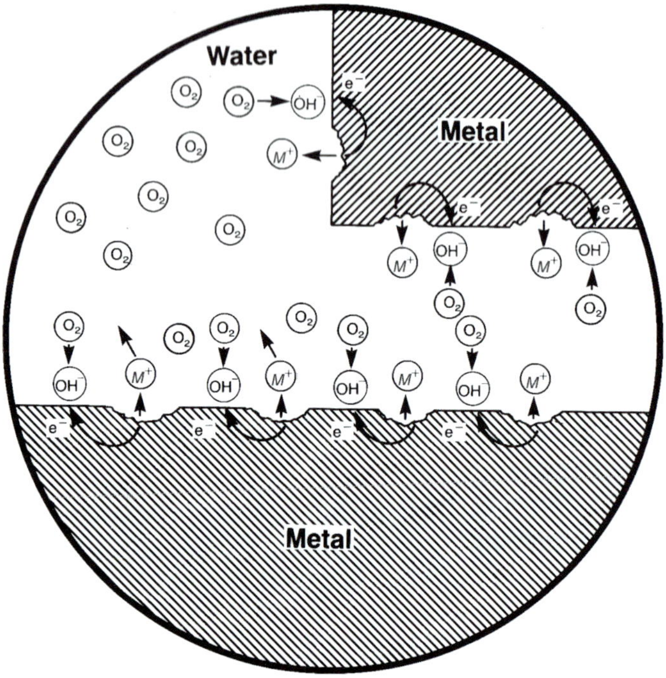

FIGURE 11.2 Initial stage of corrosion near a crevice.

Oxygen concentration outside the crevice is held almost constant by water flow. Thus, a differential oxygen concentration cell is created. A differential oxygen concentration cell is sometimes referred to as a differential aeration cell. The oxygenated water allows Reaction (11.2) to continue outside the crevice. Regions outside the crevice become cathodic, and metal dissolution ceases there.

Within the crevice, Reaction (11.1) continues (Fig. 11.3). Metal ions migrating out of the crevice react with the dissolved oxygen and water to form metal hydroxides as in Reactions (11.3) and (11.4):

$$4M^+ + 4e^- + O_2 + 2H_2O \rightarrow 4MOH \tag{11.3}$$

In the case of steel,

$$2Fe^{2+} + 4e^- + O_2 + 2H_2O \rightarrow 2Fe(OH)_2 \tag{11.4}$$

As metal ion concentration increases in the crevice, a net positive charge accumulates in the crevice electrolyte. This attracts negatively charged ions dissolved in the water. Chloride, sulfate, and other anions diffuse into and concentrate in the crevice (Figs. 11.4 and 11.5). Hydrolysis produces acids in the crevice, causing acid corrosion. Studies have shown that the crevice pH can decrease to 2 or less in

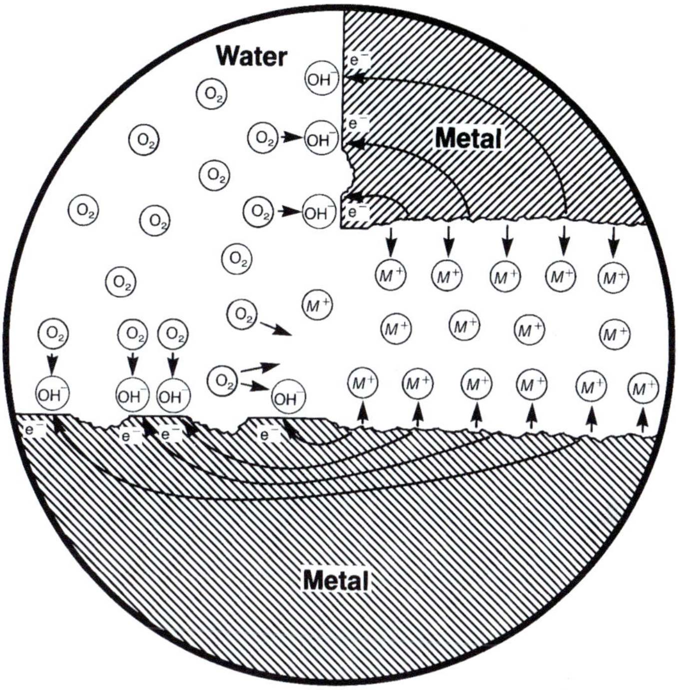

FIGURE 11.3 Later stage than in Fig. 11.2 when an oxygen concentration cell is established between regions inside and outside the crevice. Oxygen has been depleted inside the crevice.

freshwater solutions containing chloride that have a neutral pH. This mechanism is called ion concentration cell corrosion.

In seawater with a pH of 8, the pH that develops within a crevice may fall below 1, and chloride concentration can be many times greater than in the water. The crevice environment becomes more and more corrosive with time as aggressive anions concentrate within. Areas immediately adjacent to the crevice receive ever-increasing numbers of electrons from the crevice. Hydroxyl ion formation increases just outside the crevice, locally increasing pH and decreasing attack there [Reaction (11.2)]. Corrosion inside the crevice becomes more severe with time due to the autocatalytic diffusion and concentration of acidic anion into the crevice.

Underdeposit Corrosion

General Description

Deposits may cause corrosion both directly and indirectly. If the deposits that form on a surface contain corrosive substances, attack is direct. Indirect corrosion may occur as a consequence of surface

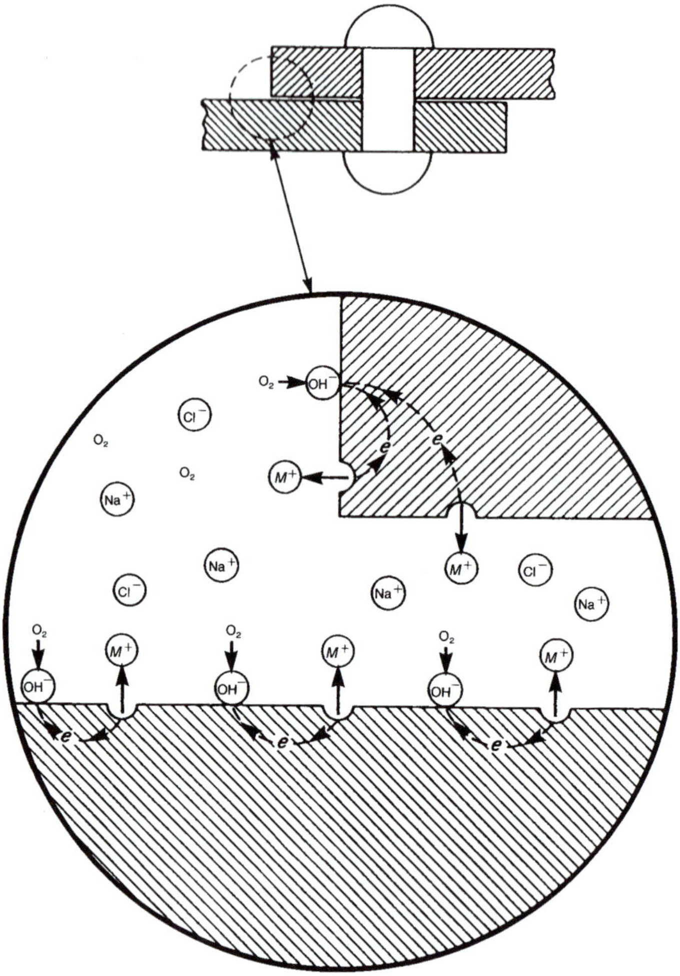

FIGURE 11.4 Crevice corrosion—initial stage in oxygenated water containing sodium chloride. (*Courtesy of Mars G. Fontana and Norbert D. Greene, Corrosion Engineering, McGraw-Hill Book Company, New York, 1967.*)

shielding provided by deposit layers or other occluding material. Both direct and indirect attack may involve concentration cell corrosion, but indirect attack almost always involves this form of corrosion. Corrosion may be caused by the formation of differential oxygen concentration cells and/or ion concentration cells.

As general oxygen corrosion proceeds on carbon steel and other alloys, the accumulation of corrosion products may also result in concentration cell formation. Concentration cell corrosion will also occur beneath iron oxide tubercles (see Chap. 12).

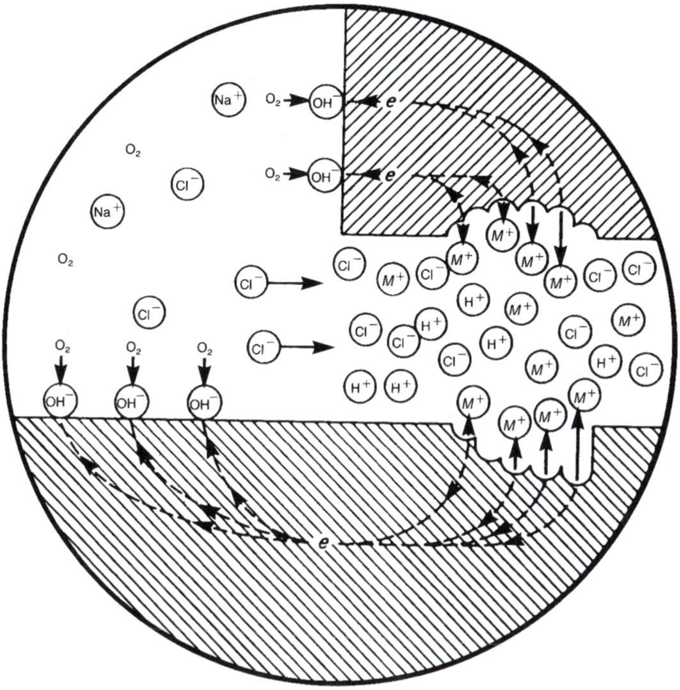

FIGURE 11.5 Crevice corrosion—final stage in oxygenated water containing sodium chloride. (*Courtesy of Mars G. Fontana and Norbert D. Greene, Corrosion Engineering, McGraw-Hill Book Company, New York, 1967.*)

Severe ion concentration cell corrosion may occur beneath deposits due to the concentration of aggressive anions. Concentrations of sulfate and chloride, in particular, are deleterious. Mineral acids, such as hydrochloric and sulfuric, form by hydrolysis.

Frequently, differential oxygen concentration and the concentration of aggressive ions beneath deposits combine to produce severe localized wastage on stainless steel (Figs. 11.6 and 11.7) and

FIGURE 11.6 Severe localized wastage on a 316 stainless steel heat exchanger tube. Attack occurred beneath deposits, which were removed to show wastage.

FIGURE 11.7 A 304 stainless steel plate from a storage tank floor, which pitted beneath a deposit layer. Crevice corrosion also occurred where bolted washers contacted the surface around the drilled holes.

other passivating alloys such as aluminum, titanium, and nickel. The depleted oxygen concentration beneath deposits reduces the ability of these alloys to form a protective oxide layer. However, this condition alone is not severe enough in most cases to initiate attack on these alloys. Yet, in the presence of depassivating agents, such as chloride ions, severe pitting may occur on stainless steels and aluminum alloys.

Underdeposit corrosion is also affected by the pH of the water. As with crevice corrosion, susceptibility to corrosion increases as a function of decreasing pH. Some deposits will dissolve when exposed to waters below a certain pH. Under such conditions, the removal of deposits will cause underdeposit corrosion to cease. However, the conditions of acidic water exposure may promote acid corrosion directly.

Certain acidic deposits that form on surfaces may contribute to attack on stainless steels. Amphoteric alloys such as aluminum experience corrosion when exposed to both acidic and alkaline deposits (Fig. 11.8).

Metal loss can also occur indirectly because of turbulence associated with flow over and around a deposit layer of appreciable thickness. Increased turbulence may initiate mechanical forms of damage (see Chap. 23, "Erosion-Corrosion," and Chap. 24, "Cavitation Damage").

Locations

As the name implies, crevice corrosion occurs between two surfaces in close proximity, such as a flanged joint. Table 11.1 gives a partial listing of common crevice corrosion sites in cooling water systems. Any region where two surfaces are loosely joined or come into close proximity qualifies as a crevice site as long as water may enter. Crevices may be produced beneath partially cracked, damaged, or peeled

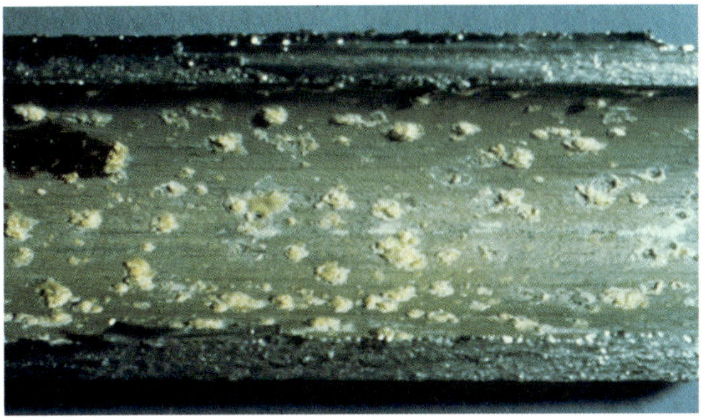

FIGURE 11.8 Corrosion product mounds covering localized areas of metal loss on an aluminum heat exchanger tube. Attack initiated beneath a thin deposit layer.

coatings. The surface that contacts the corroding metal surface may be metallic or nonmetallic. Crevices commonly are present at gaskets, flanges, washers, bolt holes, rolled tube ends, incompletely fused welds, or welds with incomplete penetration, contact points in plate-and-frame heat exchangers, threaded couplings, riveted seams,

Component	Location
Shell and tube heat exchangers	Rolled ends at tubesheet
	Defective welds at tubesheet
	Water box gaskets
	Bolt holes, nuts, washers
	Disbonded water box coatings
	Baffle contact points in exchangers using shellside cooling
Plate and frame heat exchangers	Beneath gaskets
	Plate contact points
Cooling tower and system piping	Threaded pipe joints
	Bolts, nuts, washers
	Partially exfoliated coatings
	Lap joints in sheet metal
	Between bushings and shafts on pumps
	Pump gaskets
Engine cooling	Gasketed surfaces
	Bolts, nuts, washers

TABLE 11.1 Common Crevice Corrosion Sites in Cooling Water Systems

screen wires, lap joints, beneath coatings and insulation, and anywhere surfaces abut.

Underdeposit corrosion may occur on surfaces that are covered by a deposit and/or corrosion product layer. Cooling water system deposits are ubiquitous. Deposits can be generated internally by precipitation of dissolved salts if solubilities are exceeded. Deposits may also be laid down as transported corrosion products or brought into the system from external sources. Hence, underdeposit corrosion can be found in virtually any cooling water system at any location. Especially troubled systems contain large amounts of silt, sand, grease, oil, biomass, precipitates, transported corrosion products, and other detritus.

Equipment in which water flow is slow or intermittent is subject to deposition and associated corrosion. Hence, service water system components that operate intermittently are more prone to attack. Deposits accumulate in narrow orifices, screens, long horizontal pipe runs, sumps, or at regions of constricted flow.

Components in which water temperature changes abruptly with distance, such as heat exchangers, tend to accumulate precipitates and deposits. Heater surfaces also accumulate precipitates if the dissolved species have inverse temperature solubilities. Systems in which pH excursions are frequent may accumulate deposits. Low-pH excursions may dissolve deposited material, corrosion products, and metal from surfaces. The dissolved material may precipitate when normal operating pH is restored. High-pH excursions may cause precipitation of some species due to reduced solubility. Plenum regions, such as in headboxes for heat exchangers that employ tubeside cooling water flow, and contact points of tubes with tubesheets and baffles in heat exchangers that employ shellside cooling water flow tend to be locations where deposited material will collect.

Critical Factors

Crevice Corrosion

Alloys

Metals that produce protective oxide layers (such as stainless steels) are especially susceptible to crevice corrosion. The reduced oxygen concentration in the crevice inhibits repair of the protective oxide film. This is especially true if aggressive anions such as chloride are present, which further retards oxide repair. Molybdenum alloy additions to stainless steels increase the stability of the protective oxide layer, which provides greater resistance to crevice corrosion.

Incubation Period

There is often a period before corrosion starts in a crevice for passivating metals. This incubation period corresponds to the time necessary to establish a crevice environment aggressive enough to dissolve the passive oxide layer. The incubation period is well known in stainless steels exposed to waters containing chloride. After a time period in which crevice corrosion is negligible, attack begins, and the rate of metal loss increases (Fig. 11.9).

The incubation period varies widely depending on such factors as crevice morphology, water chemistry, and temperature. However, experience in a wide variety of cooling water environments has shown that many stainless alloys develop noticeable attack within six months of first being exposed to water containing significant chloride concentrations. It is rare to see attack initiating many years after equipment commissioning unless there are changes in service conditions.

Crevice Morphology

The crevice geometry markedly affects corrosion. Crevices so tight that water may not enter are entirely immune to attack. Conversely, wider crevices are less susceptible to producing crevice corrosion. Typically, crevice corrosion tends to occur when crevices are a few thousandths of an inch in width. In misting environments or alternately wet-dry environments, the crevice holds water and may allow continued attack even when nearby surfaces are dry. In seawater, the severity of attack in stainless steel crevices depends on the ratio of the exposed surface area within the crevice to the cathodic surface area outside the crevice. If the cathodic area is large relative to crevice area, susceptibility to crevice corrosion increases.

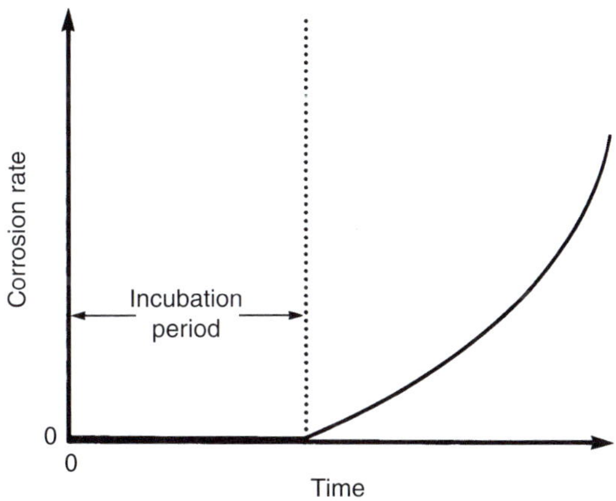

FIGURE 11.9 Schematic representation of corrosion rate as a function of time in a crevice in stainless steel exposed to chloride-containing water. The time before corrosion initiation is called the incubation period.

Water Chemistry

Acidic Anion Concentration

The amount of chloride, sulfate, thiosulfate, or other aggressive anions dissolved in water necessary to produce noticeable attack depends on many interrelated factors. Extraordinarily, if the water is quite aggressive, general corrosion may occur so rapidly outside the crevice that concentration differences cannot easily develop between the crevice interior and exterior. However, it is usually safe to assume that as the concentration of aggressive anions increases in solution, crevice attack is stimulated. As described and shown in Fig. 11.10, susceptibility of

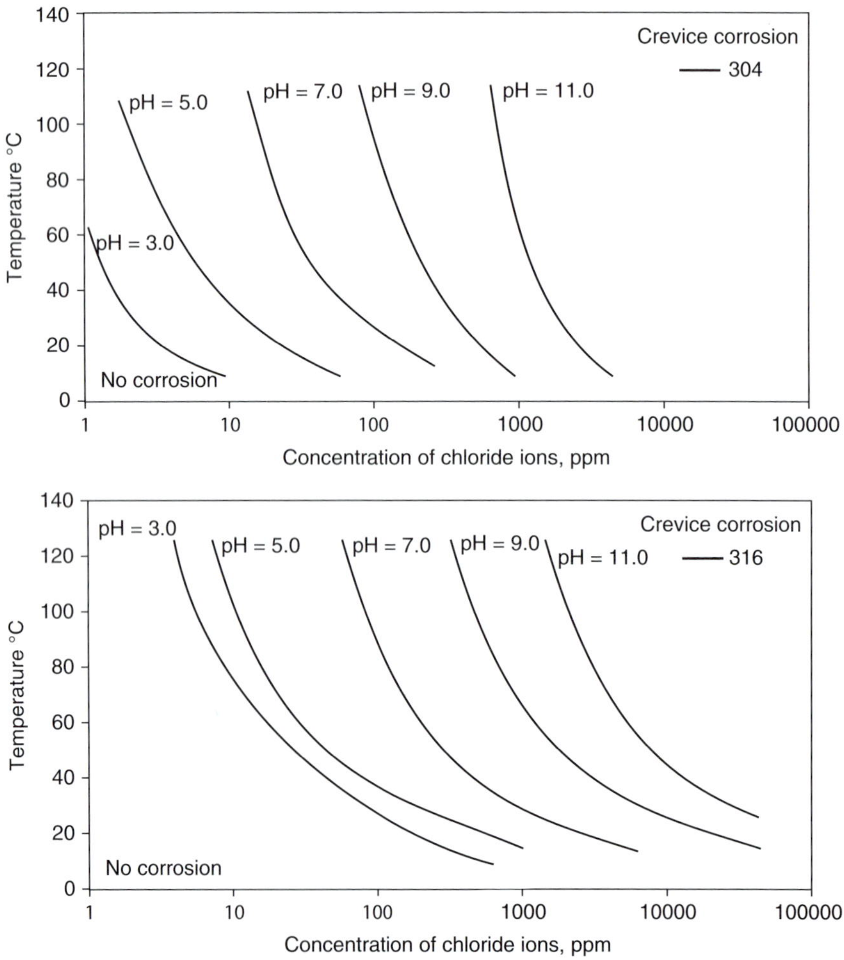

FIGURE 11.10 Curves that illustrate the recommended maximum chloride limits in water solution for 304 and 316 stainless steels as a function of pH to control crevice corrosion. (*Courtesy of Alpha Laval.*)

304 and 316 stainless steel to crevice corrosion increases as a function of increasing chloride concentration. Seawater chloride concentrations may produce severe attack in most stainless crevices in a few weeks. Figure 11.10 also illustrates how 316 stainless steel is more resistant to crevice corrosion than 304 stainless steel.

Solution Conductivity

Crevice corrosion is markedly increased as water conductivity increases.

Solution pH

Susceptibility of stainless steels to crevice corrosion increases as a function of decreasing pH. Figure 11.10 contains curves that illustrate the recommended maximum chloride limits within crevices in water solution for 304 and 316 stainless steel. Corrosion usually develops faster and is more severe as pH decreases.

Abrasion

If closely fitting surfaces within crevices rub against one another, protective oxides may be dislodged, stimulating corrosion (Figs. 11.11 and 11.12). This is especially true if temperature variation is great and if differences in the thermal expansion of abutting metals are large. Often such attack may resemble fretting, which is a mechanical wear mechanism.

Galvanic Effects

Coupling of alloys which vary widely in galvanic potential will stimulate corrosion of the less noble alloy. This is especially the case if a copper alloy is coupled to a less noble alloy such as steel. The corrosion of the steel is stimulated by the galvanic effect caused by the coupling of dissimilar metals. Hence, the sacrificial corrosion of the steel controls or prevents corrosion of the copper alloy (Fig. 11.13). See Chap. 16.

Underdeposit Corrosion

The most harmful deposits are those that are water permeable, which allows concentration cells to develop. Truly water-impermeable material

FIGURE 11.11 Wastage in a circumferential region where a 90:10 cupronickel main condenser tube passed through a baffle. In places, metal loss was greater than 25 percent of the nominal wall thickness.

FIGURE 11.12 Abrasion-related corrosion, as in Fig. 11.11. Note the more severe attack at the bottom of the tube where it rested against a support.

FIGURE 11.13 Severely corroded crevice between rolled brass heat exchanger tube and mild steel tubesheet.

is protective, since corrosion cannot occur unless water contacts metal surfaces. Innately acidic or alkaline deposits are troublesome on amphoteric alloys (those attacked at high and low pH—e.g., aluminum and zinc).

Deposits containing carbonate can be protective. The carbonate buffers acidity caused by the segregation of potentially acidic anions in and beneath deposits. However, deposits are rarely composed of only a single chemical; mixed deposits are the rule. Deposit morphology also influences corrosion. Hence, although sometimes carbonate deposits are beneficial, they often also cause underdeposit corrosion under some circumstances.

Clean surfaces are almost always easier to protect chemically from corrosion with corrosion inhibitors. The primary cause for decreased inhibitor effectiveness is that the deposits serve as barriers between the surface that needs to be protected and the treated cooling water. The inhibitor must react with the metal surface or an oxide layer to provide proper protection. The effectiveness of almost all commonly used corrosion inhibitors increases as surface cleanliness improves.

Identification

Crevice Corrosion

Crevice corrosion follows the shielded areas formed by abutting surfaces. Corrosion beneath washers and gaskets produces ghost images of the abutting surfaces (Figs. 11.14 and 11.15). At threaded joints, attack is confined to locations where water may flow between the engaged threads (Fig. 11.16). In stainless steel joints, pitting may be pronounced (Fig. 11.17). General wastage almost always occurs at cast iron and carbon steel joints (Figs. 11.18 and 11.19).

Corrosion products are almost always absent in crevices formed by abutting stainless steel components. However, areas just outside of the crevices may be stained brown and orange with iron oxides (Figs. 11.20 and 11.21). Metal ions migrate out of the crevice. Precipitation occurs by reactions similar to Reactions (11.3) and (11.4). Crevice interiors remain relatively free of iron oxide (Figs. 11.16 and 11.17). On carbon

FIGURE 11.14 Metal loss within annular regions below rubber O-rings on a cast iron valve block. Note how damage varies from hole to hole, probably due to variation in the crevice geometry.

FIGURE 11.15 Close-up of metal loss in annular regions shown in Fig. 11.14.

FIGURE 11.16 Austenitic stainless steel pipe nipple that experienced crevice corrosion at threads due to exposure to an acidic, chloride-containing solution.

steel and cast irons, iron oxide accumulates at crevice mouths. Darker oxides often are present within crevices (Figs. 11.18 and 11.19).

Chemical analysis of fluids within crevices often reveals high concentrations of chloride or sulfate. Chemical spot tests can indicate acidic pH.

Specially designed washers are used in coupon studies to monitor for susceptibility of specific metals to crevice corrosion in a particular environment (Fig. 11.22). There are several designs; most consist of a small Teflon[1] washer with radially oriented, wedge-shaped teeth. The washer is attached to the coupon surface by a mounting bolt that passes through a central hole. Crevices are produced where the teeth contact the coupon surface. Following a suitable exposure period, the coupon is removed, and the washer is taken off to determine if corrosion occurred

FIGURE 11.17 Pitting at a free-machining austenitic stainless steel threaded valve throat. The valve controlled flow of a chlorine-containing, low-pH biocide.

FIGURE 11.18 Wasted threads on small-diameter mild steel pipe.

within the crevices (Fig. 11.23). The test produces qualitative results. Crevice corrosion either occurs or does not.

Underdeposit Corrosion

Almost all cooling water system deposits are waterborne. The types of deposits that form generally depend upon water chemistry and the environment. It would be impossible to list each deposit specifically, but general categorization is possible. Deposits in cooling water systems

FIGURE 11.19 Gray cast iron elbow showing thread wastage.

FIGURE 11.20 Austenitic stainless steel plate from plate-and-frame heat exchanger. The orange oxide was formed from corrosion product originating at the regularly spaced pits. Pits are present near points of contact between adjacent plates. Corrugations run at right angles on adjacent plates.

FIGURE 11.21 Orange oxides weeping from a tight seam in martensitic (440) stainless steel surgical scissors.

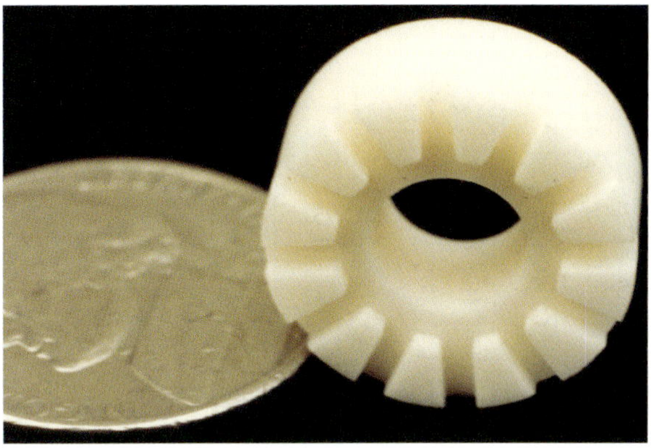

FIGURE 11.22 Crevice washer. The radially oriented, wedge-shaped washer teeth provide small areas between which crevices are formed.

FIGURE 11.23 Shallow pitting in crevice areas on a 304 stainless steel coupon exposed to a misting atmosphere. Note the relatively clean areas where the washer teeth contacted the coupon surface.

may consist of one or more of the following components: dissolved mineral salts from the water that exceed the solubility limit to form precipitates, silt and other transported suspended particles, transported corrosion products from upstream sources, biological materials, compounds produced by reaction of treatment chemicals with minerals and corrosion products, and a wide variety of contaminants such as grease, oil, and process chemicals. Associated corrosion is fundamentally related to whether deposits are innately aggressive or simply serve as an occluding medium beneath which concentration cells develop. An American Society for Testing and Materials (ASTM) list of water-formed deposits is given in Table 11.2.

Name	Formula	Name	Formula
Acmite	$Na_2O\text{-}Fe_2O_3\text{-}4SiO_2$	Hydroxyapatite	$Ca_{10}(OH)_2(PO_4)_6$
Ammonium bicarbonate	NH_4HCO_3	Hydroczincite	$2ZnCO_33Zn(OH)_2$
Analcite	$Na_2O\text{-}Al_2O_3\text{-}4SiO_2\text{-}2H_2O$		
Anhydrite	$CaSO_4$	Lazurite	$3Na_2O\text{-}3Al_2O_3\text{-}6SiO_2\text{-}2Na_2S$
Ankerite	$CaCO_3(Fe,Mg,Mn)CO_3$		
Aragonite	$CaCO_3$	Lepidocrocite	$Fe_2O_3\text{-}H_2O$
Atacamite	$CuCl_43Cu(OH)_4$	Libethenite	$4CuO\text{-}P_2O_5\text{-}H_2O$
Azurite	$2CuCO_3\text{-}Cu(OH)_2$	Maghemite	Fe_2O_3
Barite	$BaSO_4$	Magnesia	MgO
Bayerite	$Al_2O_33H_2O$	Magnesium chloride hydrate(basic)	$MgCl_25Mg(OH)_28H_2O$
Bloedite	$Na_2SO_4\text{-}MgSO_4\text{-}H_2O$		
Boehmite	$Al_2O_3\text{-}H_2O$	Magnesium hydroxyphosphate	$3Mg_3(PO_4)_2Mg(OH)_2$
Bornite	Cu_5FeS_4	Magnesite	$MgCO_3$
Brochantite	$CuSO_4\text{-}SCu(OH)_2$	Magnetite	Fe_33O_4
Brucite	$Mg(OH)_4$	Malachite	$CuCO_3\text{-}Cu(OH)_2$
Bunsenite	NiO	Meta halloysite	$Al_2O_3\text{-}2SiO_2\text{-}xH_2O$
Burkeite	$Na_2CO_3\text{-}2Na_2SO_4$	Meta thenardite	Na_2SO_4I
Calcium oxalate monohydrate	$CaC_2O_4\text{-}H_2O$	Montmorillonite	$Al_2O_3\text{-}4SiO_2\text{-}H_2O\text{-}nH_2O$
Calcium phosphate (dibasic)	$CaHPO_4$	Mullite	$3Al_2O_3\text{-}2SiO_2$
Calcium pyrophosphate	Ca_2P_2O7	Muscovite	$KAl_2(Si_3Al)O_{10}(OH_2F)_2$
Calcium sodium phosphate	$CaNaPO_4$	Natrolite	$Na_2O\text{-}Al_2O_33SiO_2\text{-}2H_2O$
Calcium sulfate	$CaSO_4$		
Calcite	$CaCO_3$	Noselite	$4Na_2O3Al_2O_36SiO_2\text{-}SO_3$
Calcium aluminate	$3CaO\text{-}Al_2O_3\text{-}6H_2O$	Oldhamite	CaS
Cancrinite	$4Na_2O\text{-}CaO\text{-}4Al_2O_3\text{-}2CO_24SiO_23H_2O$	Olivine	$2(Mg,Fe)O\text{-}SiO_2$
Celestite	$SrSO_4$	Paratacamite	$CuCl_23Cu(OH)_2$
Cementite	Fe_3C	Pectolite	$Na_2O\text{-}4CaO\text{-}6SiO_2\text{-}H_2O$
Cerrucite	$PbCO_3$	Periclase	MgO
αChalcocite	αCu_2S	Portlandite	$Ca(OH)_2$
βChalcocite	βCu_2S	Pyrrhotite	FeS
Chalcopyrite	$CuFeS_2$	Quartz	SiO_2

Copper	Cu	Sepiolite	$2MgO\text{-}3SiO_2\text{-}2H_2O$
Copper aluminate	$CuAlO_2$	Serpentine	$3MgO\text{-}2SiO_2\text{-}2H_2O$
Corundum	αAl_2O_3	Siderite	$FeCO_3$
Covellite	CuS	Smithsonite	$ZnCO_3$
Cristrobalite	SiO_2	Sodium disilicate	$\beta Na_2Si_2O_5$
Cuprite	Cu_2O	Sodium metasilicate	Na_2SiO_3
Delafossite	$Cu_2O\text{-}Fe_2O_3$	Sphalerite	βZnS
Dolomite	$CaCO_3\text{-}MgCO_3$	Syngenite	$K_2SO_4\text{-}CaSO_4\text{-}H_2O$
Favalite	$2FeO\text{-}SiO_2$	Talc	$3MgO\text{-}4SiO_2\text{-}H_2O$
Ferrous bicarbonate	$Fe(HCO_3)_2$	Tenorite	CuO
Ferrous sulfate monohydrate	$FeSO_4\text{-}H_2O$	Teschemacherite	NH_4HCO_3
Ferrous sulfate quadrahydrate	$FeSO_4\text{-}4H_2O$	Thenardite	Na_2SO_4
δ Ferric Oxide	δFe_2O_3	Thermonatrite	$Na_2CO_3\text{-}H_2O$
β Ferric oxide monohydrate	$\beta Fe_2O_3\text{-}H_2O$	Troilite	FeS
Fluorite	CaF_2	Trona	$3Na_2O\text{-}4CO_2\text{-}5H_2O$
Foshagite	$5CaO3SiO_23H_2O$	Vermiculite	$(Mg,Fe)_3(Al,Si)_4O_{10}(OH)_2\text{-}4H_2O$
Forsterite	$2MgO\text{-}SiO_2$	Vivianite	$Fe_3(PO)_28H_2O$
Gaylussite	$CaCO_3\text{-}Na_2CO_35H_2O$	Wedellite	$CaC_2O_4\text{-}2H_2O$
Gehlenite	$3CaO\text{-}Al_2O_3\text{-}2SiO_2$	Wikeite	$Ca_{10}O[(Si,P,S)O_4]_6$
Gibbsite	$Al_2O_3\text{-}3H_2O$	Willemite	Zn_2SiO_4
Glauconite	$K_2(MgFe)_2Al_6(Si_4O)_3(OH)_{12}$	Witherite	$BaCO_3$
Goethite	$\alpha Fe_2O_3\text{-}H_2O$	Whitlockite	$\beta Ca_3P_2O_8$
Gypsum	$CaSO_4\text{-}2H_2O$	Wollastonite	$\beta CaSiO_3$
Gyrolite	$2CaO\text{-}_33SiO_2\text{-}H_2O$	Wustite	FeO
Halite	$NaCl$	Xonotlite	$5CaO5SiO_2\text{-}H_2O$
Hemihydrate	$CaSO_4 \cdot ½H_2O$		
Hematite	Fe_2O_3		
Hydromagnesite	$4MgO \cdot 3CO_2 \cdot 4H_2O$		

*Source: *Manual on Water,* 3d ed., ASTM, STP-442, 1969.

TABLE 11.2 ASTM List of Common Compounds Found in Water-Formed Deposits*

Oxides of manganese and iron may be found deposited together. Similar conditions cause oxidation of both iron and manganese ions. Exposure to oxygenated water, chlorination, and some microbiological processes causes such oxidation. Often, a few percent chloride is found in deposits, possibly because of associated chlorination.

Manganese-rich deposits usually take one of three forms:

1. A loosely adherent, friable, brown, or black deposit may occur (Fig. 11.24).
2. A thin, dark, brittle, glassy manganese layer sometimes forms on heat transfer surfaces (Fig. 11.25).
3. Nodular manganese oxide deposits also occur (Fig. 11.26).

Figure 11.24 Friable brown deposit patches on a brass condenser tube. The deposit contained more than 30 percent manganese.

Figure 11.25 A thin, glassy layer of predominantly manganese oxide on the internal surface of a brass condenser tube. The many light colored spots are pits at fractures in the manganese oxide layer.

FIGURE 11.26 Small, manganese-rich nodules on a 90:10 cupronickel condenser tube. Note the small pits beneath each nodule. *(Courtesy of National Association of Corrosion Engineers; Andy Howell Public Service of Colorado, Corrosion '89 Paper No. 197 by H. M. Herro.)*

Manganese oxide deposits are more noble than the metal. Consequently, deep pits may form at breaks in such deposit layers due to galvanic corrosion effects. Even corrosion-resistant alloys such as stainless steel may be affected.

Localized attack tends to occur beneath manganese deposits on copper alloys (Fig. 11.27); it is most common beneath nodular and glassy deposits (Figs. 11.25 and 11.26).

Stainless steels may pit or generally corrode beneath deposits (Figs. 11.28–11.31). Conditions beneath the deposit become more aggressive due to oxygen depletion, decreasing pH, and/or the

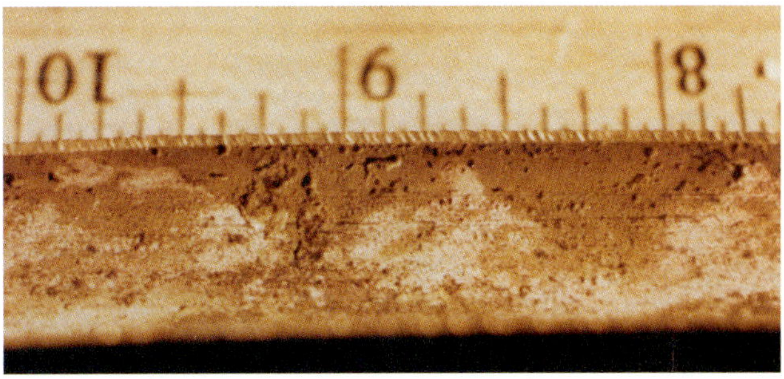

FIGURE 11.27 Pitting on internal surface of a brass condenser tube revealed by removal of manganese-rich deposit.

Figure 11.28 Deposits on a 316 stainless steel heat exchanger tube.

Figure 11.29 The tube shown in Fig. 11.28, with deposits removed to show pitting.

Figure 11.30 General wastage on a 304 stainless steel heat exchanger tube.

FIGURE 11.31 Same type of general wastage as in Fig. 11.30. Note how abruptly corrosion terminates at deposit edge.

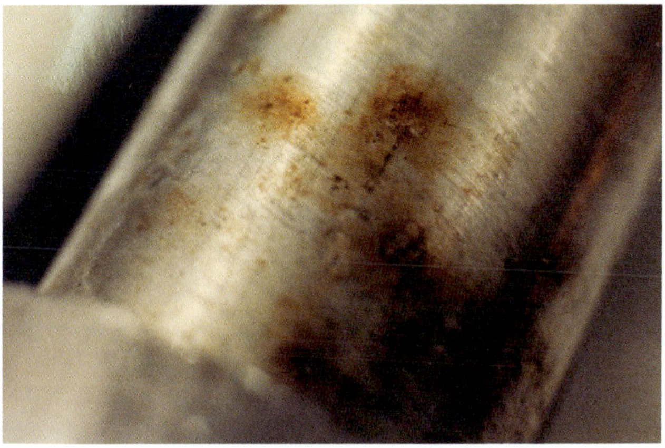

FIGURE 11.32 Rust patches surround small pits on a stainless steel shaft. The rust is formed by oxidation of ejected ferrous ions from the tiny pits. Pits were initiated by chloride and sulfate ions concentrated by evaporation.

concentration of depassivating agents. Metal loss that occurs beneath deposits is usually in the form of pits. Such corrosion is unlikely unless depassivating agents such as chloride and/or low pH are present. If conditions beneath deposits are sufficiently aggressive, corrosion may become general (Fig. 11.30). Borders separating corroded from unattacked regions are usually sharp, mirroring deposit patterns (Fig. 11.31). Usually, corrosion products are reddish brown and may surround the pits (Fig. 11.32). The corrosion products formed on stainless steels resemble orange and brown iron oxides that form on carbon steel and cast iron surfaces but are considerably less voluminous.

Contaminants

Silt, sand, concrete chips, shells, tower fill, plastic, and various other materials foul many cooling water systems. These materials produce indirect attack by establishing differential oxygen concentration cells and ion concentration cells. Attack is usually general on carbon steel, cast iron, and most copper alloys. Below silt accumulations on copper, brass, and cupronickel heat exchanger tubes, a layer of bluish-white copper carbonate often forms (Figs. 11.33 and 11.34). Localized attack is almost always confined to strongly passivating metals such as stainless steels and aluminum alloys.

Precipitates

When dissolved mineral salts in the water exceed their solubility limits, a wide variety of precipitates will form in cooling water systems. Some common precipitated compounds include carbonates, silicates, sulfates, and phosphates (see Table 11.2).

FIGURE 11.33 Silt layer on the internal surface of a 90:10 cupronickel condenser tube.

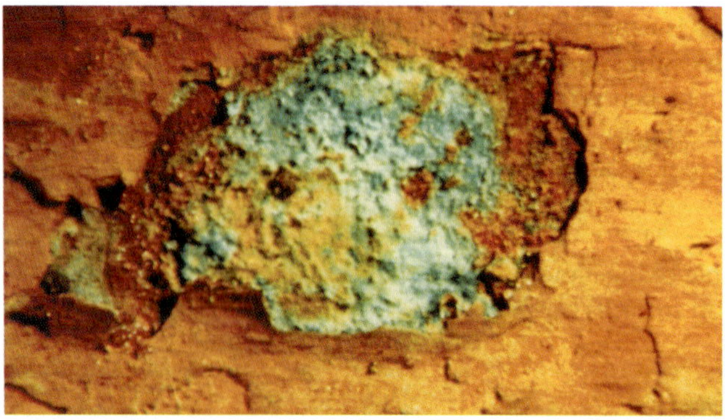

FIGURE 11.34 Same condenser tube as in Fig. 11.33 but with silt removed to reveal bluish-white copper carbonate corrosion products.

FIGURE 11.35 Thick calcium carbonate deposits on condenser tube and copper transfer pipe. Note the stratification.

Calcium carbonate makes up the largest percentage of material in bulk deposits in many cooling water systems (Fig. 11.35). Deposits are usually heavily stratified, indicating distinct periods of formation that reflect changes in water chemistry, heat transfer, and flow. Calcium carbonate deposits readily form as pH and water temperature rise. Corrosion may be slight beneath heavy accumulations of fairly pure calcium carbonate. Such layers can inhibit some forms of corrosion. When nearly pure, calcium carbonate is white. However, calcium carbonate compounds are often intermixed with silt, metal oxides, and other precipitates, leading to underdeposit corrosion. Copper carbonate can form beneath deposit accumulations on a copper alloy component, producing a friable bluish-white corrosion product (Fig. 11.36). Beneath the copper

FIGURE 11.36 Corroded copper tubes from a heating coil. Cooling water contacted the visible surface. The mounds consist of copper carbonate (see Fig. 11.37).

carbonate, sparkling, ruby-red cuprous oxide crystals will often be found (Fig. 11.37). The cuprous oxide is friable, as these crystals are small and do not readily cling to one another or other surfaces (Fig. 11.38). If chloride concentrations are high, a white copper chloride corrosion product may line the surface beneath the cuprous oxide layer.

Sulfur Compounds

Hydrogen sulfide and other sulfides are directly aggressive to many metals and are particularly aggressive to copper alloys

FIGURE 11.37 As in Fig. 11.36 but with carbonate mound removed to reveal sparkling lavender cuprous oxide crystals.

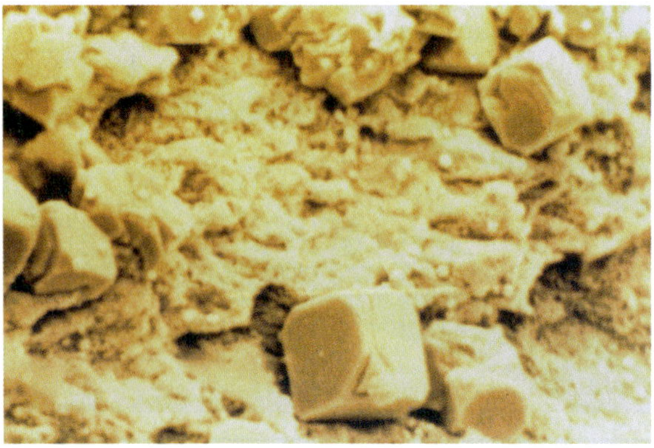

FIGURE 11.38 Scanning electron microscope picture of cuprous oxide crystals as shown in Fig. 11.37. Note the partial octahedral symmetry.

FIGURE 11.39 Sulfide deposits (dark patches) on longitudinally split brass heat exchanger tube. Note the perforation where wastage penetrated the tube wall. Sulfide was spalled after perforation by escaping fluids.

(Fig. 11.39). Sulfides are easily detected in the field by the strong odor of hydrogen sulfide, which is produced after exposure to hydrochloric acid.

Many metal sulfides produce poorly adherent corrosion product layers. This leads to rapid spalling during thermal cycling or turbulent flow. In particular, nonadherent and easily spalled sulfide layers form on steel and cast irons.

Organic Contaminants

Petroleum greases and oils can be excellent corrosion inhibitors on a variety of alloys. The hydrophobic layer produced by oil or grease can prevent water from contacting surfaces and can essentially eliminate corrosion. For this reason, it is common practice for systems to be laid up dry during idle periods using soluble oil treatments. The oil layers deposit onto metal surfaces following drainage, which repels residual water and condensed moisture to prevent corrosion.

Unfortunately, the addition of oil and grease cannot be recommended as a corrosion-reduction measure in most operating cooling water systems for three basic reasons.

1. Greases and oils may serve as binders that accumulate and hold large amounts of suspended solid material from the water and on surfaces. The accumulated material may foul surfaces and/or block flow through piping and tubing downstream.
2. Greases and oils can break down by a variety of processes to produce organic acids that accelerate corrosion.

3. Grease and oil may promote the growth of aggressive microorganisms (see Chap. 13). Attacked surfaces may resemble those associated with acid corrosion and microbiologically influenced corrosion.

Corrosion Control

Crevice Corrosion

Crevice corrosion requires the formation of a suitable, tight crevice geometry that allows water to enter the crevice. Hence, crevice corrosion may be controlled or prevented most effectively by eliminating the crevices if possible. This may be accomplished by changing the design or construction of system components or by sealing the crevices.

The techniques that are most effective are usually system specific. In general, the following steps have been shown to be effective:

- Do not use riveted joints.
- Employ sound welding practice to minimize porosity and achieve full weld metal penetration and fusion. Continuous welds are preferred to many short runs.
- Properly seal known crevices before exposure to water using appropriate resistant sealing materials such as caulk, thread sealant, and organic coating materials such as epoxy.
- Coat heat exchanger and condenser water boxes with protective materials as necessary. Special attention should be paid to the tube sheet area and any sharp projections such as nuts and bolts.
- Avoid using hydrochloric acid to clean equipment that contains crevices, particularly in systems that contain stainless steels. Chloride will concentrate in preexisting crevices during cleaning and may not be adequately removed during subsequent rinsing operations.
- In severe environments, welding tube ends into tubesheets is preferred to rolling to avoid the formation of crevices.
- Ensure that all gaskets and seals are in good repair and bolts are properly tightened.
- During prolonged outages, remove wet packing and clean gasketed surfaces thoroughly.
- In misted areas above water surfaces, exercise special care in maintaining paint layers. Regularly inspect known problem areas.

- If mechanical changes listed here are not practical, the water chemistry may be controlled, if possible to minimize crevice corrosion. Key parameters include reduction of aggressive anion concentration and conductivity, as well as maintaining alkaline pH (in systems containing carbon steel and stainless steel components). These changes must be in-line with the overall chemical treatment program.
- Suitable, more resistant alloys may be substituted for susceptible alloys.

Underdeposit Corrosion

Deposit-related corrosion may be lessened in the following basic ways:

Water Treatment

Limiting deposition related to precipitation is one of the primary goals of most treatment programs. Pretreatment processes such as clarification and softening may be used to reduce suspended and dissolved solids in the makeup water to control deposition. Chemical dispersants may be added to the cooling water, the pH may be adjusted, and cycles of concentration may be reduced to control deposition of dissolved and suspended solids. Biodispersants and biocides may be used to control biofouling. The judicious use of chemical corrosion inhibitors will control aqueous corrosion, including underdeposit corrosion. Effective corrosion inhibition relies upon maintaining clean metal surfaces. Process leaks must be identified and eliminated, since such contaminants will either interfere with or directly or indirectly cause degradation of corrosion inhibitors or the protective layers that they form. If significant amounts of deposits and corrosion products form, the water chemistry may be controlled, if possible to minimize underdeposit corrosion. Key parameters include reduction of aggressive anion concentration and conductivity, and maintaining alkaline pH (in systems containing carbon steel and stainless steel components). These changes must be in line with the overall chemical treatment program.[2]

Deposit Removal

Regular cleaning involving mechanical methods such as water blasting, air rumbling, and chemical cleaning are commonly used as regular maintenance procedures. Cleaning balls made of various materials have been used in condensers and heat exchangers during service to remove waterside deposits with varying degrees of success.

Design and Operational Changes

Deposition caused by settling of suspended particulate may be reduced by increasing flow or turbulence. Dead legs, stagnant areas, and other low-flow regions should be eliminated or redesigned to permit flow greater than 3 ft/s (1 m/s). However, excessive flow velocities should be avoided for soft metals such as copper alloys to prevent erosion. Suspended solids may be removed through the use of filters and strainers.

Material Substitution

Sometimes there is no practical, economical way to reduce deposition, either by a change in design or operation, or by treating systems chemically. Under such conditions, a material change is required to control underdeposit corrosion. The most economical solution usually is to coat existing structures with protective materials.

Application of a protective coating by painting sometimes sufficiently reduces corrosion under intermittent wetting and drying conditions. However, such coatings are less effective when constant immersion and deposition occur. Phenolic resins, epoxies, vinyl ester, and other chemically resistant organic coatings are quite effective in reducing corrosion of underlying metal surfaces when thermal expansion and contraction are slight and when surface contours are smooth. In more severe service and when temperatures are high, plating or cladding may be used.

Substituting a resistant alloy for a susceptible alloy may be the only viable solution to a specific corrosion problem. However, caution should be exercised; this is especially true in a cooling water environment containing deposits. Concentration cell corrosion is insidious. Corrosion-resistant materials in oxidizing environments such as stainless steels can be severely pitted when surfaces are shielded by deposits. Each deposit is unique, and nature can be perverse. Thus, replacement materials ideally should be tested in the specific service environment before substitution is accepted.

Cautions

Occasionally, corrosive bacteria may colonize crevices. Although it is less common than crevice and underdeposit corrosion, microbiologically influenced corrosion should always be suspected if large amounts of sulfide or biological material are present in or near corroded crevices.

Underdeposit corrosion is not so much a single corrosion mechanism as it is a generic description of wastage beneath deposits. Attack may appear much the same beneath silt, precipitates, metal oxides, and debris. However, when deposits tend to directly interact with metal surfaces, attack is easier to recognize.

Even when interaction is direct, forms of attack may be difficult to differentiate from other corrosion mechanisms. For example, corrosion beneath biological material can exhibit strikingly unique attack morphologies according to the degree of microbiological involvement (see Chap. 13).

Related Problems

See Chap. 12, "Oxygen Corrosion"; Chap. 13, "Biologically Influenced Corrosion"; Chap. 14, "Acid Corrosion"; Chap. 16, "Galvanic Corrosion"; and Chap. 17, "Dealloying."

Case History 11.1

Industry:	Steel
Specimen location:	Stainless steel plate-and-frame heat exchanger. Closed chilled water cooling loop used to cool continuous caster mold cooling water.
Orientation of specimen:	Vertical
Environment:	Cooling water side: Softened cooling water using closed system inhibitor Mold side: Softened, hot mold cooling water, 120°F (49°C) using closed system inhibitor
Time in service:	Less than 1 year
Sample specifications:	0.023 in. (0.058 cm) thick, corrugated 316 stainless steel sheet

Fine pits ran transversely across the apexes of many corrugations on the stainless sheets (Figs. 11.40 and 11.41), at points of contact with adjacent sheets in the plate-and-frame heat exchanger. Close examination of plate surfaces with a low-power stereomicroscope indicated that the attack sites were elongated holes or tunnels that originated at pit sites at the apexes of corrugations. The pits propagated through the plate wall (Fig. 11.42).

Pitting was caused by crevice corrosion at crevices formed by closely overlapping corrugations of adjacent plates. Chloride concentration was high within pits. The tunneling of the pits and their precise alignment in the stainless steel plate rolling direction indicates that the tunneling occurred along coring in the microstructure of the steel. Crevice corrosion may be controlled by reducing the concentration of chloride and conductivity in the cooling water as well as through proper pH control.

FIGURE 11.40 Corrosion site at apex of corrugation on plate from a plate-and-frame heat exchanger.

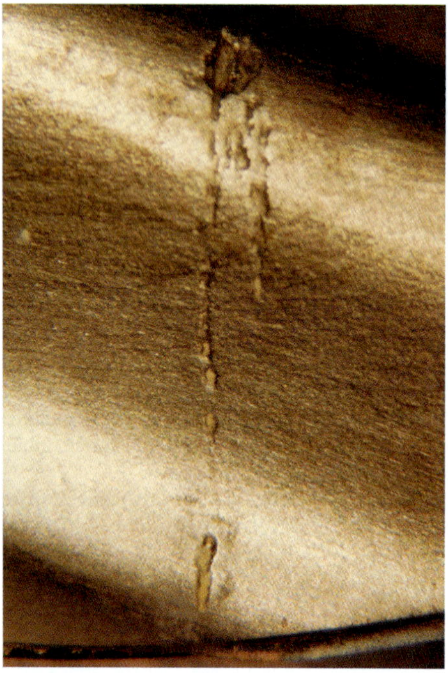

FIGURE 11.41 Pit-site and tunnel propagation down the side of the wall.

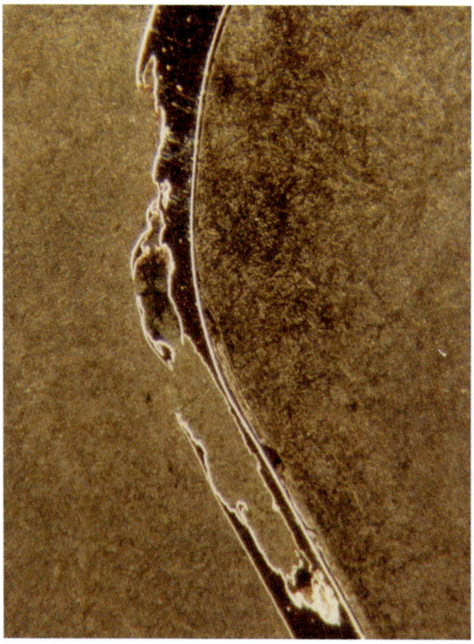

FIGURE 11.42 Same pitting as in Fig. 11.41. Longitudinal cross section through tunnel.

Case History 11.2

Industry:	Food processing
Specimen location:	Heat exchanger for air compressor
Specimen orientation:	Horizontal
Environment:	Untreated cooling water tubeside, air shellside
Time in service:	2 years
Sample specifications:	36 in. (91.4 cm) long, 8 in. (20.3 cm) diameter exchanger with carbon steel tubesheets and baffles housing ¼ in. (0.64 cm) outer diameter copper tubes

Cooling water flowed through the copper tubes in the heat exchanger to cool compressed air that flowed past their shellsides. Failures occurred in the exchanger, causing cooling water to leak into the air side. The end caps were removed from both ends of the exchanger to reveal thick, stratified iron oxide layers and tubercles on the end caps and tubesheets (Figs. 11.43 and 11.44). The tubesheet surfaces were brushed to remove thick iron oxide layers to aid in identifying leaking tubes (Fig. 11.45). The airside was filled with pressurized water, and leakage occurred readily at the location of one tube. One end of the leaking tube was open, but the other end was plugged with iron oxide (Fig. 11.46). A few other, closely surrounding tubes, along with several tubes at other locations were also plugged with iron oxide.

The copper tubes were expanded into the tubesheet to create flared ends. The flared ends were sawcut in order to remove the failed tube for closer inspection. When the tubes were expanded into the tubesheet, narrow gaps were produced between the tubes and tubesheet bores at the flared ends. Moderately deep, general metal loss occurred on the water side of the tubesheets (Fig. 11.47). Corrosion penetrated deeply within the crevices between the tubes and tubesheet due to crevice corrosion (Fig. 11.48).

Deep depressions covered with small mounds of green corrosion products formed on the internal surface of the leaked copper tube due to underdeposit corrosion. A perforation occurred at the deepest depression. Deposits formed

FIGURE 11.43 Thick, stratified iron oxide layers and tubercles on the end cap.

FIGURE 11.44 Thick, stratified iron oxide layers and tubercles on the tubesheet.

FIGURE 11.45 Tubesheet surface was brushed to remove thick iron oxide layers to aid in identifying leaking tubes.

on the internal surfaces of the copper tubes due to the migration of iron oxide corrosion products. These corrosion products formed due to oxygen corrosion on the steel tubesheet and end caps, as well as other carbon steel system components within the cooling water circuit. The underdeposit corrosion on the copper tubes resulted from the formation of differential oxygen concentration cells and ion concentration cells beneath occlusive iron oxide deposit layers. Some tubes, particularly the identified leaking tube, experienced total blockage

FIGURE 11.46 One end of the leaking tube was open, but the other end was plugged with iron oxide. Adjacent tubes were also blocked.

FIGURE 11.47 Moderately deep, general metal loss occurred on the water side of the tubesheet.

of water flow due to pluggage by iron oxide. Underdeposit corrosion will be further aggravated by stagnant flow and increased tube wall temperatures that result from tube blockage.

Waterside corrosion of the copper tubes and carbon steel components may be controlled most effectively through the use of adequate concentrations of proper corrosion inhibitors, biocides, and dispersants. Control of steel corrosion will substantially reduce the amount of iron oxide corrosion products that may cause underdeposit corrosion on the copper tubing. Corrosion inhibitors

FIGURE 11.48 Corrosion penetrated deeply within the crevice between the tube and tubesheet due to crevice corrosion.

and biocides are less effective when flow rates are low or stagnant, and when occlusive corrosion product and deposit layers cover surfaces. Therefore, it was recommended that other exchangers in the system that are heavily fouled with corrosion products should be properly cleaned to restore proper flow prior to the introduction of chemical treatment. It was also recommended that proper flow be established through the exchanger during idle periods as well as during service. If the exchanger was to be idled for an extended period under stagnant flow conditions, it was recommended that proper layup procedures should be followed.

Case History 11.3

Industry:	Hospital building
Specimen location:	HVAC chiller, condenser
Orientation of specimen:	Horizontal
Environment	Internal: Open recirculating cooling water
Time in service:	10 years
Sample specifications:	¾ in. (1.9 cm) outer diameter enhanced copper tube (externally finned, internally rifled)

The internal surface of a failed copper condenser tube was generally covered with light-colored deposit layers that built up between the grooves of the rifling. However, in scattered locations, mounds of blue-green corrosion products interrupted (Fig. 11.49). Removing some random mounds revealed shallow to moderately deep metal loss (Fig. 11.50). Acid was applied to the internal surface, which caused the deposits to effervesce and dissolve, indicating that the deposits were probably mostly in the form of calcium carbonate.

The shallow depressions that occurred beneath the scattered blue-green corrosion product mounds were caused by underdeposit corrosion. Underdeposit

FIGURE 11.49 Internal surface of an enhanced copper chiller condenser tube is mostly covered with light-colored deposit layers that built up between the grooves of the rifling. However, in scattered locations, mounds of blue-green corrosion products interrupt.

FIGURE 11.50 Removing mounds shown in Fig. 11.49 reveals shallow to moderately deep metal loss.

corrosion resulted from the formation of differential oxygen concentration cells and ion concentration cells beneath deposit and corrosion product layers. Deposit and corrosion product layers also serve as barriers, limiting corrosion inhibitors from adequately protecting the surfaces. Corrosion inhibitors are less effective when flow rates are low or stagnant, particularly if the system was laid up wet with inadequate layup practice. It was recommended that the tubes be cleaned properly on a regular basis. Chemical cleaning will be more effective, particularly for enhanced, rifled tubes that tend to entrap deposits between the grooves of the rifling. Following cleaning, the tubes should be properly passivated prior to startup on normal maintenance dosages of copper corrosion inhibitors.

Case History 11.4

Industry:	Food manufacturing
Specimen location:	HVAC chiller, condenser
Orientation of specimen:	Horizontal
Environment	Internal: Open recirculating cooling water
Time in service:	15 years
Sample specifications:	¾ in. (1.9 cm) outer diameter enhanced copper tube (externally finned, internally rifled), carbon steel tubesheet

Inspection revealed that the inlet tubesheet was covered with substantial amounts of tower fill fragments and other debris (Fig. 11.51). The tubesheet surface beneath the debris was covered with a tenacious layer of brown corrosion products and deposits. The bulk material was cleaned away to reveal a generally wasted tubesheet surface, with deeper crevices surrounding some of the tubes (Fig. 11.52). The copper tubes experienced no significant metal loss or deposition. Figure 11.53 shows how nonuniform, general metal loss occurred on the tubesheet. Tubesheet wastage was much deeper in locations where debris, iron oxide, and deposit layers collected and formed.

Chloride and sulfate diffused and concentrated to high enough levels that a very corrosive acidic environment formed beneath occlusive tubercles, corrosion product, and deposit layers, as well as within crevices between the tubes and

Figure 11.51 Inlet tubesheet is covered with substantial amounts of tower fill fragments and other debris.

Figure 11.52 Bulk material shown in Fig. 11.51 was cleaned away to reveal a generally wasted tubesheet surface, with deeper crevices surrounding some of the copper tubes.

Figure 11.53 Tubesheet wastage was nonuniform and deep in locations where debris, iron oxide, and deposit layers collected and formed.

tubesheets. It is possible that some of the carbon steel tubesheet corrosion at locations of copper tube contact was influenced by galvanic corrosion to some degree. The copper tubes, being more noble than the steel, may promote corrosion of the steel. Galvanic corrosion will tend to occur more readily when the water is high in conductivity or contains high concentrations of aggressive anions. It must be emphasized that deep metal loss did not occur around all of the copper tubes. The tubes that were covered with thick iron oxide tubercles and other material were affected most. Therefore, ion concentration cell formation beneath tubercles and deposits, as described previously, was the most significant contributor to the corrosion.

Flow through the tubes on the inlet tubesheet was partially or perhaps even completely blocked with tower fill fragments and other debris. Low or stagnant flow at locations of bypassed blocked tubes also contributed to corrosion, since corrosion inhibitors and biocides are less effective under such conditions. Buildups of debris on tubesheets should be eliminated through the use of proper filtration.

Case History 11.5

Industry:	Chemical process
Specimen location:	Heat exchanger for production of vinyl chloride monomer
Orientation of specimen:	Vertical tubes
Environment:	Shellside: Open recirculating cooling water Tubeside: Process fluids
Time in service:	2 to 3 years
Sample specifications:	2 in. (5.1 cm) outer diameter, carbon steel tube.

A tube experienced shallow, general metal loss on the external surface at a location where it contacted a baffle (Fig. 11.54). The wasted surface was covered with iron oxide corrosion products that contained small concentrations of sulfide. The external surface was cleaned to reveal a rough groove within the wasted area (Fig. 11.55).

The metal loss that occurred on the external surface resulted primarily from crevice corrosion. The occluded surface within the crevice between the tube and baffle produced conditions that were favorable for the formation of differential oxygen concentration cells. It is possible that some metal loss in this area was caused by the diffusion and concentration of aggressive anions such as chloride, sulfate, and sulfide. The presence of sulfide in the corrosion products may indicate that some corrosion was caused by sulfate-reducing bacteria (SRB).

Figure 11.54 Tube external surface experienced shallow, general metal loss at a location where it contacted a baffle.

FIGURE 11.55 External surface at location shown in Fig. 11.54 was cleaned to reveal a rough groove within the wasted area.

The oxygen-depleted region within the crevice will provide a suitable environment for the growth of anaerobic microorganisms like SRBs.

Information gathered during the investigation revealed that the cooling water contained excessive amounts of suspended solids. Corrosion inhibitors may become less effective if substantial amounts of suspended solids are present in the water. The inhibitor will tend to become depleted due to attachment to the solids, and will not be available to protect the tube surfaces. In addition, it is probable that suspended solids settled and built up on the surfaces of the horizontally oriented baffles to promote the crevice corrosion. Corrosion inhibitors are less effective when significant amounts of occlusive deposits cover surfaces, and when flow rates are low or stagnant. Periodic cleaning of built-up deposit layers is recommended, which may be difficult to achieve on the shellside of a shell-and-tube heat exchanger. In addition, proper control methods, such as filtration, should be implemented if the amount of suspended solids becomes excessive.

Case History 11.6

Industry:	Chemical process
Specimen location:	Heat exchanger–propylene refrigerant condenser
Orientation of specimen:	Horizontal
Environment:	Shellside: Propylene and process fluids Tubeside: Open recirculating cooling water containing high chloride and sulfate concentrations (>2000 ppm)
Time in service:	14 years
Sample specifications:	¾ in. (1.9 cm) outer diameter, externally finned, carbon steel

Cooling water flowed through the tubes to condense propylene on the external surfaces. It was reported that numerous and recurrent leaks occurred at tube-to-tubesheet joints in the condenser, but apparently not within the tubes themselves. Since the external surfaces of the analyzed tube sections were finned along their entire lengths, they were not removed from tube-to-tubesheet joints.

Photographs of the tubesheet after mechanical cleaning revealed severe, general metal loss on the tubesheet and tube-to-tubesheet welds that gave them highly irregular contours (Fig. 11.56). The tubesheet surfaces also appeared to have been covered in most locations by layers of orange-brown deposits and corrosion products. Some thicker, adherent mounds of corrosion products appeared to have remained in scattered locations. Since no sections of the tubesheet or welds were received, no definite statements could be made concerning metal loss at those locations.

The internal surfaces of the tube sections were uniformly covered with layers and mutually intersecting tubercles of brown iron oxides and corrosion products

Figure 11.56 Severe, general metal loss on the tubesheet and tube-to-tubesheet welds that gave them highly irregular contours.

FIGURE 11.57 Tube internal surface is uniformly covered with layers and mutually intersecting tubercles of brown iron oxides and corrosion products.

(Fig. 11.57). The tubercles had hollow interiors. Chemical spot tests of material beneath the tubercles revealed the presence of acidic substances, chlorides, sulfates, and sulfides. The metal surfaces beneath the tubercles were mostly lined with thin layers of orange-brown corrosion products. Removing all layers revealed an internal surface that experienced moderately deep, general metal loss (Fig. 11.58). Superimposed on the wasted surfaces were numerous longitudinally

FIGURE 11.58 Removal of corrosion product and deposit layers revealed an internal surface that experienced moderately deep, general metal loss. Superimposed on the wasted surfaces were numerous longitudinally oriented, narrow, straight grooves, or striations.

FIGURE 11.59 Metallographic cross section through wasted internal surface revealed undercut pits and depressions. The surfaces were overlaid with hollow tubercles of iron oxide.

oriented, narrow, straight grooves, or striations. No perforations were found anywhere within the received sections.

Metallographic examination revealed undercut pits and depressions (Fig. 11.59). The surfaces were overlaid with layers and hollow tubercles of iron oxide and corrosion products that were overlaid with thin deposit layers.

Aggressive agents such as chloride, sulfate, and sulfide diffused and concentrated beneath the tubercles due to the formation of ion concentration cells to several times that in the bulk water, to produce very corrosive, locally acidic environments. It is probable that failures at the tubesheet were caused by similar corrosion. Cleaning frequency of the exchangers was increased, and the use of a less aggressive water source was investigated.

Case History 11.7

Industry:	Food manufacturing
Specimen location:	Ammonia heat exchanger
Orientation of specimen:	Horizontal
Environment:	Internal: Open recirculating cooling water
Time in service:	15 years
Sample specifications:	1¼ in. (3.2 cm) outer diameter 304 stainless steel tubes

Cooling water flowed through the tubes to cool shellside ammonia. It was reported that contaminated water containing a high concentration of manganese from a permanganate addition was exposed to the waterside during an idle period following the most recent exchanger shutdown. The internal surfaces were partially covered with thin layers of powdery brown deposits that could be easily removed by wiping with a towel (Fig. 11.60). A deposit analysis indicated that the deposits had a high organic content and fairly high concentration of manganese. In some locations the soft deposit layer was removed to reveal transversely oriented iron oxide (rust) streaks that surrounded very tight pinhole pits (Fig. 11.61). The interiors of the pits were bare-surfaced. The depressions

FIGURE 11.60 Internal surface that was partially covered with thin layers of powdery brown deposits.

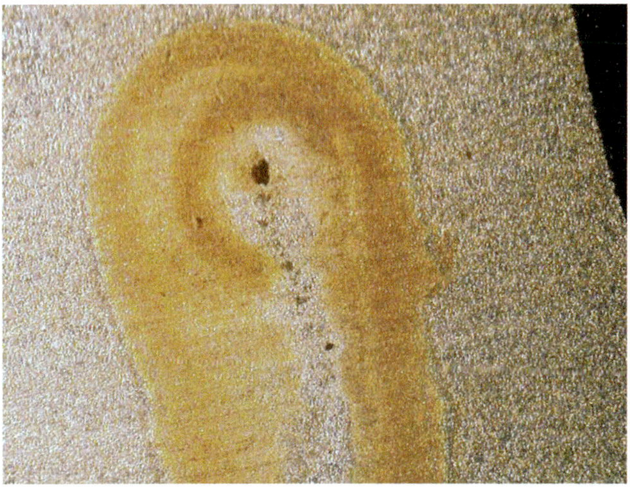

FIGURE 11.61 Soft deposit layer shown in Fig. 11.60 was removed to reveal transversely oriented iron oxide (rust) streaks that surrounded very tight pinhole pits.

may have formed in part due to the formation of ion concentration cells beneath deposit layers. Aggressive ions, such as chloride or sulfate, diffused and concentrated beneath the occlusive deposit layers in localized areas to several times that in the bulk water to form a corrosive, acidic solution.

The internal surface deposits appeared to be more like loose silt than adherent deposits, which normally form during system operation. Manganese oxide–containing material reportedly contaminated the exchanger following system shutdown.

Thus, it is probable that the internal deposits that covered the received sections came primarily from the contamination. Deposits that contain high concentrations of manganese oxide may promote pitting since they are more noble than the stainless steel, and they can promote galvanic corrosion in the form of pits at breaks in the layer. The transversely oriented rust streaks and associated lines of pits along their paths indicate that the pits formed during an idle period when water did not flow. This evidence strongly indicates that those pits formed after the contamination event.

Case History 11.8

Industry:	Commercial building
Specimen location:	Air handler coil
Orientation of specimen:	Horizontal
Environment:	Internal: Open recirculating cooling water External: Building air
Time in service:	15 years
Sample specifications:	¾ in. (1.9 cm) outer diameter copper tubes covered with aluminum fins on the external surface

Reportedly, the bundle from which the section was removed was in service for about 15 years, with little to no cooling water flow during the winter months, sometimes for weeks to months. It was also reported that cooling water leaked within the coil, prompting removal of the coil from service.

A visual inspection revealed that the internal surfaces of some tube sections within the coil were clean, bright, and unattacked, while others were covered with friable layers of deposits and corrosion products along their bottom sides (Fig. 11.62). The top sides were mostly bare and bright. In scattered locations along the bottom side beneath the deposit layers, the internal surfaces were interrupted by green corrosion product mounds that covered moderately deep to very deep depressions (Fig. 11.63). These depressions were filled with dense

FIGURE 11.62 Internal surface of tube is covered with friable layers of deposits and corrosion products along the bottom side.

FIGURE 11.63 Beneath the deposit layers shown in Fig. 11.62, the internal surfaces are interrupted by green corrosion product mounds that cover moderately deep to very deep depressions.

copper oxides and corrosion products. The localization of the deposits to the bottom sides indicated that deposition occurred during idle periods or when cooling water flow rates were very low. Under such conditions, suspended material in the water will tend to settle. In some locations, depressions appear to have formed along apparent air/water interfaces.

In all of the locations described, the metal loss was primarily caused by underdeposit corrosion. Underdeposit corrosion resulted from the formation of differential oxygen concentration cells and ion concentration cells beneath occlusive layers of deposits.

Corrosion inhibitors are less effective when occlusive deposit layers cover surfaces, and when flow rates are low or stagnant, particularly for extended periods. When the unit is laid up for long periods of time, it is recommended that proper layup practice should be implemented. Periodic cleaning is also recommended.

Case History 11.9

Industry:	Fossil utility
Specimen location:	Main condenser near inlet water box
Orientation of specimen:	Horizontal
Environment:	Tubeside: Brackish, estuarine water (polluted), 2–3 ppm tolyltriazole residual, ferrous sulfate 1 ppm as iron for 2 months for 1 h/day, dispersant 5–8 ppm Shellside: Condensing steam
Time in service:	2 years
Sample specifications:	7/8 in. (2.2 cm) outer diameter, 90:10 cupronickel tubes

The condenser was plagued by rapid corrosion on waterside surfaces. The entire internal surface was fouled with silt and other deposits, beneath which a cuprous oxide layer was present (Fig. 11.64). Localized areas of metal loss were present beneath mounds of corrosion product. Some of these localized areas were deep enough to threaten tube integrity.

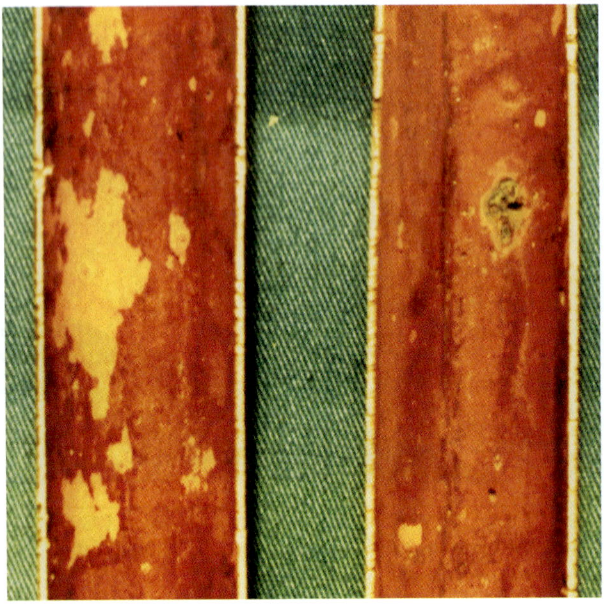

FIGURE 11.64 Thick, red cuprous oxide layer covering internal surface of 90:10 cupronickel condenser tube. Note the green mounds marking sites of localized wastage.

Ferrous sulfate was added to the condenser in hopes of controlling corrosion. A thick, tan deposit layer rapidly formed on tubes near inlets (Fig. 11.65). Corrosion continued unabated. Underdeposit corrosion caused localized areas of metal loss (Fig. 11.66). Corrosion product mounds contained up to 10 percent chloride.

FIGURE 11.65 Internal surface of condenser tube as shown in Fig. 11.64.

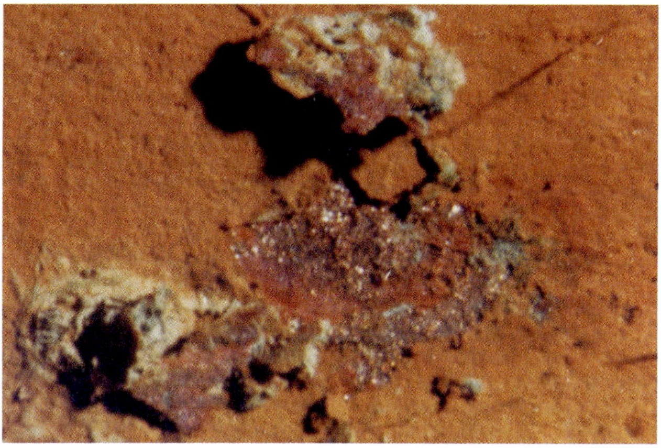

FIGURE 11.66 Same condenser tube corrosion as in Figs. 11.64 and 11.65. Corrosion product mound broken open to reveal sparkling cuprous oxide crystals.

More frequent manual cleaning of tubes was suggested as on-line sponge-ball cleaning did not appear effective. Higher concentrations of tolyltriazole were fed with no discernible effect. The use of a more corrosion-resistant tube alloy was recommended.

References

1. Teflon is a registered trademark of E. I. du Pont de Nemours & Company.
2. Chemical treatment programs that can be used to control deposition and corrosion are described in more detail in D. Flynn (ed.), *The Nalco Water Handbook*, 3d ed., McGraw-Hill, 2009.

CHAPTER 12

Oxygen Corrosion

General Description

Iron and Steel

Oxygen corrosion of carbon and low-alloy steels and low-chromium cast iron alloys results from exposure to water containing dissolved oxygen. Below 11 percent chromium in these alloys, corrosion resistance to oxygenated water is almost the same as in unalloyed iron and steel. Oxygen reacts with iron in the steel to form iron oxide corrosion products.

The major cathodic and anodic reactions in oxygenated water with slightly alkaline pH are shown in Reactions (12.1) and (12.2). The cathodic reaction (12.2) is known as oxygen reduction.

$$\text{Anode: Fe} \rightarrow \text{Fe}^{2+} + 2\,e^- \tag{12.1}$$

$$\text{Cathode: H}_2\text{O} + ½\,\text{O}_2 + 2\,e^- \rightarrow 2\,\text{OH}^- \tag{12.2}$$

The initial corrosion product that forms during oxygen corrosion is ferrous hydroxide, shown in Reaction (12.3).

$$\text{Fe} + \text{H}_2\text{O} + ½\,\text{O}_2 \rightarrow \text{Fe(OH)}_2 \tag{12.3}$$

The ferrous hydroxide is rapidly oxidized to ferric hydroxide in oxygenated waters [Reaction (12.4)]:

$$\text{Fe (OH)}_2 + ½\,\text{H}_2\text{O} + ¼\,\text{O}_2 \rightarrow \text{Fe (OH)}_3 \tag{12.4}$$

A schematic showing how the reactions occur on the metal surface is also shown in Fig. 12.1.

The corrosion rate is controlled mainly by cathodic reaction rates. Cathodic reaction (12.2) is usually much slower than anodic reaction (12.1). The slower reaction controls the corrosion rate.

If dissolved oxygen concentration is high in the water exposed to the metal surface, the driving force for Reaction (12.2) increases. Reaction (12.2) is controlled primarily by the diffusion of oxygen to

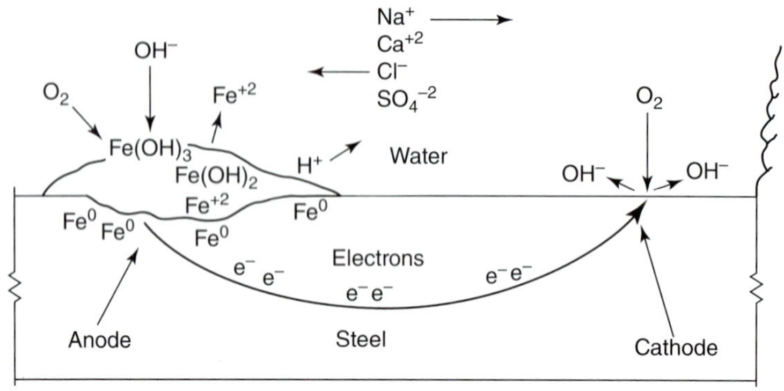

FIGURE 12.1 Schematic of oxygen corrosion reactions that occur on carbon steel surfaces.

the corroding surface. The corrosion rate of carbon and low-alloy steels, and low-chromium cast iron alloys, is nearly proportional to the concentration of dissolved oxygen at concentrations up to several parts per million (Fig. 12.2). For oxygen corrosion to become appreciable

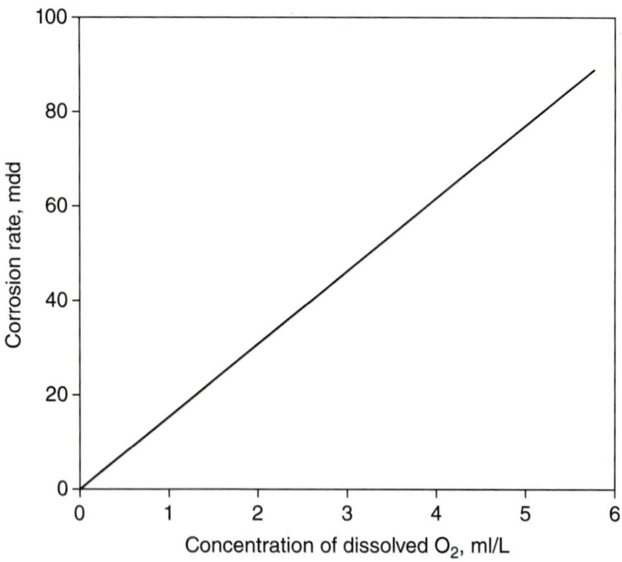

FIGURE 12.2 Effect of oxygen concentration on corrosion of mild steel in slowly moving water containing 165 ppm $CaCl_2$; 48-hour test, 250°C. (*Courtesy of H. H. Uhlig, D. N. Triadis, and M. Stern, "Effect of Oxygen, Chlorides, and Calcium Ion on Corrosion Inhibition of Iron by Polyphosphates,"* J. Electrochem. Soc. *102, p. 60 (1955). Reprinted with permission by The Electrochemical Society, Inc.*)

requires exposure of metal surfaces to oxygenated water for prolonged periods. Such conditions are met in a variety of cooling water applications.

When a bare metal surface of a carbon, low-alloy steel, or low-chromium cast iron alloy component is exposed to oxygenated water, iron oxide layers will form quickly. Wastage is usually general if oxygen concentrations remain near saturation at room temperature. Surfaces have irregular, undulating contours formed by shallow, intersecting areas of metal loss (Fig. 12.3). Corrosion rates are initially high and decrease rapidly while iron oxide layers form and thicken. The accumulated oxide layers retard diffusion of oxygen to the metal surface. Thus, Reaction (12.2) slows. Eventually, nearly steady-state corrosion is achieved.

In open recirculating cooling systems, dissolved oxygen is replenished. However, the dissolved oxygen concentration may decrease over time within a tight, closed system. As dissolved oxygen concentration decreases through consumption by corrosion and by aerobic bacteria, Reaction (12.2) slows to reduce corrosion rates. The corrosion rate of carbon and low-alloy steels in neutral pH deaerated water at room temperature is very low, less than 0.2 mil/y (0.005 mm/y). In addition, any process that impedes oxygen diffusion, such as accumulation of corrosion product and deposit layers, slows reaction (12.2), ultimately reducing corrosion rates.

Ferrous hydroxide [$Fe(OH)_2$] and ferric hydroxide [$Fe(OH)_3$] usually contain waters of hydration. A layer of hydrated magnetite typically forms between the ferric and ferrous hydroxides. The threefold layer of corrosion products is sometimes called *rust.* The bulk of rust is the ferric hydroxide layer. The topmost layer is orange and brown, whereas the underlying layers are usually black.

FIGURE 12.3 Generally corroded steel surface.

Figure 12.4 Uniform, stratified iron oxide layers covering surface that experienced oxygen corrosion.

During the initial stages of oxygen corrosion, uniform layers of iron oxide corrosion products are produced (Fig. 12.4). Where uniform, protective, cathodic iron oxide corrosion product layers crack or peel away from the metal surface in localized areas, subsequent corrosion that occurs in such locations may produce nodular mounds of iron oxide corrosion products that are called tubercles (Fig. 12.5). Regions adjacent to tubercles may remain smooth and more shallowly wasted. The outer layers shield the underlying metal surface,

Figure 12.5 Emergency service water supply line shows moderate tuberculation.

Figure 12.6 Utility service water system pipe almost plugged by heavy tuberculation.

so dissolved oxygen concentration decreases in the water beneath the layer. Thus, differential oxygen concentration cells form and more reduced forms of oxide are produced beneath the tubercle cap.

Tubercles can eventually grow to the point that they severely reduce flow, or even plug piping and tubing, leading to increased pumping costs and reduced supply of cooling water to downstream components (Fig. 12.6). Porous tubercles covering heat exchanger tubes will also substantially reduce heat transfer. Tubercles are much more than amorphous lumps of corrosion product and deposit. They are highly structured. Structure and growth are interrelated in complex ways.

All tubercles have five structural features in common:

1. Outer crust
2. Inner shell
3. Core material
4. Fluid-filled cavity
5. Corroded floor

These features and their associated compounds are shown in Fig. 12.7. Typical reactions and their locations appear in Fig. 12.8.

Outer Crust

A friable outer crust forms atop the tubercle. The crust is composed of ferric hydroxide and various other components from the water, including carbonates, silicates, other precipitates, settled particulate,

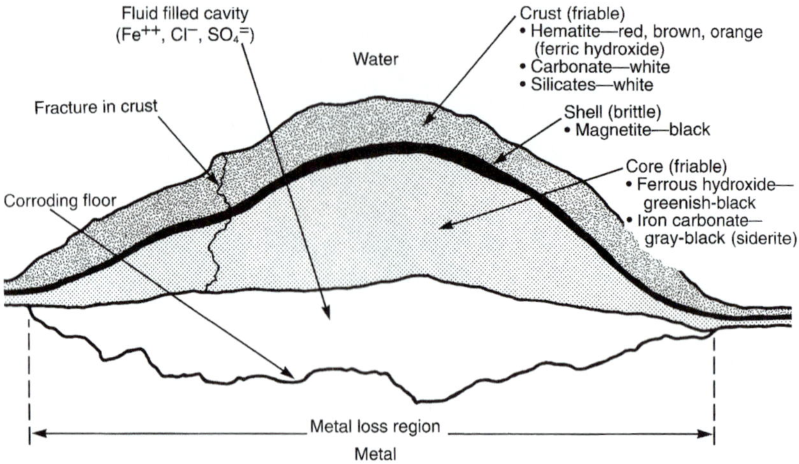

FIGURE 12.7 Well-developed tubercle shows chemical compounds and structure. (*Courtesy of National Association of Corrosion Engineers, Corrosion '91 Paper No. 84 by H. M. Herro.*)

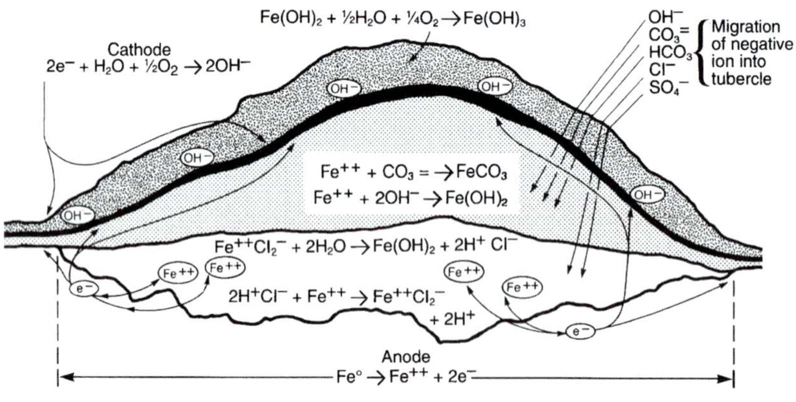

FIGURE 12.8 Chemical reactions within the tubercle shown in Fig. 12.7. (*Courtesy of National Association of Corrosion Engineers, Corrosion '91 Paper No. 84 by H. M. Herro.*)

and detritus. Components of treatment chemicals may also deposit preferentially atop tubercles in response to associated corrosion. Ferrous ions and ferrous hydroxide generated within the tubercle diffuse outward through fissures, where they encounter dissolved oxygen. Ferric hydroxide is produced and precipitates atop the tubercle.

Inner Shell

Just beneath the outer crust a dense, brittle, black magnetite shell develops. The shell separates the region of high dissolved oxygen concentration outside the tubercle from the very low dissolved oxygen regions in the core and fluid-filled cavity below. The difference in dissolved oxygen concentration promotes corrosion due to the formation of differential oxygen concentration cells (see Chap. 11). The shell is mostly magnetite and thus has relatively high electrical conductivity, which promotes electron migration. Electrons generated at the corroding floor are transferred to regions around the tubercle and to the shell, where cathodic reactions produce hydroxyl ions, locally increasing pH. Dissolved compounds with normal pH solubility, such as carbonate, deposit preferentially atop the shell where pH is elevated. A cross section of the outer crust, inner magnetite shell, and core material is shown in Fig. 12.9.

Core

Friable core material is present beneath the magnetite shell (Fig. 12.9). The core consists mostly of ferrous hydroxide formed by Reaction (12.3).

The hydroxyl ions migrate inward, attracted by the positive charge that is produced by the ferrous ions generated near the corroding surface (Fig. 12.8). Other anions such as carbonate, chloride, and sulfate also concentrate beneath the shell due to ion concentration cell mechanisms (see Chap. 11). Carbonate may react with ferrous ions to form siderite ($FeCO_3$) as in Reaction (12.5) (Fig. 12.10).

$$Fe^{2+} + CO_3^{2-} \rightarrow FeCO_3 \tag{12.5}$$

Figure 12.9 Tubercle cross section shows black magnetite-rich shell beneath deposit and hematite cap. Note the core material below the magnetite shell (Magnification: 7.5×). (*Courtesy of National Association of Corrosion Engineers, Corrosion '89 Paper No. 197 by H. M. Herro.*)

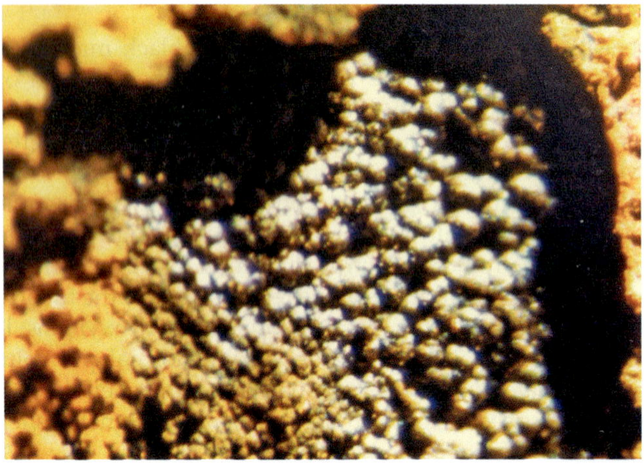

FIGURE 12.10 Siderite ($FeCO_3$) nodules inside a tubercle revealed by breaking open the magnetite shell. (*Courtesy of National Association of Corrosion Engineers, Corrosion '89 Paper No. 197 by H. M. Herro.*)

Cavity

A fluid-filled cavity is sometimes present beneath the core (Fig. 12.7). The cavity may be large (Fig. 12.11) or small (Fig. 12.12). The cavity may result, in part, from acidic conditions that develop internally. Acidic conditions may prevent precipitation of oxides and hydroxides inside the tubercle.

FIGURE 12.11 Large tubercle broken open to reveal the internal cavity. The cavity is filled with fluid in service.

FIGURE 12.12 The steel surface is at the bottom, and the red material is an epoxy mounting medium. Note the small, shallow central cavity.

Floor

The corroded tubercle floor is usually in the form of a depression that is much wider than it is deep (Figs. 12.13 and 12.14). The width of the metal loss almost exactly matches the tubercle mound width. Iron dissolves at the floor, forming ferrous ions according to Reaction (12.1).

If chlorides diffuse beneath the tubercle caps to sufficient concentrations due to the formation of ion concentration cells, acidic conditions will be produced due to hydrolysis and will cause corrosion.

Floor–core boundary:

$$Fe^{2+} + 2(Cl^-) + 2HO \rightarrow Fe(OH)_2 + 2(H^+C1^-) \tag{12.6}$$

FIGURE 12.13 Note the very shallow, broad depression beneath the tubercle. (*Courtesy of National Association of Corrosion Engineers, Corrosion '91 Paper No. 84 by H. M. Herro.*)

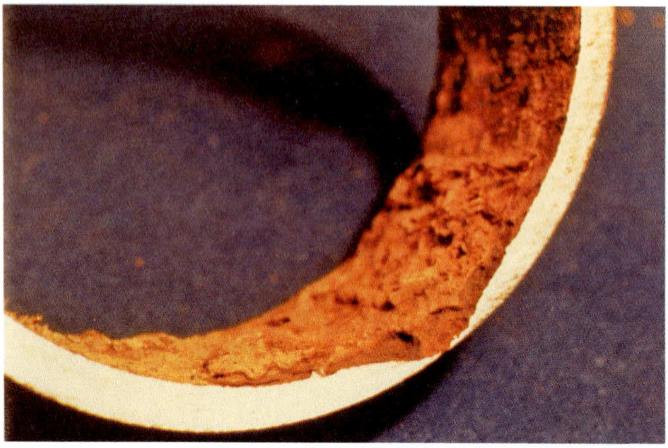

FIGURE 12.14 Perforation at a dish-shaped depression on the internal surface of a large-diameter steel pipe. A large tubercle capped the depression but was dislodged during pipe sectioning. (*Courtesy of National Association of Corrosion Engineers, Corrosion '89 Paper No. 197 by H. M. Herro.*)

Near floor:

$$2(H^+Cl^-) + Fe^{2+} \rightarrow FeCl_2 + 2H^+ \quad (12.7)$$

Similarly, sulfate may accelerate attack by depressing internal pH.

At sufficiently low pH, the concentration of hydrogen ions in the water is great enough that the following hydrogen reduction reaction (12.8) predominates at the cathode.

$$2H^+ + 2e^- \rightarrow H_2 \quad (12.8)$$

If the pH of the water within the tubercle is depressed sufficiently, the hydrogen reduction reaction is favored over oxygen reduction, and corrosion accelerates.

As a result of the concentration of anions such as chloride and sulfate, it is common for material scraped from the inside of tubercles to be acidic when mixed with water. Acidity varies not only from tubercle to tubercle but also from place to place in a given tubercle. Acidity is greatest near the corroded metal surface. The size of the fluid-filled cavity can indicate acidity. More acidic internal environments tend to produce larger cavities.

Numerous small, shallow parallel grooves (striations) will form on the corroded floor (Fig. 12.15). Undercutting is very rare. The striations are caused by preferential corrosion along microstructural features such as inclusion stringers and pearlite bands. The corroded floor usually is covered with porous, friable corrosion products containing ferrous

Figure 12.15 Striated, corroded depressions are revealed on a large-diameter pipe after tubercles were removed. Striated surfaces are caused by preferential corrosion along microstructural irregularities in the steel. Such attack can be caused by low-pH conditions.

hydroxide, which may form in place by conversion of the steel surface. If acidity is significant, the corrosion product layer lining the floor will be thin.

When first exposed to air, tubercles are often quite dark both externally and internally (Figs. 12.16 and 12.17). After prolonged exposure to air, however, tubercle color changes to red, brown, and orange (Figs. 12.18 and 12.19). The more reduced ferrous hydroxide is converted to ferric hydroxide. A very rapid conversion to lighter colors sometimes signifies acidic conditions within the cavity.

Figure 12.16 Tubercle only a few minutes after removal from a water-filled pipe.

FIGURE 12.17 Tubercle in Fig. 12.16 opened to show the dark interior.

FIGURE 12.18 Tubercle in Figs. 12.16 and 12.17 after several hours of exposure to air. Note the reddening.

General Tubercle Features

When tubercles dry, the waters of hydration are removed; this can cause them to become weak and brittle. As such, they can be crushed by gentle pressure with a finger. Therefore, when a system is restarted following long-term dry layup, tubercles may break away from surfaces to be swept downstream, and they may significantly occlude or plug narrow piping or orifices. Tubercle caps can be dislodged whole with a hard implement (Figs. 12.20–12.22). The physical strength of the tubercle is related to the thickness of the magnetite

FIGURE 12.19 Tubercle in Fig. 12.18 opened to show color change of interior.

FIGURE 12.20 Knife dislodging a dry tubercle from a steel pipe surface.

FIGURE 12.21 Tubercle in Fig. 12.20 with cap completely dislodged.

FIGURE 12.22 Intact tubercle from Figs. 12.20 and 12.21 showing physical integrity after removal from surface.

shell and densities of crust and core material. A typical magnetite shell in cross section is shown in Fig. 12.9. Dense, thick shells harden structures. These harder, denser tubercles have usually grown at a slower rate than lower-density tubercles. Thin magnetite shells are usually indicative of fast growth. Thus, thin magnetite shells indicate an undesirable condition.

Multiple magnetite shells may form by successive fracture. Ferrous species spew out of the fractured shell and are quickly oxidized to form a new ferric hydroxide crust. Beneath the new crust, another magnetite shell forms (Fig. 12.23). As tubercles age, their internal structure and outer morphology are altered.

Tubercles may become very large [up to 6-in. (15 cm) diameter structures have been reported]. The volume of metal loss beneath tubercles is usually much smaller than the accumulated volume of the tubercles themselves. This is due to the typical substantial amount of porosity within the cavities of tubercles. Tubercle height may be

FIGURE 12.23 Multiple magnetite shells in a small tubercle. Multiple shells form due to successive fracture during growth.

from 5 to 30 times as great as the metal loss depth below. Tubercles typically only form on surfaces that are submerged in water.

Atmospheric Oxygen Corrosion

Uniform iron oxide layers tend to form in misted areas (Figs. 12.24 and 12.25). Stratification of corrosion products is common under intermittent wetting and drying conditions.

FIGURE 12.24 Atmospheric oxygen corrosion on structural beam. Uniform iron oxide layers cover the corroded surfaces.

FIGURE 12.25 Close-up of stratified iron oxide layers.

Other Alloys

Copper Alloys

Copper is not ordinarily corroded in water unless dissolved oxygen is present. However, dissolved oxygen in water has a dual role in corrosion of copper. It stimulates attack by an interaction at the cathode, and it retards corrosion by producing a protective corrosion-product layer. In nearly pure aerated water, a thin, protective layer of cuprous oxide and cupric hydroxide forms. Oxygen must diffuse through the film for corrosion to occur. However, corrosion will not be appreciable if exposure is to dissolved oxygen alone. It is necessary for significant concentrations of other aggressive agents to be present in the water along with dissolved oxygen for significant corrosion to occur. Such agents may include dissolved carbon dioxide, acids, chlorides, sulfates, ammonia, sulfur dioxide, sulfides, and oxidizing species, including halogens such as chlorine and bromine.

Aluminum

The corrosion resistance of aluminum relies upon the presence of protective oxide layers on the metal surface. Passive aluminum oxide layers are produced on bare aluminum alloys due to exposure to oxygenated waters at neutral pH. The corrosion resistance of aluminum is reduced in deaerated solutions. Under reducing conditions and exposure to aggressive anions, particularly chloride, the oxide layer may be compromised.

Stainless Steels

Dissolved oxygen in water solution combines with chromium in stainless steels to form a highly adherent and protective oxide film that gives stainless steels their typically high resistance to corrosion. Therefore, stainless steels are generally not susceptible to oxygen corrosion per se. Sensitized stainless steel may experience oxygen corrosion and tuberculation, but such cases are rare. Sensitized stainless steel is described in more detail in Chap. 18, "Intergranular Corrosion," and Chap. 26, "Weld Defects." The corrosion resistance of stainless steel is reduced in deaerated solutions. Under reducing conditions, and by exposure to aggressive anions, particularly chloride, the oxide film may be compromised.

Zinc and Galvanized Steel

Galvanized steel is simply carbon steel that is coated, either by plating or hot dipping, with a thin layer of zinc. In cooling water systems, the use of galvanized steel panels and piping is common. Protective zinc oxides and hydroxides and basic zinc carbonate corrosion product layers form when zinc is immersed in oxygenated water. Corrosion is usually general if oxygen supply is adequate. Localized attack is favored when oxygen concentration falls below saturation levels. Thus, zinc is susceptible to attack from oxygen concentration cells. Shielded areas or areas depleted in oxygen concentration tend to corrode, forming voluminous, white, friable corrosion products. Once the zinc layer is breached, the underlying steel becomes susceptible to oxygen corrosion (Figs. 12.26 and 12.27). Additional information about corrosion of zinc and galvanized steel are presented in Chaps. 15 and 16.

FIGURE 12.26 Severely corroded galvanized steel pipe supplying water to a cooling tower fan-bearing system. Where the galvanized metal has been consumed, only a brown rust patch is visible.

FIGURE 12.27 The steel pipe from Fig. 12.26 in cross section, showing severe wastage where the galvanized layer is absent.

Locations

Oxygen corrosion and tubercle formation only occur on unprotected carbon and low-alloy steel, galvanized steel that experiences loss of the protective zinc layer, and on low-chromium cast iron surfaces exposed to oxygenated waters. Coating layers that are breached will expose the underlying metal to corrosive environments. Metal loss and associated iron oxide corrosion product formation will occur generally across the entire exposed surface. The corrosion product layers formed by oxygen corrosion are a precursor to tubercle development.

Carbon steel heat exchangers, cooling water distribution and service water system piping, storage tanks, standpipes, fire protection systems, galvanized steel cooling tower decks and basins, cast iron pump and engine components, and virtually all nonstainless ferrous components are susceptible to oxygen corrosion when exposed to oxygenated water.

Surfaces must be submerged in oxygenated water for extended periods in order for tubercles to form and grow. Deposits stimulate tubercle formation. Hence, regions where foreign material accumulates are common tubercle breeding grounds. Locations where flow rates are low or stagnant, such as deadlegs, also promote tubercle growth. For the same reason, oxygen corrosion and tubercle formation and growth occur readily during extended idle periods.

The relationship between flow and tubercle growth is dichotomous. Low or stagnant flow of oxygenated water may stimulate oxygen corrosion and tubercle growth. However, if the water is completely depleted of dissolved oxygen, oxygen corrosion will cease. Tubercles that form under conditions of significant flow can be elongated along the flow direction (Fig. 12.28). However, if flow rates

FIGURE 12.28 Tubercles elongated by flow in a mill water supply line.

are very high, turbulence may dislodge incipient tubercles. Thus, pump impellers and other components experiencing severe turbulence almost never show tubercular growth unless they have been out of service for an extended period.

Identification

Chemical analysis using x-ray fluorescence or energy dispersive spectroscopy using an electron microscope will be useful in identifying the material that covers a corroded surface as iron based; x-ray diffraction may be able to identify various oxides and hydroxides that are typically associated with oxygen corrosion. However, it is important to distinguish whether the material is in the form of corrosion products that formed in place due to corrosion at the site, or if it is deposited material that was transported from sources upstream. Corrosion products that form in place will have definite structure as dense layers that may be stratified as in Fig. 12.4. The stratification of the iron oxide layers indicates that oxygen corrosion occurred during many separate episodes.

It is important to conduct a metallographic examination through a surface that has experienced corrosion. Metallographic analysis of a surface that has experienced oxygen corrosion may reveal burrowing depressions that are filled with dense iron oxide, as shown in Fig. 12.29. In addition, the iron oxide may form in multiple, distinct, stratified layers as shown in (Fig. 12.30). Iron oxide layers may also contain remnants of pearlite colonies or carbide from the original steel microstructure (Fig. 12.31). The presence of such remnants provides definite proof that the iron oxide formed in place due to oxygen corrosion.

FIGURE 12.29 Metallographic cross section of burrowing depressions that are filled with dense iron oxide.

FIGURE 12.30 Iron oxide corrosion products formed in multiple, distinct, stratified layers.

Remnant carbides are often suggestive of more recent oxygen corrosion due to the relatively slow process of converting iron carbide to iron oxide, as compared to the more rapid oxidation of the ferrite phase. An absence of remnant carbides suggests older attack.

A tubercle will contain a distinct, dense outer shell of iron-based corrosion products that will be in the form of a raised mound, as shown in Fig. 12.12. It is common for porous material to be present inside the cavity of a tubercle due to localized acidic conditions that

FIGURE 12.31 Iron oxide layers containing remnants of pearlite colonies from the original steel microstructure.

may be produced beneath the tubercle cap. Metallographic examination may also reveal cavities above the corroded floor. This indicates that locally acidic conditions developed beneath the tubercle cap.

Iron oxide deposits that form when material is transported from a source upstream will not have the distinct structures of iron oxide layers and tubercles that form in place due to oxygen corrosion. Deposited iron oxide will be in the form of an agglomeration of granular, nonstructured, highly porous material as shown in Fig. 12.32.

FIGURE 12.32 Agglomerated granular, nonstructured, highly porous iron oxide deposits.

FIGURE 12.33 Atmospheric oxygen corrosion that occurs on surfaces following drainage, also referred to as "flash corrosion," produced orange and brown, shiny iron oxide layers.

Atmospheric oxygen corrosion that occurs on surfaces following drainage, also referred to as "flash corrosion," tends to produce orange and brown, shiny iron oxide layers (Fig. 12.33). Metal loss due to flash corrosion tends to be shallow due to short-term exposure to residual oxygenated water prior to evaporation. In some cases oxygen corrosion that occurs during idle periods may be substantial on low-lying surfaces, such as on the bottom sides of horizontally oriented piping and tubing. Metal loss also may be substantial below the interfaces of shallow pools of water that repeatedly collect during idle periods.

Critical Factors

Water Chemistry

The most important factor contributing to oxygen corrosion and tubercle growth is dissolved oxygen concentration. Differential oxygen concentration between the bulk water and in the water beneath occlusive iron oxide layers, tubercles, and deposits stimulates localized corrosion (see Chap. 11, "Crevice Corrosion"). Local corrosion rates may be much higher than general corrosion rates due to unfavorable anode-to-cathode area ratios established by differential oxygen concentration. This can result in localized corrosion.

At zero dissolved oxygen concentration in the bulk water, oxygen corrosion and tubercle growth associated with differential oxygen concentration cell activity ceases.

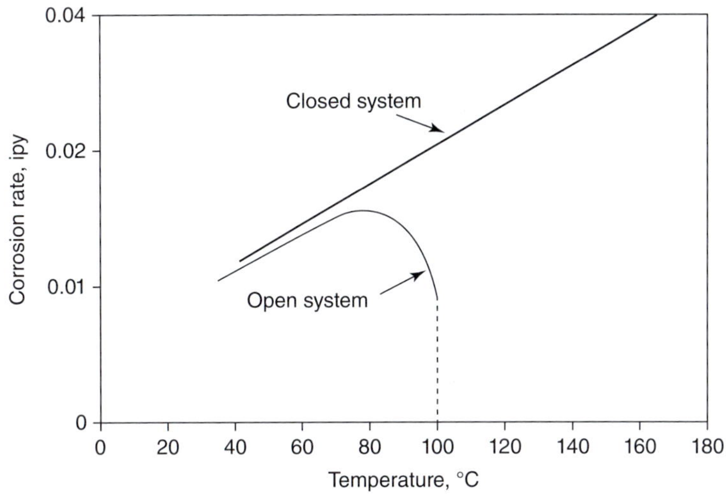

FIGURE 12.34 Effect of temperature on corrosion of iron in water containing dissolved oxygen. (*Courtesy of McGraw-Hill Inc.,* Corrosion, Causes and Prevention, *by F. Speller, p. 168, McGraw-Hill, 1951.*)

Corrosion rate is nearly proportional at temperatures up to about 180°F (80°C) in systems open to the air. Although reaction rates increase with temperature, dissolved oxygen is driven from solution as temperatures increase. As temperatures approach boiling, corrosion rates fall to very low values, since dissolved-oxygen concentration also decreases as water temperature rises (Fig. 12.34).

Soft waters with high bicarbonate alkalinity stimulate tubercle formation, as do high concentrations of sulfate, chloride, and other aggressive anions. High concentrations of aggressive anions will promote the formation of locally acidic conditions due to the production of ion concentration cells as tubercles develop. When water pH is 4 to 10 at room temperature, oxygen corrosion rates of carbon steel are nearly constant (Fig. 12.35). Below a pH of 4, protective corrosion products are dissolved. A bare carbon steel surface contacts the acidic water, and the corrosion rate increases sharply. Above a pH of about 10 at room temperature, surface alkalinity increases and corrosion rates fall sharply.

Alloy Effect

For carbon and low-alloy steels and low-chromium cast irons, differences in alloy and carbon concentration, heat treatment, and mechanical forming usually produce only small differences in corrosion rate in a pH range of 4 to 10.

Biological Interactions

It is a common belief that the presence of tubercles on surfaces in cooling water systems provides evidence of microbiological activity or

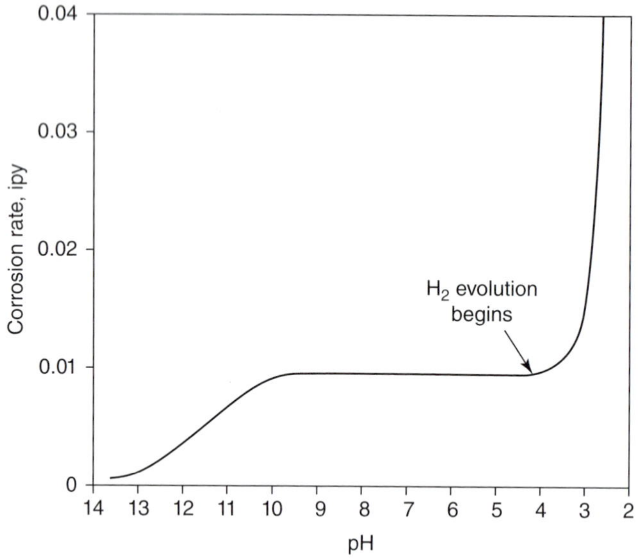

FIGURE 12.35 Effect of pH on corrosion of iron in aerated water at room temperature. (*Reprinted from* Industrial & Engineering Chemistry *16, 665, 1924, W. Whitman, R. Russell, and V. Altieri, published by the American Chemical Society.*)

even microbiologically influenced corrosion. This assumption is certainly not absolute. For instance, tubercles may form on certain boiler components when exposed to waters containing excessive dissolved oxygen concentration during high-temperature service. Bacteria that may grow in cooling water systems will not survive at the operating temperatures in a boiler. This reveals that microorganisms are not required for tubercle formation and growth. However, it is common for microorganisms, or even macroorganisms, to colonize tubercles in cooling water systems, though they are not required for tubercle formation and growth. The degree to which such organisms influence tubercle development is variable (see Chap. 13, "Microbiologically Influenced Corrosion"). The oxygen-depleted regions beneath tubercle caps provide environments that may be suitable for the growth of anaerobic bacteria such as sulfate-reducing bacteria. Sulfate-reducing and acid-producing bacteria may contribute to corrosion within tubercles to stimulate their growth. In-depth analysis and assessment of system conditions is necessary in order to determine the contribution of microorganisms to corrosion and tubercle growth.

Flow

Tubercle formation is promoted by both high- and low-flow conditions. Flow directly influences tubercle morphology. Tubercles will elongate in the direction of the flow (Fig. 12.28). Flow also affects growth by replenishing dissolved oxygen, aggressive anions, chemical inhibitors, and suspended particulates. If flow is very high, turbulence will dislodge tubercle structures.

Corrosion Control

Chemical Treatment

Proper chemical treatment may reduce general corrosion rates associated with oxygen corrosion and tuberculation on carbon and low-alloy steels, and low-chromium cast irons. The exact treatment program that will be most effective is dependent upon the type of cooling water system, its design, water chemistry, and operation. Water chemistries and operating practices may differ widely even among similar industries using similar equipment. Chemical treatment programs that are effective for the control of oxygen corrosion and tuberculation in cooling water systems most commonly employ corrosion inhibitors as the primary means of corrosion control. Chemical corrosion inhibitors, when maintained in the water at proper concentrations, reduce corrosion by affecting cathodic and/or anodic processes. A wide variety of corrosion inhibitors may be used.

Anodic corrosion inhibitors interfere with the anodic reaction by promoting the formation of protective films at anodic sites on carbon and low-alloy steel and low-chromium cast iron surfaces. Some common anodic corrosion inhibitors include nitrites, molybdates, orthophosphates, and phosphonates. These inhibitors either promote the formation of protective iron oxide layers or produce precipitates that improve the resistance of the oxide layer. Chromates are very effective anodic inhibitors; they were commonly used in the past, but they have been essentially banned from use in most countries due to environmental concerns. If anodic inhibitors are used alone, it is vital that the concentration be sufficient, or the lack of protection at anodic sites may produce severe localized corrosion or pitting.

Some common cathodic corrosion inhibitors, which promote the formation of protective films at cathodic sites on the metal surfaces, include polyphosphates and zinc. Cathodic inhibitors form films that limit the transport of dissolved oxygen to prevent oxygen reduction and electron transfer, which are necessary for corrosion reactions to occur. It is common practice to use a combination of both anodic and cathodic corrosion inhibitors in open recirculating cooling water systems to provide the most effective protection.

The effectiveness of the program will depend upon water chemistry parameters such as pH, alkalinity, calcium hardness, and concentration of aggressive anions such as chloride and sulfate. Temperature and flow rates also affect corrosion inhibition. It is common for suitable dispersant chemicals to be used along with the corrosion inhibitors in order to stabilize dissolved calcium in the water.

This controls the deposition of calcium salts and prevents the loss of phosphate-based inhibitors due to precipitation. The use of dispersants is highly recommended in systems containing silt, sand, oil, grease, biological material, and/or other foreign material. Not only does increased dispersion generally increase the effectiveness of chemical inhibition, it also prevents the nucleation of differential oxygen concentration cells beneath deposits.[1]

In addition, corrosion inhibitors are less effective when flow rates are low or stagnant. This is due to limited or possible insufficient replenishment of corrosion inhibitors to the surfaces. In addition, low or stagnant flow tends to promote deposition of suspended solid materials. No matter what chemical treatment is used, it will almost always be more effective on an essentially clean metal surface. Corrosion inhibitors will be less effective when thick iron oxide tubercles cover surfaces. Large tubercles occupy a much larger surface area than the surface area of the metal beneath. Therefore, there is much more surface area to protect with inhibitors, and more cracks to repair in the tubercle shells. Tubercles, with rare exceptions, grow more rapidly when young than when old. Older tubercles may have slow growth rates, and may even experience dormant periods. Nevertheless, experience has shown that removal of tubercular encrustations is beneficial when chemically treating a system. There are numerous methods of removal, including water blasting, mechanical abrasion, air rumbling and, most effective of all, chemical cleaning.

Oxygen corrosion during idle periods may be controlled through the use of proper layup procedures. In closed systems that employ wet layup practice, the use of very high concentrations of appropriate corrosion inhibitors, such as nitrite and molybdate, have produced excellent results. Successful dry layup practice involves removal of liquid water and/or the use of water repellent soluble oil coating treatments or filming amines.

Altering System Operation

Because of the diversity of cooling water systems, only general guidelines for altering system operation can be given.

Flow is very important in controlling tuberculation. It is usually best to maintain a flow of at least 3 ft/s (1 m/s) to allow treatment chemicals to provide effective protection. Increased flow also retards fouling and thus reduces tubercle initiation sites. If water flow is stagnant, dissolved oxygen concentration in regions far from air-liquid boundaries would eventually fall to zero, and oxygen concentration cell activity would cease. However, it is *not* recommended that flow be suspended in any cooling water system, since attendant corrosion, fouling, and biological activity may make tuberculation problems seem tame by comparison. Low flow [less than 3 ft/s (1 m/s)] can produce serious tubercle growth in a short time.

Material Substitution

Certain conditions, ultimately dictated by economics, make the substitution of more resistant materials a wise choice. Stainless steels and corrosion-resistant alloys do not experience oxygen corrosion or the formation of tubercles in oxygenated water, as long as they are not in a sensitized condition. Copper alloys and aluminum also do not experience oxygen corrosion. However, each of these alloys may suffer other problems that would preclude their use in a specific environment.

A frequently useful and more economical solution to oxygen corrosion and tuberculation involves the use of protective coatings. Galvanized sheet steel is immune to tubercle growth until the zinc-rich layer is consumed. Epoxy and other field- and factory-applied coatings have been used extensively in cooling water systems to retard corrosion. Most utility and larger condenser water boxes are coated with a protective layer. Application skill is critical to good service. Sharp corners, threads, edges, crevices, and projections are troublesome areas to coat effectively.

Cautions

Tubercles grow at about the same rate in large- and small-diameter piping used in the same system when flow rates are similar. Small-diameter pipes are likely to plug with tubercles before perforation. The reverse is true of large-diameter pipes. Hence, replacement or cleaning of plugged small-diameter pipe may be necessary, but large-diameter pipes should not be ignored in a troubled system. Nondestructive techniques such as ultrasonic testing can be useful for assessing pipe wall thickness losses.

Related Problems

See also Chap. 11, "Crevice and Underdeposit Corrosion (Concentration Cells)"; Chap. 13, "Biologically Influenced Corrosion"; and Chap. 19, "Graphitic Corrosion."

Case History 12.1

Industry:	Steel
Specimen location:	Continuous caster spray chamber
Specimen orientation:	Horizontal and vertical
Environment:	Open recirculating cooling water spray for secondary cooling of steel slabs
Time in service:	10 years
Sample specifications:	Carbon steel structural beams, ½ in. (12.7 mm) cross-sectional thickness

FIGURE 12.36 The original protective organic coating covering structural beams degraded after about 2 to 3 years of service. The underlying steel beam surfaces were exposed to the spray chamber environment for an extended period of time, allowing substantial atmospheric oxygen corrosion to occur.

Beams installed in the spray chamber for the continuous caster were originally covered by an organic coating that was applied to provide corrosion protection. The original coating degraded after about 2 to 3 years of service, causing it to become brittle (Fig. 12.36). The brittle nature of the coating allowed numerous cracks to form, which exposed the underlying steel beam surfaces to the spray chamber environment for an extended period of time. The spray chamber environment consisted of cascading and rebounded spray water and condensed steam that issued from direct spray water cooling of the steel slab. Atmospheric oxygen corrosion occurred on the exposed steel beam surfaces, forming thick, stratified layers of dense, black and brown iron oxide corrosion products (Fig. 12.37). Substantial beam thickness loss occurred, including perforations of beam webs. The metal loss compromised the strength of the beams and required repairs and replacement. The stratification of the iron oxide layers indicated that oxygen corrosion occurred during many separate episodes that were associated with the operation of the caster. The corrosion inhibitors used in the water largely deposited onto the hot slab along with other dissolved solids. Consequently, they were essentially unavailable to protect the steel structural components in the spray chamber. Water flow was also highly variable and was not uniformly distributed across surfaces within the spray chamber. In addition, condensed steam in the spray chamber contained no corrosion inhibitors. Essentially, corrosion inhibitors used in the spray water cannot be relied upon to control atmospheric corrosion. Atmospheric oxygen corrosion is controlled most effectively through the use of proper corrosion-resistant coatings. Installation of a more resistant coating material than was used during original construction was recommended. Proper coating installation requires suitable surface preparation and adequate curing time.

FIGURE 12.37 Thick, stratified layers of dense, black and brown iron oxide corrosion products covering beam surface.

Case History 12.2

Industry: Pharmaceutical
Specimen location: Process water supply line
Specimen orientation: Horizontal
Environment: Condensate and demineralized water containing no corrosion inhibitors
Time in service: 2 years
Sample specifications: 3 in. (76 mm) outer diameter, 1/8 in. (3.175 mm) thick galvanized steel piping

Several failures occurred throughout the line within only 2 years of service. The internal surface was covered with numerous thick, longitudinally elongated, flow-oriented iron oxide tubercles (Fig. 12.38). Removing the tubercles uncovered deep depressions. Metallographic examination revealed that the internal surface was partially covered with a layer of zinc plating (Fig. 12.39) that was interrupted by several depressions filled with dense iron oxide. Metallic copper particles were found within the corrosion product layers. The external and internal surfaces were originally covered completely by zinc galvanizing layers. Much of the galvanizing on the internal surface was lost. The presence of metallic copper particles in the corrosion products indicate that some of the zinc experienced accelerated galvanic corrosion due to the plating of copper from water solution. Zinc is very susceptible to corrosion in demineralized water.

Once the galvanizing layer on the surface was consumed, the exposed steel became more susceptible to corrosion. The deep metal loss on the underlying

FIGURE 12.38 Internal surface covered with numerous thick, longitudinally elongated, flow-oriented iron oxide tubercles.

FIGURE 12.39 Metallographic cross section showing internal surface partially covered with a layer of zinc plating that is interrupted by several depressions filled with dense iron oxide.

steel was caused by oxygen corrosion resulting from exposure to water containing high concentrations of dissolved oxygen. The longitudinal orientation of the tubercles indicated that oxygen corrosion was active during service while water flowed through the pipe. Oxygen corrosion in the piping may be controlled most effectively through the use of adequate concentrations of proper corrosion inhibitors. However, it was reported that no chemical treatment was used in the system since the water was being used to supply a process. Consequently, the use of a more resistant material, such as an appropriate grade of stainless steel, was recommended for the piping.

Case History 12.3

Industry:	Steel
Specimen location:	Basic oxygen furnace (BOF) hood
Specimen orientation:	Vertical
Environment:	Closed recirculating cooling water
Time in service:	3 years
Sample specifications:	Carbon steel tubing, 1.5 in. (38 mm) outer diameter, 0.202 in. (51.3 mm) thick tube wall, attached in parallel by welded membranes into a panel.

Periodic inspection of the internal surfaces of tubes within the panel showed that they were mostly smooth and even, and covered with thin, uniform layers of black iron oxide (Fig. 12.40). However, in scattered locations, irregularly shaped, longitudinally aligned, narrow strips of shiny, dark brown iron oxide corrosion products interrupted along the cold side of the tube. Sandblast cleaning revealed shallow metal loss with a maximum depth of 0.009 in. (0.23 mm) occurred beneath the strips of corrosion products that formed along the cold side (Fig. 12.41). The location, orientation, distribution, and nature of the corrosion products were consistent with idle time oxygen corrosion, which is also referred to as "flash corrosion." This corrosion occurred during idle periods after cooling water was drained from the hood, and a thin, narrow stream of residual cooling water dripped down the cold side. The depth of metal loss was very shallow, and did not significantly threaten tube integrity. The metal loss and corrosion products formed only on the cold side of the tube. There was no reduction of heat transfer, and no internal surface metal loss on the hot side that could promote the formation and growth of corrosion fatigue cracks. Therefore, the tube was considered to be in a serviceable condition. Even so, idle time oxygen corrosion was henceforth controlled through use of proper layup procedures.

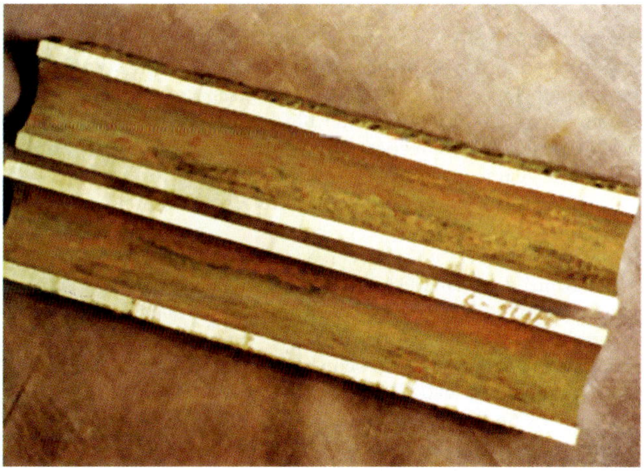

FIGURE 12.40 Internal surfaces of BOF hood panel tubes are mostly smooth and even and covered with thin, uniform layers of black iron oxide. In scattered locations, irregularly shaped, longitudinally aligned, narrow strips of shiny, dark brown iron oxide corrosion products interrupted along the cold side.

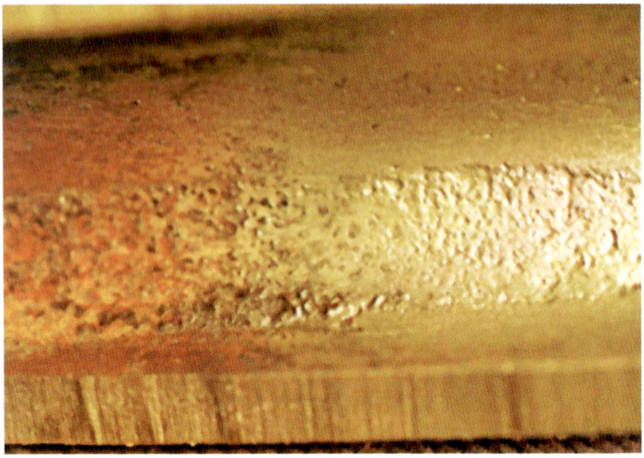

FIGURE 12.41 Sandblast cleaning revealed shallow metal loss beneath the strips of corrosion products that formed along the cold side.

Case History 12.4

Industry:	Steel
Specimen location:	Hot strip mill water supply main
Specimen orientation:	Horizontal
Environment:	Lake Michigan water treated with dispersant; heavily fouled with grease and oil
Time in service:	23 years
Sample specifications:	54 in. (137 cm) outer diameter, 0.330 in. (8.38 mm) thick mild steel

After almost 20 years of service, many leaks developed throughout the mill water supply system. Upon inspection, it was found that perforations had occurred beneath large tubercles. Some tubercles were up to 6 in. (15 cm) in diameter. Accessible pipe was water blasted to remove tubercles, and the areas where metal loss was deepest (several thousand) were weld repaired.

After 5 more years of service, a few leaks began to develop. Window sections cut from the main contained numerous tubercle encrustations (Fig. 12.42).

The entire internal surface was covered with a layer of dark, greasy deposit and slime intermixed with reddish-brown oxides; oxides were dark immediately upon removal from the surface (Figs. 12.16 and 12.17). Analysis showed that the material within the tubercles was fairly acidic.

Beneath each tubercle was a dish-shaped depression. Some depressions were as deep as half the intact wall thickness. Many shallow striations were present in the depressions on the internal surfaces (Fig. 12.15). Circumferential welds were preferentially attacked (Fig. 12.43). All of this evidence revealed that acid corrosion occurred beneath the tubercles.

Recommendations were made to begin treatment with corrosion inhibitors. However, due to the high system volume, this was cost prohibitive. Other recommendations were made to reduce grease and oil fouling and to pursue biocidal treatment.

Figure 12.42 Typical internal surface of heavily tuberculated mill water supply line about 16 hours after removal from system. Note the weld where corrosion product and deposit were removed.

Figure 12.43 Supply line of Fig. 12.42 showing preferential weld attack.

Case History 12.5

Industry:	Utility
Specimen location:	Steel baffle plate from a test water box containing flowing cooling water
Specimen orientation:	Vertical
Environment:	Cooling water containing more than 2000 ppm of chlorides and sulfates, pH 7.2 to 7.6, approximately 80°F (27°C), no corrosion inhibitors used, flow 3 ft/s (1 m/s)
Time in service:	2 weeks
Sample specifications:	4-7/8 in. by 1-3/4 in. by 1/8 in. (12.4 cm by 4.4 cm by 0.32 cm), carbon steel, cold rolled

FIGURE 12.44 Many incipient tubercles formed on a steel Dame baffle in 2 weeks. (*Courtesy of National Association of Corrosion Engineers, Corrosion '89 Paper No. 197 by H. M. Herro.*)

A test water box was installed during a 2-week trial to monitor corrosion and fouling in a utility cooling water system. A baffle plate from the test box was removed after the test. Small, hollow incipient tubercles dotted surfaces (Fig. 12.44). Small amounts of carbonate were present atop and around each tubercle. Each tubercle capped a small depression no deeper than 0.005 in. (0.013 cm) (Fig. 12.45). This indicated local average corrosion rates were as high as 130 mpy (3.3 mmpy). This high corrosion rate indicated fast formation and growth rates of tubercles combined with the production of acidic conditions beneath them.

Each tubercle exhibited small clam-shell marks or growth rings at high magnification (Fig. 12.46). Each ring was formed by fracture at the tubercle base

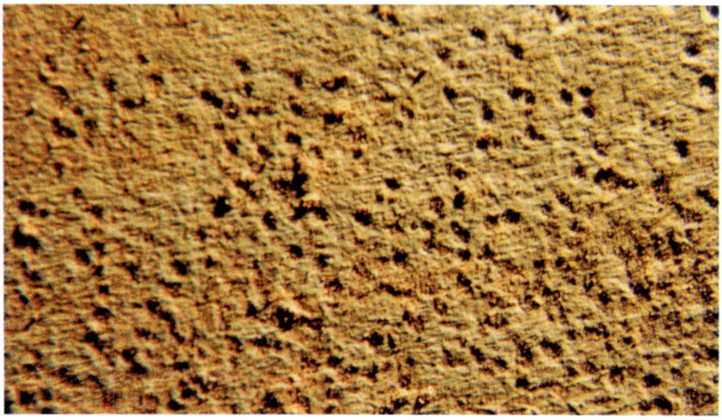

FIGURE 12.45 The baffle plate of Fig. 12.44 with tubercles removed to show pit-like depressions beneath each mound. (*Courtesy of National Association of Corrosion Engineers, Corrosion '89 Paper No. 197 by H. M. Herro.*)

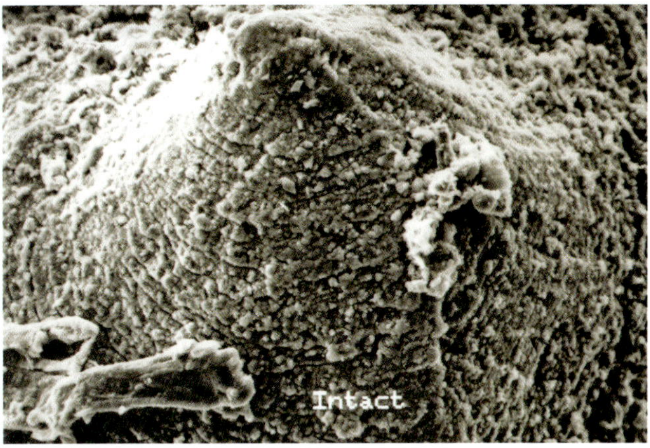

FIGURE 12.46 Scanning electron micrograph of tubercles in Fig. 12.45. Note the clam-shell growth steps formed by successive fractures at the tubercle base. Tubercle is about 200 μm in diameter. (*Courtesy of National Association of Corrosion Engineers, Corrosion '91 Paper No. 84 by H. M. Herro.*)

during growth. Ejected internal contents rapidly deposited when contacting oxygenated waters. Tubercles were hollow (Fig. 12.47). Surfaces below the cap contained concentrations exceeding 10 percent of chloride and sulfate due to the formation of ion concentration cells, producing localized acidic conditions that accelerated corrosion.

Oxygen corrosion and the associated tubercle initiation may be controlled through the use of adequate concentrations of proper corrosion inhibitors. Reduction of the combined high chloride and sulfate concentration in the water was also proposed.

FIGURE 12.47 Tubercles in Fig. 12.47 broken open to reveal hollow interiors. (*Courtesy of National Association of Corrosion Engineers, Corrosion '91 Paper No. 84 by H. M. Herro.*)

Case History 12.6

Industry:	Nuclear utility
Specimen location:	Emergency service water pipe
Specimen orientation:	Horizontal
Environment:	Near the end of a deadleg with low flow rates, about 1 ft/s (0.3 m/s). Dispersant, tolyltriazole, silicate 15 to 25 ppm residual, bromine/chlorine, 1 h/day
Time in service:	7 years
Sample specifications:	3 ½ in. (8.9 cm) outer diameter, mild steel pipe

An emergency service water system pipe section was removed because of reduced flow. In places, as much as 40 percent of the internal cross-sectional area was blocked by tubercle growth (Fig. 12.6).

Tubercles consisted of hard, black iron oxide shells overlaid with friable carbonate-containing deposits. In places, several laminated black magnetite shells existed. The outer crust could be crushed by gentle pressure with a finger. Tubercles were riddled with white crystalline fibers. Other detritus was incorporated into the tubercle core and crust. Metal loss was less than 0.030 in. (0.76 mm) below each tubercle. Wall thickness was almost 0.25 in. (6.4 mm).

Chemical cleaning was suggested. A tannin-citric acid process was tested and produced good results. Water chemistry, chemical treatment, and operating changes were suggested, but were deemed inadvisable due to discharge restrictions of chemicals into local waters.

Case History 12.7

Industry:	Nuclear utility
Specimen location:	Turbine cooling water system piping
Specimen orientation:	Horizontal and vertical
Environment:	Water from cooling tower pump suction, pH 8.6 to 8.8, flow 2 to 5 ft/s (0.6 to 1.5 m/s). Dispersant, 1 to 3 ppm tolyltriazole, sodium hypochlorite at 0.8 free residual chlorine for 2 hours per day; 0.6 to 0.8 ppm total zinc and 0.1 to 0.2 ppm soluble zinc. Free chlorine maintained at 1 ppm for 5 consecutive days per month during the summer. Chemical treatment started after 2 years of no treatment.
Time in Service:	3 years
Sample specifications:	L-shaped pipe section of 2 in. (50.8 mm) carbon steel pipe socket welded to a 90° carbon steel elbow

Internal surfaces were covered with moderate-sized tubercles (Fig.12.48). The tubercles had thick, hard magnetite shells that capped large internal cavities (Fig. 12.12). The cross-sectional area of the pipe bore was reduced by at least 30 percent in some places due to tubercle intrusion. Tubercles were aligned with flow, indicating that growth occurred during service. No failure occurred, and deepest metal loss was 0.093 in. (0.33 mm) from the nominal pipe wall thickness of 0.225 in. (5.72 mm).

X-ray analysis of material scraped from internal surfaces indicated that it was 88 percent iron, 7 percent silicon, and 1 percent each of magnesium, aluminum, chlorine, and sulfur.

This system was not treated except for chlorination during its first 2 years of existence. Corrosion rates were reduced during the following year, due to proper

FIGURE 12.48 Small, hard tubercles on an essential service water system pipe in a nuclear utility.

treatment with corrosion inhibitors. Chemical cleaning was considered since small-diameter piping had been plugged in deadleg regions.

Case History 12.8

Industry:	Building cooling system
Specimen location:	Closed cooling water system piping
Specimen orientation:	Horizontal
Environment:	Primarily closed loop cooling water (softened, using nitrite corrosion inhibitor). Intermittent use of open recirculating condenser water (free cooling). Water treated with low-level phosphate corrosion inhibitor
Time in service:	7 years
Sample specifications:	Externally threaded steel pipe nipple 13/16 in. (20.6 mm) inside diameter, 1 in. (25.4 mm) outside diameter, attached to a valve and fittings. Deadleg piping with flow rates about 1 ft/s (0.3 m/s).

The internal surface was partially covered along the bottom side with fairly thick iron oxide tubercles (Fig. 12.49). Metal loss occurred within a few scattered patches on the internal surface, including at the threaded end where perforations penetrated to the external thread roots (Fig. 12.50). Chemical spot tests indicated that components of the corrosion products covering the depressions were acidic in a distilled water solution, and contained chloride. Under normal operating conditions when the system operated as a closed loop, the system contained softened water treated with high concentrations of nitrite inhibitor with lower dissolved oxygen concentration. During such normal operation, corrosion rates were low, even under low-flow conditions.

It was determined that significant oxygen corrosion occurred when condenser water flowed through the piping on an intermittent basis to provide energy savings. Corrosion rates measured under higher flow rates [>3 ft/s (1.0 m/s)] in the condenser water system were low [<2 mpy (0.05 mmpy)]. At low flow rates, corrosion rates in condenser water are higher. This is due to

FIGURE 12.49 The internal surface of piping from a closed loop in an HVAC system is partially covered along the bottom side with fairly thick iron oxide tubercles.

FIGURE 12.50 Metal loss occurred within scattered patches on the internal surface, including at the threaded end where perforations penetrated to the external thread roots.

the higher concentrations of dissolved oxygen, lower inhibitor concentrations, and the presence of higher concentrations of aggressive agents and dissolved and suspended solids that are typically present in condenser water. The use of condenser water was discontinued.

Reference

1. More detailed descriptions concerning the proper use of anodic and cathodic corrosion inhibitors and their combinations, as well as dispersants, in cooling water systems are contained in D. Flynn (ed.), *The Nalco Water Handbook*, 3d ed., McGraw-Hill, 2009.

CHAPTER 13

Biologically Influenced Corrosion

General Description

One of the most complex mechanisms to accurately diagnose in cooling water systems is corrosion related to biological interactions. This is due to the wide diversity of organisms, both large and microscopic in size, which can be found in cooling water and other water systems. Organisms are generally introduced to the cooling water system from water sources such as oceans, lakes, rivers, ponds, wells, and treated potable water. The natural water sources include fresh, brackish, and seawater. Each type has a wide variety of organisms that can be associated with it. In addition, organisms can also be introduced by contaminants, such as airborne materials that are incorporated into the system at open cooling towers. Biological organisms are ubiquitous, and their possible contribution to corrosion in cooling water systems must be considered.

Corrosion due to biological material has generally been categorized into two influences on metal loss. The categories are based on whether biologically produced substances cause corrosion actively or passively. Each mechanism either accelerates preexisting corrosion or establishes a new form of metal loss. However, the distinction between active and passive corrosion is often vague.

For passive corrosion, biological material acts as a chemically inert substance, and wastage is an indirect consequence of the biological mass or by-products present on the metal surface. Biological material acts as any deposit accumulation. This material can provide shielded areas in which concentration cells develop (see Chap. 11, "Crevice and Underdeposit Corrosion"). Passive attack involving underdeposit corrosion tends to involve large system surface areas and, hence, accounts for the greatest amount of metal loss, by weight, in cooling water systems. Appreciably sized masses may also create flow disturbances and cause erosion.

Passive corrosion caused by chemically inert substances is the same whether the substance is living or dead. The substance acts as an occluding medium, changes heat transfer, and/or influences flow. However, as living organisms die, decomposition may generate substances that can cause corrosion, such as ammonia and sulfur compounds.

Active biological corrosion may be defined as the chemical interaction of living organisms with materials to directly produce corrosion and/or appreciably accelerate preexisting corrosion processes. Most of what is referred to as *microbiologically influenced corrosion* (MIC) fits into this category. Ultimately, active biological corrosion directly accelerates or establishes electrochemical corrosion reactions. Factors influencing electrochemistry include pH, area effects, temperature, polarization, flow, oxygen concentration, and electrolyte conductivities. Also, corrosive substances such as acids and ammonia may be produced by biological activity.

Active corrosion may be caused by microorganisms. Potentially troublesome bacteria are either aerobic or anaerobic, although facultative bacteria may grow under either high or low oxygen conditions, changing their metabolic processes and concomitant chemistries accordingly. Active attack may produce intense localized corrosion and perforations. Microbiologically influenced corrosion should be considered if corrosion is not normally predicted by the environmental conditions.

One important aspect of microbiologically influenced corrosion is that individual organisms are not isolated from each other. In fact, bacteria grow in a consortium, in which several varieties of organisms coexist in an energy-efficient community. Synergistic effects are common. This consortium often develops into a diverse biofilm. A schematic of biofilm formation is shown in Fig. 13.1. Initially, individual cells of certain microorganisms are attracted to a surface and attach themselves to it. Once attached, the cells can secrete extracellular polymeric substances such as polysaccharides, which cause the film to grow with a slimy, glue-like consistency. Figure 13.2 shows an example of a biofilm on a surface, with the individual cells and the surrounding extracellular polymer matrix revealed at high magnification. The environment produced by the film allows for the incorporation of particulate matter such as suspended solids and other microorganisms. Consumption of oxygen by aerobic microorganisms in the outer portion of the layer will produce a favorable oxygen-depleted environment for anaerobic organisms beneath the biofilm at the metal surface.

Generally, greater than 90 percent of the microorganisms in cooling water systems are sessile, meaning they are attached to surfaces. Microorganisms in the water are referred to as planktonic.

Biofilm dynamics frequently change with time. For example, one type of microorganism may break down a particular molecule common to the system. A second microorganism may further degrade the molecular by-products. A third microorganism may use such by-products to obtain energy.

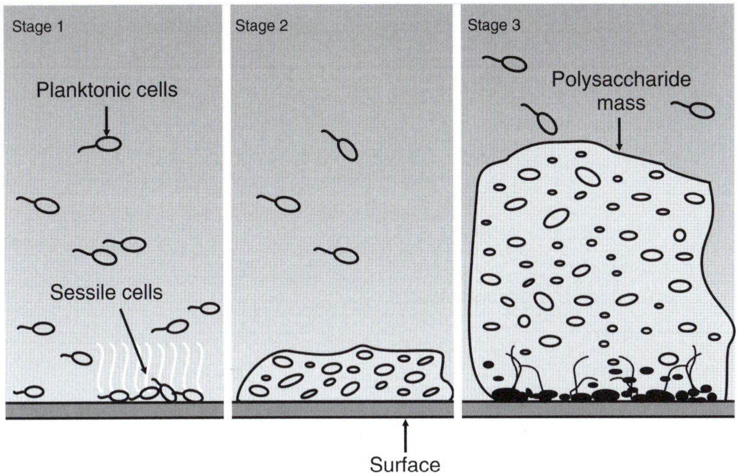

FIGURE 13.1 Typical stages of biofilm formation. Initially, individual cells attach to the metal surface, then the cells surround themselves with extracellular material (polysaccharide), and finally other substances and microorganisms are incorporated into the material.

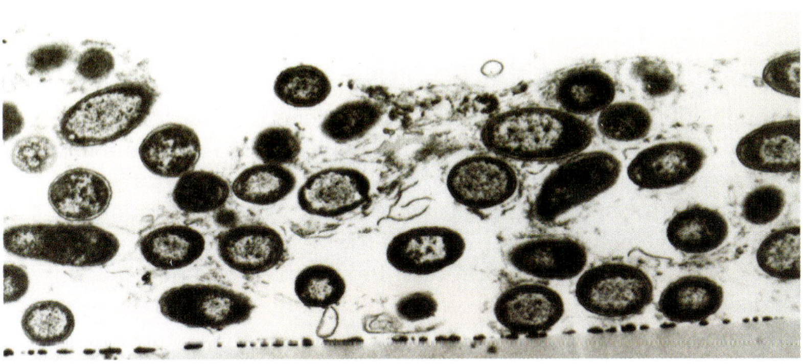

FIGURE 13.2 Microscopic view of a biofilm, showing individual cells of *Pseudomonas aeruginosa* surrounded by extracellular polymer.

One of the major effects that a biofilm layer will have on heat exchanger equipment is substantial reduction of heat transfer and increasing flow resistance. The thermal conductivity of a biofilm and water are essentially identical, as biofilms are largely composed of water. Table 13.1 lists the thermal conductivities for different layers that typically form on metal surfaces in cooling water systems, such as mineral scales and biofilm. A biofilm is only about one-fourth as thermally conductive as calcium carbonate and only about half as conductive as analcite. Therefore, for a given layer thickness, biofilm

Scale	Thermal Conductivity (watt m^{-1} K^{-1})
Calcium carbonate	2.26–2.93
Calcium sulfate	2.31
Calcium phosphate	2.60
Magnesium phosphate	2.16
Magnetic iron oxide	2.88
Analcite (sodium aluminum silicate)	1.27
Biofilm	0.63
Water	0.63

TABLE 13.1 Thermal Conductivities of Scales and Biofilm[1]

material will have a significantly more detrimental effect on heat transfer than common mineral scales. In critical cooling applications, such as continuous caster molds, decreased heat transfer may lead to the formation of defects in the cast steel product.

Characteristics of Biological Organisms

Although environments in cooling water systems that may cause biologically influenced corrosion can, and typically do, include many organisms, it is important to understand individual features and characteristics of specific organisms or groups of microorganisms. This knowledge helps in determining if the microorganisms played a distinct role in the corrosion process. Proper diagnosis depends on establishing likely contributions from specific organisms to the attack on the metal surface.

Each variety of bacteria causes corrosion by changing localized environments on the metal surface, thus influencing reactions occurring at the anode and/or cathode in an electrochemical corrosion cell. Surprisingly, there has been considerable disagreement among researchers as to specific reactions involved in microbiologically influenced corrosion. Theories have changed, been modified, or expanded over the years as more scientific information becomes available. Biological pathways to corrosion can be intricate, and some reactions are often open to speculation because of the perceived complexities of microbiological processes. In addition, the chemical composition of the electrolyte will influence the impact of microorganisms on corrosion, such as their type and numbers in the system. Factors affecting the aggressiveness of corrosive bacteria include temperature, total organic carbon and nitrogen concentrations, flow, oxygen or ammonia concentrations, chemical treatment, pH, and other influences, many of which are unknown.

Four main types of bacteria that have been linked to corrosion in cooling water systems include:

1. Slime formers
2. Sulfate reducers
3. Acid producers
4. Metal oxidizers (depositors)

Table 13.2 summarizes some information for several common microorganisms associated with microbiologically influenced corrosion, their processes, and preferred environments for activity. Some specific microorganisms, as well as larger biological organisms associated with passive corrosion, will be discussed separately in the following sections.

Genus or Species	pH Range	Typical Temperature Range	Oxygen Requirement	Action
Pseudomonas	4–9	70–105°F (20–40°C)	Mainly aerobic	Slime former
Desulfovibrio	4–8	50–105°F (10–40°C)	Anaerobic	Sulfate reducer
Desulfotomaculum	6–8	70–105°F (20–40°C) Some 115–165°F (45–75°C)	Anaerobic	Sulfate reducer
Desulfotomonas	-	50–105°F (10–40°C)	Anaerobic	Sulfate reducer
Acidithiobacillus thiooxidans	0.5–8	50–105°F (10–40°C)	Aerobic	Acid producer
Acidithiobacillus ferrooxidans	1–7	50–105°F (10–40°C)	Aerobic	Iron oxidizer
Gallionella	7–10	70–105°F (20–40°C)	Aerobic	Metal oxidizer
Leptothrix	6.5–9	50–95°F (10–35°C)	Aerobic	Metal oxidizer
Sphaerotilus	7–10	70–105°F (20–40°C)	Aerobic	Metal oxidizer

Table 13.2 Common Microorganisms Associated with Corrosion in Cooling Systems and Some Characteristics, Including Typical Ranges of Activity.[2] (*Reprinted with permission of ASM International. All rights reserved. www.asminternational.org.*)

Macroscopic Biological Organisms and Material

Organisms and inert material are mainly associated with passive corrosion, as they create conditions favorable for concentration cell corrosion mechanisms. Shells, clams, wood fragments, and other biological materials can provide environments suitable for concentration cell corrosion. Additionally, fragments can lodge in heat exchanger inlets, locally increasing turbulence and altering flow rates, promoting the tendency for erosion-corrosion. If deposits are of substantial size on a surface exposed to flowing water, then turbulence and associated erosion-corrosion can occur downstream.

Zebra mussels and Asiatic clams are particularly troublesome. Because of their prodigious reproduction and relatively rapid growth rates under favorable conditions, components such as water intakes, transfer piping, grates, and screens can be rapidly blocked. Portions of service water systems in nuclear power plants have been rendered temporarily inoperative due to clams, mussels, and similar organisms. Besides gross occlusion, these organisms increase rates of silt, sand, and detritus accumulation, also accelerating concentration cell corrosion.

Small organisms frequently become embedded within corrosion products and deposits. In some cases, the organisms may make up a sizable fraction of the deposits and corrosion products. Seeds and fibrous material often blow into cooling towers (Fig. 13.3), where they are transported throughout the system. This material sticks to surfaces, acting like sieves by straining particulate matter from the water.

Figure 13.3 Drain screen in a cooling tower, with the holes in the mesh mostly occluded by deposit layers containing inorganic and biological material, such as small sprouts from seeds blown into the cooling tower.

FIGURE 13.4 Stress corrosion cracking of a brass condenser tube caused by ammonia from decomposing slime masses lodged on the internal surface.

Deposit mounds form, reinforced by the fibers (see Case History 23.2 for an example where biological material resulted in erosion-corrosion).

Finally, when living organisms die, decomposition may generate ammonia. In some cases, locally high concentrations of ammonia can be produced, causing stress corrosion cracking of brass condenser tubes (Fig. 13.4).

Slime Formers

The development of slime formation can be very rapid on a surface, with partial coverage occurring in hours to days. Slime is basically a network of secreted strands (extracellular polymeric substances—EPS) produced by microorganisms. This typically provides the matrix for biofilms, which are intermixed with water, other bacteria, gases, detritus, and extraneous matter. Commonly, 99 percent of the slime layer is water, although much silt and debris may also become entrapped in it. Dried microorganisms themselves consist mostly of carbon, nitrogen, oxygen, and hydrogen.[3]

Most slime formers are aerobic, although some, such as *Pseudomonas*, are facultative and can also grow in low-oxygen environments. Closed recirculating cooling water systems containing low oxygen concentrations are usually free of significant slime masses, as long as makeup water additions to the system are not appreciable.

Slime layers can contribute to corrosion both actively and passively. First, since the indigenous slime formers are aerobic, they consume oxygen, stimulating the formation of differential oxygen concentration cells. The slime layer forms an occluding mass, also contributing to aggressive ion concentration cell formation and passive corrosion (see Chap. 11, "Crevice and Underdeposit Corrosion").

Second, and again because most slime formers are aerobic, these bacteria are present atop corrosion products and deposits in proximity to oxygenated water. Slime layers produce a stagnant zone next to the surface that retards convective oxygen transport and increases diffusion distances. Surface shielding is further accelerated by the gathering of dirt, silt, sand, and other materials incorporated into the biofilm. The depletion of oxygen that develops beneath the biofilm provides an environment suitable for the growth of anaerobic bacteria. Sulfate-reducing bacteria and acid-producing bacteria may proliferate beneath slime layers (Fig. 13.5).

Sulfate-Reducing Bacteria

Probably the microorganisms most widely associated with microbiologically influenced corrosion are sulfate-reducing bacteria (SRB). Sulfate reducers cause most localized industrial cooling water corrosion associated with bacteria. Desulfovibrio, Desulfomonas, and Desulfotomaculum are three genera of sulfate-reducing bacteria.

These bacteria are anaerobic. They may survive, but not actively grow, when exposed to aerobic conditions. They occur in most natural waters including fresh, brackish, and seawater. Most soils and sediments also contain sulfate reducers. Sulfate or sulfite must be present for active growth. The bacteria may tolerate pH ranges from about 4 to 9, and some species can withstand temperatures as high as about 176°F (80°C).

There are several reported theories to describe how sulfate reducers cause corrosion. One theory indicates that these bacteria cause

Figure 13.5 Severely pitted aluminum heat exchanger tube that was exposed to cooling water on its external surface. Pits were caused by sulfate-reducing bacteria colonies that developed beneath a slime layer. Note the distinct edge of the slime layer along the side of the tube, highlighting that attack was localized to the top of the tube.

cathodic depolarization by removing hydrogen from cathodic sites. This was based on undisputed evidence that several sulfate reducers possess enzymes that are capable of converting hydrogen as part of their metabolism. In this process, inorganic sulfates are reduced to sulfides in the presence of hydrogen, with the principal reaction of conversion of sulfate by the bacteria as:

$$8H\ (\text{adsorbed}) + SO_4^{2-} \rightarrow S^{2-} + 4H_2O \quad (13.1)$$

The reduced sulfur produced from the reaction can then combine with iron, with the overall reaction on steel stated as:

$$4Fe + SO_4^{2-} + 4H_2O \rightarrow 3Fe(OH)_2 + FeS + 2OH^{2-} \quad (13.2)$$

Alternative theories regarding the involvement of SRB in corrosion have been developed over the years, with some that consider additional cathodic reactions. Other theories emphasize the effect of the sulfide films that form on metal surfaces, with interactions between the sulfide layers and the development of localized anodic regions on the metal surface. Recent reviews highlight that SRB can influence a number of corrosion mechanisms simultaneously.[4]

Acid-Producing Bacteria

Many bacteria produce acids. Acids may be organic or inorganic depending on the specific bacterium. In either case, the acids that are produced will lower the pH. This typically causes corrosion to occur or accelerates corrosion rates if the pH is appreciably depressed. Although many kinds of bacteria may generate acids, *Thiobacillus thiooxidans* and *Clostridium* species have most often been linked to accelerated corrosion on steel. Activity of *Thiobacillus thiooxidans* in wastewater systems is also known to create acidic environments capable of causing deterioration of concrete.

Thiobacillus thiooxidans is an aerobic organism that oxidizes various sulfur-containing compounds to form sulfuric acid. These bacteria are sometimes found near the tops of tubercles (see Chap. 12, "Oxygen Corrosion," for details on tubercle growth). As stated previously, often interactions between microorganisms will result in increased localized corrosion. As an example, there is a symbiotic relationship between *Thiobacillus thiooxidans* and sulfate-reducing bacteria; *Thiobacillus thiooxidans* oxidize sulfide to sulfate, whereas the sulfate reducers convert sulfate to sulfide. It is unclear to what extent *Thiobacillus thiooxidans* directly influences corrosion processes inside tubercles. It is more likely that they indirectly increase corrosion by accelerating sulfate-reducer activity deep in the tubercles.

Clostridia are anaerobic bacteria that can produce organic acids. For instance, *Clostridium aceticum* metabolism results in the formation of acetic acid. Short-chain organic acids can be quite aggressive to carbon steel. *Clostridia* are sometimes found deep beneath deposit and corrosion-product accumulations near corroding surfaces and within tubercles. Increased acidity directly contributes to wastage.

Metal-Oxidizing Bacteria

Metal-oxidizing bacteria oxidize dissolved ferrous ions (Fe^{2+}) to ferric ions (Fe^{3+}). Reaction of these ions with dissolved oxygen produces ferric hydroxide. Some bacteria oxidize manganese and other metals. As such, these bacteria contribute to the deposition of metal oxides, and they are sometimes referred to as metal depositors.

Gallionella bacteria, in particular, have been associated with the accumulation of iron oxide–based deposits. Other metal oxidizers include *Sphaerotilus*, *Crenothrix*, and *Leptothrix* species. Each bacterium is filamentous, often with a segmented appearance (Fig. 13.6).

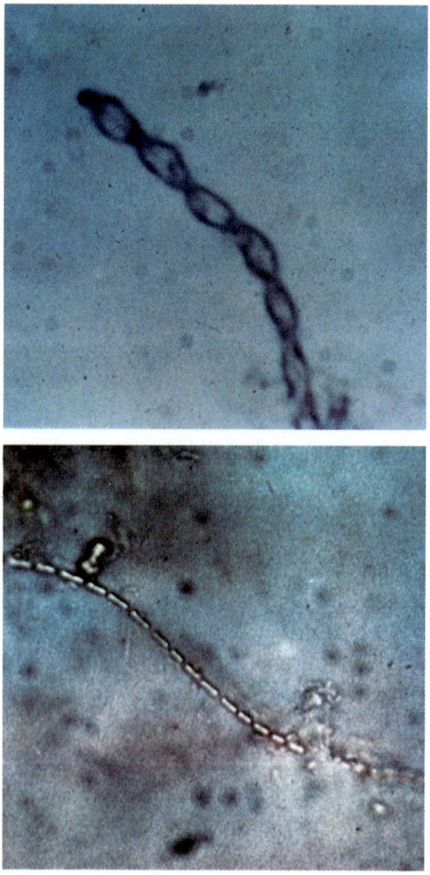

FIGURE 13.6 Segmented, filamentous structure of *Gallionella* (top) and *Sphaerotilus* (bottom) revealed by microscopy.

The metal oxide accumulates along very fine "tails" or excretion stalks generated by these organisms. In fact, up to 90 percent of the dry weight of the cell mass can be iron hydroxide.

Deposition attributed to *Gallionella* in tubercles must be carefully considered, as iron oxide that forms tubercle shells and underlying iron oxide and hydroxide-based corrosion products will be directly produced by oxygen corrosion. The amount of microbiologically induced deposition is often much less than that caused by electrochemical corrosion and deposition due to precipitation or settling. In addition, oxidizing conditions of the water can promote deposition on surfaces.

Gallionella, Leptothrix, Siderocapsa, and other microorganisms have been associated with biological deposition of manganese and iron oxides on metal surfaces, which can result in underdeposit corrosion (see Chap. 11). Manganese dioxide (MnO_2) has been shown to shift the electrochemical potential on stainless steel surfaces such that pitting attack is promoted, a process called manganese ennoblement. Manganese-containing deposits can also promote corrosion of copper and copper alloys. However, manganese dioxide deposition is not uniquely caused by microbiological activity. Chemical treatment to control microorganisms, such as chlorination, also causes oxidation of iron and manganese ions. Thus, careful consideration of both processes is required to properly determine the most appropriate remedial actions for the system.

Nitrifying and Denitrifying Bacteria

Nitrifying bacteria are capable of oxidizing ammonia (NH_3) to nitrate (NO_3). The best-known nitrifying bacteria are *Nitrosomonas* and *Nitrobacter*.

Leakage of ammonia into cooling water stimulates the growth of *Nitrosomonas.* Hence, ammonia plants often contain such bacteria. Biological activity reduces pH, and oxygen concentration falls. A drop in pH is a common sign of nitrifiers. The drop is usually sharp but transient, and corrosion effects are usually slight.

Nitrobacter, an aerobic bacterium, can materially depress pH by oxidizing nitrite (NO_2^-) to nitrate (NO_3^-), in effect producing nitric acid. Acidity may increase until the pH decreases to the range between 3 and 5. Such bacteria require high concentrations of oxygen and cause problems only in oxygenated systems.

Denitrification is a process that can affect closed cooling water loops treated with nitrite as a corrosion inhibitor. Denitrifiers are microorganisms that can convert nitrate or nitrite to nitrogen. If the nitrite concentration in cooling water continually decreases in a nitrite-treated closed cooling water system over time, activity from denitrifying bacteria should be considered, especially if the conductivity of the cooling water remains fairly constant.

Other Microorganisms

As stated previously, a wide variety of microorganisms can be present that influence processes in a cooling water system. Some additional microorganisms that deserve mention for activity in cooling water systems include algae, metal-reducing bacteria, and fungi.

Some algae produce dense, fibrous mats in sunlit areas (Fig. 13.7). The mats, when established on the surface, act as occluding mediums for passive corrosion. Differential oxygen concentration cells and ion concentration cells may form. The mats can also provide environments where corrosive anaerobes may grow.

Additionally, high concentrations of dissolved oxygen can be created by vigorously growing algae. In at least one case, dissolved oxygen concentrations measured near cooling water intakes at a steel mill were as high as 10 ppm, well above the calculated oxygen saturation concentration at the inlet temperature and pressure. Oxygen concentration in the water decreased at night when algae growth stopped. Corrosion rates measured with electric polarization devices followed a similar diurnal cycle.

Metal-reducing bacteria, such as those that convert ferric to ferrous ions, have been suggested as an accelerant for steel corrosion in oxygenated waters in some experimental studies. However, consistent evidence of these bacteria influencing corrosion in industrial systems is scarce. Some instances also suggest that these microorganisms may have corrosion-inhibiting properties.[5]

Fungi, such as yeasts and molds, may also be present in cooling water systems. These microorganisms tend to grow in moist, water-wetted areas rather than submerged surfaces. Fungi can be responsible for degradation of wood components in the system, such as in

FIGURE 13.7 Thin algae mats that developed on the basin of a cooling tower. The algae were localized to areas exposed to direct sunlight.

wooden cooling towers. Degradation is caused by certain microorganisms that attack cellulose fibers and lignin in wood. Chapter 8, "Nonmetallic Materials," contains a more detailed description of biological degradation of wood.

Locations

Virtually any cooling water system may be affected by biologically influenced corrosion. Oil- and grease-fouled systems generally contain high bacteria counts. Due to the tolerance of microorganisms for certain environments, their activity may be restricted under certain conditions. For instance, microbiological attack usually occurs where water temperature is below about 180°F (82°C). However, some organisms can tolerate even harsher conditions.

Because low-flow or stagnant conditions generally promote environments favorable for many microorganisms, deadlegs or locations that experience intermittent flow in piping are prone to MIC. Regions of piping at far distances from chemical treatment feed points may also be more susceptible to MIC. Many cases of attack due to microorganisms are related to hydrotesting components or portions of a water system. Microbiological activity can be significant if the water used for hydrotesting was allowed to remain stagnant for a long period of time in the system. Failures may not immediately happen during hydrotesting, but corrosion that began during that period may continue during service.

Slime masses foul water boxes, heat exchanger tubing, and tubesheets. Large organisms, such as clams, mussels, and other shellfish, frequently lodge in heat exchanger tube inlets (Fig. 13.8), water intakes, valves, and service water piping. These larger organisms often feed off smaller organisms.

FIGURE 13.8 Small clams lodged in heat exchanger tube inlet ends. (*Courtesy of Rick Ruckstuhl, Nalco Chemical Company.*)

Algal growth can occur in open sunlit areas, and it can cover surfaces in cooling towers. Appreciable masses can plug strainers and water intakes. Large organisms (those visible to the naked eye) are a sure indicator of bacterial presence and may indirectly cause attack.

Potentially corrosive anaerobic bacteria may be present beneath biofilm layers due to the locally oxygen-depleted regions that develop there. These organisms may also thrive in systems containing little oxygen. Cases of MIC causing or contributing to metal loss in fire protection systems are well-known (Fig. 13.9). Localized metal loss predominately occurs along the sides and bottoms of horizontally oriented pipe that experiences extended periods with stagnant conditions.

Tubercles that form on carbon steel surfaces provide localized oxygen-depleted regions suitable for anaerobic microorganisms. The floors of tubercles sometimes contain sulfate-reducing and acid-producing bacteria. However, the presence of tubercles is not definitive evidence of MIC, as tubercle structures can develop on steel surfaces without direct contribution from microbiological activity. Studies have shown that many microorganisms are attracted to actively corroding sites. Determining the cause for corrosion initiation is often difficult.

Corrosion most commonly occurs where biological matter is deposited on surfaces. However, attack may also occur away from attached organisms and settled biomass. For example, biologically produced chemicals such as ammonia, hydrogen sulfide, and acids can increase general corrosion rates in a system, near and away from generating organisms.

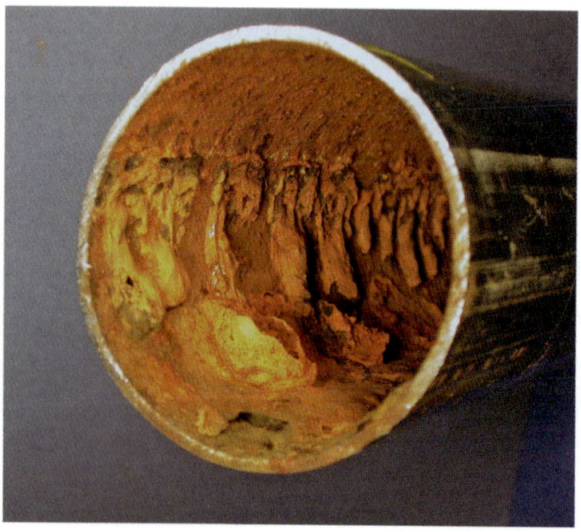

FIGURE 13.9 Internal surface of a troubled fire protection system. Corrosion was deepest at and below a waterline in the horizontally oriented pipe.

Microbiologically influenced corrosion, most often associated with sulfate-reducing bacteria, can affect the external surfaces of buried pipelines. Regions where the surrounding soil is saturated with water are most susceptible.

Virtually all metallurgies can be attacked by corrosive bacteria. However, cases of MIC on highly alloyed stainless steels (typically containing greater than 6 wt% molybdenum) are very rare. Titanium is generally considered to be immune to direct attack, although surface fouling and underdeposit corrosion may still occur. Although copper alloys are resistant to fouling and copper ions are toxic to many organisms, copper and its alloys have experienced corrosion in some rare cases due to direct microbiological activity, such as that of sulfate-reducing bacteria.

Critical Factors

In order to properly diagnose biologically influenced corrosion, biological matter or specific offending organisms that may promote corrosion must be present in a system suffering active corrosion. However, not every system containing such material or organisms is adversely affected. A biological presence (current or past) is virtually certain in any cooling water system, but not every cooling water system is significantly attacked by biological organisms.

Four factors must be present for a complete diagnosis of MIC:

1. Presence of microorganisms or their by-products
2. Compatible environmental conditions for the specific microorganisms
3. Specific corrosion products and deposits that are characteristic of the microbiological interaction on the surface
4. Corrosion morphologies that are consistent with the type of microorganism or microorganisms responsible for attack

Identification

Diagnosis is not based on a preponderance of evidence; it is based on complete consistency of all evidence. It should be vigorously stressed that all four factors just stated must be consistent for a specific mechanism. The presence alone of potentially corrosive bacteria or other organisms is not proof of corrosion related to these organisms.

Combined relations between biological, metallurgical, and chemical evidence is needed for proper diagnosis. Often, MIC is implicated based on the appearance of damage or rapidity of the attack. In many cases, consideration of other corrosion mechanisms will provide a more likely explanation for the observed metal loss.

Note that evidence related to the preceding four factors is unique for specific bacteria. Each form of biologically influenced corrosion can be recognized by examining wastage morphologies, corrosion-product and deposit composition, biological analysis, and compatible environmental conditions.

Biological Analysis

Specific mention needs to be made regarding biological analysis. Biological analysis is used to determine what microorganisms are present either in the system or on a surface, and their relative amounts in some cases. Some common methods for analysis may include culture techniques, biochemical assays, microscopic examination, and genetic techniques. These methods are discussed briefly in this section, and they do not represent all available techniques for biological analysis.

Culturing

Culturing, also known as plating, has been a primary method used over the years to detect and enumerate microorganisms using specially prepared media that allow for the growth of some specific groups of bacteria. As culturing requires that the microorganisms are viable, the results are very dependent on proper sampling technique and transportation, as well as prompt analysis of the samples. For this reason, culturing does not always provide an accurate assessment of the numbers and types of microorganisms in an environment, and it will always underestimate the microorganism population of the system.[6]

Culturing techniques cannot be used on samples that have been removed from the cooling water system for an extended period of time. If microbiologically influenced corrosion is suspected, it is imperative to obtain samples from the metal surface for analysis immediately after the component is removed from the system. It is good practice to take multiple samples from the surface, both at and away from corrosion damage for comparison. When significant amounts of material, such as tubercles, cover metal surfaces, it is often useful to collect swab samples from beneath the occlusive material—for example, beneath tubercle caps. As previously discussed, the oxygen-deficient environment in an occluded location can allow for the growth of significantly different species than the bulk water environment. Therefore, sampling material in both occluded and nonoccluded areas may prove valuable in an investigation. Some guidelines regarding microbiological control in a system based on differential microbiological analysis of water samples and swabs from surfaces are listed in Table 13.3.

Biochemical Assays

Biochemical assays can be used to identify the activity of organisms associated with MIC. These methods measure different constituents associated with the microorganisms, such as adenosine triphosphate (ATP).

Microbiological Analysis	Good Control	Poor Control
Total aerobic (bulk water)	$<10^3$–10^4 cfu/mL	$>10^6$ cfu/ml
SRB (bulk water)	<1 cfu/mL	>10 cfu/ml
Pseudomonas (bulk water)	$<10^2$–10^3 cfu/mL	$>10^5$ cfu/ml
Fungi, yeast, mold (bulk water)	<10 cfu/mL	$>10^2$ cfu/ml
Filamentous algae (bulk water)	None	Many
Total aerobic (swab)	$<10^5$–10^6 cfu/swab	$>10^7$ cfu/swab
SRB (swab)	<10 cfu/swab	$>10^2$ cfu/swab
Clostridia AP bacteria (swab)	<10 cfu/swab	$>10^2$ cfu/swab
Pseudomonas (swab)	$<10^3$–10^4 cfu/swab	$>10^6$ cfu/swab
Fungi, yeast, mold (swab)	<10 cfu/swab	$>10^2$ cfu/swab
Filamentous algae (swab)	Few	Many

Note: cfu = colony-forming units.

TABLE 13.3 Guidelines for Microbiological Activity in Recirculating Cooling Water Systems Based on Differential Microbiological Analysis for Bulk Water and Swab Samples

ATP is a biomolecule that is required for cell activity and is a key indicator of the viability of an organism. Because all microorganisms contain ATP, measurements provide an indication of the total number of viable microorganisms in a sample. However, this method does not distinguish between ATP produced by different microorganisms.

Microscopy

Numerous microscopic methods are used to identify microorganisms in samples. Some common methods use light microscopy in conjunction with staining techniques to highlight certain microorganisms. Scanning electron microscopy can be very useful to identify morphologies associated with microorganisms in corrosion products. Microscopic methods can often identify filamentous-type structures in corrosion products that can be associated with metal-oxidizing bacteria (Fig. 13.10).

Genetic Techniques

In recent years, genetic techniques have been developed to identify and quantify microorganisms. One method capable of providing information on microbiological populations is the polymerase chain reaction (PCR), which amplifies target sequences of nucleic acid (DNA or RNA) in a sample through replication. The amplified sequences can be used to identify specific bacteria species within a sample. A major benefit of this technique is that it can be used to quantify microorganisms in dry samples after their removal from the system. Because of

FIGURE 13.10 Filamentous microorganisms encrusted with iron oxide present in corrosion products contained in a tubercle.

this, genetic methods have become more widely used for analysis. However, the technique does not necessarily provide information about whether the microorganisms were active at the time the component was removed from the system, although some specialized methods can be used to distinguish active microorganisms in some cases. Again, it is best practice to obtain samples both at and away from corrosion sites for comparison to assess the likelihood of microorganisms contributing to the metal loss.

Organism Types

The following sections provide details on some specific organism types, regarding factors other than biological analysis, which are required for diagnosis of biologically influenced corrosion. Organisms associated with passive corrosion are discussed first, followed by microorganisms responsible for active corrosion.

Large Organisms

A great variety of large organisms foul cooling water systems. Many creatures are frequently found in cooling tower basins, condensers, heat exchangers, and water supply systems, as conditions can be very favorable for growth at these locations.

Clams, mussels, and other bivalves are a serious problem in many cooling water systems (Fig. 13.11). Due to a lack of natural predators and prodigious reproduction rates, they have rapidly become a serious threat to the operation of many cooling water systems. Shells foul many cooling water systems on seacoasts and tend to lodge near tube

FIGURE 13.11 A main mill water strainer basket caked with clams.

inlet ends in shell-and-tube heat exchangers (Fig. 13.8). Objects slightly smaller than tube inner diameters are swept through tubes, and larger ones cannot fit into openings. Copper, brass, and cupronickel tubing may suffer severe, localized erosion-corrosion where shells touch tube surfaces.

Massive accumulations of shells may obstruct pipes, screens, valves, and other water passages. Often, obstruction occurs immediately after or during high-flow conditions in normally stagnant systems. Large numbers of weakly adhering and/or dead shells are detached from surfaces during high flow, causing valve malfunction and obstructions in the system.

Small Organisms

Admittedly, the distinction between large and small organisms is somewhat arbitrary—all large organisms must have been small at one time. From an empirical viewpoint, however, small organisms may be loosely defined as those creatures less than 0.4 in. (1 cm) in length. Into this category falls a wide variety of invertebrates and plants. Diatoms, larval bivalves, and other small organisms may be found attached to surfaces in cooling water systems (Figs. 13.12–13.14). Vegetative fibers, seeds, insects, and other biological materials are blown into cooling towers and make their way throughout the cooling system.

Small organisms may accelerate deposition. If fibrous, organisms may act as sieves and/or become encased in particulate. Small organisms may become incorporated into deposits and corrosion products (Fig. 13.15). Often, organisms are encased in or are cloaked by deposit and corrosion products, making visual recognition difficult.

FIGURE 13.12 Pinhead-sized, shell-like organisms covering a carbon steel pipe surface.

FIGURE 13.13 A 1 in. (2.5 cm) diameter mild steel service water pipe containing many small cone-shaped organisms.

FIGURE 13.14 Close-up of organisms shown in Fig. 13.13.

FIGURE 13.15 A corrosion product and deposit mound on a mild steel service water pipe that is honeycombed by small tube-like organisms. More than 50 percent by volume of the surrounding material is composed of these casts. Each hole is approximately 0.01 in. (0.025 cm) in diameter.

In some cases, metal -oxidizing bacteria cause manganese and/or iron enrichment in deposits. These filamentary organisms can be seen upon close inspection of affected surfaces by microscopic techniques. However, many small organisms without exoskeletons shrivel upon drying. Only creatures containing substantial inorganic constituents and other "skeletal" remnants not subject to degradation by desiccation are easily visible in dried material (Fig. 13.16).

Slime Layers

Slime layers are easily recognized by the gelatinous mass that accumulates on surfaces (Fig. 13.17). The material generally feels greasy or slick to the touch. When dry, slime will form a thin, brittle layer,

FIGURE 13.16 Small organisms found inside a tubercle. Each is about 0.05 in. (0.13 cm) high. The organisms have segmented, fibrous stalks and bivariate heads.

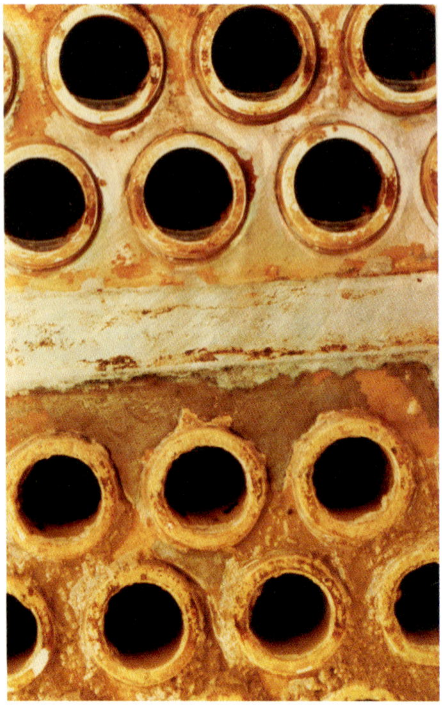

FIGURE 13.17 A gelatinous slime layer on the inlet side (bottom) of a heat exchanger tubesheet. Note that the outlet side is mostly clean.

which easily peels from the surface, often exfoliating in large patches (Fig. 13.18). Slime may be colored by dirt and other debris that accumulates in the gooey mass. A ninhydrin spot test can be used to qualitatively indicate the presence of proteins associated with slime.

Tubesheets and headboxes are frequently affected. Sliming is often more severe on inlet tube sheets than on outlets (Fig. 13.17). Passive attack beneath slime is usually general, and rusting on steel may color surfaces brown and red (Fig. 13.19).

Bacterial counts obtained from proper culturing of the material in slime layers are usually quite high, exceeding tens of millions (Table 13.4). In waters taken from such systems, bacteria counts are often several orders of magnitude lower.

Locations covered by a slime layer will provide an environment conducive to the growth of anaerobic microorganisms. If sulfate reducers are present, localized areas with pitting may be superimposed on the generally corroded surface (Fig. 13.5).

Sulfate-Reducing Bacteria

Sulfate-reducing bacteria require anaerobic environments to flourish. These environments may be highly localized, such as inside a tubercle,

FIGURE 13.18 A dried slime layer peeling off water box surfaces on a small heat exchanger.

beneath a spotty deposit, or within a crevice. Locations beneath slime layers and established biofilms provide environments suitable for SRB activity.

An example of a microbiological analysis in a troubled carbon-steel service water system is given in Table 13.5. When biological

FIGURE 13.19 A generally corroded carbon steel surface revealed after a slime layer was removed.

	Water	Slime Layer
Total aerobic bacteria	8000	46,000,000
Enterobacter	<1000	100,000
Pigmented	<1000	None
Mucoids	<1000	None
Pseudomonas	<1000	9,100,000
Spores	None	9200
Total anaerobic bacteria	<10	33,000
Sulfate reducers	None	30,000
Clostridia	None	3000
Total fungi	1000	3100
Yeasts	<10	<100
Molds	<10	3000
Iron-depositing organisms	None	None
Algae	None	None
Other organisms	None	None

TABLE 13.4 Typical Microbiological Analysis of the Slime Layer in Fig. 13.18 and from the Water in the Cooling System. All counts expressed as colony-forming units (cfu) per milliliter or gram.

	Water	Deposits and Corrosion Products
Total aerobic bacteria	490,000	1,100,000
Enterobacter	30	<1000
Pigmented	<70	<1000
Mucoids	<10	<1000
Pseudomonas	100,000	210,000
Spores	12	8700
Total anaerobic bacteria	35	150,000
Sulfate reducers	30	120,000
Clostridia	5	30,000
Total fungi	2	<100
Yeasts	2	<100
Molds	None	<100
Iron-depositing organisms	None	None
Algae-Filamentous	Very few	None
Algae-Nonfilamentous	Few	Few
Protozoa	Few	None

TABLE 13.5 Typical Microbiological Analysis in a Service Water System Pipe Experiencing Microbiologically Influenced Corrosion. All counts expressed as colony-forming units (cfu) per milliliter or gram.

counts of sulfate reducers in solid materials scraped from corroded surfaces are more than about 10^4, significant attack is possible. Counts above 10^5 are common only in severely attacked systems. It is important to note that SRB are probably not uniformly distributed throughout the deposit and corrosion-product mass (especially in aerated systems).

In addition, planktonic counts (in water samples) are usually unreliable as an indicator of active corrosion. The presence of any sulfate reducers in the water, however, indicates much higher concentrations of viable sessile bacteria somewhere in the system.

Table 13.6 shows a similar analysis for a different cooling water system, which had significant levels of sulfate reducers present in the water. The deposits on the metal surface of a cupronickel utility main condenser indicated appreciable levels of aerobic bacteria such as *Pseudomonas* and some accumulation of SRB. However, the metal at this location showed no significant corrosion associated with sulfate reducers.

Tests have been developed for the detection of sulfate reducers without the need for culturing. These tests are based on detecting certain compounds produced by SRB and have applicability (in some cases) even if the producing organisms have died. Laboratory studies have shown adequate agreement between such tests and live culture analyses when viable organisms are present.

	Water	Deposits
Total aerobic bacteria	310,000	14,000,000
Enterobacter	10	<1000
Pigmented	12,000	<1000
Mucoids	<10	<1000
Pseudomonas	30,000	4,200,000
Spores	None	1000
Total anaerobic bacteria	350	1300
Sulfate reducers	350	1000
Clostridia	None	300
Total fungi	20	1000
Yeasts	None	<100
Molds	20	1000
Iron-depositing organisms	None	None
Algae	None	None
Other organisms	None	None

Table 13.6 Typical Microbiological Analysis at a Main Condenser Outlet, which Showed No Significant Corrosion by Sulfate-Reducing Bacteria on Cupronickel Tubes. All counts expressed as colony-forming units (cfu) per milliliter or gram.

FIGURE 13.20 Pitting on the waterside surface of a Carpenter 20 heat exchanger tube caused by sulfate reducers.

Corrosion Morphologies Sulfate-reducing bacteria frequently cause localized corrosion. Discrete hemispherical depressions form on most alloys, including Carpenter 20 (Fig. 13.20), stainless steels (Figs. 13.21 and 13.22), aluminum (Fig. 13.5), and carbon steels (Fig. 13.23). Metal loss morphology on copper alloys is not well defined.

Pit interiors are characteristically smooth and distinctly hemispherical, but become rougher on less-noble alloys. Pits tend to cluster together, consistent with the formation of colonies on metal surfaces. Many overlapping pits form irregularly dimpled surfaces. Frequently, a lightly etched aureole surrounds the pit clusters. These etched areas are often produced by shallow corrosion beneath deposit and slime masses that covered the sulfate reducers in service.

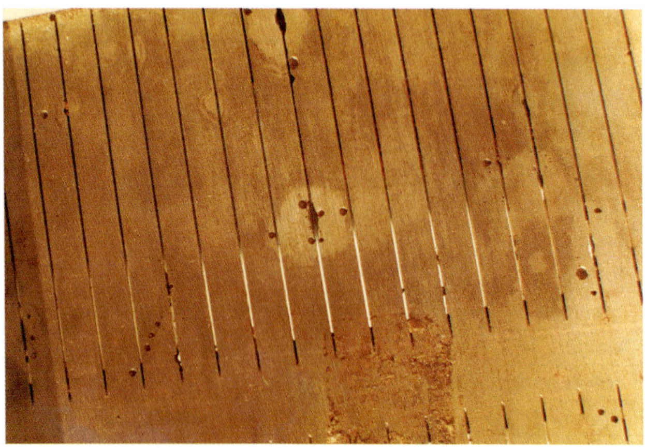

FIGURE 13.21 Small pits on a 316 stainless steel plate. A light area covers the clustered pits, marking ghost images of deposit mounds. Note the smooth hemispherical pit interiors covered by a black layer and the tendency of the pits to cluster together.

Figure 13.22 Many small hemispherical pits on a 304 stainless steel heat exchanger tube end. The heat exchanger was removed from service and stored vertically for an extended period. Deposit accumulated at the lower tube ends where sulfate reducers flourished.

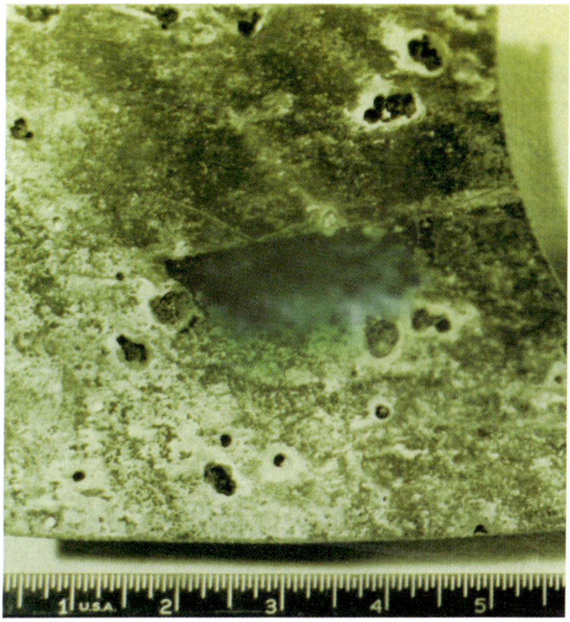

Figure 13.23 Clustered sulfate-reducer pits on a carbon steel tank bottom.

In metallographic cross sections, pits may exhibit scalloping and mild undercutting, especially on stainless steels (Fig. 13.24). On carbon steels, pit interiors are somewhat less regular. Interiors may contain terrace steps, giving the pits a peculiar bulls-eye pattern when viewed

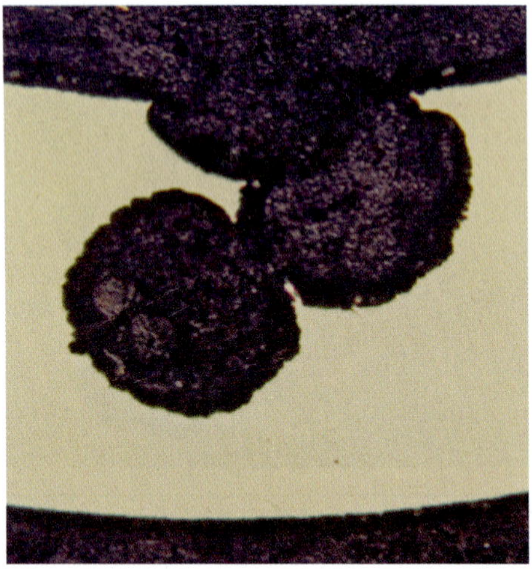

FIGURE 13.24 Scalloped, partially undercut pits in a metallographic cross section of a 316 stainless steel tube.

from above (Fig. 13.25). It is tempting to speculate that each terrace marks a corrosion-arrest stage in pit development corresponding to changes in biological activity.

Attack at welds due to bacteria is possible, but it is not nearly as common as is often supposed. Welds show a predisposition to corrode

FIGURE 13.25 Metal loss areas on the internal surface of steel fire protection piping with significant contributions from sulfate-reducing bacteria and acid-producing bacteria. The areas have terraced edges and interiors have a dimpled contour due to clusters of mutually intersecting superimposed depressions. Note also the striated features due to acidic conditions that developed within the areas.

FIGURE 13.26 Irregular deposit and corrosion-product mounds containing concentrations of sulfate-reducing bacteria on the internal surface of a 316 stainless steel transfer line carrying a starch-clay mixture used to coat paper material. Attack only occurred along a weld that had incomplete penetration, and it resulted in many perforations.

compared to other locations in the system because of factors such as residual stresses, microstructural differences, compositional variation, and surface irregularities. These factors will promote preferential corrosion of the welds by most corrosion mechanisms, including MIC. Corrosion is common along locations of incomplete penetration of butt welds that join piping (Figs. 13.26 and 13.27) and along incompletely closed pipe and tube weld seams. The crevices at these locations can develop anaerobic conditions suitable for SRB. Figure 13.28 shows a severely corroded carbon steel pipe from a service water system. Wastage began at a circumferential weld employing a backing ring, which also created a crevice (see Case History 13.1).

Corrosion Products and Deposits Sulfate-reducing bacteria produce metal sulfides as corrosion products. Sulfide usually lines pits or is entrapped in material just above the pit surface, with the sulfide appearing as a black layer on most metals.

Numerous spot tests can be used to indicate the presence of sulfide. When freshly corroded surfaces are acidified, the rotten-egg odor of hydrogen sulfide is easily detected. Lead acetate paper will produce a color change when exposed to hydrogen sulfide. More sensitive spot tests using sodium azide solutions are often successful at detecting metal sulfides at very low concentrations in localized areas on surfaces.

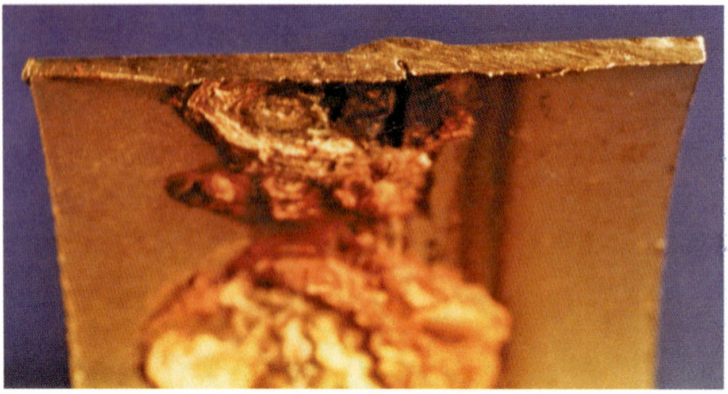

FIGURE 13.27 Longitudinal cross section of a pipe, showing a pit tunneling into location of incomplete penetration of a weld and overlying corrosion products with a black layer covering the metal.

FIGURE 13.28 A perforated carbon steel pipe at a weld-backing ring. The gaping pit was caused by sulfate-reducing bacteria.

Rapid, spontaneous decomposition of metal sulfides occurs after sample removal, as water vapor in the air adsorbs onto metal surfaces and reacts with the metal sulfide. The metal sulfides are slowly converted to hydrogen sulfide gas, and may eventually remove all traces of sulfide. Therefore, in the absence of significant amounts of occlusive corrosion products and/or deposits, freshly corroded surfaces will contain appreciably more sulfide. Pseudomorphically oxidized sulfide, which appears as distinct hexagonal platelets, can sometimes be found in tubercles that had SRB present (Fig. 13.29).

Sulfides are typically intermixed with iron oxides and hydroxides on carbon steels and cast irons. Carbon steel pits are commonly capped

FIGURE 13.29 Scanning electron microscopy image of pseudomorphically oxidized sulfide platelets found in the interior of a tubercle in a fire protection system.

with voluminous, friable brown iron oxide mounds (tubercles), sometimes containing black iron sulfide plugs (Fig. 13.28). Additionally, corrosion product stratification is sometimes very distinct, with sulfide concentration generally being highest in the corrosion products near metal surfaces. Cross-sectional elemental mapping by energy dispersive spectroscopy can be used to show areas with elevated sulfur levels. The corrosion products filling the pits sometimes will have a mottled appearance, consisting of sulfides mixed with oxides (Fig. 13.30).

FIGURE 13.30 Metallographic cross section through a depression, showing mottled corrosion product containing iron oxides and iron sulfide.

Stainless steels attacked by sulfate reducers show well-defined pits. Black metal sulfides are generally present within pits on freshly corroded surfaces (Fig. 13.21). Rust stains may surround pits or form streaks running in the direction of gravity or flow from attack sites. Sometimes, pits will have a bare-surfaced, shiny appearance (Fig. 13.22).

Acid-Producing Bacteria

Corrosion usually is moderate and localized. Almost all significant attack is associated with anaerobic bacteria, as aerobic acid-producing varieties usually reside near the top of deposits and corrosion products contacting oxygenated waters. Thus, the direct effect on corrosion at the metal surface from the aerobic microorganisms is limited. Additionally, although acidic products may be expected to increase corrosion rates, acidity cannot be extremely low in deposits since the deposits and corrosion products would dissolve at sufficiently low pH.

Clostridia frequently are also often found where sulfate-reducing bacteria are present, sometimes at high levels inside tubercles. A microbiological analysis of tubercle material removed from the corroded surface in a troubled service water system main is given in Table 13.7. *Clostridia* counts above 10^2 cfu/g of material are high

	Deposit and Corrosion Products
Total aerobic bacteria	1,300,000
Enterobacter	<1000
Pigmented	<1000
Mucoids	<1000
Pseudomonas	150,000
Spores	900
Total anaerobic bacteria	24,400
Sulfate reducers	22,200
Clostridia	2200
Total fungi	<100
Yeasts	<100
Molds	<100
Iron-depositing organisms	None
Algae	None
Other organisms	None

Table 13.7 Microbiological Analysis of Tubercular Material Containing High Concentrations of *Clostridia*. All counts expressed as colony-forming units (cfu) per gram.

enough to cause concern. When acid producers markedly accelerate corrosion, counts are often greater than sulfate reducer counts.

Surfaces beneath affected tubercles often have a striated contour due to increased acidity (Fig. 13.25). Striated surfaces are caused by preferential attack along microstructural and microcompositional irregularities that have been elongated during steel rolling (see Chap. 14, "Acid Corrosion").

Short-chain and low-molecular-weight organic acids, such as acetic acid and formic acid, can be formed by certain bacteria. The resulting organic acid salts are not easily detected without techniques such as infrared spectroscopy.

Elimination

Best practice for avoiding biologically influenced corrosion in a cooling water system is to start with a clean system and avoid the development of fouling conditions. Once biological material develops in locations, it may be difficult to remove and chemically treat. Often, troubled areas may reinoculate other parts of the system. Once established, multiple contributing factors in the system often need to be addressed. Biological corrosion and deposition may be prevented by system design, system operation, and chemical treatment.

Perhaps the single greatest enemy of good biological control in cooling water systems is low flow. Stagnant regions are almost always the first places in which MIC occurs. Deadlegs should be eliminated if possible. Proper flow not only prevents settling of detritus, but also replenishes biocides and corrosion inhibitors. The costs of chemical treatment can usually be reduced by appropriate system design and operation.

Water treatment can include a wide variety of biocides to control almost any biological problem. Oxidizing biocides, such as chlorine, hypochlorite, chlorine dioxide, bromine, peroxide, and ozone, are primarily used in open and once-through systems. A variety of nonoxidizing biocides, which attack microorganisms via different mechanisms, are also available for treatment. Proper choice of biocides is system dependent. Biodispersants may also be used to aid in the removal of existing biofoulants from surfaces and allow them to be removed from the system.[7] Discharge limitations, associated corrosion, and other problems will often restrict chemical use.

In addition to proper selection of biocides for a system, their effectiveness will depend on dosage. Periodic slug feeding is often utilized. Shocking with massive amounts of biocides may be effective in treating some systems, but not all systems will respond identically. Shocking heavily fouled systems may produce sloughing of large biological mats that plug components. After shocking, bacterial growth may be rapid, and the system can return to its previous state quickly. It is imperative that biological control not be erratic. It is much easier and decidedly less costly to maintain good control than to bring a seriously troubled system back into control.

Large organisms such as clams and mussels can be removed by strainers and grates. Bivalves are relatively resistant to some biocides but will succumb if treatment is persistent. Larval forms must be eliminated by chemical treatment. Knowledge of life cycles and related information can be necessary in controlling large organisms. Treatment is much more effective at certain times in each organism's development.

Dissolved oxygen is the enemy of anaerobic bacteria. Oxygen corrosion consumes oxygen. Without convective replenishment, oxygen-depleted zones form and corrosion by anaerobes may be worse.

Finally, microbiological issues during service may have begun due to preoperational conditions. Hydrostatic testing of equipment and piping should follow proper procedures. The procedures typically require the use of treated water of suitable quality. In addition, the equipment should be drained if it is not going to be placed into service soon after testing to avoid prolonged stagnant periods.

Cautions

Bacteria are present in virtually all cooling water systems. Indeed, corroded systems usually contain a diverse group of microorganisms (as do unattacked systems). The presence of organisms (large or small) in proximity to corrosion by itself is not proof of biologically influenced corrosion. It should be stressed vigorously that all evidence must be consistent with any single corrosion mode before a definitive diagnosis can be made (see "Critical Factors"). Further, all alternative explanations must be carefully examined, as many forms of corrosion resemble each other. However, this book highlights how in many cases a corrosion mechanism produces a unique "fingerprint" by which it can be differentiated from all other forms of attack.

Many cases of attack in cooling water systems can be misdiagnosed as biologically influenced corrosion if not carefully investigated. In addition, microbiological influences are often not appropriately considered in some instances. The critical issue to address when investigating any case of suspected MIC is the extent to which organisms have influenced attack. For instance, the following questions should be answered: Is the corrosion exclusively associated with biological activity, or would essentially the same corrosion have occurred in the absence of organisms? Do other coexisting or competing forms of corrosion better represent observed damage morphologies, deposits, and corrosion products? Such questions can be answered, but only by careful, unprejudiced, critical examination by informed investigators well versed in cooling water system failure analysis.

Related Problems

See also Chap. 11, "Crevice and Underdeposit Corrosion (Concentration Cells)"; Chap. 12, "Oxygen Corrosion"; Chap. 14, "Acid Corrosion"; and Chap. 22, "Stress Corrosion Cracking."

Case History 13.1

Industry:	Nuclear utility
Specimen location:	Emergency service cooling water system
Orientation of specimen:	Horizontal
Environment:	Internal: Cooling water near ambient temperature, intermittent low flow, daily chlorination for 1 hour External: Ambient air
Time in service:	5 years
Sample specifications:	3.5 in. (8.9 cm) outer diameter, carbon steel pipe, wall thickness 0.225 in. (0.572 cm)

A weeping leak developed in a carbon steel emergency service water pipe at a circumferential weld employing a weld-backing ring. A rubberized saddle clamp was used to plug the leak temporarily. After several weeks the section was cut out of the system and the failure was examined.

The tube was received while still wet, less than 48 hours after removal from the system. Though the sample was not shipped cold as recommended, the timeliness of the microbiological analysis ensured that the detected species were most likely representative of those present during service. Materials taken near the failure site and water samples from the system were analyzed. The resulting microbiological analyses are given in Table 13.8.

	Water	Deposits and Corrosion Products
Total aerobic bacteria	2,400,000	1,700,000
Enterobacter	<1000	<1000
Pigmented	<10,000	<10,000
Mucoids	<10,000	<10,000
Pseudomonas	20,000	150,000
Spores	2000	9500
Total anaerobic bacteria	35,000	32,000
Sulfate reducers	30,000	25,000
Clostridia	5000	7000
Total fungi	<100	300
Yeasts	<100	<100
Molds	<100	300
Iron-depositing organisms	None	None
Algae	None	None
Other organisms	None	None

TABLE 13.8 Microbiological Evaluations of Water and Deposits in Case History 13.1. All counts expressed as colony-forming units (cfu) per milliliter or gram.

FIGURE 13.31 A longitudinally split carbon steel emergency service water pipe. Note the circumferential weld-backing ring, which is consumed on one side (top).

Internal surfaces were heavily tuberculated (Fig. 13.31). The circumferential weld backing ring was severely corroded. Pits were filled with deposits, friable oxides, and other corrosion products. Black plugs embedded in material filling the gaping pit at the perforation contained high concentrations of iron sulfide. Bulk deposits contained about 90 percent iron oxide.

The perforation shown in Fig. 13.28 occurred at the backing ring, which was entirely consumed by corrosion at this location. At intact areas the weld was sound, showing adequate penetration and no unusual characteristics. The weld was riddled with mildly undercut, gaping pits. Deep attack was confined to the fusion and heat-affected zones, with a pronounced lateral or circumferential propagation (Fig. 13.28). The resulting perforation at the external surface was quite small.

The large amounts of sulfide, high bacterial counts, pit morphology, and other factors strongly indicated corrosion was accelerated by sulfate-reducing bacteria. Microscopic evidence also implicated *Clostridia.* The crevice between the backing ring and the tube wall provided a shielded location and created an environment favorable for anaerobes.

The pit morphology was particularly interesting. The major depression was cavernous and showed lateral propagation. However, only a small weeping perforation was present. It is tempting to speculate that once the pipe wall was breached, air in the vicinity of the perforation limited anaerobic activity, thus producing the laterally propagating pit.

Case History 13.2

Industry:	Steel
Specimen location:	Thermocouple housing in a cooling water supply line
Orientation of specimen:	Vertical
Environment:	90–110°F (32–43°C), pH 8.5, flow 5–6 ft/s (1.4–1.8 m/s), conductivity 60 μmhos
Time in service:	1 year
Sample specifications:	3.5 in. (8.9 cm) long, 0.5 in. (1.3 cm) diameter, mild steel, wall thickness 0.12 in. (0.30 cm)

Surfaces exposed to water contained many localized areas of wastage (Fig. 13.32). The localized areas of wastage consisted of irregular metal-loss regions covered by brittle black or brown caps. Many small fibers were embedded in corrosion

FIGURE 13.32 A thermocouple housing showing localized wastage. Small fibers containing high concentrations of sodium chloride were found in corroded areas.

products, and many more were wrapped around surfaces. Most fibers were about 0.001 in. (0.025 mm) in diameter and were hollow. Chemical analysis showed that up to 20 percent by weight of the fibers was sodium chloride.

The fibers were vegetative in origin, possibly seed hairs. They concentrated salt from the brackish water in which their parent plant grew. These fibers blew into cooling towers and were circulated throughout the cooling water system. Some became attached to rough surfaces. The high concentration of salt in the fibers resulted in underdeposit corrosion due to salt-laden biological material.

Case History 13.3

Industry:	Alcohol production
Specimen location:	Condenser
Orientation of specimen:	Horizontal
Environment:	Internal: Cooling water, exit temperature 120°F (29°C), calcium 270 ppm, magnesium 150 ppm, sodium 94 ppm, manganese 1.2 ppm, chloride 94 ppm, fluoride 9.3 ppm, pH 7.3 External: Ethyl alcohol
Time in service:	18 months
Sample specifications:	304 stainless steel tubing, 0.75 in. (1.9 cm) outer diameter

The section was perforated in several locations due to severe, localized wastage on the internal surface. The cooling water had a history of low-pH excursions, with documented excursions to a pH below 5. The system also had been plagued with high sulfate-reducing bacteria counts.

Internal surfaces were covered with brown and tan deposits and corrosion products (Fig. 13.33). Pits were present beneath small mounds of reddish-brown

FIGURE 13.33 A longitudinally split 304 stainless steel condenser tube covered with deposits.

FIGURE 13.34 The condenser tube shown in Fig. 13.33, but with deposits removed to show pits.

FIGURE 13.35 Metallographic cross section, showing a smaller undercut pit growing off the large hemispherical depression.

oxide. When deposits and corrosion products were completely removed, many pits and depressions were revealed (Fig. 13.34). Most pits and depressions were shallow, but some were deep. Chemical spot tests revealed appreciable indications of sulfides in depressions. Metallographic cross sections revealed the larger depressions had distinctly hemispherical contours (Fig. 13.35). In places, smaller undercut pits extended from the interiors of the larger hemispherical depressions.

The differences in morphologies of the depressions and pits indicated two distinct causes for metal loss. Sulfate reducers had formed the large hemispherical depressions. The more undercut pits were formed during a low-pH excursion involving mineral acid after the sulfate-reducing bacteria became inactive. It is likely the low-pH excursion deepened preexisting sulfate-reducer depressions, causing final perforation.

Case History 13.4

Industry:	Nuclear utility
Specimen location:	Emergency service water system piping
Orientation of specimen:	Vertical and horizontal (90° bend)
Environment:	Internal: Ambient river water, unclarified External: Ambient air
Time in service:	10 years
Sample specifications:	1 in. (2.5 cm) outer diameter, mild steel pipe

FIGURE 13.36 Low-magnification scanning electron micrograph of small tube-like organisms in Figs. 13.13 and 13.14. Note how all organisms (except one) point into the direction of flow.

After almost 10 years of service, several pipes were removed because of restricted flow associated with accumulation of corrosion products and macrofouling (Fig. 13.13). The most heavily fouled sections contained unclarified river water and remained stagnant for as long as 30 days at a time. Eighteen months before removal of the pipes, parts of the system had been cleaned using high-pressure water jets. No failure occurred in the piping. The deepest observed wastage was only approximately 0.030 in. (0.076 cm). However, small-diameter tubes were severely obstructed. Up to half the volume of obstructing material consisted of small, tube-like organisms (Fig. 13.14) similar to bryozoans.

Chemical analysis showed that each organism contained up to 50 percent silica by weight. Each was coated with iron oxides, silt, and other deposits and corrosion products. In places, large deposit accumulations were clearly correlated with large numbers of organisms.

From the orientation of the tube-like organisms (Fig. 13.36), it was clear that they were flourishing even during periods with sufficient flow of chlorinated water.

Case History 13.5

Industry:	Light manufacturing
Specimen location:	Cooling water recirculation line
Orientation of specimen:	Horizontal
Environment:	Untreated cooling water from a pond
Time in service:	About 4 years
Sample specifications:	Type 304L stainless steel pipe, 6.5 in. (15.4 cm) outer diameter

The cooling water piping system was developing leaks with increasing frequency. The majority of the leaks occurred at circumferential butt welds joining piping. Examination of the internal surface at a leaking circumferential weld revealed large deposit and corrosion product nodules on the surfaces at many locations (Fig. 13.37). In addition to the large corrosion product mounds, the internal surfaces were covered by many small, black organisms and other evidence of biological material (Fig. 13.38).

FIGURE 13.37 Large corrosion product mound located along the circumferential weld. Note the many black spots scattered across the surface that are small organisms.

FIGURE 13.38 Close-up of biological material on the internal surface of the stainless steel pipe.

The mounds at the welds contained friable, filamentary corrosion products, typical morphologies associated with metal-oxidizing bacteria. Chemical analysis of material on the surface using a scanning electron microscope equipped with an energy dispersive spectrometer indicated areas that had elevated concentrations of manganese. It was likely that metal-oxidizing bacteria contributed to manganese oxide deposition on the surface, as the water was not treated with

FIGURE 13.39 Small pit adjacent to the circumferential weld (left side of the image) and an area of shallow metal loss having a wavy morphology at a location beneath a large corrosion product mound.

oxidizing biocides. In addition, compositional analysis of the corrosion products indicated elevated sulfur levels in the material. Spot tests identified the presence of sulfides. Evidence of chlorides was not observed. The underlying surface beneath the corrosion product mounds had shallow, wavy metal loss areas and deep pits along the welds, some of which perforated the pipe wall (Fig. 13.39).

Analysis of the pond water at different times indicated chloride levels between 60 and 120 ppm (as $CaCO_3$). These concentrations of chlorides would not be expected to result in pitting attack on 304 stainless steel for the reported temperatures of the water. Microbiological analysis indicated 20 cfu/mL of sulfate-reducing bacteria in a sample of the pond water. The various pieces of evidence in this case consistently indicated MIC was primarily responsible for the corrosion in this system, with contributions from sulfate-reducing bacteria and metal-oxidizing bacteria that deposited manganese oxide. Importantly, the pond water was not treated to control microbiological activity during service. Numerous weld-related issues, such as appreciable heat tint oxide, also increased the susceptibility to corrosive attack.

Case History 13.6

Industry:	Chemical processing
Specimen location:	Coupon rack in an open cooling water system
Orientation of specimen:	Horizontal
Environment:	Treated cooling water
Time in service:	33 days
Sample specifications:	Mild steel coupon

Corrosion monitoring by mild steel coupons typically indicated corrosion rates over 15 mpy calculated by weight loss, with localized metal loss that penetrated as deep as 0.016 in. (0.406 mm).

The coupons removed from the system were generally covered by unusual corrosion products that were oriented in the direction of flow across the coupon

Figure 13.40 Mild steel coupon surface that is generally covered by moderately thick layers of deposits and corrosion products.

surface (Fig. 13.40). Chemical analysis of the material indicated it was not crystalline and that it consisted primarily of iron with minor amounts of phosphorus. Trace amounts of calcium, silicon, and sulfur were also present. The loss at 925°C from the sample was 23 percent by weight, which included appreciable contributions from water of hydration and likely organic material.

Microscopic examination of the material on the surface revealed distinct platelets in the corrosion products. Localized chemical analysis using a scanning electron microscope equipped with energy dispersive spectroscopy indicated the platelets consisted of iron with sulfur at appreciable levels (13 wt%). The presence of sulfides was also confirmed by spot tests.

Removal of the deposits and corrosion products revealed distinct, localized metal loss areas (Fig. 13.41). The areas had clusters of hemispherical depressions and growth rings that emanated from the clusters along the direction of flow.

The evidence presented on the coupon, including the characteristics of the corrosion products and morphology of the metal loss areas, strongly indicated that MIC was primarily responsible for the high corrosion rates and localized attack on the coupons. Microbiological analysis of the material confirmed the diagnosis, as swabs from the surface of the coupons immediately after removal from the system indicated the presence of viable sulfate-reducing bacteria and acid-producing bacteria. This case history illustrates how analysis of corrosion coupons

Figure 13.41 Wide areas of localized metal loss that spread out from a central point that contains clusters of many small, hemispherical depressions.

can provide valuable information about the cause of corrosion as well as the extent and type of attack.

Case History 13.7

Industry:	Building
Specimen location:	Basin of a cooling tower
Orientation of specimen:	Horizontal
Environment:	Treated cooling water
Time in service:	Unknown
Sample specifications:	Galvanized steel, overall sheet thickness 0.100 in. (2.54 mm)

Deep localized metal loss occurred on the basin floor of a cooling tower, at a location where makeup water was added (Fig. 13.42). These areas were reportedly covered by thick layers of sludge.

Microbiological analysis of swabs from the cooling tower basin in the areas of localized metal loss indicated the presence of sulfate-reducing bacteria (at 1000 cfu/swab) and appreciable amounts of slime-forming aerobic bacteria (26,000,000 cfu/swab) that included *Pseudomonas* (1,200,000 cfu/swab).

Numerous wide, deep depressions on the surface reduced the sheet thickness by nearly 70 percent. The depression interiors were generally bare-surfaced with an irregular contour having a smoothened, terraced morphology (Fig. 13.43). Spot tests indicated the presence of sulfides in the reddish-brown corrosion products surrounding the depressions.

The layers of sludge that accumulated in areas of the cooling tower basin shielded the underlying metal surface and provided suitable environments for anaerobic organisms. Sulfate-reducing bacteria were at least partly responsible for the MIC on the galvanized steel surface. Acid-producing bacteria were

FIGURE 13.42 Area in the corner of a cooling tower that was covered by a thick sludge layer. Removal of the sludge layer revealed many wide, moderately deep depressions.

Figure 13.43 Wide depressions having a smoothened, terraced morphology surrounded by brown corrosion products that contain sulfide.

suspected to have also contributed, based on the bare surface and appearance of most depressions. Frequent cleaning of the cooling tower basin and chemical treatment with biocides was implemented to reduce the attack.

Case History 13.8

Industry:	Laboratory
Specimen location:	Cooling water piping
Orientation of specimen:	Horizontal
Environment:	Untreated water from various sources including wells
Time in service:	Several months
Sample specifications:	Type 304 stainless steel, schedule 10 pipe

Numerous leaks occurred at circumferential welds throughout sections of the piping system after hydrotesting. A section submitted for analysis had corrosion product mounds on the internal surface overlying the circumferential weld at multiple locations (Fig. 13.44). A leak was present along the reported bottom of the pipe, and deposits and corrosion products extended along longitudinal directions from this location. There was appreciable misalignment of the pipes at the weld on this side (Fig. 13.45).

In addition to misalignment in places, the circumferential weld had a band of very dark heat tint oxide centered on the weld along the entire circumference (Fig. 13.44). Metallographic examination also indicated locations of incomplete penetration of the weld. The microstructure of the weld fusion zone was dendritic with interconnected ferrite. Measurements indicated high ferrite content in the weld, a condition related to high heat input. Thus, the circumferential weld had numerous characteristics that appreciably reduced the corrosion resistance, and deep, cavernous pits extended from locations at and near the weld into the metal.

The water used for hydrotesting had a relatively low chloride content that would not be expected to initiate corrosion on type 304 stainless steel. However, microbiological analysis revealed that some water sources used for hydrotesting were found to contain sulfate-reducing bacteria. The corrosion products

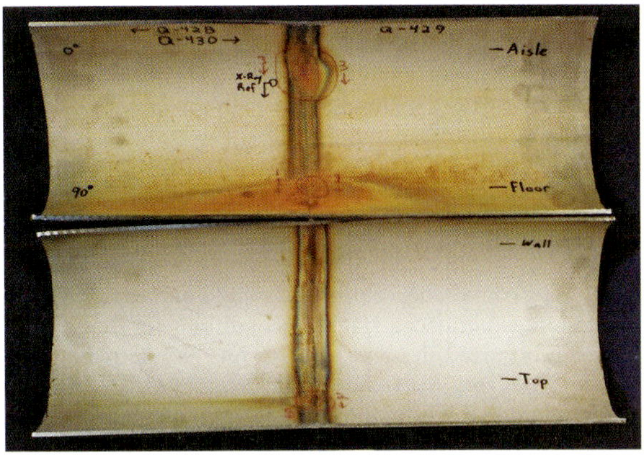

Figure 13.44 Multiple corrosion product mounds on the internal surface along the circumferential welds. A leak was noted at the location along the bottom of the pipe (floor). Note the dark oxide band (heat tint) straddling the circumferential weld.

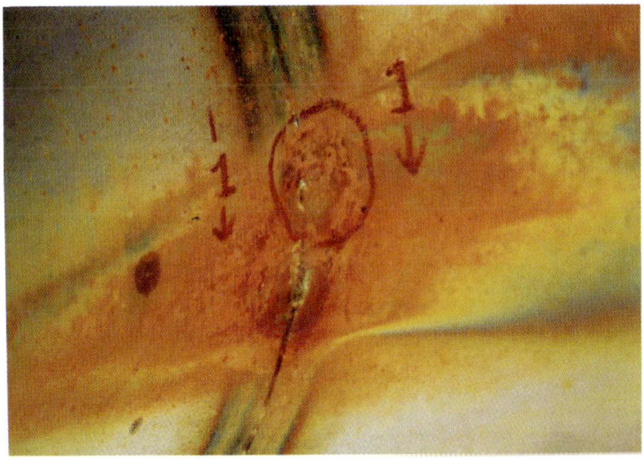

Figure 13.45 Corrosion along the bottom of the piping at a location of significant misalignment at the butt weld. A leak occurred at this site.

overlying pit sites had elevated sulfur levels in places, and spot tests indicated the presence of sulfides and sulfates.

The use of untreated water containing sulfate-reducing bacteria, and extended stagnant periods after hydrotesting, combined to initiate corrosion along a circumferential weld which contained defects that made the surface highly susceptible to corrosive attack (see Chap. 26, "Weld Defects"). Incomplete penetration of the weld resulted in crevices that provided environments suitable for anaerobic organisms such as SRB. Weld repairs were required using proper welding procedures. The use of untreated water was discontinued for hydrotesting. It was also recommended that the water be drained and not allowed to remain stagnant in sections for long periods of time.

Case History 13.9

Industry:	Pharmaceutical
Specimen location:	Chilled process loop
Orientation of specimen:	Horizontal
Environment:	Closed cooling water system treated with nitrite
Time in service:	25 years
Sample specifications:	Carbon steel piping, outer diameter 4-1/2 in. (11.4 cm), wall thickness 0.234 in. (5.9 mm)

A section of piping containing a perforation was removed from a closed cooling water system that experienced stagnant conditions for long periods of time. The perforation occurred in a wasted area on the internal surface that was covered by a large, very thick mound of material (Fig. 13.46). Microscopic examination of material within the corrosion product mound revealed pseudomorphically oxidized sulfide platelets and cellular structures in places (Fig. 13.47). In addition, spot tests indicated the presence of sulfides in the material beneath the corrosion product mounds. Removal of material on the surface revealed areas of metal loss and numerous depressions, some of which were deep (Fig. 13.48). The deeper depressions were hemispherical and had smooth interiors.

The presence of sulfides in material beneath the corrosion product mounds on the internal surface, combined with the morphology of the deep depressions, suggested that MIC due to sulfate-reducing bacteria contributed to the metal loss. Low flow and stagnant conditions promoted microbiological activity and underdeposit corrosion, due to the accumulation of deposition and microbiological material. Corrosion inhibitors and biocides are much less effective under low and stagnant flow conditions, particularly for extended periods of time.

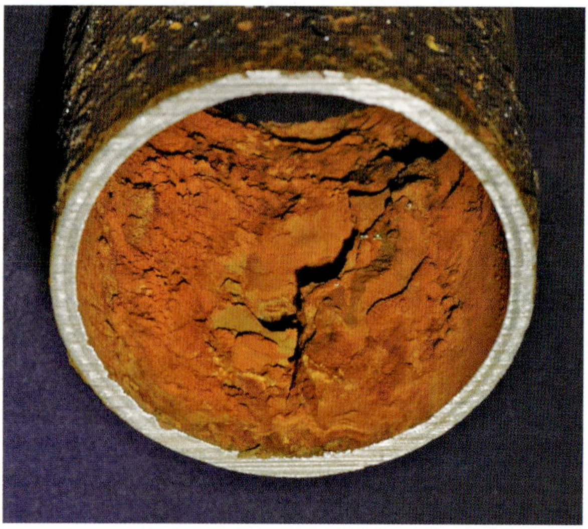

Figure 13.46 Internal surface, showing the remnant of a large corrosion product mound of material near the perforation.

FIGURE 13.47 Corrosion products in the material beneath the large mound containing sparkling crystals of pseudomorphically oxidized sulfide.

FIGURE 13.48 Internal surface of the pipe before (top) and after (bottom) cleaning by glass bead blasting. Note the clusters of moderately deep, hemispherical depressions beneath the layers of deposits and corrosion products.

In addition, monitoring water chemistry of the system revealed low nitrite corrosion inhibitor levels for a period of time. This was likely associated with the activity of denitrifying bacteria. Microbiological analysis detected the presence of denitrifying bacteria in the water of the closed loop. Changes to the biocide treatment program were able to reduce microbiological activity, but cleaning the surfaces of deposits and corrosion products was needed to reduce corrosion by underdeposit mechanisms.

Case History 13.10

Industry:	Pulp and paper
Specimen location:	Stock piping after alum injection point
Orientation of specimen:	Horizontal
Environment:	Papermaking stock 125°F (52°C), pH between 6.2 and 6.6.
Time in service:	25 years
Sample specifications:	Type 316 stainless steel piping, wall thickness 0.625 in. (15.9 mm)

Piping from the stock system experienced corrosion confined to locations at and slightly downstream from the injection point for alum. A section in this area was removed from the bottom of the pipe, and the internal surfaces were reportedly covered with thick layers of deposits that were determined to be 99 percent alum.

The internal surface contained areas with moderately deep metal loss. Metal loss was in the form of depressions that were generally hemispherical with smooth interiors, and many were clustered together (Fig. 13.49). A thin layer of black corrosion products lined most of the depressions, and spot tests indicated the presence of sulfates and sulfides in the corrosion product layers.

FIGURE 13.49 Metal loss areas consisting of clusters of hemispherical depressions lined by black sulfide-containing layers on the internal surface of stainless steel paper stock piping.

FIGURE 13.50 Metallographic cross section through a depression, showing overlying corrosion products that contain a network of sulfide.

Metallographic cross sections reveal mottled corrosion products (Fig. 13.50). Analysis using energy dispersive spectroscopy indicated a network of sulfur.

Although the evidence on the sample strongly suggested that metal loss was caused by microbiologically influenced corrosion due to sulfate-reducing bacteria, microbiological analysis of the water and deposit was not available at the time the sample was removed from the system. However, analysis of the corrosion products by the polymerase chain reaction (PCR) technique indicated genetic evidence of species of sulfate-reducing bacteria in the sample, allowing for complete diagnosis.

The thick layers of deposited alum that accumulated along the bottom of the pipe provided an oxygen-depleted environment conducive to SRB activity and shielded them from biocides added to the stock for microbiological control. In addition, the high concentrations of sulfate in the deposits provided a supply of nutrients for the metabolic reactions of SRB. Cleaning the deposition from the surfaces and proper control of the alum injection to avoid deposit buildup was suggested to prevent corrosion from recurring.

Case History 13.11

Industry:	Light industry
Specimen location:	Bottom of a heat exchanger shell
Orientation of specimen:	Horizontal
Environment:	Cooling water treated with phosphates, azole, HEDP, glutaraldehyde, and slug feeds of oxidizing biocides. Acid feed was used to maintain pH around 6.8.
Time in service:	7 years
Sample specifications:	Carbon steel, 6.5 in. (15.4 cm) outer diameter shell

The shell of a heat exchanger used to cool nitrogen gas experienced multiple perforations (Fig. 13.51). The exchanger used shellside cooling, and low flow of the cooling water was reported.

FIGURE 13.51 Thick, iron-based deposits covering the internal surface of a heat exchanger shell with shellside cooling. Note the perforation at the center of the section.

Bulk chemical analysis of the material covering the surface by x-ray fluorescence and x-ray diffraction indicated it was predominately iron as iron oxide hydroxide (Lepidocrocite: FeOOH and Goethite: FeOOH) and iron oxide (Magnetite: Fe_3O_4). Some silicon (6 percent) and sulfur (4 percent) were also found in the material. Qualitative spot tests also indicated the presence of microbiological residue (slime) within the deposit and corrosion products. Microscopic analysis also indicated fine platelets in the corrosion products (Fig. 13.52), and spot tests indicated the material contained sulfides. In addition, an acidic solution was produced when the material was mixed with water. This evidence

FIGURE 13.52 Fine, platelet-shaped particles observed microscopically in the corrosion products on the surface.

strongly suggested that MIC caused by sulfate-reducing bacteria and acid-producing bacteria contributed to metal loss on the section. Oxygen corrosion and underdeposit corrosion were also indicated as mechanisms that contributed to the wastage.

Microbiological analysis was not performed on the material on the surface of the received section. However, analysis of deposited material from the cooling tower basin indicated the presence of high levels (30,000 cfu/g) of sulfate-reducing bacteria and some acid-producing bacteria, indicating the potential for significant MIC elsewhere in the system.

This case history demonstrates that conditions in exchangers with shellside cooling, especially with low flow of cooling water to the exchanger, can develop environments favorable for anaerobic microorganisms. Deposits tend to accumulate along the bottom of the shell and along the tops of tubes, especially near baffles.

References

1. N. Zelver, W. G. Characklis, and F. L. Roe, CTI Paper No. TP239A, 1981, Annual Meeting of the Cooling Tower Institute, Houston, Texas.
2. S. C. Dexter, "Microbiologically Influenced Corrosion," in *ASM Handbook*, Volume 13A: *Corrosion: Fundamentals, Testing, and Protection*, ASM International, Materials Park, OH, 2003, p. 400.
3. S. W. Borenstein, *Microbiologically Influenced Corrosion Handbook*, Industrial Press Inc., New York, 1994, p. 17.
4. B. J. Little and J. S. Lee, *Microbiologically Influenced Corrosion*, John Wiley & Sons Inc., Hoboken, NJ, 2007, p. 28.
5. R. Javaherdashti, *Microbiologically Influenced Corrosion: An Engineering Insight*, Springer, London, 2008, pp. 60–62.
6. B. L. Little and J. S. Lee, *Microbiologically Influenced Corrosion*, pp. 56–57.
7. An in-depth discussion concerning the use of biocides and biodispersants and their advantages and disadvantages is provided in D. J. Flynn (ed.), *The Nalco Water Handbook*, McGraw-Hill, 2009, pp. 21.17–21.26.

CHAPTER 14

Acid Corrosion

General Description

Acid corrosion in cooling water systems is usually caused by an upset. Acid spills, process leaks, airborne gas contamination, and unintentional overfeed of acid during chemical water treatment are the major causes of attack. Acid corrosion may occur to some degree during chemical cleaning operations, even when exposure times are short, and proper amounts of corrosion inhibitors are used to limit metal loss. Metal loss caused by acid conditions can be substantial, and the severity of attack quickly becomes apparent through rapid failure. Even slightly depressed pH may increase corrosion rates to produce significant wastage during long exposure times.

Many factors influence acid corrosion. Physical and mechanical factors include metallurgy, temperature, water flow, surface geometry and fabrication, as well as deposit and corrosion-product composition and thickness. Key water chemistry factors include dissolved oxygen concentration, metal-ion concentration, chemical treatment, and of course, the kind of acid that is produced (oxidizing or nonoxidizing, strong or weak).

Carbon and Low-Alloy Steel

The corrosion rate of carbon and low-alloy steel (and low-chromium cast irons) in pure water is about constant between a pH of 4 to 10 (Fig. 14.1). In solutions containing strong acids such as hydrochloric and sulfuric, normally protective iron oxides are dissolved below a pH of 4. Dissolution of the oxide layer occurs at slightly higher pH (5 or 6) in weaker acids such as carbonic acid and a variety of organic acids. Once the protective oxide layer is removed, the acid solution will react directly with the metal. The main anodic reaction in acid solutions is given in Reaction (14.1); iron is dissolved at a rate determined by the cathodic process. Thus, hydrogen gas is generated in most acid solutions. If dissolved oxygen is present, the cathodic Reaction (14.2) can be accelerated by depolarization as in Reaction (14.3).

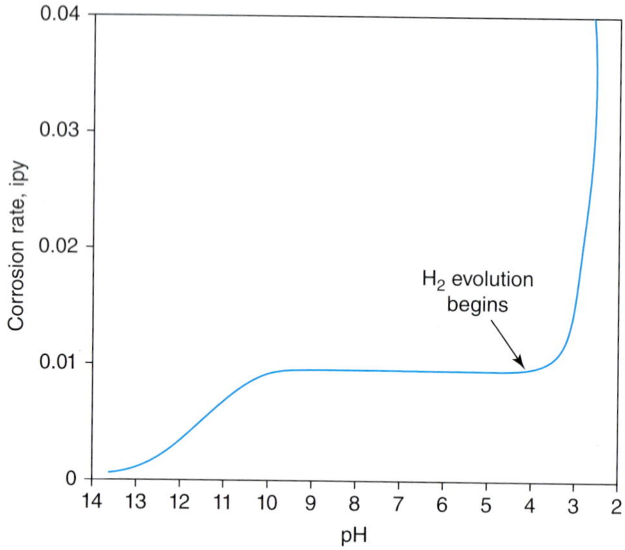

FIGURE 14.1 Effect of pH on corrosion of iron in aerated water at room temperature. (*Reprinted from Industrial & Engineering Chemistry 16, 665, 1924, W. Whitman, R. Russell, and V Altieri, published by the American Chemical Society.*)

Anode:

$$Fe \rightarrow Fe^{2+} + 2e^- \tag{14.1}$$

Cathode:

$$2H^+ + 2e^- \rightarrow H_2 \tag{14.2}$$

Cathode depolarization:

$$4H^+ + O_2 + 4e^- \rightarrow 2H_2O \tag{14.3}$$

Acid strength depends on the tendency of the acid to dissociate into a hydrogen ion and counter ion. Weak acids, such as carbonic and organic acids, dissociate only slightly compared to strong mineral acids such as sulfuric, hydrochloric, and nitric. Hydrogen ions consumed in corrosion cannot be replaced in strong acid solutions since dissociation of these acids is almost complete. In weak acid solutions, however, dissociation is incomplete at a consistent pH, and more hydrogen ions can be produced by dissociation as the hydrogen ions in solution are consumed during corrosion.

Depolarization effects are important in weak and non-oxidizing acids. Non-oxidizing acids include hydrochloric, hydrofluoric,

Acid	Under O_2 mpy (mmpy)	Under H_2 mpy (mmpy)	Ratio
6% acetic	550 (14.0)	6 (0.15)	92
6% H_2SO_4	360 (9.1)	30 (0.76)	12
4% HCl	480 (12.2)	31 (0.79)	16
0.04% HCl	390 (9.9)	5.5 (0.14)	71
1.2% HNO_3	1820 (46.2)	1570 (39.9)	1.2

TABLE 14.1 Effect of Dissolved Oxygen on Corrosion of Mild Steel in Acids Corrosion Rate[1]

phosphoric, and weak sulfuric. The synergistic effects of dissolved oxygen with carbonic acid and a variety of organic acids are significant. The effects of dissolved oxygen are also significant for weak and dilute inorganic acids (Table 14.1). Access of oxygen to carbon steel surfaces during corrosion influences the wastage process in non-oxidizing acids. Fluid velocity can influence the amount of oxygen reaching the metal surface and, therefore, the corrosion rate. In deaerated acid solutions, steel corrosion rate is constant with fluid velocity. If dissolved oxygen is present, however, the corrosion rate is proportional to fluid velocity.

Oxidizing acids serve as oxidizing agents, and they contain oxygen in the ionic structure. Oxidizing acids such as nitric acid and strong sulfuric acid solutions show virtually no synergism with oxygen. Thus, the addition of oxygen has little or no effect on corrosion rates in oxidizing acids since they act as depolarizers. Concentrated sulfuric acid (>93 percent) at room temperature is handled routinely in carbon steel drums and tanks, since a protective iron sulfate layer forms on surfaces under such conditions. However, if flow rates exceed about 3 ft/s (1 m/s), or impinging or turbulent flow occurs, the protective sulfate layer will be removed, and substantial corrosion may occur. In addition, sulfuric acid solutions become extremely corrosive as concentration decreases, particularly at concentrations below 50 percent where protective sulfate layers do not form. Corrosion rates decrease to much lower levels at low sulfuric acid concentrations.

During acid cleaning or severe acid upsets, ferric ion concentration may increase to high levels, albeit much more slowly than ferrous ion concentration. As ferric ion concentration increases, corrosion can be severe. Iron is oxidized, and ferric ions reduced, as in Reaction (14.4), on bare steel surfaces that develop during system cleaning or upsets:

$$Fe + 2Fe^{3+} \rightarrow 3Fe^{2+} \tag{14.4}$$

Stainless Steel

Acid corrosion on stainless steels differs from corrosion on nonstainless steels in two important respects. The differences are based upon the interaction of the acid with the passive oxide layer that forms on stainless steel surfaces. First, corrosion by non-oxidizing acids on stainless steels is usually more severe in deaerated solutions. Second, oxidizing acids attack stainless steel far less strongly than carbon steel. Hence, nitric acid solutions at low temperatures cause only superficial damage on stainless steels. In fact, nitric acid solutions are used for passivation treatments on stainless steels. Comparatively, certain non-oxidizing acids, particularly hydrochloric acid, cause truly catastrophic damage. Essentially, due to its inherent corrosivity to stainless steels, there is no safe concentration of hydrochloric acid that can be recommended for exposure to stainless steels.

Alloys whose corrosion resistance depends on forming a protective oxide layer, such as stainless steel, are susceptible to severe localized attack when pH falls as a result of exposure to non-oxidizing acid solutions. Such solutions may be produced by acid excursions and in-leakage of process contaminants. Pits form local areas of metal loss associated with breakdown of a protective oxide layer. However, there is usually a period of immunity from attack corresponding to the time necessary to breach the protective oxide layer. Hence, stainless steels usually experience very little attack during brief low-pH excursions. During long-term acid exposure, however, pits may initiate and grow at an ever-increasing rate. Breakdown is stimulated by low pH as well as by the decrease of dissolved oxygen in occluded regions. Small pit sites form and remain stable because of the large ratio of cathodic surface area (unattacked metal surface) to the small anodic pitted areas. If conditions become very severe, pitting will give way to general attack as more and more of the surface becomes actively involved in corrosion.

Copper Alloys

Copper corrosion by mineral and organic acids is promoted to a large degree by the presence of oxidizing agents, most commonly, dissolved oxygen. In addition, oxidizing acids such as nitric acid cause severe attack. Non-oxidizing acids and organic acids cause very little corrosion if the oxygen concentration is very low. Copper is vigorously attacked by acetic acid at low temperatures; at temperatures above boiling, no attack occurs because no dissolved oxygen is present.

A specific type of organic acid corrosion may occur on copper alloy tubing, which is known as formicary corrosion or "ant nest corrosion". The term "ant nest corrosion" is attributed to the tunneling nature of metal loss that occurs. Ant nest corrosion requires the simultaneous presence of moisture, oxygen, and a specific corrodent, usually organic acids. In many cases, the organic acids are produced by the decomposition or hydrolysis of residual self-evaporating lubricants, organic detergents, or organic solvents used to manufacture tubing. The aggressive organic acids are produced during long-term storage. Formicary corrosion refers to a specific organic acid, formic acid, which has been identified during investigations of this type of corrosion. Other organic acids have also been identified as contributors to such corrosion.

Zinc brasses and cupronickels behave in much the same way as copper with regard to acid corrosion.

Aluminum

Similar to other metals, the type of acid, its concentration, and the solution temperature influence corrosion rate. Aluminum corrodes at a fairly low rate between a pH of 4 and 8.5 at room temperature. Non-oxidizing acids such as hydrochloric acid are quite corrosive in all but very dilute concentrations. The corrosion rate in a 10 percent hydrochloric acid solution will increase one hundredfold as temperature increases from 50°F (10°C) to 176°F (80°C). Phosphoric acid can be used to etch aluminum uniformly at dilute concentrations. At higher concentrations, attack is rapid. Aluminum is also particularly susceptible to corrosion by organic acids.

Aluminum is resistant to oxidizing nitric acid at concentrations above 80 percent due to the production of stable aluminum oxide layers. However, at 50 percent nitric acid concentration at room temperature, corrosion rates are as high as 80 mpy (2.0 mmpy).

Titanium and Corrosion-Resistant Alloys

Titanium and corrosion-resistant alloys, particularly nickel alloys, are highly resistant to acidic conditions that may be produced in cooling water systems under upset conditions. Such alloys are commonly used because of their resistance to corrosion by acidic environments on the process side. Therefore, they will tend to be very resistant to leaks of process fluids into cooling water that produce acidic conditions.

Locations

Virtually any cooling water system component that is made of a susceptible alloy that contacts low-pH water may experience acid corrosion. However, there may be appreciable differences in corrosion severity and initiation times depending on alloy composition, type of acid, temperature, and location.

FIGURE 14.2 Brass condenser tube inlet end severely corroded by an acid upset involving sulfuric acid.

The condition of the metal may also influence the corrosion rate. In general, the severity of acid corrosion increases as a function of increasing residual or applied stress at a given acidic pH. Tube ends that have been rolled or welded often contain high residual stress. This may cause heat exchanger tube ends to experience severe metal loss that produces rough surface contours (Fig. 14.2). Screens, rolled sheet metal, and other heavily deformed metals that are not stress relieved are also prone to attack. A similar type of corrosion that occurs on boiler components due to acid exposure is referred to as stress-assisted corrosion.[2]

Crevices are sometimes present in which acidic species may concentrate [see Chap. 11, "Crevice and Underdeposit Corrosion (Concentration Cells)"]. Surfaces containing pits, crevices, and other shielded areas are more likely to be attacked by low-pH conditions. Occluded sites allow the concentration of aggressive anions and assist the autocatalytic pitting process.

Corrosion involving non-oxidizing acids can be highly sensitive to flow. Thus regions of high flow and turbulence are often more severely attacked than more quiescent regions (Figs. 14.3 and 14.4) (Chap. 23, "Erosion-Corrosion"). Weirs, lips, and other flow obstructions increase turbulence and thus corrosion. Pipe elbows, tees, and joints are also more susceptible to corrosion. Outer curves at pipe bends often are more severely wasted than inner bends.

FIGURE 14.3 Locally wasted region at an inlet pipe. See Fig. 14.4. Note the severe attack at the discharge end of the straight pipe.

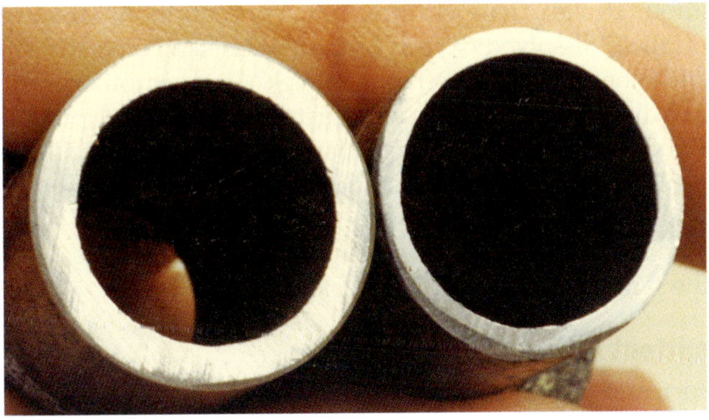

FIGURE 14.4 Generally wasted pipe (to the viewer's right) adjacent to a new pipe. Wastage was general near a pump discharge due to high flow.

Weld Attack

Welds are often more susceptible to corrosion than other areas (see Chap. 26, "Weld Defects"). Welds may contain porosity, crevices, and other imperfections that favor corrosion. Elevated residual stresses that are inherent to the welding process may promote localized corrosion at welds. Carbon steel weld heat-affected zones may be ditched by acid attack (Fig. 14.5).

FIGURE 14.5 Severe acid corrosion at weld on a large-diameter steel pipe caused by an acid upset.

Critical Factors

Acid corrosion damage is dependent upon many factors. Metallurgy, deposition, design, temperature, pH, concentration, dissolved-ion concentration, dissolved oxygen concentration, acid type and strength, flow, and other factors all influence attack. As ferric ion concentration increases during acid cleaning of industrial systems, the corrosion rate of steel increases rapidly. Damage may become widespread if the bulk cooling water solution pH is sufficiently depressed. In some cases, acidic conditions can develop beneath occlusive deposits and corrosion products. The appearance of metal loss beneath the layers will have the same features as acid corrosion.

Intermittent Attack

Acid corrosion in cooling water systems is often caused by upset conditions and chemical cleaning, which may occur intermittently. Consequently, periods of attack may be separated by intervals of virtually no acid corrosion.

In stainless steels such distinctions are not as clear, since little to no corrosion will occur in intervening periods due to the high corrosion resistance in normal oxygenated cooling waters at proper cooling water pH. Dissolved oxygen, water, acid, and metal-ion concentrations can have a pronounced effect on acid corrosion.

Mixed Acids

The corrosivity of solutions containing more than one acid type may be quite different than the corrosivity of either acid alone. Mixed-acid solutions are used widely in a variety of industrial applications.

Corrosion Products

Acids typically cause damage directly by reacting with the protective oxide layers and then the metal itself. Acid exposure may also cause corrosion indirectly by removing deposits and corrosion products from one location to promote deposition and underdeposit corrosion elsewhere. For example, during a severe acid excursion, large amounts of iron, copper, zinc, or other materials may be solubilized. Dissolved iron may reach concentrations as high as 500 ppm at pH below 2.5. The resulting suspended and dissolved material will be moved downstream. Suspended solid material can settle, and dissolved salts can oxidize or become insoluble when pH returns to normal operating levels. Subsequent deposition may foul critical cooling surfaces such as heat exchanger tubing. Blockage of pipes, screens, and other passages may also occur.

Identification

Attack by strong acids tends to produce intense localized wastage. Weak acids cause more general corrosion. However, numerous exceptions to these general rules exist. Fortunately, corrosion damage caused by acids has many unique features that allow easy recognition in most cases. Fresh acid attack is recognized by the absence of corrosion product in wasted areas and the sharpness or roughness of the corroded surface. Oxide layers are usually easily stripped by a test drop of hydrochloric acid in freshly corroded areas. Deposits are almost always absent. Edges of attacked areas are sharp and angular.

Strong Acids

Most metals that are exposed to strong acids experience pitting (Fig. 14.6). In strong mineral-acid corrosion, pits may be produced that penetrate to depths that are greater than their widths (Fig. 14.7). Cavernous chambers with pronounced undercutting form just below the surface due to concentration of aggressive anions within the pits.

When corrosion begins, a small amount of positive metal ion is put into solution at the corroding surface. The positively charged ions attract negatively charged ions from the acidic water solution. Hydrolysis in the growing pit produces ever-decreasing internal pH, leading to more metal dissolution. The pit begins to undercut, reflecting the increasingly acidic conditions within. The pH within active pits may fall to much lower values than bulk water pH. Since pH remains low, corrosion products do not readily precipitate within the pit. Hence, pits usually contain only small amounts of corrosion products. However, areas surrounding pits may be covered by corrosion products formed by normal pH precipitation processes or by oxidation of metal ions leaking out of pits.

FIGURE 14.6 Carbon steel pipe attacked by a strong mineral acid.

FIGURE 14.7 Close-up of pits shown in Fig. 14.6. Note the narrow pit mouths and the pronounced undercutting.

A striking feature of mineral-acid corrosion is the directionality of pit growth. Pits tend to grow in the direction of gravity, even when flow is present. Because pits entrap metal ions and acidic anions, the density of fluid within can be much greater than bulk water density. Pits forming on roof or ceiling surfaces tend to drain by convection. This process tends to decrease acidity within such pits.

Pits growing in the direction of gravity can form grooves or furrows on wall surfaces on stainless steels (Fig. 14.8). Contents leak out of pit interiors and depassivate areas immediately below the active corrosion sites. Grooves form, following the direction of gravity or flow.

FIGURE 14.8 Pits on an austenitic stainless steel pipe. Note that the grooves formed by leakage of pit contents travel down the pipe wall in the direction of gravity. (Pipe wall was vertical in service.) Note the multiple parallel striations within the larger grooves following stress lines remaining from the original pipe manufacturing processes.

Pit morphology varies widely with acid strength, metallurgy, and many other factors. As acids become weaker, pit interiors tend to become less jagged and more rounded, wider, and smoother. Also, flow tends to smooth pit edges. It is interesting to note that all acid corrosion begins as localized attack since some area on the corroding metal surface must be cathodic for corrosion to occur. Thus, in early stages of acid attack it is not at all clear that pits will eventually develop, even when localized attack is present.

Strong acid corrosion will often highlight microstructural features in steel and other formed and rolled alloys. Striations and grooves appear as the acid follows microstructural irregularities, such as manganese sulfide inclusion "stringers" elongated by rolling or forming processes (Fig. 14.9), or more highly cold-worked regions (Fig. 14.10). Mutual intersection of pits can lead to a jagged surface contour. Straight grooves may also form at 45 and 135° angles to the longitudinal axis of the tube where residual shear stresses are elevated due to tube-forming operations (Fig. 14.11). Metallographic examination will reveal undercut pits and depressions that are bare or lined with thin

Figure 14.9 Surface striations and shallow grooves on a carbon steel coupling. Acid etched the metal along microstructural irregularities, producing the grooves. Compare to Fig. 14.10.

Figure 14.10 Striations on the internal surface of admiralty brass condenser tube after acid cleaning.

FIGURE 14.11 Straight grooves formed at 45 and 135° angles to the longitudinal axis of the tube where residual shear stresses were elevated due to tube-forming operations.

FIGURE 14.12 Metallographic cross sections of undercut pits and depressions that are bare or lined with thin corrosion product layers.

corrosion product layers (Fig. 14.12). Elongation of pit mouths may occur because of preferential attack along microstructural irregularities such as inclusion stringers, second phase, and coring (Fig. 14.13).

Weak Acids

Carbonic Acid

Carbonic acid corrosion commonly occurs in steam and condensate systems,[3] but it rarely occurs in cooling water systems. However, the

FIGURE 14.13 Preferential acid corrosion along inclusion stringers.

metal loss that is produced by carbonic acid corrosion illustrates damage that tends to be produced by weak acids. Weak acids such as carbonic acid produce smoother metal loss than strong acids. Pitting and surface roughness on carbon steel may increase as the amount of dissolved oxygen increases (Figs. 14.14 and 14.15). In steam and condensate systems, carbonic acid may be produced when high concentrations of carbon dioxide and oxygen dissolve into condensed steam. Corrosion may be significant at locations of initial condensation where grooving frequently occurs (Figs. 14.16–14.18). Grooves are

FIGURE 14.14 Smooth attack on carbon steel by carbonic acid. Note the vivid red color of hematite, indicating high oxygen concentration.

FIGURE 14.15 Jagged fir-tree pattern of metal loss caused by condensing steam with high concentrations of carbon dioxide and oxygen. Pipe was slanted, causing narrow streams of carbonic acid to channel to central collection point.

FIGURE 14.16 Vertical carbon steel gas cooler with grooving at bottom of the last three coils. Carbonic acid–containing condensate attacked.

FIGURE 14.17 Close-up view of coil from Fig. 14.16. Note the presence of orange iron oxides and carbonate. Also note how attack starts as small pits due to dropwise condensation.

FIGURE 14.18 General thinning along the bottom of a steel conduit carrying carbonic acid–laden water.

caused by the flow of acidic condensate across surfaces. Friable, light-colored iron corrosion products containing carbonate are often present. Similar corrosion products may surround leaks and failure sites. Carbonate-containing material effervesces strongly when exposed to mineral acids.

Organic Acids

Organic acid strength tends to increase as molecular weight decreases. Hence, low-molecular-weight organics such as formic acid are quite

FIGURE 14.19 Cast iron pipe elbow corroded by organic acid.

corrosive relative to longer-chain acids. Additionally, and just as in carbonic acid corrosion, oxygen tends to increase the corrosiveness of many organic acid solutions on carbon steels.

If pits form, they will tend to be shallower, rounder, and less severely undercut than surfaces attacked by strong acids (Fig. 14.19). Wasted surfaces are typically rough to jagged, but they become smoother and more undulating as acid strength, oxygen concentration, and potentially corrosive metal-ion concentration decrease.

Corrosion Control

Acid corrosion may be controlled by preventing acid excursions and process fluid in-leakage, and by minimizing chemical cleaning exposure times. Acid excursions may be prevented through careful system monitoring and control. Acid- and chemical-feed equipment, storage equipment, and especially pumps, should be inspected on a regular basis to ensure that they are in proper working order. Possible sources of acid in-leakage, such as perforated process heat exchangers or makeup water contamination, should be identified and isolated. Windborne acid-gas contamination of cooling towers should also be eliminated. Chlorine feed should also be monitored and controlled properly to ensure that acidic conditions are not produced. Carbonic acid corrosion in steam and condensate systems is controlled by the

pH Range	Estimated Potential Corrosion Rate, mpy (μm/y)	Time to Reach Increased Corrosion Rate
6.0–7.0	3–10 (75–250)	Hours to days
4.5–6.0	10–50 (250–1250)	Hours
3.0–4.5	~100 (~2500)	Minutes
2.0–3.0	200–500 (5000–12,500)	Seconds
1.0–2.0	≥ 1000 (≥ 25,000)	Instantaneous

TABLE 14.2 Relative Corrosion Rates that May Occur on Carbon Steel Components Within Specific pH Ranges during Upset Conditions. Time Frames to Initiate Corrosion are Included.

use of condensate corrosion inhibitors that usually consist of neutralizing amines. The use of amines is uncommon in cooling water applications.

If an acid excursion occurs during service, high corrosion rates may be experienced, depending upon system pH. Table 14.2 shows the relative corrosion rates that may occur on carbon steel components within specific pH ranges during upset conditions. This table also reveals the time frame in which corrosion will initiate to produce corrosion. At very low pH (<4.5), corrosion will initiate within several minutes, and when pH is less than 3, within seconds. Consequently, time spans for acid excursions should be minimized. Acid corrosion and resultant fouling that may occur during upset conditions may be controlled by taking specific steps, depending upon the specific pH ranges that are reached. Control measures that are recommended for moderate, severe, and extreme low-pH excursions in open recirculating cooling water systems are presented next.

Moderate Low-pH Excursions (pH 4.5–6.5)

Acid feed should be stopped and the acid feed system should be blocked off until the acid source is identified. Feed of oxidizing biocide should also be suspended. The acid feed system should be inspected to ensure that it is in proper working order. If there are no leaks or problems with the acid feed system, then other sources of acid contamination, such as the feedwater, process leaks, excessive feed of biocides that produce acidic conditions, or intrusion of airborne acid sources into the cooling towers should be investigated. Blowdown rate should be increased appropriately, since soluble iron, copper, and other system metals will increase. Moderate pH excursions may be corrected through blowdown, depending upon system size and available blowdown rate. The pH monitoring system should be standardized to ensure proper operation. The system pH should be monitored until the desired levels are reached. If the system pH cannot be controlled within a reasonable time frame, appropriate alkalinity supplements may be used. When the pH returns to normal, chemical and acid feed and blowdown rate can be returned to normal. Higher concentrations of corrosion inhibitors will be required. Depending upon the length of the pH excursion, or if the pH dropped below 6, a passivation treatment may be required. Fouled critical heat exchangers may need to be back-flushed.

Severe Low-pH Excursions (pH 2.5–4.5)

Acid feed should be stopped and the acid feed system should be blocked off until the acid source is identified. Feed of corrosion inhibitors, dispersants, and oxidizing biocide should also be suspended. The acid feed system should be inspected to ensure that it is in proper working order. If there are no leaks or problems with the acid feed system, then other sources of acid contamination, such as the feedwater, process leaks, excessive feed of biocides that produce acidic conditions, or intrusion of airborne acid sources into the cooling towers should be investigated. Blowdown should be increased to the maximum allowable rate. Makeup will be increased accordingly. Drain valves on heat exchangers and at deadlegs should be opened to allow drainage. The pH monitoring system should be standardized to ensure proper operation. The system pH should be monitored until the desired levels are reached. Iron concentration should be monitored several times during the recovery process. During recovery, sufficient iron dispersant should be added to stabilize or disperse the iron. Alkalinity supplements must be used. When the pH is 2.5 to 4.5, sodium carbonate should be used. When the pH reaches 4.5, and the water is clear, with less than 3 ppm iron, the blowdown may be reduced, and the drains on heat exchangers and deadlegs may be closed. Alkalinity supplements should continue to be added until the proper pH is reached. When the pH returns to the appropriate range for the treatment program, chemical and acid feed and blowdown rate can be returned to normal. Fouled critical heat exchangers may need to be back-flushed. Higher concentrations of corrosion inhibitors will be required. Depending upon the length of the pH excursion, a passivation treatment may be required.

Extreme Low-pH Excursions (pH <2.5)

Acid feed should be stopped and the acid feed system should be blocked off until the acid source is identified. Feed of corrosion inhibitors, dispersants, and oxidizing biocide should be suspended. The acid feed system should be inspected to ensure that it is in proper working order. If there are no leaks or problems with the acid feed system, then other sources of acid contamination, such as the feedwater, process leaks, excessive feed of oxidizing biocide, or intrusion of airborne acid sources into the cooling towers should be investigated. Blowdown rate should be increased to the maximum allowable rate.

Makeup will be increased accordingly. Drain valves on heat exchangers and at deadlegs should be opened to allow drainage. The pH monitoring system should be standardized to ensure proper operation. The system pH should be monitored continuously until the desired levels are reached. Sample locations for pH should be at the return headers. Iron concentration should be monitored several times during the recovery process. Sufficient iron dispersant should be added to stabilize or disperse the iron. Alkalinity supplements must be used. Below pH 2.5, it is necessary to use caustic to achieve more rapid neutralization. However, this step should not be conducted too quickly to raise the pH above 2.5 to 3 or substantial iron oxide deposition may occur. Between pH 2.5 and 4.5, sodium carbonate should be used. When the pH reaches 4.5, and the water is clear, with less than 3 ppm iron, the blowdown may be reduced, and the drains on heat exchangers and deadlegs may be closed. Alkalinity supplements should continue to be added until the proper pH is reached. When the pH returns to the appropriate range for the treatment program, chemical and acid feed and blowdown rate can be returned to normal. Fouled critical heat exchangers may need to be back-flushed and should be inspected. A passivation treatment is required. Following passivation, normal concentrations of corrosion inhibitors can be resumed.

Cautions

Oxygen corrosion, cavitation, and erosion-corrosion frequently produce damage resembling acid attack. Pits and depressions that form due to oxygen corrosion can usually be differentiated from acid corrosion damage by analyzing surface corrosion products and deposits and by carefully examining pit geometry. Pits formed due to oxygen corrosion have less undercut interiors than acid pits and are usually filled or lined with reddish-brown corrosion products. The corrosion products formed by oxygen corrosion are much thicker and more difficult to remove than those found in acid pits.

Localized acidic conditions may be produced beneath occlusive deposit and corrosion product layers due to the formation of ion concentration cells.

Pitting on stainless steels is commonly associated with acid conditions that develop within pits by autocatalytic processes. Solution pH outside the pit may be only mildly acidic or not acidic at all.

Cavitation produces spongy, porous-appearing surfaces, strongly resembling acid corrosion. However, cavitation usually causes highly localized areas of metal loss that are surrounded by unwasted surfaces, unlike acid corrosion, where metal loss occurs over a much wider area, or even the entire exposed surface.

Related Problems

Also see Chap. 11, "Crevice and Underdeposit Corrosion (Concentration Cells)"; Chap. 12, "Oxygen Corrosion"; Chap. 23, "Errosion-Corrosion"; and Chap. 24, "Cavitation Damage."

Case History 14.1

Industry:	Building cooling system
Specimen location:	Centrifugal chiller condenser
Orientation of specimen:	Horizontal
Environment:	Cooling water
Time in service:	8 years
Sample specifications:	Carbon steel

A divider plate for a heat exchanger in a centrifugal chiller reportedly experienced significant metal loss (Fig. 14.20). The plate tore away at the welded edges that contacted the bell.

Within scattered patches occupying approximately half of the total surface area on one side of the plate, the surface was shallowly wasted and was covered by friable layers of light-colored deposits that overlaid iron oxide layers. Between these patches, moderately deep to very deep, mutually intersecting depressions interrupted. The depressions were covered with thin layers of friable brown iron oxide that stripped away easily when acidified to reveal bare metal surfaces (Fig. 14.21). Perforations occurred where metal loss was deepest (Fig. 14.22). Substantial thinning also occurred along the edges of the plate. The deep metal loss that occurred on both sides of the plate was primarily caused by acid corrosion. The presence of thin iron oxide layers on the depressions indicated that acid corrosion was not active shortly before the plate was removed from service. It is also probable that acid corrosion occurred intermittently. Acid corrosion may have occurred to some degree due to exposure to

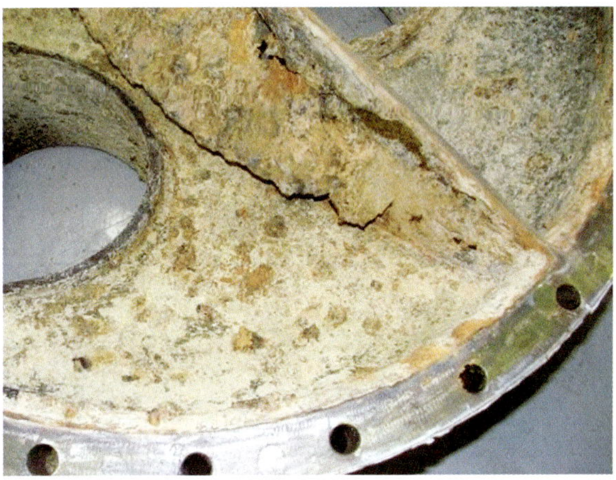

FIGURE 14.20 A divider plate for a centrifugal chiller reportedly experienced significant metal loss. The plate tore away at the welded edges that contacted the bell.

cooling water that became acidic during service. In order for iron oxide layer and subsequent metal removal to occur, the cooling water pH would have to drop below about 4. Reportedly, acid excursions to pH 2 occurred in the system periodically during its service life.

Figure 14.21 Depressions were covered with thin layers of friable brown iron oxide that stripped away easily when acidified to reveal bare metal surfaces.

Figure 14.22 Perforations occurred where metal loss was deepest.

Case History 14.2

Industry:	Steel
Specimen location:	Outlet piping of a sulfuric acid storage tank
Orientation of specimen:	Horizontal
Environment	Internal: Concentrated sulfuric acid, 70 to 100°F (21–38°C)
Time in service:	About 1 year
Sample specifications:	½ in. (12.7 mm) outer diameter, 304 stainless steel tubing.

FIGURE 14.23 Austenitic stainless steel pipe that carried sulfuric acid. Failure was caused by severe general wastage from internal surfaces.

Feedlines from a concentrated sulfuric acid storage tank failed. The tubing contained large, ragged perforations due to severe internal surface wastage (Fig. 14.23).

The internal surface was generally covered with dark, tenacious corrosion products. Traces of greenish-white corrosion product were also present. Many small, intersecting pits pockmarked internal surfaces. Thinning was pronounced within a wide area near the failure where the remaining tube wall was paper thin.

The failure was caused by corrosion due to exposure to sulfuric acid that had an improper concentration. Nearly pure sulfuric acid (>93 percent concentration) is very mildly corrosive. However, when the concentration drops to about 80 percent at 100°F (38°C), corrosion rates on 304 stainless steel may reach 1000 mils (25 mm) per year. At 70 percent concentration, corrosion rates may be an order of magnitude greater.

It was likely that water entered the storage tank due to condensation of moisture-laden air that was sucked into the tank. As night fell and temperature decreased, moisture would condense and dilute the acid. Air-drying systems were not in good operating order. Suggestions were made to test acid water content, and if the acid was found to fall below 90 percent concentration, concentrated acid was to be added to raise the concentration above 93 percent. It is also possible that wastage of the tubing was promoted by excessive flow rates or turbulent flow through the line, which would remove protective sulfate corrosion product layers.

Case History 14.3

Industry:	Chemical process industry
Specimen location:	Process heat exchanger
Orientation of specimen:	Horizontal
Environment	External: Process fluids
Internal:	Open recirculating cooling water
Time in service:	6 years
Sample specifications:	¾ in. (19.1 mm) carbon steel tube

The internal surfaces of the tubes in most areas experienced fairly uniform, general metal loss. However, deep, bare surfaced, undercut pits and depressions interrupted in scattered locations (Figs. 14.24 and 14.25), indicating that some metal loss occurred due to acid corrosion. Acid corrosion resulted from a chemical cleaning or exposure to cooling water that became acidic due to upset conditions, such as process fluid in-leakage.

FIGURE 14.24 Deep, bare-surfaced, undercut pits and depressions interrupted in scattered locations due to acid corrosion.

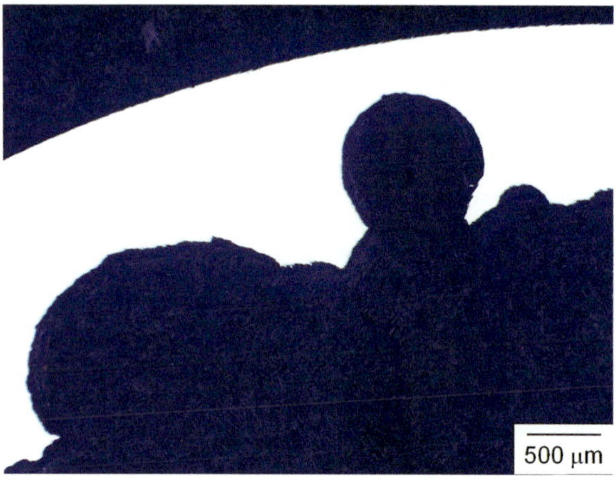

FIGURE 14.25 Metallographic cross sections of deep, undercut, bare-surfaced pits and depressions on internal surface. Unetched.

Case History 14.4

Industry:	Industrial gas production
Specimen location:	Heat exchanger
Orientation of specimen:	Horizontal
Environment:	External: Open recirculating cooling water (maximum 104°F, 40°C), Internal: Nitrogen gas [268°F (131°C) cooled to 95°F (35°C)]
Time in service:	2 years
Sample specifications:	¾ in. (19.1 mm) outside diameter, 304 stainless steel

It was reported that thin layers of light-colored deposits and corrosion products which covered the external surfaces of some tubes were chemically removed using a hydrochloric acid solution. The external surface was mostly bare and bright with a metallic luster. The external surface in most areas experienced no metal loss and had a smooth and even contour. However, in a few scattered locations, undercut pits interrupted (Fig. 14.26). The microstructure consisted of twinned grains of austenite that contained some ferrite phase stringers as well as some intergranular carbides. Corrosion penetrated laterally along the ferrite stringers (Fig. 14.27). The pits formed due to acid corrosion

FIGURE 14.26 Undercut pits interrupting mostly unattacked stainless steel surface.

FIGURE 14.27 Microstructure consisting of twinned grains of austenite that contained some ferrite phase stringers. Acid corrosion penetrated laterally along the ferrite stringers. Etchant: Electrolytic oxalic acid.

resulting from exposure to the hydrochloric acid solution. Essentially, there is no safe concentration of hydrochloric acid to which austenitic stainless steels can be exposed. It was recommended that other, less aggressive acids be used for chemical cleaning, including oxidizing acids such as nitric acid or weaker organic acids.

Case History 14.5

Industry:	Chemical process industry
Specimen location:	Vinyl chloride monomer heat exchanger
Orientation of specimen:	Horizontal
Environment:	Internal: Vinyl chloride monomer External: Open recirculating cooling water
Time in service:	3 years
Sample specifications:	1 in. (25.4 mm) outer diameter, carbon steel tube

The external surfaces were interrupted by numerous scattered, shallow to moderately deep, rough-surfaced depressions, most of which were undercut (Fig. 14.28). The depressions and surrounding surfaces were lined with layers of iron oxides and deposits that were acidic in a distilled water solution, and contained small amounts of chloride. The layers stripped quickly when exposed to concentrated hydrochloric acid, indicating that they were quite thin (Fig. 14.29). Most of the wastage at the pits and depressions was caused by acid corrosion that was active shortly before the sections were removed from the system. Acid corrosion occurred when aggressive, chloride-containing process fluids (vinyl chloride monomer) leaked into the cooling water to produce acid solutions. It is possible that acid corrosion occurred on an intermittent basis.

Figure 14.28 External surfaces interrupted by numerous scattered, shallow to moderately deep, rough-surfaced, mostly undercut depressions.

FIGURE 14.29 Iron oxide layers stripped quickly when exposed to concentrated hydrochloric acid, indicating that they were quite thin.

Case History 14.6

Industry:	Chemical process industry
Specimen location:	Debutanizer overhead condenser
Orientation of specimen:	Horizontal
Environment:	External: Process fluids
	Internal: Open recirculating cooling water
Time in service:	2 years
Sample specifications:	¾ in. (19.1 mm) carbon steel tube

Moderately deep depressions and straight grooves oriented at 45 and 135° angles to the longitudinal tube axis formed on the internal, waterside surface (Fig. 14.30). The internal surface was covered in most areas by thin corrosion

FIGURE 14.30 Straight grooves oriented at 45 and 135° angles to the longitudinal tube axis.

product and deposit layers. Chemical spot tests indicated the presence of substantial amounts of sulfide in the layers. The internal surface metal loss was primarily caused by acid corrosion due to in-leakage of process contaminants into the cooling water system. Acid corrosion was not active shortly before the section was removed from the system. It is possible that acid corrosion occurred intermittently. The presence of sulfide in the corrosion products on the internal surface indicated that some corrosion was caused by aggressive sulfur compounds, probably hydrogen sulfide. Waterside corrosion may be controlled by minimizing cooling water contamination and by taking appropriate remedial action if and when contamination occurs.

Case History 14.7

Industry:	Plastics and glass manufacture
System type:	Condenser tube in a process chiller
Specimen:	Externally finned, internally rifled tube
Material:	90/10 cupronickel
Time in service:	Less than 2 years; problems began about 6 months after installation.
Environment:	Internal: Filtered raw river water, no chemical treatment External: Freon

After months of intermittent operation, a condenser tube failure resulted in Freon loss and considerable downtime and loss of production. The unit was idle for 2 months prior to the failure.

Inspection of tubes removed from the system revealed extensive localized corrosion within portions of the tube that were not internally rifled. These short regions did not have fins or rifling formed into the surface to allow a tight fit with tube supports and baffles. The rifled surfaces did not exhibit significant corrosion.

Examination of a section revealed internal surfaces that were covered everywhere by a thin layer of tan deposits that contained major amounts of silicon, iron, and aluminum (Fig. 14.31). Silicon and aluminum are typical of silt from river water, and some of the iron may be transported corrosion products from elsewhere in the system. Although the thin deposit layer might suggest that an underdeposit

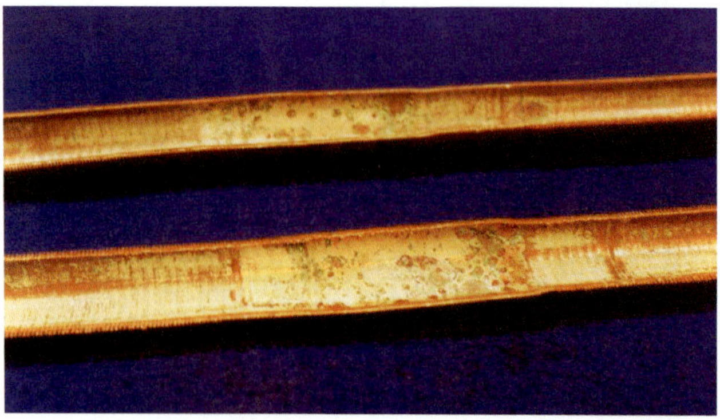

FIGURE 14.31 Internal surface of cupronickel enhanced tube covered by a thin layer of tan deposits.

FIGURE 14.32 Green corrosion product mounds in scattered locations.

corrosion mechanism was responsible for the attack, the localization of the attack indicates that the corrosion was likely initiated by another factor.

Green corrosion product mounds formed in scattered locations in the unrifled regions of the tubes (Fig. 14.32). Depressions were present beneath the mounds. Close examination revealed clusters of small, deep pits beneath the corrosion product mounds (Fig. 14.33). Purple-colored stains surrounded the pitted area. Cross sections through the pits revealed a unique morphology of interconnected cavities (Fig. 14.34). This morphology is not typical of underdeposit corrosion, but it is characteristic of ant nest corrosion. As discussed previously under "Copper Alloys," ant nest corrosion, or formicary corrosion, is a specific mechanism

FIGURE 14.33 Depressions beneath corrosion product mounds.

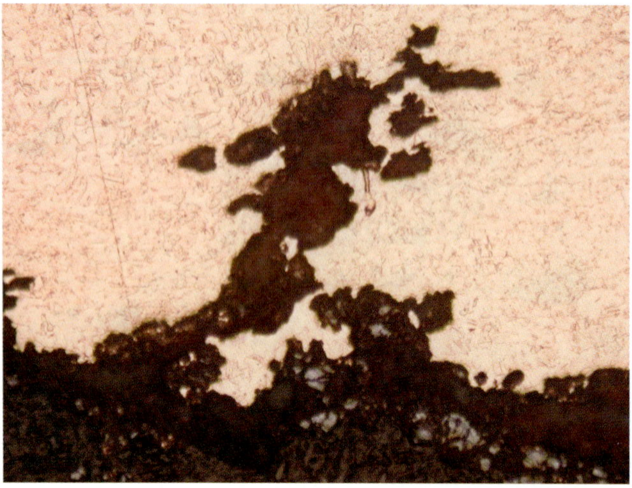

FIGURE 14.34 Cross sections through the pits revealing a unique morphology of interconnected cavities. This morphology is not typical of underdeposit corrosion, but it is characteristic of ant nest corrosion.

caused by the presence of organic acids on the surface of copper-based alloys. Infrared spectroscopy showed evidence for the presence organic acid salts, although the exact type (such as formate or acetate) was not identified. This confirmed that the attack was likely initiated by ant nest corrosion.

Ant nest corrosion requires the simultaneous presence of moisture, oxygen, and a specific corrodent, usually an organic acid. In many cases, the organic acid is associated with the decomposition or hydrolysis of self-evaporating lubricants, organic detergents, or organic solvents used in the manufacture of the enhanced tubing. The localization of the attack at the unrifled areas was probably related to residual tube manufacturing chemicals left on the internal surface followed by long-term storage prior to installation and system startup.

References

1. L. W. Whitman and R. Russell, *Industrial & Engineering Chemistry* 17, 1925, p. 348.
2. J. Dillon, P. Desch, and T. Lai, Chapter 11, *The Nalco Guide to Boiler Failure Analysis*, 2d ed., McGraw-Hill, 2011.
3. Refer to Dillon, Desch, and Lai, *The Nalco Guide to Boiler Failure Analysis*, 2d ed.

CHAPTER 15

Alkaline Corrosion

General Description

Corrosion of industrial alloys in alkaline waters is not as common or as severe as corrosion associated with acidic conditions. High-pH solutions produce little corrosion on steel, stainless steel, cast iron, nickel, and nickel alloys under most cooling water conditions. This is at least in part explained by inherent alloy corrosion behavior and the interaction of specific ions on the surface of these metals. High temperatures and elevated concentrations of caustic are required to cause alkaline corrosion on steel components.

Another effect that reduces the likelihood of alkaline corrosion is that many dissolved minerals in cooling water have normal pH solubility and thus deposit at faster rates when pH increases. Precipitated minerals such as phosphates, carbonates, and silicates, for example, can reduce corrosion on many alloys by producing layers that provide some protection.

However, certain metals that are used in cooling water environments, notably aluminum and zinc, can be vigorously attacked at elevated pH. As they are also significantly corroded at low pH, these metals are said to be *amphoteric*. A plot of the corrosion behavior of an aluminum alloy as a function of pH when exposed to solutions containing specific ions is illustrated in Fig. 15.1. The graph also shows that the influence of various ions is often more important than solution pH in determining corrosion on aluminum. The mechanism of alkaline corrosion on aluminum and zinc is generally associated with dissolving or breakdown of protective corrosion product layers that form on the metal surface.

The interaction of ammonia on copper alloys is different. Ammonia produces an alkaline solution when dissolved in water at high concentration. However, the corrosion mechanism on copper alloys is not attributed specifically to alkaline corrosion. Rather, corrosion of copper alloys by ammonia solutions is caused by complexation reactions. Details on some specific metals and alloys are presented in the following sections.

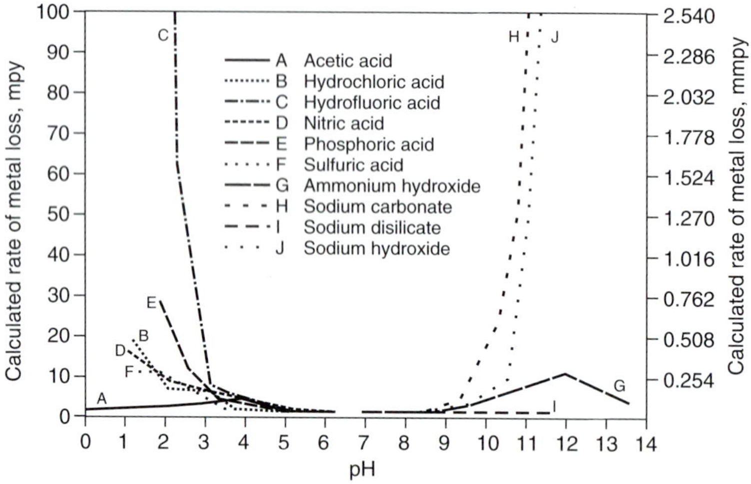

Figure 15.1 Effect of pH on corrosion of aluminum (specifically 1100-H14 alloy) by various chemical solutions. Observe the minimal corrosion in the pH range of 5 to 8.5. The relatively low corrosion rates in acetic acid and ammonium hydroxide demonstrate that the nature of the individual ions in solution is more important than the degree of acidity or alkalinity. (*Courtesy of Alcoa Laboratories; from* Aluminum Properties and Physical Metallurgy, *ed. John E. Hatch, American Society for Metals, Metals Park, Ohio, 1984, Figure 19, page 295.*)

Aluminum Alloys

Thermodynamically, aluminum should be a highly reactive metal. However, reactivity is limited in most natural environments due to its ability to develop a protective oxide layer on the surface, which quickly forms when exposed to water and/or air. Once formed, the oxide slows further corrosion. This oxide layer may be as thin as about 5 nm (5×10^{-9} m) when formed naturally in air, but it is thicker (about 20–200 nm) when formed in water.[1] The oxide layer can be made much thicker by anodizing, which is an electrochemical metal finishing technique (for details see Chap. 5, "Aluminum").

Aluminum corrodes in water as in Reaction (15.1):

$$2Al + 6H_2O \rightarrow 2Al(OH)_3 + 3H_2 \quad (15.1)$$

The final corrosion product, aluminum oxide trihydrate (also written as $Al_2O_3 \bullet 3H_2O$), is called bayerite.

At a pH above about 9 and in the absence of suitable aluminum corrosion inhibitors such as silicate, the protective oxide layer will rapidly dissolve, and significant corrosion may occur (Fig. 15.1). Aluminum exposed to an aqueous solution containing sodium hydroxide corrodes as in Reaction (15.2):

$$Al + NaOH + H_2O \rightarrow NaAlO_2 + 3/2H_2 \quad (15.2)$$

Figure 15.1 shows that aluminum is corroded by alkaline solutions, albeit at different rates, when water pH exceeds 9. Note the very low corrosion rates for sodium disilicate between pH 10.0 and 11.5; silicates are used as corrosion inhibitors for aluminum alloys in some cooling water systems. Corrosion by most inorganic salts between a pH of 5.0 and 8.5 is very slow near room temperature. Of course, natural waters vary widely in composition, and exceptions do occur.

Zinc

Like aluminum, zinc is amphoteric and is therefore attacked at low and high pH (Fig. 15.2). In weakly alkaline solutions near room temperature, corrosion is actually very slight, being less than 1 mpy (0.025 mmpy), even at a pH as high as 12. The corrosion rate increases rapidly at higher pH, approaching 70 mpy (1.8 mmpy) at a pH near 14. Just as in aluminum corrosion, the corrosion rate of zinc depends on the ionic species in the water.

Since zinc is a very reactive metal, corrosion protection is primarily provided by a stable corrosion product layer that forms spontaneously when the surface is exposed to water. The corrosion product layer that forms depends upon dissolved gases and ions in the water and temperature. In many conditions, a zinc hydroxide layer will form on the surface, as indicated in Reaction (15.3).

$$Zn + 2H_2O \rightarrow Zn\,(OH)_2 + H_2 \tag{15.3}$$

Highly alkaline conditions dissolve the corrosion product layer, leading to rapid general attack.

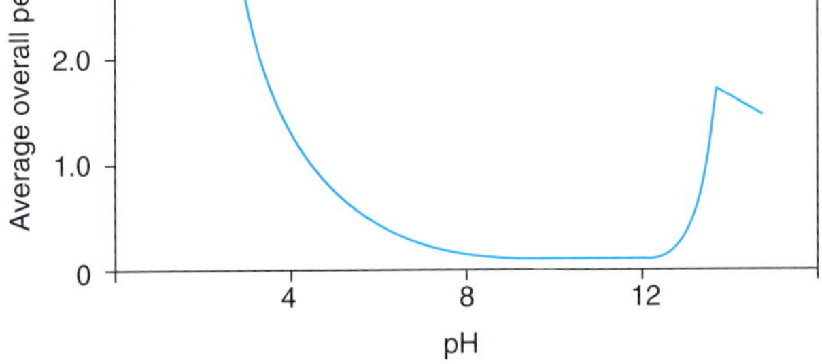

Figure 15.2 Corrosion rate of zinc in distilled water as a function of pH (pH adjustments were made by additions of HCl and NaOH). (*Courtesy of Springer.*)

The development of corrosion resistance for zinc and zinc-coated (galvanized) components in cooling water environments depends on forming and maintaining a stable passive corrosion product layer. Although very high pH values can result in the dissolution of zinc corrosion products, some cooling water conditions at slightly elevated pH values can limit the ability of the zinc surface to form the protective corrosion product layer or may cause its breakdown. Instead of dense protective corrosion products forming on the surface, the surface develops a white, voluminous, porous layer of basic zinc carbonate and hydrated zinc oxide that is not protective. Such nonprotective corrosion products are known as white rust. White rust formation on galvanized steel surfaces in a cooling tower is shown in Fig. 15.3.

White rust on galvanized steel can be avoided by maintaining water conditions that allow for the formation of the protective zinc hydroxide layer. The pH of the water is one of the primary factors that influence the development of white rust. Typically, a passivation period is employed in which the water conditions are carefully controlled. During this period, pH values are generally maintained between 7 and 8 for a period of 4 to 12 weeks to promote the formation of the protective corrosion product layer. During normal operation, pH values above 9 should be avoided.[2] Other parameters, such as hardness, alkalinity, chloride content, sulfate content, and free chlorine are also important.

Copper Alloys

Copper alloys can experience corrosion at high pH under certain conditions. However, attack is not usually caused by elevated pH alone, but by the presence of ammonia or substituted ammonium compounds.

FIGURE 15.3 Spotty white rust on a galvanized steel support beam in a cooling tower.

	Corrosion Rate	
Alloy	**mpy**	**mmpy**
Phosphorized copper	6.7	0.17
Admiralty brass	11.0	0.28
Aluminum brass	3.9	0.10
90-10 Cupronickel	3.9	0.10
70-30 Cupronickel	0.4	0.01
Alloy 400 (Monel)	0.4	0.01

Table 15.1 Corrosion Rates of Some Copper Alloys Exposed to an Ammonia-Containing Water Mist[3]

Ammonia (NH_3) dissolves in cooling water to form a weak base (ammonium hydroxide, NH_4OH) which will increase pH. This environment will be aggressive to copper and its alloys if dissolved oxygen is also present. Wastage is caused by complexation of the copper. Corrosion increases as oxygen concentration and/or temperature increase. High concentrations of ammonia are generally not present in cooling water systems. However, high levels of dissolved ammonia can occur in condensed moisture on some process-side applications, such as steam on condenser tube surfaces. Corrosion rates approaching 240 mpy (6.1 mmpy) have been measured at room temperature in 2N ammonium hydroxide solutions.

Copper-alloy corrosion behavior depends on the alloying elements and their concentrations in the metal. Thus, corrosion rates depend on alloy composition. Table 15.1 shows the corrosion rates of some copper alloys in an environment of misted water containing 1000 mg/L ammonia, known as a fog test. Note that nickel additions to copper, as in cupronickel alloys, significantly reduce corrosion due to ammonia dissolved in water. Monel, which is a nickel-copper alloy, is also resistant to attack in ammonia-containing waters and vapors.

Copper resists corrosion in caustic solutions. For example, corrosion rates in hot caustic soda may be less than 1 mpy (0.025 mmpy). Alloying copper with zinc increases corrosion rates in caustic solutions, whereas nickel additions decrease corrosion rates. Caustic solutions attack zinc brasses at rates of 2 to 20 mpy (0.051–0.51 mmpy). Concentrated caustic solutions are generally not encountered in cooling water systems, but cleaning solutions for some processes may be caustic-based. Care must be taken to avoid exposing certain alloys to caustic solutions.

Locations

Wastage is pronounced in equipment contacting high-pH fluids. Chemical process equipment, heat exchangers, water-cooled process reactors, valving, transfer pipes, and heating and cooling systems can be attacked. Equipment that is exposed to high-pH cleaning solutions may also be affected.

General corrosion may occur throughout a cooling water system if pH becomes excessive for susceptible alloys. Areas of high turbulence or flow may experience more severe attack.

Evaporation can concentrate alkaline species. More severe attack may occur at locations such as waterlines. Similarly, alkaline species can become concentrated in areas that are intermittently wetted, promoting attack.

Process sides of certain water-cooled heat exchangers may experience condensation, with aggressive gases such as ammonia at very high concentrations in water solution. Such exchangers that are tubed with certain copper alloys may experience substantial wastage.

Critical Factors

Aluminum alloys will corrode at high pH in the absence of corrosion inhibitors. However, corrosion rates will vary depending upon the corrodent. Perhaps one of the most corrosive solutions contains caustic, which causes corrosion rates to increase substantially as pH rises above a value of 9. The Al^{+3} ion reacts vigorously with OH^- to produce AlO_2^-. Ammonium hydroxide only moderately increases corrosion rates of aluminum alloys.

Conversely, copper alloys are appreciably attacked by ammonium hydroxide, but corrosion rates increase only moderately with caustic concentration.

Just as at low pH, concentration mechanisms can substantially increase attack due to alkaline species. The two principal mechanisms of concentration are evaporation and condensation. Whatever the water composition, evaporation increases solute concentrations of compounds with vapor pressures lower than water (such as caustic compounds). This is the reason cooling systems that experience boiling and/or large evaporative losses without sufficient makeup water additions may be especially prone to attack. Cycling may also increase dissolved corrodent concentrations. Condensation allows for aggressive gases such as ammonia to dissolve and concentrate.

Identification

Aluminum Alloys

Aluminum that is exposed to concentrated alkaline solutions tends to experience broad areas of general metal loss (Fig. 15.4). Intersecting areas of shallow metal loss produce generally wasted surfaces. Localized attack may occur during early stages of corrosion and on castings (Figs. 15.5 and 15.6). Wastage progresses so that individual

FIGURE 15.4 Corrosion of an aluminum water manifold from a diesel engine cooling system due to concentrated caustic. Note the generally wasted internal surface and the presence of white deposits and corrosion products.

FIGURE 15.5 Corroded aluminum alloy casting valve seat. The seat was in contact with water having a pH between 9 and 10 for three months. Red marks were drawn on the surfaces to assist metallographic preparation.

FIGURE 15.6 The valve seat of Fig. 15.5 magnified to show surface detail. Note the intersection of depressions to form more general metal loss.

areas of attack eventually merge, and well-defined pits are generally rare.

White, friable corrosion products composed of bayerite ($Al_2O_3 \cdot 3H_2O$), caustic, and $NaAlO_2$ cover corroded areas (Fig. 15.4). The white corrosion product and deposit layers usually test as distinctly alkaline when mixed with distilled water. Corrosion products usually cling tenaciously to the underlying metal and do not form voluminous lumps.

Corroded surfaces may have wide depressions with interiors having smooth contours. Darker corrosion product layers may line depressions, especially if cooling water is at elevated temperatures (Fig. 15.7).

Zinc

Corroded surfaces are typically covered by voluminous white corrosion products, known as white rust. These corrosion products may appear gelatinous or waxy when wet, and powdery when dry. The corrosion products often start as small nodules that spread to cover the surface (Fig. 15.8). These corrosion products consist of combinations of ZnO, $Zn(OH)_2$, and $ZnCO_3$. The white rust corrosion products can be quite similar chemically to the protective dull-gray passive oxide that forms on the zinc surface. The major difference between the corrosion products is that white rust is porous and generally nonprotective, while the passive corrosion product layer is dense and effectively protects the underlying metal from exposure to the environment. The surface beneath the corrosion products may have a

FIGURE 15.7 Corroded aluminum surface exposed to cooling water of pH as high as 11.

FIGURE 15.8 Spotty white corrosion product nodules on the surface of a galvanized steel cooling tower.

darkened appearance with shallow metal loss. On galvanized steel, reddish brown corrosion products may be surrounded by white corrosion products if the zinc coating has been penetrated and the underlying steel is attacked (as shown in Fig. 7.6).

Copper Alloys

Wastage may appear as general attack or it may be localized. The attack morphology is typically characteristic of the concentration mechanism associated with corrosion. Pitting almost never occurs.

Corrosion caused by ammonia is much more likely on the process side of heat exchangers, such as in condensers with steam on the shellside. Grooving-type attack is common. Grooves are produced by rivulets of aggressive ammonia-containing condensate that run in the direction of gravity. Broad annular grooves are cut into tube surfaces abutting the tube sheets and baffles (Fig. 15.9). These grooves are caused by condensed vapors that contain high concentrations of ammonia. The condensed moisture containing ammonia runs down the tubesheets and baffles, collecting where tubes are rolled into the sheets or pass through baffles. In extreme cases, attack can become general, producing undulating surface contours (Fig. 15.10).

If condensation occurs inside piping, grooving can be pronounced. Grooves are formed as rivulets of condensed vapor, laden with ammonia, run down vertical surfaces. Frequently, sinuous grooves which might superficially resemble stress corrosion cracks may occur (Fig. 15.11). (see Chap. 22, "Stress Corrosion Cracking").

FIGURE 15.9 Severe grooving by ammonia-containing condensate on an admiralty brass condenser tube. Condensate flowed down the tubesheet and collected at the tube end, cutting an annular groove. Note the perforation just outside the tube roll.

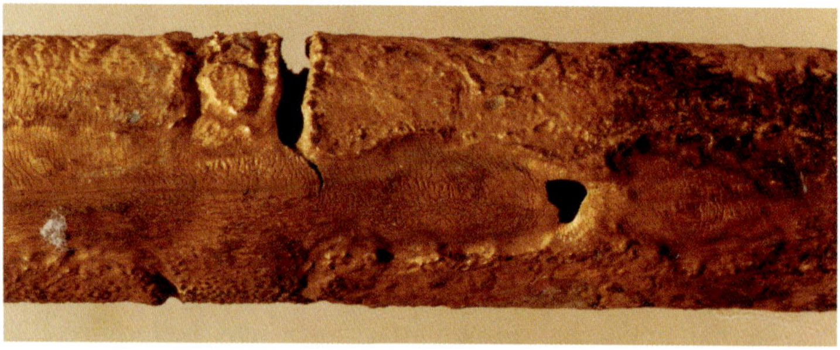

FIGURE 15.10 A brass condenser tube severely wasted by condensing vapors containing ammonia.

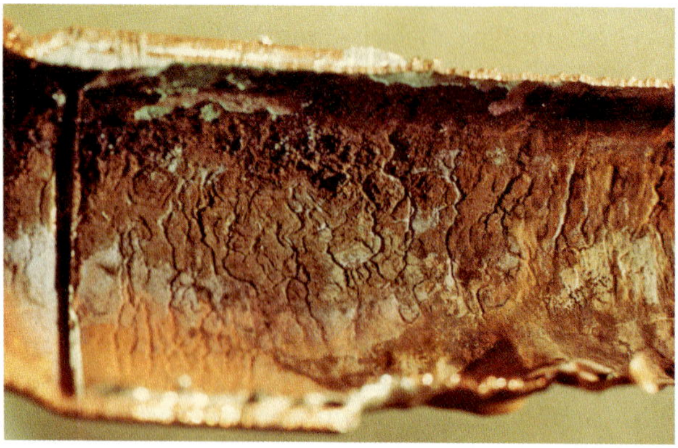

FIGURE 15.11 Severe grooving on the internal surface of a copper pipe carrying condensate. Grooves were cut by condensing vapors containing ammonia running down pipe walls.

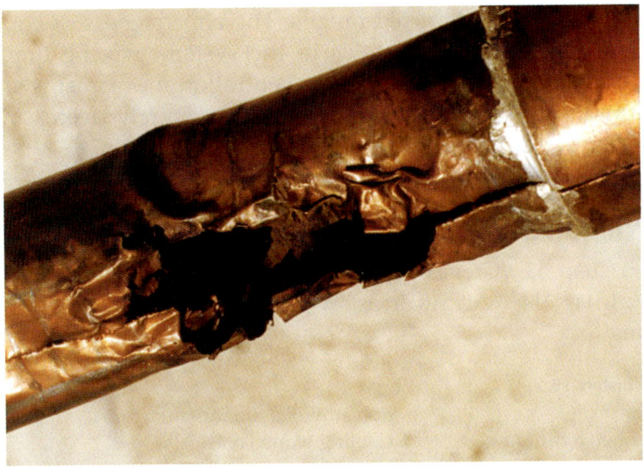

FIGURE 15.12 A large, ragged perforation in a region of severe internal wastage on a copper tube. Corrosion was caused by ammonia-containing condensate collecting along the bottom of the tube.

A broad band of general metal loss may form along the bottom of horizontal piping where ammonia-rich condensate collects (Fig. 15.12).

Other than grooving, localized attack due to ammonia is relatively rare. Patches of metal loss can occur in conjunction with biological fouling and decomposition of organic materials, which generate ammonia.

Voluminous corrosion products are usually absent, as most copper complexes are quite soluble. Adjacent to corroded areas, small amounts of corrosion products and deposits may form that are colored a vivid blue-green by compounds containing liberated copper ions.

Elimination

Minimizing attack requires control of pH to acceptable levels. Concentrations of alkaline-producing species known to promote attack should be reduced where possible.

Some corrosion inhibitors can be used to limit corrosion. For instance, sodium silicate markedly improves the corrosion resistance of aluminum alloys in many alkaline aqueous solutions.

If water chemistry cannot be altered to limit corrosion, then metal alloys more resistant to corrosion in alkaline environments should be considered for substitution.

Cautions

Reducing pH usually has a beneficial effect on corrosion caused by alkaline substances. However, this seemingly obvious solution could have some drawbacks. Chemical treatment programs work most effectively within certain pH ranges. The effectiveness of corrosion inhibitors for other metallurgies in the system may be limited at lower pH values. In addition, if concentration mechanisms such as evaporation or condensation are operating, merely decreasing pH may prove ineffective in controlling attack.

Related Problems

See also Chap. 11, "Crevice and Underdeposit Corrosion (Concentration Cells)"; Chap. 14, "Acid Corrosion"; and Chap. 22, "Stress Corrosion Cracking."

Case History 15.1

Industry:	Polyethylene film manufacture
Specimen location:	Cooling water conduit
Orientation of specimen:	Vertical
Environment:	Cooling water, 70–100°F (21–38°C)
Time in service:	Less than 1 year
Sample specifications:	1 in. (2.5 cm) diameter, aluminum alloy tubing, anodized

An aluminum alloy cooling water conduit section was removed from a machine that made polyethylene sheeting. A large perforation formed just below a ¾ in. (1.9 cm) diameter cooling water orifice (Fig. 15.13). Metal loss was also observed around the edges of the cooling water orifice.

The orifice allowed water to exit and move down an annular space between the conduit and a sheathing pipe. The elliptical perforation was caused by erosion-corrosion, which was due to impinging water flow. The relatively high water pH (9.5–10.5) caused corrosion that accelerated the erosion.

The localization of the wastage that perforated the conduit demonstrates the accelerating effect of turbulence in alkaline attack on aluminum alloys.

FIGURE 15.13 An aluminum cooling water conduit severely attacked by caustic in service. The large circular hole of the cooling water orifice is corroded around its edges. The elliptical hole between the 6- and 7-in. ruler markings was formed by corrosion penetrating the conduit wall.

Case History 15.2

Industry:	Metals fabrication
Specimen location:	Annealing machine sleeve
Orientation of specimen:	Vertical
Environment:	Water containing neutralizing amine, oxygen scavenger, and caustic
Time in service:	1 month
Sample specifications:	3 in. (7.6 cm) outer diameter aluminum tubing

Aluminum sleeves used in an annealing machine to heat-treat copper rod failed rapidly (Figs. 15.14 and 15.15). Attack was caused by caustic that concentrated due to evaporation of water in the annealing sleeve.

Water taken from below these sleeves had a pH of 10.2. It was estimated that the water pH may have been near 14.0 at the metal surface.

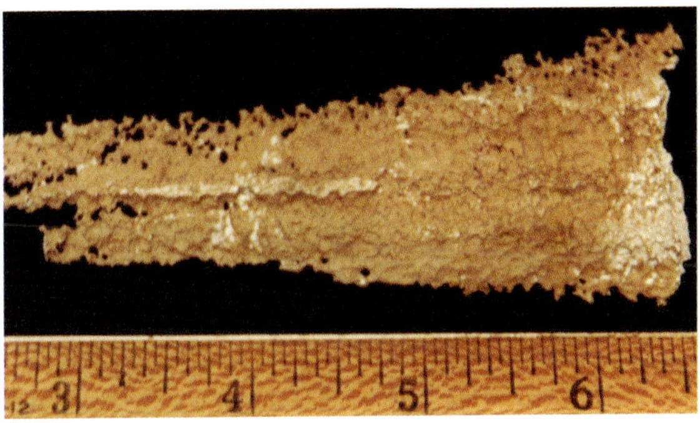

Figure 15.14 A fragment of a severely corroded aluminum annealing sleeve. Attack was by caustic, which caked the specimen.

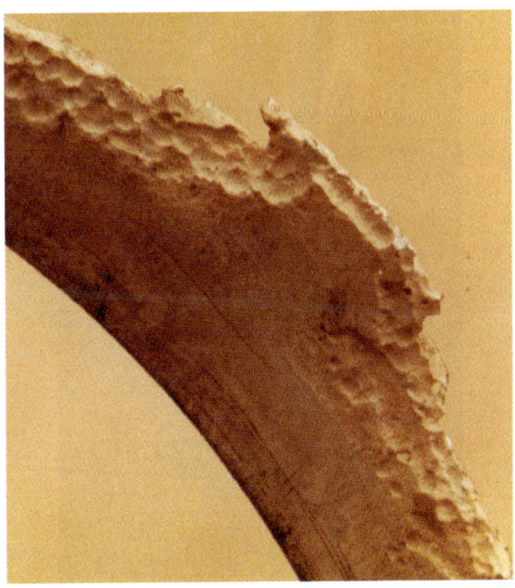

Figure 15.15 A threaded portion of the annealing sleeve shown in Fig. 15.14. Note the heavy caustic deposits.

Case History 15.3

Industry:	Utility
Specimen location:	Low-pressure turbine condenser inlet just outside tubesheet
Orientation of specimen:	Horizontal
Environment:	Internal: Cooling water, pH 7.8–8.2, phosphate-zinc, chlorine, 65–100°F (18–38°C), hardness 300–600 ppm, total alkalinity 45–60 ppm External: (a) Typical service conditions: Saturated steam 105–130°F (41–54°C), NH_3 <0.3 ppm, pH 8.7–9.0, dissolved oxygen 10–20 ppb (b) Startup conditions, once or twice a year: NH_3 up to 75 ppm, pH 8.7–9.4, dissolved oxygen up to 3500 ppb
Time in service:	14 years
Sample specifications:	1 in. (2.5 cm) outer diameter, admiralty brass tubing

The tube section was one of several that contained perforations within circumferential bands of severe external surface metal loss (Fig. 15.9). The maximum measured wall thickness was 0.040 in. (0.10 cm) in the region of the tube that was rolled into the tubesheet. Just outside the rolled end within the circumferential band, the wall thickness was reduced to 0.030 in. (0.076 cm) or less.

Close visual inspection in the thinned band revealed small, longitudinally oriented grooves and gullies running from the rolled tube end across the thinned region (Fig. 15.16). The grooves were deepest at the rolled tube end and branched as they moved away from the tubesheet. The external surface was generally

Figure 15.16 Small grooves on the tube shown in Fig. 15.9. Grooves are condensate rivulet paths cut into the tube surface. Note the spots covered by bluish-green corrosion products.

covered with a thin, dark copper oxide layer. The layer was covered by spotty bluish-green corrosion products in a few places. In the thinned region, the surface had an undulating contour.

Wastage was caused by corrosion due to exposure to condensate containing high concentrations of dissolved ammonia. Corrosion was likely more severe during startup of the system, when ammonia and oxygen levels were significantly higher. Steam containing high levels of ammonia condensed, ran down the tubesheet, and collected at tube ends, causing pronounced localized wastage and grooving.

Case History 15.4

Industry:	Printing
Specimen:	Chilled roll
Orientation of specimen:	Horizontal
Environment:	Cooling water with pH typically around 9.5
Time in service:	2 years
Sample specifications:	5-7/8 in. (14.9 cm) outer diameter, Aluminum alloy

The received section of an aluminum alloy roll contained portions of spiraling, deep grooves on the internal surface (Fig. 15.17). The grooves reportedly occurred at locations along the internal surface on the downstream side of a spirally wound coil. The coil was installed to enhance flow through the roll. The thickness of the roll was reduced by as much as 70 percent, relative to the maximum measured wall thickness. Close examination of the grooves revealed smooth, rounded interiors lined with white corrosion products (Fig. 15.18).

The internal surface was covered by layers of deposits and corrosion products. X-ray fluorescence of material removed from the surface indicated it was predominately aluminum with small amounts of iron, silicon, calcium, magnesium, and phosphorus. X-ray diffraction indicated the presence of aluminum hydroxide, specifically bayerite, $Al(OH)_3$, and boehmite, AlO(OH). Spot tests indicated that some of the material near the grooves produced a slightly alkaline solution when mixed with water. It was reported that the untreated cooling

Figure 15.17 Chilled roll from a printing press that contained deep grooves that formed adjacent to a spirally wound coil. Deposits and corrosion products cover the internal surface.

FIGURE 15.18 Close-up of the aluminum roll in Fig. 15.17, showing deep grooves that are covered by a layer of white corrosion products in most places.

water maintained a pH around 9.5. This evidence indicated that alkaline corrosion contributed to the attack.

However, attack was not general across the surface of the roll, but localized adjacent to the coil on the internal surface. It is likely that turbulent conditions developed immediately downstream from the coil, significantly promoting wastage at these locations due to a contribution from erosion.

The cooling water was reportedly untreated for corrosion protection of aluminum. The use of an aluminum corrosion inhibitor, such as silicate, was recommended to help reduce wastage of aluminum components in the cooling system.

Case History 15.5

Industry:	Building
Specimen location:	Cooling tower
Environment:	Treated cooling water, pH 7.7 to 8.8, total hardness 360 to 420 mg/L, conductivity 650 to 1525 μmhos
Time in service:	4 months
Sample specifications:	Hot mill galvanized steel sheet

The water-contacting surfaces of the cooling tower developed spotty patches or nodules of friable white material in locations, including the hot deck distribution pans and the lower basin (Fig. 15.19). Thicker layers were present along the waterline of the lower basin. A dark corrosion product layer and extremely shallow metal loss were found beneath the white material (Fig. 15.20). Metallographic cross sections revealed that the entire zinc coating thickness was converted to corrosion products in places (Fig. 15.21). No attack of the underlying steel was noted.

Chemical analysis of the light-colored material at all locations examined indicated that it predominantly contained zinc with much lower levels of silicon, calcium, and phosphorus. The chemical composition of the corrosion products

FIGURE 15.19 White corrosion product layers on the basin of a cooling tower and along the waterline on the sides.

FIGURE 15.20 Close-up of the spotty white corrosion products with an area partially cleaned, revealing extremely shallow metal loss with a darkened appearance beneath them.

FIGURE 15.21 Metallographic cross section through a location of white corrosion products on a galvanized sheet. Note that the voluminous corrosion products overlie an area where the entire galvanized coating was converted to corrosion products. The underlying steel is unattacked.

and their consistency indicated that white rust had formed on the galvanized surfaces.

Review of the water chemistry indicated that pH values of the cooling water gradually increased over time, as acid feed was not available to control pH. Operating the cooling tower at higher pH levels than recommended for initial passivation was primarily responsible for the formation of white rust. Recommended remedial actions for the extent of damage may include cleaning and coating with zinc-containing paints, followed by repassivation at appropriate water conditions.

Case History 15.6

Industry:	Electronic component manufacturing
Specimen location:	Tubing in a closed cooling system
Orientation of specimen:	Varies
Environment:	Low-conductivity cooling water, untreated
Time in service:	Unknown
Sample specifications:	¼ in. (0.6 cm) outer diameter copper tubing

Analysis of water from a cooling system revealed total copper concentrations as high as 800 mg/L, with as much as 690 mg/L dissolved in the water. This indicated severe corrosion of copper components in the system. Examination of copper tubing from the system revealed sections that experienced deep internal surface metal loss (Fig. 15.22). The smooth shapes of the wasted surfaces indicate that the metal loss at those locations was caused by erosion-corrosion, with high flow rates and turbulence in the small-diameter tubing contributing to the attack.

The wasted surfaces were covered by spotty blue corrosion products in places. A strong scent of ammonia was emitted from the surfaces of the tubing samples, which was a major indicator that the material on the surface contained amines or ammonia that volatilized into the air. Water analysis confirmed the presence of

Figure 15.22 Internal surface of a copper tube that experienced smooth metal loss in most areas. Note the spotty blue corrosion products in the wasted areas.

significant concentrations of dissolved ammonia in the cooling water. Erosion-corrosion was significantly aggravated by the presence of ammonia, which is very corrosive to copper and will release copper ions into water solution.

In order to control corrosion of copper in the pure water, the source of ammonia into the cooling water needed to be eliminated. The chemical treatment should include adequate concentrations of proper corrosion inhibitors, such as azoles. It was recommended that flow rates in the system be checked to assure the water velocities are not excessive for copper (see Chap. 23, "Erosion-Corrosion"). If changes to the chemistry and operation of the system could not limit attack, then alternative metallurgies, such as stainless steel, would need to be considered for replacement. Stainless steel tubing and fittings were not corroded in the system.

References

1. J. R. Davis (ed.), *Corrosion of Aluminum and Aluminum Alloys*, ASM International, 1999, pp. 25–27.
2. White Rust Prevention: An Industry Update and Guide Paper – 2012, Presented By: Association of Water Technologies (AWT).
3. L. Caruso and H. T. Micheals, "The Resistance of Copper-Nickel Alloys to Ammonia," *Corrosion in Simulated Steam Condenser Environments*, Proceedings of the American Power Conference, 1980.

CHAPTER 16

Galvanic Corrosion

General Description

Of the many types of corrosion that can affect cooling water systems, one of the most familiar is galvanic corrosion. The basic factors involved in galvanic corrosion are relatively simple and the concepts comparatively uncomplicated. Nevertheless, it is common to find galvanic corrosion incorrectly diagnosed as the primary mechanism for the cause of deterioration of one metal simply because a different metal is nearby. In addition, certain features of the galvanic corrosion process occasionally can become complex and subtle, making correct diagnosis in these cases a challenge.

One of the main factors required for galvanic corrosion is contact between dissimilar metals or conductive materials. This is generally associated with metals of different compositions. The metals will have a galvanic potential in a specific environment, which can be ranked from noble to active in a galvanic series. Table 16.1 provides a galvanic series of some metals in seawater. The location of coupled metals in the galvanic series is particularly important, as corrosion of the active metal is more likely due to galvanic effects if there is a large potential difference between the metals.

The basic features involved in galvanic corrosion can be clearly illustrated by considering a simple dry cell (Fig. 16.1). A dry cell consists of a noble material (graphite in the conventional dry cell) and an active material (zinc in the conventional dry cell) embedded in a common, electrically conductive environment (electrolyte). The noble and active materials are linked by an external electrical pathway, which can be opened or closed. When the circuit is closed (such as by switching on a light), the active metal begins to corrode and the noble material remains intact. The corrosion of the active material furnishes electrons that flow to the noble material through the external electrical pathway. A corresponding counterflow of positive ions moves through the electrolyte to complete the circuit.

Noble
Gold
Graphite
Titanium
Silver
Hastelloy C
(316 stainless steel (passive)
304 stainless steel (passive)
410 stainless steel (passive))
Nickel (passive)
Monel (Alloy 400)
(70:30 cupronickel
90:10 cupronickel
Silicon bronze
Copper
Red brass (84–86% copper)
Admiralty brass (70–73% copper)
Yellow brass (64–69% copper)
Naval brass (62–65% copper)
Muntz metal (59–63% copper))
Nickel (active)
316 stainless steel (active)
304 stainless steel (active)
410 stainless steel (active)
(Cast iron (not graphitically corroded)
Carbon steel)
Aluminum alloys
Zinc (and galvanized coatings)
Magnesium
Active

TABLE 16.1 Galvanic Series of Selected Metals in Seawater

Hence, the basic requirements of a galvanic corrosion process are:

- An electrochemical interaction of two or more materials (materials 1 and 2 in Fig. 16.1) having a sufficient galvanic potential difference
- An electrolyte (3) common to the surface of both materials, through which an ionic current flows
- An electrically conductive pathway (4) physically linking the two materials

In any galvanic couple, the corrosion rate of the active material (anode) will typically increase, and the corrosion rate of the noble material (cathode) will typically decrease or possibly even cease,

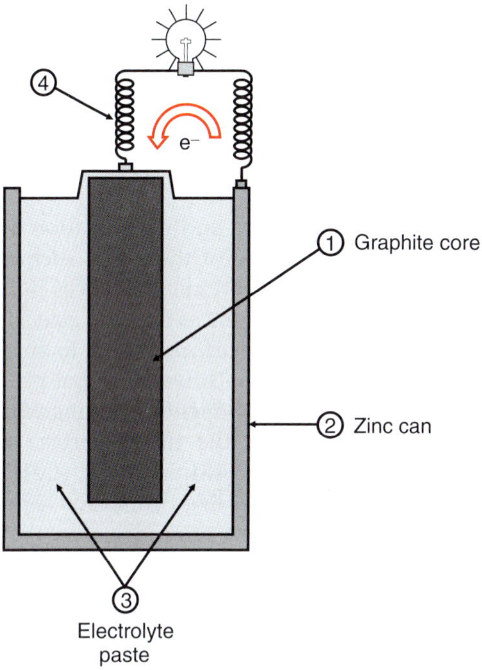

FIGURE 16.1 Simple dry cell battery. Electrons are conducted along the external circuit (4), which physically connects the active (2) and noble (1) materials. An equivalent ionic countercurrent is conducted through the electrolyte (3), thereby completing the circuit.

relative to their corrosion rates in the environment when they are not electrically coupled together.

As corrosion proceeds, reaction by-products may form on the metal surfaces, creating a resistance to electrical exchanges at these surfaces. Consequently, the reaction rate diminishes correspondingly. If the corrosion reaction is stopped, then a time-dependent recovery can occur. When the reaction is restarted, the initial corrosion rate is reestablished. This effect is often observed in conventional dry cells and automobile batteries. The magnitude of this resistance phenomenon can vary from metal to metal. Hence, the rate and extent of galvanic corrosion for any particular couple may be less than anticipated when these resistance effects are operating.

Galvanic corrosion is typically associated with the coupling of two or more dissimilar metals. However, sufficient variations that can produce changes in other parameters, either environmental or physical, can also result in galvanic potential differences that can drive the electrochemical corrosion reaction.

Some environmental parameters include fluid chemistry such as differential oxygen concentration and ion concentration cells (see Chap. 11, "Crevice and Underdeposit Corrosion (Concentration Cells"), temperature, and flow velocity. Appreciable environmental differences can produce a corrosion current even within the same metal. For instance, sufficient temperature differences between sites on the same component can result in galvanic current flow. In these cases, the site with the higher temperature is usually the corrosion site. Galvanic corrosion of this form can potentially affect heat exchangers and refrigeration equipment.

Similarly, physical parameters of the metal such as differential strain, differences in surface condition, differences in thermal treatment or microstructure, and differences in grain size and orientation can make one area on the metal surface anodic to an adjacent area. One of the more common physical differences that promotes preferential corrosion in areas on a component in cooling water systems is differential strain. Locations on the metal surface that have an increased degree of cold work or residual tensile stress are preferred areas to experience corrosion. Localized corrosion, primarily due to residual stresses, is often observed around welds. Welds also have microstructural differences in the fusion and heat-affected zones that can also promote corrosion in some cases.

Locations

Galvanic corrosion is location-specific in the sense that it occurs at a coupling of dissimilar metals (Fig. 16.2). It is metal-specific in the sense that corrosion typically affects the metal that has less resistance in the environment to which the couple is exposed. Hence, in principle, we would anticipate galvanic corrosion of relatively reactive metals wherever they are in physical contact with relatively noble metals in a sufficiently aggressive, common environment. Experience has shown, however, that all such couples do not necessarily result in unsatisfactory service. This is because of the interplay of various critical factors that influence galvanic corrosion. These critical factors are discussed in the next section.

The coupling of dissimilar metals can be broken down into categories based on how the couple is formed. The categories include direct

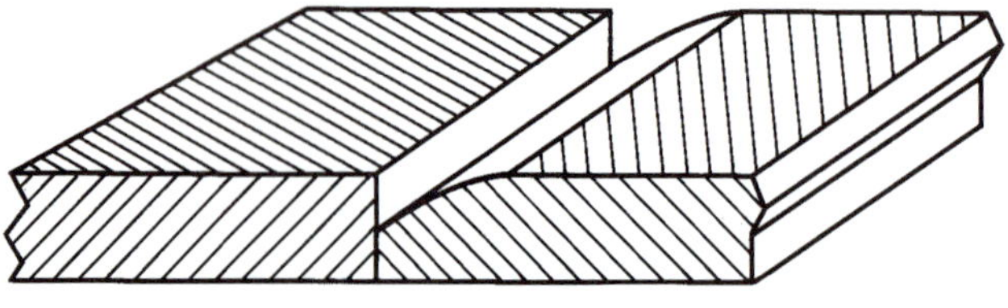

FIGURE 16.2 Galvanic corrosion along a plane of bimetallic contact.

coupling of different components, application of metallic coatings, deposition of metallic ions, and microstructural differences in the metal.

Direct Coupling

The design of cooling water equipment such as cooling coils, heat exchangers, and condensers often involves direct coupling of dissimilar metals, offering the possibility of galvanic corrosion. Channels, heads, tubesheets, and gasket surfaces are targets of galvanic corrosion. Tubes in heat exchangers may be significantly noble with respect to the tubesheet metal or to baffles, causing localized metal loss in the tubesheet around the tubes (Fig. 16.3). Similarly, steel baffles may corrode where they contact copper tubes. Metallic inserts installed to prevent inlet-end erosion may induce galvanic corrosion of less-noble tubes. Systems that have different metals used for piping, fittings, and other components can be subject to galvanic corrosion (Fig. 16.4). Cast iron components such as floating head covers may galvanically corrode if coupled to relatively noble metals such as copper or copper-alloy tubes and tubesheets. Interestingly, if the deterioration of the cast iron produces graphitic corrosion (see Chap. 19, "Graphitic Corrosion"), a potential reversal may occur. The graphitically corroded cast iron surfaces that are covered by iron oxide containing graphite become more noble than the copper or copper alloy, resulting in galvanic corrosion of these metals.

Coatings

Some components can be coated with metallic layers for a variety of reasons, including appearance, corrosion protection, and wear resistance. A sacrificial coating is one that is anodic to the underlying

Figure 16.3 Example of grooving around brass tubes rolled into a steel tubesheet in a heat exchanger exposed to high-conductivity cooling water.

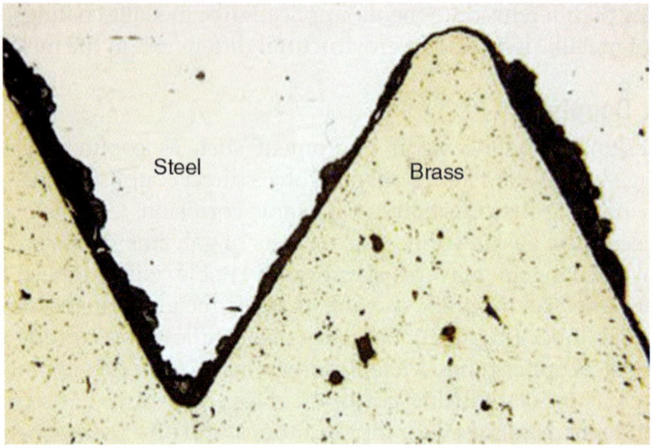

FIGURE 16.4 Micrograph of a cross section through a threaded joint where water penetrated, showing corrosion of the steel component and no corrosion of the brass fitting it was coupled to.

substrate (base metal). A common example is galvanized steel, where the zinc and/or aluminum-based coating can be protective to the underlying steel (Figs. 7.3 and 7.4). If a noble coating is applied over a component of a more reactive metal, then severe attack may occur at breaks in the coating due to the galvanic coupling (see Figs. 7.1 and 7.2 for examples). In some cases, brass or copper alloys can be plated onto zinc die-cast components for appearance. If these components are exposed to a water-based environment, they can degrade quickly. Chromium and nickel plating on steel or aluminum are other examples. Additional information on the interaction of coatings with the substrate metal is provided in Chap. 7.

Note that a conductive nonmetal that is formed in place by reaction of the metal surface with chemical species can also act as the noble material in a galvanic couple. For example, galvanic corrosion may occur on the surface of steel components that are covered or partially covered with mill scale (magnetite, Fe_3O_4) or iron sulfide corrosion product layers. Both mill scale and iron sulfide are noble with respect to steel. Significant galvanic corrosion can occur where breaks or holidays in these layers expose unprotected metal.

Deposition

Galvanic corrosion may also occur by transport of relatively noble metals, either as particulate or as dissolved ions, to the surface of an active metal. For example, ions of copper, perhaps resulting from corrosion or erosion elsewhere in the cooling water circuit, may be carried by cooling water to surfaces of aluminum, steel, or even stainless

FIGURE 16.5 Micrograph showing a nearly continuous layer of deposited elemental copper that resulted in galvanic corrosion of the steel surface in a cooling water environment.

steel components. If the ions are reduced and plated onto the component surfaces, localized galvanic corrosion may result, as shown in Fig. 16.5. Due to the large potential difference, galvanic corrosion caused by deposited elemental copper is more frequently encountered on aluminum alloy and galvanized steel components in cooling water systems. As little as 0.02 to 0.05 ppm of copper in cooling water can initiate pitting attack on aluminum due to galvanic effects.[1]

Microstructure

The solution potential difference between microstructural constituents of a particular metal alloy can affect its corrosion resistance. An appreciable potential difference between the phases in the microstructure and the surrounding matrix can result in galvanic corrosion on the microscopic scale, which can subsequently lead to widespread metal loss. Significant alloy segregation can produce similar effects. The exact interaction depends on the metal under consideration. Some common examples for different alloys are briefly discussed next.

Aluminum alloys often show preferential attack along precipitates in the microstructure or grain boundaries due to elemental segregation. Typical alloying elements in aluminum will change the solution potential relative to pure aluminum. Magnesium and zinc additions will lower the potential of aluminum solid solution (make it more anodic), while silicon, copper, and manganese will make the potential more cathodic. Aluminum alloys, especially high-strength wrought alloys and cast alloys, are particularly susceptible to microgalvanic corrosion when exposed to water of sufficient conductivity.

These alloys typically have second phase constituents that have appreciably different solution potentials than the aluminum matrix. Preferential attack due to microgalvanic corrosion can occur in the aluminum matrix along the boundaries of more noble precipitates, such as Al_2Cu (a constituent that may be present in 2000 series aluminum alloys) and silicon (typically present in cast aluminum alloys). An example of microgalvanic corrosion along more noble precipitates in a cast aluminum alloy is shown in Fig. 16.6.

A form of microgalvanic corrosion may occur at stainless steel welds if sensitization has taken place or if welding has produced dissimilar phases with an unfavorable morphology (see Case History 26.1). These forms of microgalvanic corrosion can arise due to microstructural changes during heating and solidification of the weld. Sensitization is caused by the development of microscopic zones along grain boundaries with reduced chromium content (see Chaps. 18 and 26). These chromium-depleted zones are more susceptible to corrosion.

Another microgalvanic effect may occur in stainless steel alloys that have a two-phase microstructure. A two-phase dendritic microstructure of austenite and ferrite typically develops during solidification of austenitic stainless steel welds. Similarly, certain cast alloys will contain residual ferrite, and ferrite stringers may be present in wrought austenitic stainless steel alloys. Differences in composition of the two phases result in potential differences that can cause preferential corrosion of the more anodic phase. Preferential attack can also

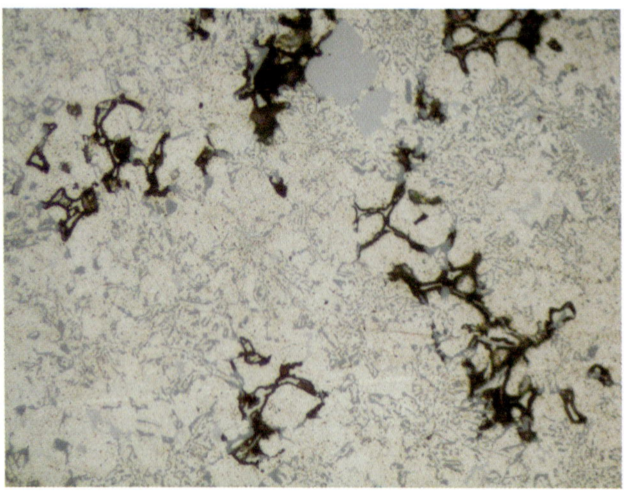

FIGURE 16.6 Micrograph illustrating preferential attack penetrating along second phases, specifically the Al_2Cu phase in the microstructure of a cast type 385 aluminum alloy. The aluminum matrix is anodic to the Al_2Cu phase in the microstructure, and microgalvanic attack occurred in locations exposed to cooling water.

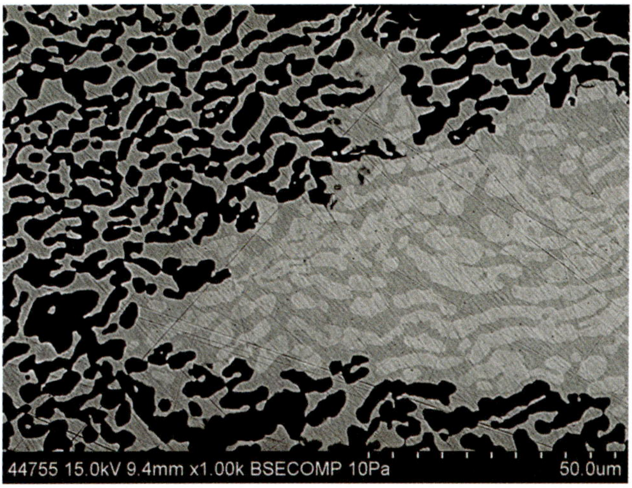

FIGURE 16.7 Scanning electron microscope backscattered electron image showing preferential attack in a type 2205 duplex stainless steel heat exchanger tube. Spot chemistries indicated that the corroding austenite phase (light gray) was lower in chromium and molybdenum but higher in nickel than the ferrite phase (dark gray).

occur in some duplex stainless steel alloys if they are exposed to severe cooling water conditions (Fig. 16.7). The phase that is anodic is dependent on the environmental conditions.

Another form of microstructural galvanic corrosion, graphitic corrosion, is unique to gray and nodular cast irons. It may be encountered in cast iron pumps and other cast iron components. Graphitic corrosion does not require connection to a different metal, as the graphite in the microstructure is considerably more noble than the surrounding iron-based matrix. Details on graphitic corrosion are presented in Chap. 19.

Some two-phase brass alloys experience selective attack on one phase in the microstructure. A common example in water systems is dezincification in brass alloys typically containing more than 32 percent zinc. Brass alloys with zinc levels greater than 32 percent will generally have a second phase, known as the beta phase. The beta phase is surrounded by an alpha phase matrix, which has a lower zinc content. Preferential attack on the higher zinc beta phase occurs because it is the more anodic phase in the microstructure. Examples of dezincification of beta phase–containing brass alloys are presented in Chap. 17, "Dealloying."

Critical Factors

The necessary conditions for galvanic corrosion are interaction of electrochemically dissimilar materials that are exposed to a common conductive fluid and are physically linked so that current can flow from one material to the other. Once these necessary conditions are met, both the determination of the anodic (corroded) and cathodic (protected) materials are affected by various environmental factors, along with the magnitude of the corrosion of the anodic metal.

Galvanic corrosion in typical industrial cooling water systems is the net result of the interplay of these factors. Some factors may accelerate the corrosion process; others may retard it. In their approximate order of importance, the more influential factors are discussed next.

Galvanic Potential

In simple terms, galvanic potential is related to the magnitude of the current induced by coupling dissimilar materials exposed to a common conductive fluid. The magnitude of the potential depends on the materials that are coupled and on the characteristics of the fluid to which the metals are exposed.

As mentioned previously, a galvanic series provides a ranking of different materials according to their galvanic potential in a specific environment. One of the most commonly known galvanic series is a ranking of materials in a seawater environment. Table 16.1 is such a listing for numerous industrial metals and alloys often found in cooling water systems. Strictly speaking, this arrangement is precise only for seawater under controlled laboratory conditions.

In other conductive fluids under industrial cooling water conditions, the magnitude of the potential, and possibly the position of the metals on the list, could change. In typical cooling water environments, however, the order of the metals as listed for seawater would not be expected to change significantly. Studies to produce a galvanic series for tap or freshwater have shown this to be true for most metals.[2,3] Some notable differences in the alloy rankings between fresh and seawater are a shift to a more noble potential for aluminum and austenitic stainless steels in freshwater. However, the potential shift does not appreciably alter the rankings, at least for the conditions of the experiments. It must be noted that other experimental conditions can dramatically change the rankings. For instance, solutions of concentrated salts, acidic species, and alkaline species have been shown to have appreciably different rankings of corrosion potentials for metals.[4]

The relative positions of the materials in Table 16.1 indicates which metal in a couple will probably be noble and which will be active and will thus experience accelerated attack due to galvanic corrosion, at least in a seawater environment. The difference in potential between two materials is a major factor that affects corrosion rates. Hence, galvanic coupling of metals widely separated on the table, such as copper and aluminum, would be expected to result in relatively rapid corrosion of the anodic member (aluminum in this example). Galvanic coupling of metals having little separation on the table, such as yellow brass and copper, would be expected to produce relatively low corrosion rates of the anodic member (yellow brass in this example).

In general, bracketed groups of metals in the table would not be expected to produce significant galvanic corrosion if coupled with each other.

Some materials in the table require more detailed explanation of their placement and possible interactions. For instance, titanium is a very reactive metal but forms a protective passive film that is considered extremely noble. Stainless steels behave similarly. Note the dual positions of stainless steels in Table 16.1 corresponding to their active and passive states. Passive stainless steel (its normal condition in well-aerated, flowing water) is quite noble and can potentially induce galvanic corrosion on aluminum, carbon steel, and even copper and copper alloys. In its active condition, however, stainless steel is less noble than copper and may suffer galvanic corrosion if coupled to copper or its alloys.

Stainless steel may lose its passivity and assume an active state under certain environments, such as low-flow or stagnant conditions, when oxygen and oxidizing ions are depleted, and especially if chloride ions are present at appreciable levels. Note also that a galvanic couple can be established between passive regions and active regions of the same stainless steel component. For example, active sites may form at crevices or beneath deposits, resulting in rapid corrosion rates, especially if the surrounding passive metal has a relatively large area (unfavorable area ratio). This caution applies to other passive metals, such as aluminum, as well. (See Chap. 11.)

Aluminum alloys, as mentioned previously, can have appreciable differences in solution potential depending on alloy composition and heat treatment.[5] Alloying additions of magnesium and zinc (elements ranked lower in potential on the galvanic series than aluminum) reduce the solution potential of aluminum (that is, make it more anodic), while silicon and copper additions raise the solution potential. Copper alloys such as brass and cupronickel behave similarly. Additions of zinc to copper, as in brass alloys, lower the solution potential, and higher-zinc brass alloys have a more anodic ranking in the galvanic series. Conversely, nickel additions to copper raise the potential.

Temperature is a parameter that can affect the solution potentials of metals and alloys. This is illustrated in Fig. 16.8, which shows potential ranges for some metals and alloys in seawater at two different temperatures. Note that the potential can be increased or decreased with increasing temperature, depending on the metal or alloy. For instance, note that the potential range for zinc increases slightly with increasing temperature, but decreases slightly for carbon steel. These shifts can produce what is known as polarity reversal of galvanized steel. At ambient temperatures steel is noble to zinc. However, a potential reversal may occur at temperatures above 140°F (60°C) in waters of certain chemistries, and the zinc becomes noble to the steel. For this reason galvanized steel should be avoided for use in many hot water systems.

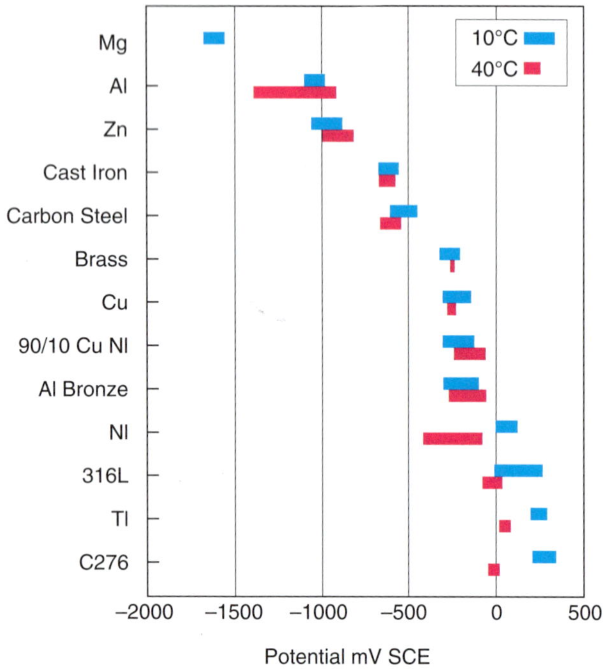

FIGURE 16.8 Chart showing potential ranges of metals and alloys in seawater at two different temperatures, 10°C (blue boxes) and 40°C (red boxes). (*Source: M. Davies and P. J. B. Scott,* Guide to the Use of Materials in Waters, *NACE International, 2003, p. 60. © NACE International 2003.*)

Conductivity of Fluid

Exposure of coupled dissimilar metals to a common fluid, or electrolyte, is necessary to complete the galvanic circuit. Cooling water is typically a suitable electrolyte. Electrons flow through the metal to form one-half of the circuit (hence the requirement that the metals be electrically connected), and ions move through the electrolyte to provide the corresponding countercurrent. Consequently, the ionic conductivity of the electrolyte exercises a substantial influence on the galvanic corrosion process, affecting both its severity and its extent. For example, galvanic corrosion in relatively low-conductivity water may be moderate and confined to a narrow zone at the junction of the metals involved. In high-conductivity water, the corrosion will be more extensive and may be severe near the junction. Thus, the likelihood and severity of galvanic corrosion increases going from conductive freshwater to brackish water to seawater. On the other hand, dissimilar metal couplings exposed to ultrapure, high-resistivity water should not experience appreciable galvanic corrosion.

Area Effect

The exposed areas of the noble (cathode) and active (anode) metals can have a significant effect on most galvanic corrosion processes. The corrosion rate of the active metal is proportional to the area of exposed noble metal divided by the area of exposed active metal. A favorable area ratio (large anode, small cathode) can permit the coupling of dissimilar metals. An unfavorable area ratio (large cathode, small anode) of the same two metals in the same environment can be costly. Table 16.2 highlights this effect for a variety of metal alloy couples in seawater. The area effect is important to consider when choosing fasteners that are a different alloy than the metals they are joining (see Case History 16.1). If dissimilar metal fasteners must be used, they should be noble relative to the metals being joined.

Correct application of this principle can lead to what would appear to be peculiar recommendations. For example, if just one member of a couple is to be coated, it should be the noble member. Most coating systems leave holidays or tiny openings where the metal is exposed. If the active metal is coated, the area of exposure at the holidays can be quite small compared to the area of the noble metal, resulting in an unfavorable area ratio. An example of severe localized attack due to galvanic corrosion on small areas of exposed coated steel in a water box of a heat exchanger is shown in Fig. 16.9. On the other hand, if the noble metal is coated, the holidays provide a small cathodic area and hence a highly favorable area ratio with respect to the active metal.

Surface Effects

The galvanic series in Table 16.1 is generally useful for predicting the tendency toward galvanic corrosion between coupled metals. The arrangement of this series is, however, based on data generated under specific, controlled laboratory conditions on clean, bare metals.

In industrial cooling water systems, metal surfaces may vary from clean to completely covered with deposits and corrosion product layers. These layers may interfere with and retard the exchange of current between the electrolyte and the metals of the electrochemical circuit. Since the electron flow through the coupled metals must be equivalent to the ionic flow through the fluid, reduction in ionic flow on the fluid side will result in a corresponding decrease in electron flow through the metals. In other words, the corrosion rate will diminish or possibly cease. Initial corrosion rates may diminish to a fraction of their original value if a corrosion product layer that reduces electrical exchange forms and develops on surfaces. In this way, corrosion inhibitors that produce protective layers on the metal surface can reduce galvanic corrosion.

Fluid Velocity

The possible effects of fluid velocity on galvanic corrosion are sometimes overlooked. Fluid velocity can affect the apparent potential of

Metal Considered		Galvanized steel	Aluminum 1100	Aluminum 2024	Mild steel	Low-alloy steels	Cast iron	12–14% Cr (410 SS)	Lead tin solder	16–18% Cr steel (430 SS)	Lead	Tin	Muntz metal	Manganese bronze	Naval brass	Nickel	Yellow brass	Admiralty brass	Aluminum bronze	Red brass	Copper	Silicone bronze	70-30 Copper-nickel	70-30 nickel-copper	Cr-Ni stainless steel (304 SS)	Cr-Ni-Mo stainless steel (316 SS)	Graphite
Zinc, galvanized steel	S																										
	E																										
	L																										
Aluminum 1100	S																										
	E																										
	L																										
Aluminum 2024	S																										
	E																										
	L																										
Mild steel	S																										
	E																										
	L																										
Cast iron	S																										
	E																										
	L																										
12-14% Cr steel (410 SS)	S																										
	E																										
	L																										
Lead tin solder	S																										
	E																										
	L																										
16-18% Cr steel (430 SS)	S																										
	E																										
	L																										
Austenitic Cr-Ni stainless steel (304 SS)	S																										
	E																										
	L																										

Table 16.2 Table Indicating the Likelihood of Corrosion of Galvanic Couples in Seawater, Highlighting the Area Ratio Effect

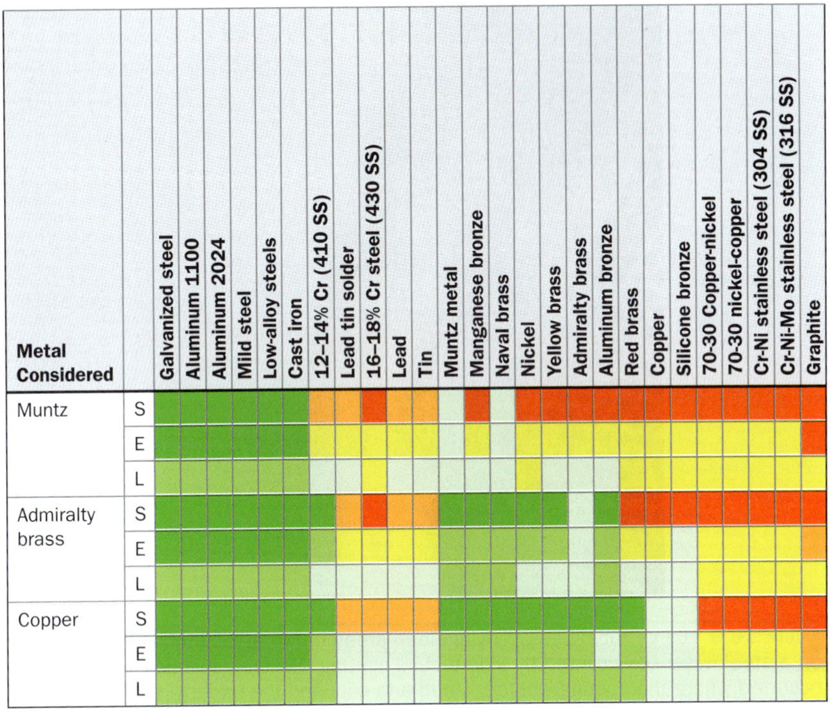

- (dark green) The corrosion of the metal under consideration will be reduced considerably in the vicinity of the contact.
- (light green) The corrosion of the metal under consideration will be reduced slightly.
- (pale) The galvanic effect will be slight with the direction uncertain.
- (yellow) The corrosion of the metal under consideration will be increased slightly.
- (orange) The corrosion of the metal under consideration will be increased moderately.
- (red) The corrosion of the metal under consideration will be increased considerably.

S Exposed area of the metal under consideration is small compared with the area of the metal with which it is coupled.

E Exposed area of the metal under consideration is approximately equal to that of the metal with which it is coupled.

L Exposed area of the metal under consideration is large compared to that of the metal with which it is coupled.

Source: H. H. Uhlig (ed.), *The Corrosion Handbook*, John Wiley & Sons, 1948, pp. 420–425.

Table 16.2 *(Continued)*

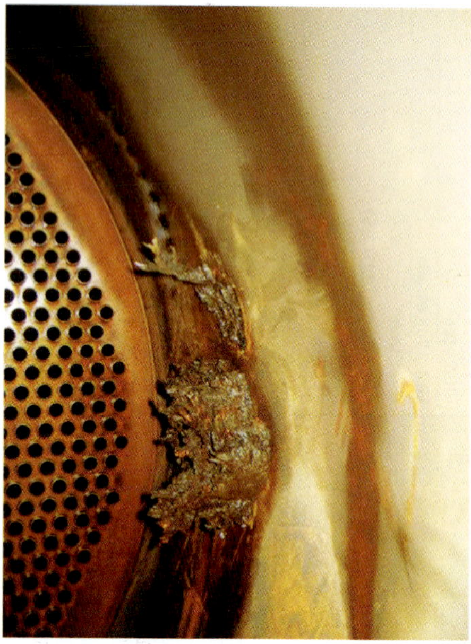

FIGURE 16.9 Large tubercles overlying deep areas of metal loss in locations on a steel water box covered by a protective coating. Severe galvanic attack occurred where the coating was degraded and exposed the underlying steel, due to the small exposed area of steel coupled to the yellow metal tubesheet.

metals in a given environment. For example, a metal under the influence of relatively rapid flow may assume either a more noble or a more active character than that indicated by the galvanic series. Occasionally, this shift in potential may result in galvanic corrosion that would not occur under stagnant or low-flow conditions.

Identification

Diagnosis of galvanic corrosion that is based on the deterioration of one metal simply because a second metal is nearby can often be incorrect. This error can generally be avoided if the required elements for the galvanic corrosion cell, as stated previously, are duly recognized. Hence, corrosion of one member of a couple when both metals are galvanically similar is probably not due to galvanic corrosion. Corrosion of one member of a couple exposed to a common, nonconducting fluid is not galvanic corrosion, since ionic flow through the fluid is required to complete the galvanic circuit. If dissimilar metals are coupled, but are not exposed to a common electrolyte, they cannot experience galvanic corrosion. Corrosion of one metal that is not coupled to a dissimilar metal is not galvanic corrosion, since a common current must pass through both metals to complete the galvanic circuit.

Galvanic corrosion will generally have the following features:

- The more active of the coupled metals will corrode.
- Corrosion of the noble metal will be slight or nonexistent, even though it would corrode in the given environment if it were not coupled to the active metal.
- In general, corrosion of the active metal in fresh cooling water will be most severe at its junction with the noble metal (Fig. 16.2), and will decrease with increased distance from the junction.
- The extent of the attack will vary depending on the conductivity of the fluid; the greater the conductivity, the more extensive the corrosion.

Elimination

The nature of galvanic corrosion is such that successful avoidance generally requires implementing preventive rather than corrective techniques. Therefore, consideration of galvanic corrosion problems must be integrated into the design of equipment. Corrective techniques applied to existing equipment can be expensive and less than satisfactory.

Because a galvanic potential difference is the driving force for corrosion, reducing the magnitude of this difference can reduce or prevent galvanic corrosion. Also, since the corrosion rate is directly influenced by the magnitude of the galvanic current flowing through the electrolyte and metals, increasing the resistance to ionic flow in the electrolyte and/or the metal will stifle galvanic corrosion. Similarly, galvanic corrosion can be prevented altogether by breaking the electrical circuit through the use of nonconductive materials at the junctions of the metals. Finally, the importance of area ratios should not be overlooked in equipment design, since favorable ratios may actually permit the use of materials having significantly dissimilar galvanic potentials.

Based on these considerations, specific preventive or corrective steps may be taken to minimize or eliminate galvanic corrosion.

Preventive Techniques

- Completely insulate the materials from one another at all junctions exposed to a common fluid. Insulation involves using dielectric couplings, nonconductive washers, inserts, sleeves, and coatings. Thread sealants may provide some degree of insulation but may not be completely reliable.
- When possible, avoid coupling materials having widely dissimilar galvanic potentials.

- If dissimilar metal couplings cannot be avoided, make use of favorable area ratios in design to avoid the active metal having a small exposed area relative to the noble metal.
- Fasteners, such as bolts and rivets, which have a significant potential difference from the materials being fastened should be relatively noble to these materials.
- If materials widely separated on the galvanic series must be joined, avoid threaded connections.
- Specify appropriate weld filler materials and welding techniques to avoid small areas of the weld or heat-affected zone being appreciably anodic to the base metal being joined.

Corrective Techniques

- Redesign to completely insulate the materials from one another at all junctions exposed to a common fluid. Insulation involves using dielectric couplings, nonconductive washers, inserts, sleeves, and coatings.
- Alter the chemistry of the common fluid to render it less conductive and/or less corrosive. Generally, water corrosivity increases with an increase in conductivity, temperature, oxygen content, and a decrease in pH. Inhibitors may be effective at limiting attack by forming protective films on the metal surfaces to reduce corrosion reactions. Note that in mixed-metal systems, higher dosages of inhibitors may be required than would be necessary in single-metal systems in the same environment.
- Use cathodic protection techniques, such as installing sacrificial anodes. An example of a sacrificial zinc anode attached to a chiller end cap is shown in Fig. 16.10. The number and placement of the sacrificial anodes for effective cathodic protection is system dependent.
- Coat both materials (preferred) or just the noble material. *Do not coat just the active material* (see "Area Effect"). Coatings should be maintained, especially the one covering the active member of the couple. Use of cathodic protection systems in association with coatings is often recommended.
- Metal ions from the corrosion of relatively noble equipment may plate out on more active metals downstream, setting up highly localized galvanic cells. This situation can be especially troublesome, for example, when copper ions are conveyed to aluminum, galvanized steel, carbon steel, and even stainless steel equipment. Corrosion of copper alloys must be controlled with azole corrosion inhibitors. If sufficient levels of azole are used, they can react with dissolved copper ions to limit or prevent the copper from plating out on active metals.

FIGURE 16.10 Sacrificial zinc anode attached to the end cap of a chiller. Note the white corrosion products covering the anode surface.

Cautions

Avoid the common tendency to conclude that galvanic corrosion is responsible for deterioration of one metal simply because another dissimilar metal is nearby. The conditions necessary for galvanic corrosion are specific, and all must be operating simultaneously for it to occur. The required conditions are outlined in the "General Description" and "Critical Factors" sections of this chapter.

Remember that a galvanic series is based on a specific set of conditions regarding exposure to a solution, such as the commonly known galvanic series from laboratory data using seawater as the exposure fluid. When there is a question about galvanic corrosion tendencies in actual industrial environments involving fluids substantially different from seawater, appropriate testing of candidate metals in these fluids may be warranted.

Caution is also necessary when changing the metallurgy of tubes in a heat exchanger. Choosing more noble tubes may induce galvanic corrosion of the tubesheet.

Graphitically corroded cast iron (see Chap. 19) can assume a potential approximately equivalent to graphite, thus inducing galvanic corrosion of components of steel, uncorroded cast iron, and even copper-based alloys. Hence, special precautions must be exercised when dealing with graphitically corroded pump impellers and pump casings (see "Cautions" in Chap. 19).

Care must be exercised when installing stainless steel inserts in the inlet or exit end of copper or copper-alloy tubes, since galvanic corrosion can occur at the tube-insert junction.

Direct coupling of dissimilar metal components is not always required for galvanic corrosion. Aluminum components, as well as galvanized steels, are sensitive to deposition of metallic ions of heavy metals, especially copper, from the water. To avoid localized galvanic corrosion of the aluminum by metallic copper reduced from copper ions, care must be exercised to prevent water containing appreciable levels of copper ions from contacting the surfaces of aluminum components. Concentrations of copper in the water as low as 0.02 to 0.05 ppm can initiate attack.

Lastly, the cooling water flow through a component must be considered when assessing the likelihood of galvanic corrosion at a dissimilar metal coupling. In some cases, localized metal loss at the coupling may be related to erosion caused by increased water flow, localized turbulence, and/or impingement conditions at the coupling. See Chap. 23, "Erosion-Corrosion," for more detail on this mechanism.

Case History 16.1

Industry:	Municipal cooling system
Specimen location:	Threaded fasteners in cooling tower basins
Orientation of specimen:	Horizontal
Environment:	Treated cooling water, alkaline program, pH 8.5–9.5, 50–105°F (10–40°C), total hardness 150 ppm, calcium 90 ppm, chlorides 70 ppm
Time in service:	7 months
Sample specifications:	3½ in. (8.9 cm) outer diameter, threaded coupling fabricated from cast zinc

Threaded zinc fasteners coupled heaters to the basins of cooling towers (the heaters were used to prevent freezing of cooling towers during the winter). Failure of the fasteners was extensive, affecting every coupling in each of several towers.

The fasteners contacted sheets of stainless steel. Threaded regions of the fasteners that were exposed to water were severely wasted. Threaded regions that were not exposed remained intact. Some exposed fastener surfaces were covered by a nonmetallic coating. Those areas were also unaffected.

Corrosion of the fasteners occurred due to galvanic corrosion, with a large potential difference between the zinc fasteners and the passive stainless steel sheet of the basin. Deterioration was rapid because of the unfavorable area ratio due to the large areas of stainless steel and the small area of the fasteners. The incomplete coverage of the fastener surfaces by the nonmetallic coating further reduced the exposed portion of the fasteners.

Note that zinc anodes are often used to cathodically protect carbon steel and other relatively noble metals. In this case, the fasteners were acting as unintentional sacrificial anodes. The simple solution to the problem would be to use stainless steel fasteners, thus eliminating the appreciable galvanic difference between zinc and stainless steels that caused the galvanic corrosion of the zinc.

Case History 16.2

Industry:	Chemical process — plastics
Specimen location:	Inner ring of cooling system following the extruder
Orientation of specimen:	Axis of ring is horizontal
Environment:	Treated cooling water 90–155°F (32–68°C), 35 psi (241 kPa)
Time in service:	12 years
Sample specifications:	15 in. (38.1 cm) diameter aluminum ring with a rectangular cross section measuring 1 × 2 in. (2½ × 5 cm); the ring has a series of brass fittings attached at cooling water inlets and outlets

Failures of the inner ring of the extruder cooling system were occurring with increasing frequency. Six failures had occurred over a two-month period.

A half-section of a ring is shown in Fig. 16.11. The brass fittings have been removed, but two insertion holes are apparent at the center of the water channel of the ring.

An irregular trough of metal loss is apparent along the circumference of the ring (Fig. 16.12). Metal loss is severe near the fitting holes (Fig. 16.13). The corroded

Figure 16.11 Section of the inner aluminum ring of an extruder cooling system. Note the inlet and outlet for cooling water.

Figure 16.12 Appearance of corrosion in the cooling water channel along the circumference of ring.

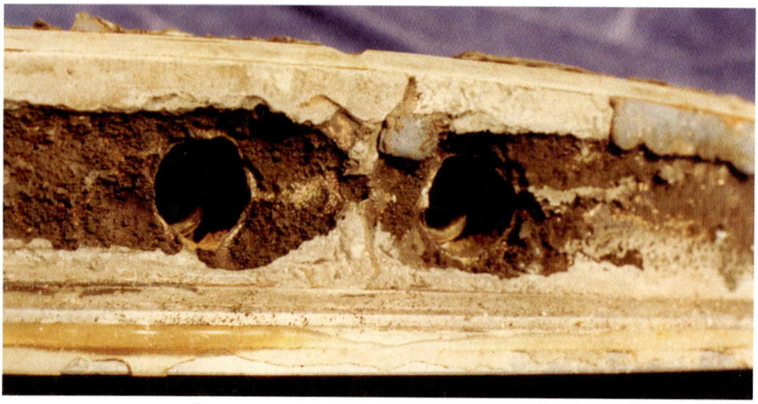

FIGURE 16.13 Severe deterioration adjacent to the cooling water inlet and outlet where brass fittings had been inserted.

zone is covered with light and dark corrosion products and deposits. Analysis of this material revealed substantial quantities of copper. Microscopic examinations revealed exfoliation of the aluminum ring in corroded regions. For more detail on exfoliation corrosion, see Chap. 18, "Intergranular Corrosion."

The large amounts of copper found on corroded surfaces distant from the brass fittings indicated corrosion of copper-based alloy components in other regions of the cooling system. Ions of copper were transported via the cooling water to the aluminum ring, where metallic copper was deposited onto the aluminum. The reduced metallic copper then established a galvanic couple with the aluminum ring, producing additional loss of aluminum.

The more severe metal loss adjacent to the brass fittings was apparently due to a direct galvanic interaction between the aluminum ring and the brass fittings. The type of exfoliation observed microscopically in corroded areas was consistent with galvanic attack due to microstructural differences of the aluminum alloy.

Case History 16.3

Industry:	Electronics
Specimen location:	Cooling plate with cooling water channels
Orientation of specimen:	Horizontal
Environment:	Closed system chemical treatment containing molybdate
Time in service:	1 year
Sample specifications:	Type 6061 aluminum plate, square ¼ in. (0.64 cm) wide cooling channels

The aluminum cooling plate was removed from service because the flow through the unit decreased appreciably. Cross sections through the plate revealed partially occluded water channels (Fig. 16.14).

The cooling water channels were generally corroded and covered by deposit and corrosion product layers (Fig. 16.15). Blue-green material was present in places. Bulk chemical analysis of the material covering the surfaces indicated it consisted predominately of aluminum as aluminum hydroxide with minor

Figure 16.14 Partially occluded square water channels near the center of the aluminum cooling plate.

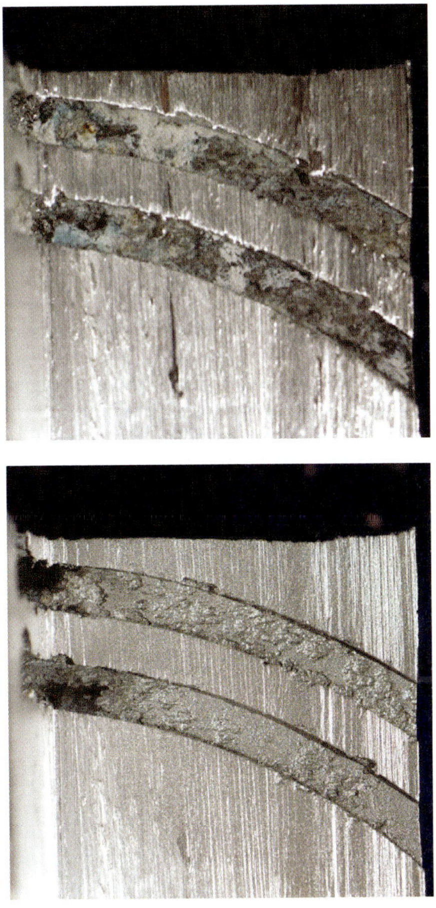

Figure 16.15 Cooling channels shown before and after cleaning. Note the blue-green colored material on the surface and the generally corroded underlying surface contour.

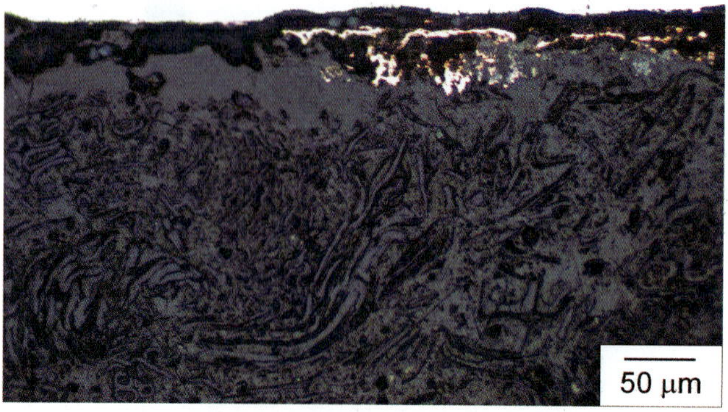

FIGURE 16.16 Micrograph of the internal surface in a water channel partially covered by deposits containing elemental copper. Note the shallow metal loss beneath the areas overlaid by the copper deposits.

amounts of molybdenum. Trace levels of copper were also present. Microscopic examination revealed the cooling water channel surfaces were covered in many places by elemental copper deposits (Fig. 16.16).

Attack on the cooling water channels was primarily due to galvanic corrosion. Galvanic corrosion was caused by the plating out of elemental copper on the metal surface, setting up localized galvanic cells. A water analysis from the system indicated copper levels as high as 0.45 ppm, well above the typical copper ion concentrations considered to be threshold values that can initiate pitting attack on aluminum. Preventing attack on aluminum from metallic copper deposition requires either removal or isolation of copper-based components in the closed water circuit, or eliminating the release of copper ions from components. Wastage of copper may be caused by erosion or corrosion mechanisms. Minimizing copper levels in the water requires maintaining sufficient concentrations of copper corrosion inhibitors, particularly azoles, in the system. These inhibitors can reduce corrosion and tie up dissolved copper in the water. Excessive flow and turbulent conditions that can cause erosion must also be avoided. Corrosion monitoring should be implemented for the metallurgies of mixed-metal systems.

Case History 16.4

Industry:	Food and beverage
Specimen location:	Wet surface air cooler
Orientation of specimen:	Horizontal
Environment:	External surface: sprayed cooling water Internal surface: ammonia
Time in service:	Unknown
Sample specifications:	Hot dipped galvanized steel tubes

Multiple steel tubes with a zinc coating were received for examination, with tube sections from different locations in the cooler. The sections from locations near the top experienced widespread conversion of the zinc coating to white corrosion products and spotty wastage of the underlying steel (Fig. 16.17). Bulk chemical analysis of the material on the surface indicated corrosion products of zinc and

FIGURE 16.17 Generally corroded external surface of a galvanized tube from a wet surface air cooler, with spotty areas of brown-orange corrosion products where the underlying steel was attacked.

iron, and about 2 percent copper. Microscopic examination revealed elemental copper deposits on the external surface, especially in locations where the metallic zinc coating was completely converted to corrosion products (Fig. 16.18). Corrosion of the zinc layer was caused by galvanic corrosion due to deposited metallic copper.

FIGURE 16.18 Hot dipped zinc galvanizing layer wasted completely through in a location where metallic copper deposits formed. Note the initial corrosion of the underlying steel tube near the center of the break in the coating.

Zinc, much like aluminum, is very susceptible to galvanic corrosion from deposited elemental copper. Corrosion of copper in mixed-metal cooling systems must be controlled to avoid galvanic corrosion due to noble deposits on active metals. Sufficient concentration of copper corrosion inhibitors, particularly azoles, is needed to minimize corrosion and release of copper ions. It is good practice to monitor for corrosion of copper and the active metal in these mixed-metal systems.

Case History 16.5

Industry:	Pulp and paper
Specimen location:	Cooling water return piping from an air compressor
Orientation of specimen:	Curved from vertical to horizontal
Environment	Open cooling water system, zinc-phosphate cooling water program, pH 8.1, and operating temperature 100–110°F (38–43°C)
Time in service:	1 year
Sample specifications:	Cast aluminum piping

Leaks in the cooling water piping system to the compressor resulted in equipment downtime. The leaks were confined to the cast aluminum piping connected to the unit. The piping was bolted to mild steel at flanges that made metal-to-metal contact (Fig. 16.19). Figure 16.20 shows brown deposit layers overlying white corrosion product layers on the internal surface of the piping, along with a generally irregular surface contour. Wastage was more severe near the flanged ends of the piping. Microscopic examination revealed preferential corrosion along the interdendritic phase in the aluminum-silicon alloy casting (Fig. 16.21).

Wastage occurred primarily due to galvanic corrosion, which was caused in part by direct coupling of the aluminum alloy piping to carbon steel. The total surface area of the cast aluminum piping was likely very small with respect to the surface area of the carbon steel to which it was connected. This unfavorable anode (cast aluminum) to cathode (carbon steel) area ratio likely accelerated corrosion of the piping.

Figure 16.19 Cast aluminum cooling water piping attached to steel components of an air compressor.

Figure 16.20 Generally corroded internal surface of aluminum cooling water piping from an air compressor. Note the irregular surface contour covered by deposit and corrosion product layers.

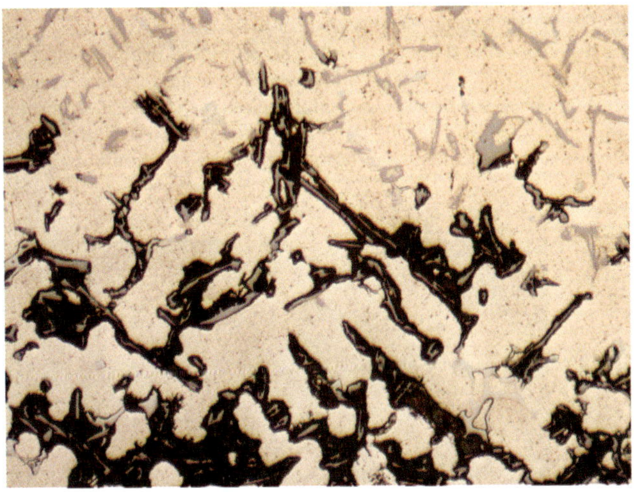

Figure 16.21 Metal loss following preferentially along the interphase boundary between interdendritic silicon and the aluminum matrix. The silicon is noble to the surrounding aluminum matrix.

Metal loss was also influenced by a microstructural galvanic corrosion mechanism, as corrosion was observed to penetrate preferentially into the pipe wall along the boundaries between microstructural phases and the aluminum matrix. Cast aluminum alloys generally contain appreciable levels of silicon, which produce a silicon-based phase in the dendritic microstructure. The silicon-based phase is cathodic to the aluminum matrix, thus producing a microscopic galvanic couple. In the presence of a suitable electrolyte, preferential corrosion of the surrounding aluminum results, leaving behind a skeletal network of the cathodic microstructural phases.

Corrosion can be controlled most effectively by avoiding the dissimilar metal coupling. Mild steel and/or cast iron would be far less susceptible to galvanic corrosion than the cast aluminum alloy with a microstructure predisposed to microgalvanic corrosion.

Case History 16.6

Industry:	Industrial laser system
Specimen location:	Water-cooled switch for power generation
Orientation of specimen:	Horizontal
Environment:	Closed-system chemical treatment containing molybdate, silicate, and azoles. pH 8.3–8.6
Time in service:	300 hours of intermittent operation
Sample specifications:	Nickel-plated aluminum alloy heat sink blocks.

The switch failed in service when temperatures became excessive. It was suspected that excessive temperatures were reached due to insufficient cooling by the water. The switch consisted of aluminum heat sinks with small diameter, about 1/8 in. (0.3175 cm), water channels. The water channels were covered in most places by a bright, shiny coating layer, which was determined to be electroless nickel plating.

Thin layers of light-colored corrosion products and deposits were found on the threads of the ports (Fig. 16.22). Scanning electron microscopy/energy dispersive spectroscopy (SEM/EDS) analysis indicated this material consisted predominately of silicon and aluminum, likely as layers of aluminum oxide corrosion products and silicate deposits from the corrosion inhibitor. The silicate in the water treatment will normally deposit on aluminum surfaces in a

FIGURE 16.22 Light-colored deposit and corrosion products in the cooling water outlet port of the aluminum heat sink.

FIGURE 16.23 Area in a water channel where the nickel plating layer was not bonded to the underlying aluminum, allowing for localized galvanic corrosion and associated formation of white corrosion products.

layer to provide corrosion protection. Within the water channels, the plating was cracked and bubbled in a few scattered locations. In areas where flakes of the plating spalled, light-colored corrosion product and deposit layers formed on the exposed aluminum surface (Fig. 16.23). Microscopic examination at locations where the nickel plating was compromised revealed shallow to moderately deep metal loss of aluminum.

Localized corrosion of water contacting surfaces of the aluminum heat sink was caused by galvanic corrosion. If the coating of nickel had been completely intact across the entire surface, it would have provided a barrier coating to protect the underlying aluminum. However, breaks in the electroless nickel plating layer promoted galvanic corrosion of the aluminum, as the electroless nickel coating is much more noble than the underlying aluminum. Locations of incomplete coverage were related to defective application of the coating or mechanical degradation in the threaded areas. The galvanic effect was intensified due to the very large cathode (nickel-coated surfaces) to anode (exposed aluminum) surface area ratio. The small areas of the exposed aluminum at breaks in the plating caused much deeper metal loss to occur at a more rapid rate than if there had been no plating covering the surfaces. Corrosion inhibitors are less effective for control of corrosion of plated metals than for nonplated metals. Tight crevices at breaks in a coating layer limit the ability of the inhibitor to reach the base metal surface.

References

1. *ASM Handbook,* Volume 13B, *Corrosion: Materials,* Materials Park, OH, ASM International, p. 104.
2. R. Francis, *Galvanic Corrosion: A Practical Guide for Engineers,* NACE, 2001, pp. 94–95.

3. Y. Matsukawa, H. Chuta, M. Miyashita, M. Yoshikawa, Y. Miyata, and S. Asakura, "Galvanic Series of Metals Conventionally Used in Tap Water With and Without Flow and Its Comparison to That in Seawater," *Corrosion*, December 2011.
4. R. W. Revie (ed.), *Uhlig's Corrosion Handbook*, 3d. ed., John Wiley and Sons, 2011, p. 129.
5. J. R. Davis (ed.), *Corrosion of Aluminum and Aluminum Alloys*, ASM International, 1999, pp. 27–29.

CHAPTER 17

Dealloying

General Description

Dealloying occurs when one or more alloying elements are preferentially removed from the metal. This depletion phenomenon differentiates dealloying from other forms of corrosion. Dealloying is also sometimes referred to as *selective leaching* or *parting.* As the word *dealloying* implies, attack occurs only in metals containing two or more alloying elements. Various alloys are susceptible to dealloying. Copper alloys such as brasses, cupronickels, and bronzes are commonly used and are thus most often associated with dealloying in cooling water environments. See Tables 4.1 and 4.2 in Chap. 4, "Copper Alloys," for compositions of copper alloys commonly used in cooling water systems. Most other alloys require conditions such as high temperatures, molten salts, acids, sulfides, or other very aggressive environments in order for dealloying to occur. In discussing dealloying, the specific terminology used comes from the alloying element removed. For example, the preferential leaching of zinc from brass is called *dezincification.* If nickel is removed, the process is called *denickelification,* and so forth. Removal of an element from the metal creates a dealloyed area that is structurally altered. Corroded areas are weak and porous, leaving components vulnerable to fracture and perforations.

A number of mechanisms have been proposed for dealloying of binary (two-element) alloys, one of which involves dissolving of both the constituents and redepositing of the more-noble element. Other proposed mechanisms involve dissolving of only the less-noble element. The details of the possible mechanisms are beyond the scope of this book, and are reviewed elsewhere.[1] Different conditions (i.e., electrochemical potential, pH) likely influence the predominant dealloying mechanism.

Dezincification

Commonly used brasses containing 15 percent or less zinc typically do not experience dezincification. The likelihood of dezincification generally increases with increasing zinc content. Small concentrations of inhibiting elements (arsenic, antimony, or phosphorus) are often added to single-phase brasses to prevent dezincification. Although uninhibited brass alloys were used in the distant past, heat exchangers have almost always been tubed with inhibited grades of brass in more recent years. Admiralty brass (nominally 71% Cu, 28% Zn, 1% Sn, < 0.10% As, Sb, or P) and arsenical aluminum brass (76% Cu, 22% Zn, 2% Al, 0.05% As) are resistant to dezincification in most cooling water environments.

Brasses containing more than about 32 percent zinc are quite susceptible to attack due to their typically duplex (two-phase) microstructure. Brasses containing up to 32 wt% zinc only contain alpha phase (Fig. 17.1). An example of a typical duplex microstructure is shown in Fig. 17.2. Alloys containing more than approximately 32 wt% zinc are particularly vulnerable to dezincification due to the usual presence of the beta phase, which contains a higher concentration of zinc than the alpha phase. Such two-phase alloys are often referred to as beta brass. Alpha-beta brasses are often selected for their low cost (relative to brasses with higher copper content) and high hot ductility, which is desirable in applications involving casting or hot working. Since zinc is less noble than copper, galvanic effects promote preferential removal of zinc from the zinc-rich beta phase first, leaving behind the alpha phase and porous, weak, copper-rich corrosion products (Fig. 17.3). Note that subsequent dezincification of the alpha phase may still occur. Dezincification-resistant (DZR) brasses have zinc contents over 32 percent, but are heat treated to

FIGURE 17.1 Typical single-phase microstructure of admiralty brass tubing, consisting of twinned grains of alpha phase copper-zinc solid solution. Etchant: Potassium dichromate.

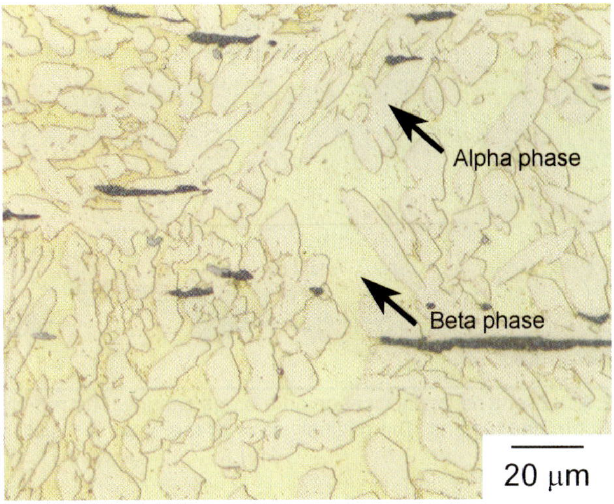

Figure 17.2 Duplex microstructure of a brass casting containing more than 32 wt% zinc, consisting of alpha phase solid solution in a matrix of beta phase solid solution. Etchant: Potassium dichromate.

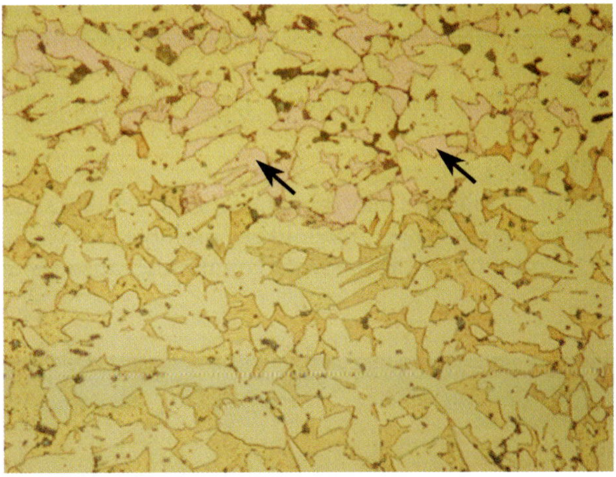

Figure 17.3 Duplex brass, showing the normal microstructure as well as dezincification of the zinc-rich beta phase toward the top of the photo. Arrows indicate areas of metallic copper corrosion products. Magnification: 500×. Etchant: Potassium dichromate.

contain only alpha phase at room temperature, and they contain a dezincification-inhibiting element, such as arsenic.

The two major forms of dezincification are "layer-type" and "plug-type." The names are taken from characteristic corrosion-product morphologies. In *layer-type dezincification,* the component surface is

FIGURE 17.4 Layer-type dezincification on a brass casting. The red layers are uniformly corroded regions. The original yellow of the brass is visible in between.

converted to metallic copper corrosion products along a zone of roughly uniform depth (Figs. 17.4–17.6). *Plug-type dezincification* produces small pockets or plugs of almost pure copper (Figs. 17.7 and 17.8). Surfaces may have to be cleaned of deposits to reveal damage, as dealloying frequently occurs beneath such material. Metallographic cross sections may be necessary for positive identification of dealloying (see the section "Cautions," later in this chapter).

Denickelification

Selective removal of nickel may occur in cupronickel alloys. Low-flow conditions and high temperatures likely promote denickelification. However, denickelification does not commonly cause the affected

FIGURE 17.5 Layer-type dezincification of a thin brass plate. The 0.019 in. (0.048 cm) plate is shown in cross section. The dezincified layers converge toward the plate edge. Note the porosity of the dezincified metal.

Figure 17.6 Layer-type dezincification on the internal surface of a cast brass pipe 90° elbow.

Figure 17.7 A large plug of dezincified metal beneath a deposit. Plugs often form on the bottom of horizontal pipes, beneath deposits.

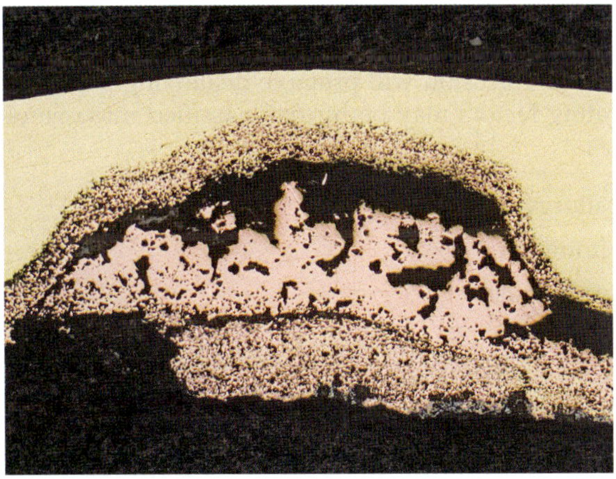

Figure 17.8 Very deep, plug-type dezincification on the cooling water side of a brass condenser tube. Unetched.

component to fail, as cupronickel wastage is usually slight compared to attack in brasses. Rather, the liberated nickel may deposit downstream and/or be released into the environment. Cupronickels are generally more prone to layer-type denickelification than to plug-type deterioration.

Dealuminification

Aluminum bronzes containing greater than approximately 9 wt% aluminum have microstructures containing more than a single phase. Those alloys are particularly vulnerable to dealuminification due to their aluminum-rich phases, in the same manner that the duplex microstructure of high-zinc-content brasses promotes dezincification. Dealuminification has been reported in waters having both high and low pH.

Locations

Because of excellent thermal conductivities, suitable mechanical properties, and superior corrosion resistance, brasses, cupronickels, and to a lesser extent bronzes are often used in heat exchangers. Thus, heat exchanger and condenser tubing and other components within cooling water systems, along with tubesheets, can experience dealloying. Piping and plumbing fittings, such as valves and connectors, as well as strainers and screens may be affected. Copper-alloy castings containing high concentrations of zinc are susceptible to dealloying.

Many copper alloys, such as bronzes and brasses, are often used in pumps as bushings, bearings, impellers, and gaskets. These pump components can be subjected to dealloying attack.

Critical Factors

Dealloying is influenced by many factors. In general, any process that increases general corrosion will promote dealloying. However, specific accelerating factors may be further classified into one of three categories:

1. Metallurgy
2. Environment
3. Water chemistry

Metallurgy

As mentioned previously, inhibited zinc brasses containing phosphorus, antimony, and/or arsenic additions dezincify much less readily than brasses free of these elements. Tin also tends to reduce the occurrence of dealloying. Commonly used brasses containing less than

15 percent zinc are usually not subject to dezincification, while those containing more than 32 percent zinc are quite vulnerable to dealloying. In some cases, cast alloys with zinc contents below 15 percent may exhibit dealloying due to compositional segregation that occurs during solidification, a phenomenon known as *coring* (see Case History 17.7).

Denickelification generally produces less wastage in cupronickels than dezincification in brasses.

Aluminum bronzes that contain more than 9 percent aluminum are particularly susceptible to dealloying.

Environment

High temperature, heat transfer, stagnant and crevice conditions, and stress accelerate dealloying. Deposits, especially porous or granular layers, promote attack. Surface shielding produced by deposition and crevices generally promotes concentration cell corrosion and allows concentration of aggressive chemical species. However, calcium carbonate deposits can actually reduce the occurrence of corrosion. High metal temperatures accelerate corrosion processes, and thermal gradients directly increase local metal dissolution. Frequent or prolonged outages that result in stagnant conditions contribute to severe attack. Highly stressed components are more susceptible to all forms of attack.

Water Chemistry

Soft waters, acidic or high-pH conditions, high concentrations of dissolved carbon dioxide, and high chloride concentrations promote dezincification. High concentrations of oxidizing biocides (e.g., chlorine and bromine) also increase susceptibility to attack. As previously mentioned, calcium carbonate precipitation in many instances reduces yellow metal corrosion, including dealloying. Calcium carbonate that precipitates at cathodic sites can actually form a barrier film that prevents the reduction of oxygen at the cathode, thus reducing the corrosion rate. Though calcium carbonate scaling can be beneficial under controlled conditions, excessive precipitation is undesirable and should be avoided. Acid waters generally accelerate corrosion on a variety of alloys. Very high pH can also accelerate corrosion and thus promote dealloying. High chloride concentration has a particularly deleterious effect on brasses and cupronickels (though less so on cupronickels). Hence, brasses and cupronickel components contacting brackish waters or seawater are at special risk. One study of potable water systems indicated a relationship between dezincification (particularly meringue-type dezincification, which refers to the formation of voluminous white corrosion products) and a high ratio of chloride concentration to carbonate hardness.[2]

In the case of dezincification, layer-type attack is often observed in conjunction with slightly acidic pH, low salt concentrations, and room temperature water. Plug-type attack is often promoted by neutral or alkaline pH, high salt concentrations, and water above ambient temperature.

Identification

Layer-type dezincification is easy to recognize visually. The original component shape and dimensions are usually preserved, but the metal color changes from the golden yellow for many copper-based alloys to the reddish color typical of elemental copper (Figs. 17.9 and 17.10). The corroded metal has little mechanical strength, and if stressed it fractures easily.

Figure 17.9 Red, dezincified gasketed surface of a cast valve throat. Note the partial spalling of the copper corrosion products.

Figure 17.10 Cross section through the valve throat in Fig. 17.9. Note how the dezincification is confined to the gasket region.

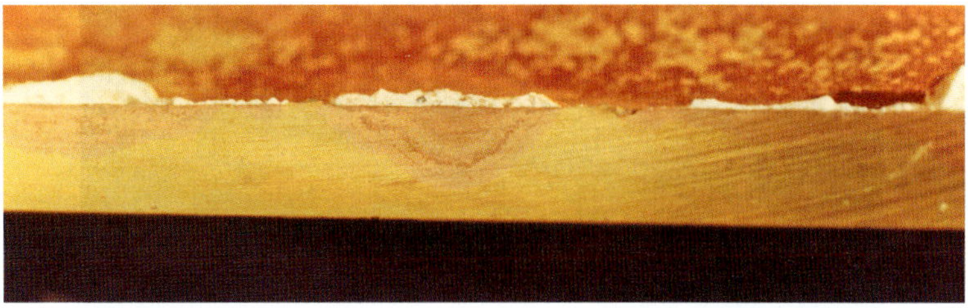

FIGURE 17.11 Brass tube wall in longitudinal cross section. Plugs of dezincified metal are present on the bottom of the horizontally oriented tube, beneath deposits. Stratification within the copper plugs is evident.

Plug-type dezincification produces localized areas filled with metallic copper corrosion products (Fig. 17.8). Such attack is common on horizontal pipe sections. Attack is usually confined to areas beneath deposits lying along tube bottoms (Figs. 17.7 and 17.11).Water will generally weep through the porous plugs that penetrate the tube or pipe wall thickness, leaving stains or mounds of dissolved solids concentrated by evaporation on external surfaces (Fig. 17.12). Unless internal pressures are high, such plugs may remain in place for long periods with no significant water loss. In some cases, the weak, porous corrosion products in the plugs are blown out of tube walls due to internal pressure, producing perforations. Close visual inspection of

FIGURE 17.12 A brass tube showing plug-type dezincification. The white patch above the plug on the external surface was caused by dissolved solids, concentrated by evaporation of water leaking through the porous dezincified plug.

FIGURE 17.13 Stratified copper corrosion product in plug-type dezincification.

both plug- and layer-type corrosion products often reveals stratification of corroded metal (Figs. 17.11 and 17.13). These layers probably represent changes in temperature, water chemistry, and/or corrosion processes.

Though plug-type dezincification is not uncommon in brasses, plug-type denickelification attack in cupronickel has never been observed at this laboratory. Similar to dezincified brass, dealloyed cupronickel surfaces have a reddish color due to the accumulation of corrosion products consisting of porous elemental copper.

Corrosion products rich in the selectively removed element may form above dealloyed surfaces. For example, in brasses, light-colored corrosion products that are rich in zinc often cover dezincified areas. In some cases of plug-type attack, the zinc-based corrosion products are very voluminous, a condition sometimes referred to as meringue-type dezincification. This has been observed on beta brass fittings in potable hot water lines, especially those that experience extended stagnant conditions with waters of appreciable chloride content. These voluminous corrosion products can restrict flow in the water line.

Elimination

Dealloying can be reduced (as can any other form of corrosion) by good system operation and the judicious use of appropriate materials and chemical treatment. Specific categories needing attention are described next.

Materials Substitution

In any specific environment, only certain alloys are affected. Substitution of more resistant materials does not always necessitate major alloy compositional changes. Adding as little as a few hundredths of a percent of arsenic, antimony, or phosphorus, for example, can markedly reduce dezincification in admiralty brass. For applications in which higher zinc contents are desirable, dezincification-resistant brasses are available for replacement of beta brass alloys having a duplex microstructure, which should generally be avoided where possible.

Surface Cleanliness

The propensity toward dealloying decreases as surface cleanliness increases. Flow should be maintained at rates high enough to prevent settling of suspended particles and to reduce biological attachment and growth. Also, chemical treatment effectiveness improves with proper flow. However, it should be mentioned that excessive water velocities can result in erosion-corrosion of brasses and other copper alloys in particular (see Chap. 23, "Erosion-Corrosion"). Surfaces on which fouling is apparent should be regularly cleaned.

Chemical Treatment

Chemical corrosion inhibition can reduce all forms of corrosion, including dealloying. In particular, film-forming chemicals such as azoles are effective in reducing the corrosion of yellow metals. However, such chemicals cannot be effective if occlusive material covers the metal surfaces. Therefore, clean surfaces should be maintained in order for treatment chemicals to function properly.

Biological Control

Any deposit material can increase the risk of dealloying. Biological accumulations, such as slime layers, are no exception to this rule. Good biological control is always beneficial, but restrictions on the use of certain biocides and their possible release into the environment must be considered.

Cautions

Identifying dealloying involves recognizing the variation of material composition caused by corrosion. Redeposition of alloy elements can occur during acid cleaning, misleading the examiner. Frequently, copper is redeposited on brasses, bronzes, and similar copper alloys during acid cleaning of deposits. Very small amounts of metallic copper are sometimes also redeposited on metal surfaces as a result of corrosion mechanisms other than dealloying. Such deposits should not be confused with corrosion products resulting from dealloying. Metallographic cross sections through suspected locations will confirm if dealloying occurred.

Related Problems

See also Chap. 11, "Crevice and Underdeposit Corrosion (Concentration Cells)."

Case History 17.1

Industry:	Mining
Specimen location:	Turbine condenser
Specimen orientation:	Horizontal
Environment:	Shellside: Condensing steam Tubeside: Cooling water, 60–80°F (16–27°C), high hardness and silica, treated with a dispersant and zinc
Time in service:	6 years
Sample specifications:	¾ in. (1.9 cm) outer diameter, CDA 442 uninhibited brass tubing

A brass condenser tube contained a 1/16 in. (0.16 cm) diameter perforation on the apparent bottom side (Fig. 17.14). Many other shallow depressions pockmarked the internal surface. Most depressions were filled with reddish, copper-colored corrosion products.

The tube was transversely sectioned and polished. Large plugs of porous corrosion products extended entirely through the 0.050 in. (0.13 cm)–thick wall (Fig. 17.15). Chemical analysis by energy dispersive x-ray spectroscopy of the

Figure 17.14 A perforation originating from the internal surface of a brass condenser tube. The perforation was caused by plug-type dezincification.

Figure 17.15 A transverse cross section through the tube wall in Fig. 17.14. Note the through-wall plug-type dezincification.

copper-colored corrosion products in the depressions indicated a composition of 94 percent copper, 5 percent zinc, and 1 percent tin.

The depressions and plugs of corroded metal were caused by dezincification. The perforation shown in Fig. 17.14 was caused when one such plug blew out of the wall due to the pressure difference between internal and external surfaces.

The condenser was mechanically cleaned every six months. However, heavy deposition due to silt, sand, and mud plagued this system and significantly promoted plug-type dezincification.

Case History 17.2

Industry:	Refinery
Specimen location:	Overhead condenser
Specimen orientation:	Horizontal
Environment:	Shellside: Atmospheric pressure, pH 5.5–6.5 (ammonia used to control pH), 240–110°F (116–43°C), overhead vapors Tubeside: Pressure 30 psi (207 kPa), pH 8.0, 87–120°F (30–49°C), cooling water (no treatment)
Time in service:	1 year
Sample specifications:	¾ in. (1.9 cm) outer diameter, uninhibited brass, 72% copper, 26.7% zinc, 1.3% tin

Severe wastage was evident on both external and internal surfaces of the tube section. There were two round perforations at depressions on one side of the external surface (Fig. 17.16). Metallographic cross sections at these locations revealed deep pockets of internal surface corrosion that penetrated the tube wall (Fig. 17.17). Spot tests suggested that material in the depressions contained sulfides, mainly concentrated near the external surface. Below were large amounts of porous metallic copper.

The internal surface was covered with a tan deposit layer up to 0.033 in. (0.084 cm) thick. The deposits were analyzed by energy dispersive x-ray spectroscopy (EDS) and were found to contain high concentrations of calcium, silicon, zinc, phosphorus, and magnesium. The porous corrosion products shown in Fig. 17.17 contained more than 90 percent copper, and small concentrations

FIGURE 17.16 What appear to be deep, hemispherical depressions on the external surface of an overhead condenser tube are actually through-wall perforations caused by severe dezincification from internal surfaces (see Fig. 17.17).

FIGURE 17.17 Porous copper corrosion products filling an internal surface depression. The edge of one crater seen in Fig. 17.16 is to the right.

of zinc, tin, and iron. Traces of sulfur and aluminum were also found. Near the external surface, high concentrations of sulfur (up to 27 percent in one location) were detected in the corrosion products.

Though visual inspection suggested that the perforations were caused by both external and internal surface metal loss, microscopic examination indicated that deep dezincification on the internal surface was the sole cause of the failures. The porous corrosion product plugs had enough structural integrity so that only a few perforations formed due to internal pressurization.

The dezincification was caused by underdeposit corrosion. The fact that the brass was not an inhibited grade was a major contributing factor. Chemical cleaning had not been done since this exchanger was installed. No chemical corrosion inhibitors were used on either external or internal surfaces.

Case History 17.3

Industry:	Utility
Specimen location:	Main condenser
Specimen orientation:	Horizontal
Environment:	Tubeside: 60–80°F (16–27°C), pH 8.3–8.7; treatment—tolyltriazole, dispersant; brackish water
Time in service:	3 years
Sample specifications:	7/8 in. (2.2 cm) outer diameter, 90:10 cupronickel, 0.031 in. (0.079 cm) wall thickness

Several sections of 90:10 cupronickel condenser tubing were received. The sections were from tubes that had been plugged because of earlier failures. No tube section contained the original tube failure, however.

Internal surfaces were covered by loosely adherent corrosion product and deposit layers. Much of the corrosion product was cuprous oxide. Substantial amounts of iron, silicon, aluminum, zinc, and nickel were also found. Not unexpectedly, chlorine concentrations up to 2 percent by weight were present; sulfur concentrations of about 1 percent were also found.

Removal of deposits and corrosion products from internal surfaces revealed irregular metal loss. Additionally, surfaces in wasted areas showed patches of

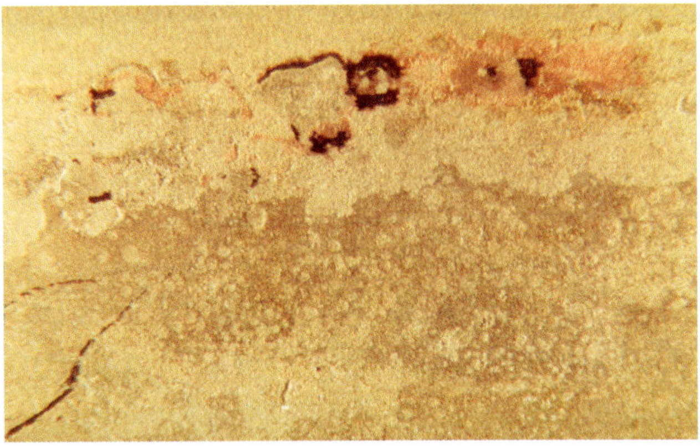

FIGURE 17.18 Close-up of a 90:10 cupronickel condenser's internal surface after acid cleaning. Elemental copper patches mark regions of shallow denickelification.

elemental copper (later confirmed by EDS) (Fig. 17.18). These denickelified areas were confined to regions showing metal loss. Microscopic analysis confirmed that dealloying, not just redeposition of copper onto the cupronickel from the acid bath used during deposit removal, had occurred.

Surfaces of plugged tubes were carefully compared to surfaces of tubes from the same condenser that were never plugged. Only plugged tubes showed denickelification. Stagnant conditions, as well as the deposits and chloride, caused the dealloying. The deepest dealloying was up to 5 percent of the remaining wall thickness.

Case History 17.4

Industry:	Utility
Specimen location:	Condenser tubes
Specimen orientation:	Horizontal
Environment:	Tubeside: 62–80°F (17–27°C), pH 8.3–8.6; treatment—tolyltriazole, dispersant
Time in service:	1 year and 12 years (two different samples)
Sample specifications:	ASTM B395 (71.4% copper, 0.97% tin, <0.005% phosphorus, <0.01% arsenic, others total <0.15%, balance zinc)

Two sections of utility condenser tubing were received. One of the sections had deep plug-type dezincification on internal surfaces (Fig. 17.19); the other showed only superficial corrosion on internal surfaces (Fig. 17.20).

The sections were removed from two similar condensers. Both condensers were fed with a common cooling water source, were of identical design, and experienced virtually identical operating conditions. However, the first condenser had been in service for 12 years, whereas the other was only online for 1 year. The dezincified tube (Fig. 17.19) was from the condenser seeing short-term service and was fabricated from an uninhibited grade of brass. The older, superficially corroded section contained 0.021 percent arsenic, an inhibiting agent.

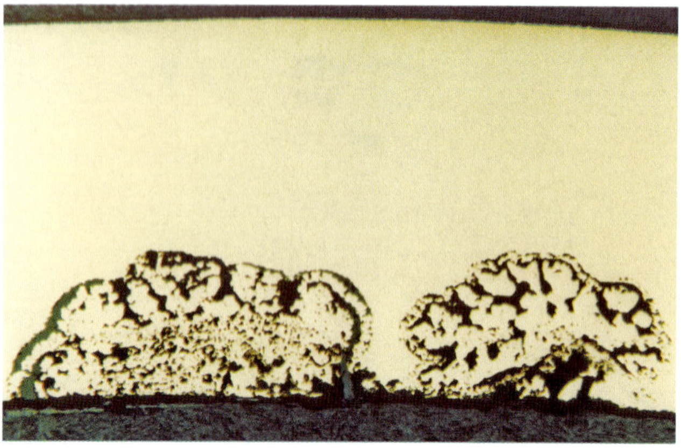

FIGURE 17.19 Plug-type dezincification on the internal surface of a brass condenser tube. Note the extreme porosity of the copper plugs. Tube wall thickness was 0.040 in. (0.10 cm). Compare to Fig. 17.20. (*Courtesy of National Association of Corrosion Engineers, Corrosion '89 Paper No. 197 by H. M. Herro.*)

FIGURE 17.20 An inhibited brass condenser tube in cross section, showing only superficial wastage on internal surfaces. This tube was exposed to virtually the same conditions as the tube shown in Fig. 17.19. However, because of the inhibition, dezincification did not occur on this tube. (*Courtesy of National Association of Corrosion Engineers, Corrosion '89 Paper No. 197 by H. M. Herro.*)

This case history dramatically illustrates the value of proper alloy choice. It was found that the failed condenser tube had been ordered many years before but was not installed until the year before the sections were removed for analysis. Today, uninhibited grades of brass are almost never used for condenser and heat exchanger service in the United States.

Case History 17.5

Industry:	Hotel
Specimen location:	Pump impeller and shaft bushing
Specimen orientation:	Vertical impeller, horizontal bushing
Environment:	Cooling water (brine well), chloride 22,000 ppm, sulfate 2400 ppm, calcium 1700 ppm, magnesium 4300 ppm, iron 0.05 ppm, silica 14 ppm, pH 7.4, 70°F (21°C), pressure 60–70 psi (0.4–0.5 MPa)
Time in service:	~1 year
Sample specifications:	6½ in. (16.5 cm) diameter bronze impeller, 1 in. (2.5 cm) diameter leaded brass bushing

A pump impeller and a shaft bushing from a small cooling water pump assembly were generally corroded. Reddish surface discoloration revealed layer-type dezincification (Figs. 17.21 and 17.22).

Figure 17.21 A generally dezincified bronze pump impeller. Surfaces were dealloyed to a depth of about 0.020 in. (0.050 cm).

Figure 17.22 The tricolor internal surface of a leaded-brass bushing. Greenish-blue corrosion product overlies the red of the generally dezincified surface. To the extreme right, only bare metal is visible due to erosion.

Attack occurred because of substantial idle periods in which flow was limited. Also, very high concentrations of chloride and sulfate in cooling waters accelerated attack.

Case History 17.6

Industry:	Utility
Specimen location:	Pump impeller
Environment:	Cooling water
Time in service:	Unknown
Sample specifications:	Aluminum bronze impeller (~85% copper, 11% aluminum, 4% iron)

Two small plugs from an aluminum bronze cooling water pump impeller exhibited reddish areas of metal loss. Microscopic examinations indicated dealuminification of the interdendritic phase, which was most likely richer in aluminum than the dendritic phase (Fig. 17.23). Perforations in the impeller were primarily caused by cavitation, though metal loss due to dealloying contributed significantly. Reportedly, the impeller was not exposed to stagnant conditions, high temperatures, soft water, or most of the other factors known to promote dealloying. High concentrations of oxidizing biocides most likely promoted dealuminification.

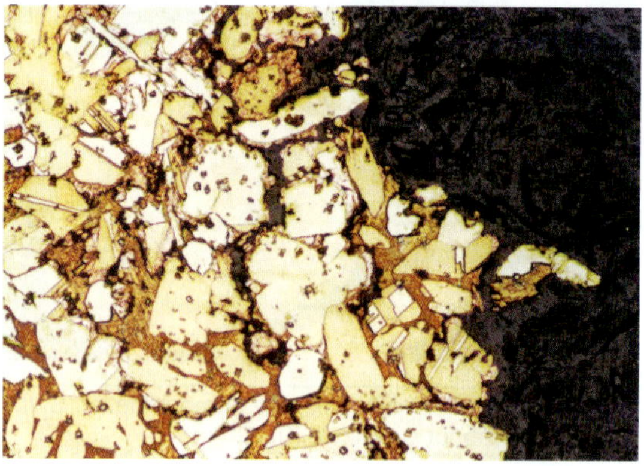

FIGURE 17.23 Aluminum bronze cooling water pump impeller, showing dealuminification along the aluminum-rich interdendritic phase. Etchant: Potassium dichromate.

Case History 17.7

Industry:	Utility
Specimen location:	Pump impeller
Environment:	Cooling water
Time in service:	Unknown
Sample specifications:	Silicon red brass (82% copper, 14% zinc, 4% silicon)

Visual examinations indicated reddish corroded areas on the impeller. Microscopic examinations clearly indicated that dealloying occurred in the interdendritic areas of the microstructure (Fig. 17.24). SEM-EDS element mapping showed a slightly elevated concentration of silicon in the unattacked interdendritic areas, relative to the dendritic areas (Figs. 17.25*a* and *b*). Analysis of the reddish corrosion products indicated that they primarily consisted of copper and were almost completely depleted of zinc and silicon (Figs. 17.25*b*–*d*, Table 17.1).

Figure 17.24 Silicon red brass pump impeller, showing dealloying in the interdendritic areas. Etchant: Potassium dichromate.

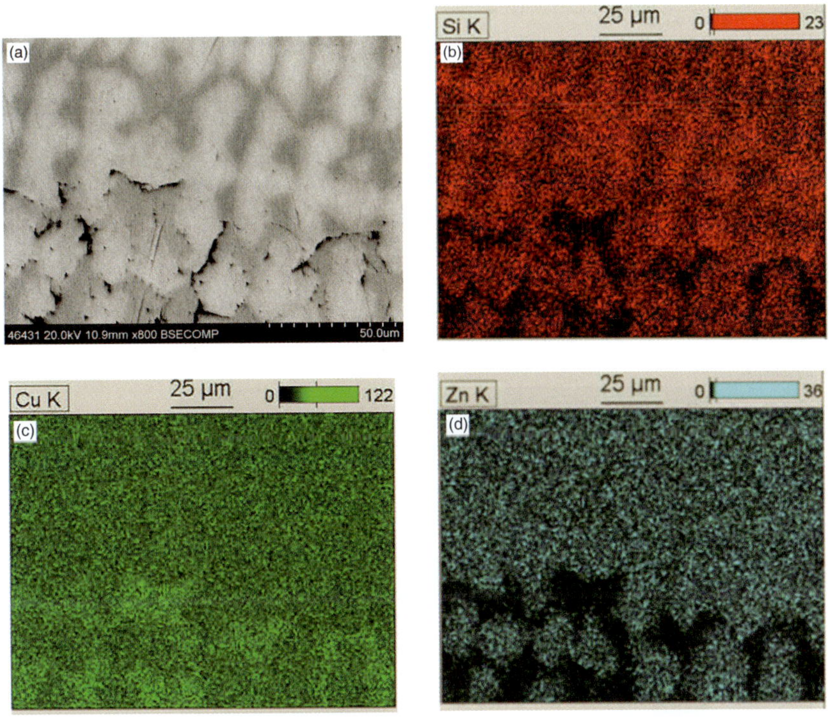

Figure 17.25 (*a*) Backscatter SEM image, showing the unattacked dendritic areas (light-colored areas) and dealloying in the interdendritic areas, toward the bottom of the photo (dark-colored areas). (*b*–*d*) Corresponding SEM-EDS maps for silicon (*b*), copper (*c*), and zinc (*d*), showing elevated silicon concentration in the unattacked interdendritic areas and near-total depletion of silicon and zinc in the dealloyed areas.

Element	Impeller Alloy	Reddish Corrosion Products
Cu	82	98
Zn	14	2
Si	4	trace

TABLE 17.1 Comparison of Bulk Impeller Alloy Composition and SEM-EDS Analysis of Corrosion Products in Dealloyed Areas

Dealloying occurred despite an overall zinc concentration in the alloy below 15 percent. Segregation of alloying elements occurred during solidification of the casting, resulting in compositional variations that promoted dealloying of the interdendritic areas. The depletion of silicon in the corrosion products suggested that desiliconification occurred in addition to dezincification. Desiliconification is not commonly encountered; however, it has been known to occur in the presence of acidic species and in high-temperature steam.[3] In this case, however, the attack occurred in a relatively low-temperature environment. It was not clear what particular chemical species in the water might have promoted dealloying, but high chloride content is a common culprit.

Case History 17.8

Industry: Electronics
Specimen location: Ball valve and fittings
Environment: Chilled water
Time in service: 8 years
Sample specifications: Failed connector: leaded forging brass (~59% copper, ~39% zinc); valve body: leaded semi-red brass (~82% copper, ~8% zinc)

Numerous pipe fittings in a chilled water system were failing. The received section exhibited a thick-edged fracture in an end connector that joined the valve body and a pipe nipple (Fig. 17.26). The fracture surface was reddish in color.

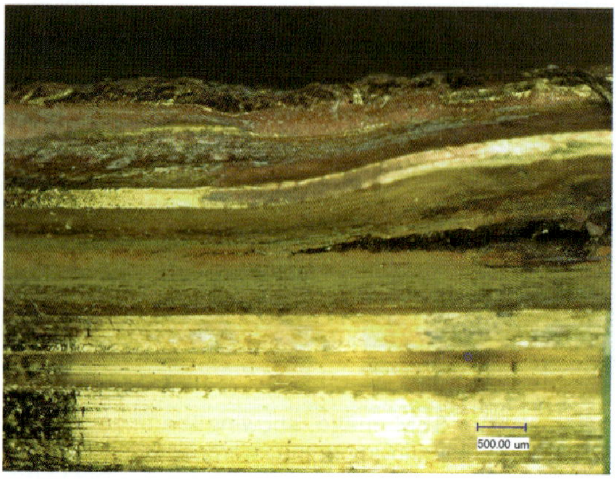

FIGURE 17.26 Reddish metallic copper corrosion products along the fracture surfaces in a threaded region of a brass fitting.

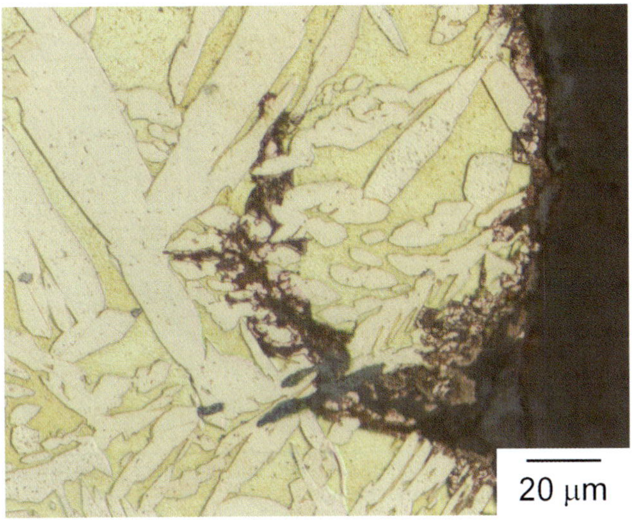

FIGURE 17.27 Metallic copper corrosion products and resulting fracture along the beta phase. Etchant: Potassium dichromate.

The failed end connector had a duplex microstructure that consisted of alpha phase in a matrix of beta phase solid solution (see Fig. 17.2). Alloy identification indicated that the failed connector contained approximately 39 percent zinc. The unfailed valve body, which did not exhibit any metal loss, was made from a semi-red brass alloy containing approximately 8 percent zinc.

Microscopic examination along the fracture edge indicated metallic copper corrosion products that were thicker toward the external surface. The fracture and corrosion products penetrated along the beta phase (Fig. 17.27). Dezincification was observed away from the fracture edge on the external surface, but not the internal surface. The connector failed due to the combination of several factors: a duplex microstructure susceptible to dezincification, condensation on the external surface, and tensile stresses due to tightening of the fittings.

Case History 17.9

Industry:	Utility
Specimen location:	Condenser
Environment:	Cooling water
Time in service:	Unknown
Sample specifications:	Brass tube plug

During an inspection of a condenser, some brass plugs used to seal tubes were found to be covered by white corrosion products (Fig. 17.28). The most severely damaged components of the plug were the bolt and washers, which exhibited fractures and reddish corrosion products in areas that were exposed to cooling water during service (Figs. 17.29 and 17.30). Examination of a washer fracture surface showed reddish corrosion products through the entire thickness (Fig. 17.31). Spot tests of white deposits and corrosion products on the bolt indicated the presence of sulfates, chlorides, and possibly carbonates. SEM-EDS analysis indicated that the white material consisted of mostly zinc, along with

FIGURE 17.28 Brass tube plug, as installed in a condenser tube, showing white corrosion products on the surface.

FIGURE 17.29 Brass bolt from a condenser tube plug, showing a transverse crack and white deposits and corrosion products on the end exposed to cooling water during service.

FIGURE 17.30 Washer from a condenser tube plug, showing cracks and white corrosion products.

Figure 17.31 Cross section of the fracture surface of a washer, showing reddish corrosion products on the entire surface.

small concentrations of sulfur, chlorine, and silicon. Elevated levels of chlorides and sulfates were indicated by spot tests. SEM-EDS analysis of the bolt away from attack indicated zinc contents up to 37 percent in the alloy.

Microscopic examination of the bolt indicated a single-phase microstructure consisting of twinned grains of alpha phase solid solution (similar to Fig. 17.1). The other examined components also had single-phase microstructures. Thick, porous metallic copper corrosion products were observed in attacked areas on the bolt (Fig. 17.32).

Despite the single-phase microstructure, severe dezincification occurred due to the high zinc content of the alloys. Stresses due to tightening of the bolt likely promoted dezincification and subsequent fracture. Zinc-based corrosion products that formed on the metal surface and silicon-containing deposits shielded

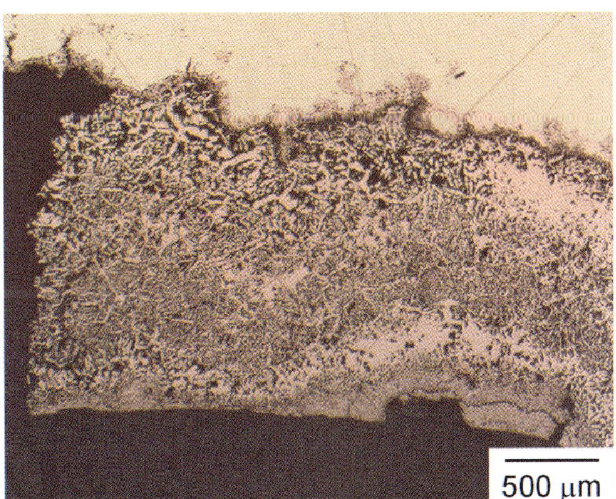

Figure 17.32 Thick, porous copper corrosion products on the surface of the bolt. Unetched.

the underlying metal and resulted in the formation of differential aeration and ion concentration cells, which contributed to dezincification significantly.

Dezincification attack was localized to specific brass plugs. Cupronickel plugs used to seal tubes did not show evidence of dealloying. Inhibited admiralty brass tubing of the condenser also did not show attack. Reportedly, the tubesheets of the condenser did not exhibit signs of dezincification despite being made from Muntz metal, which nominally contains about 40 percent zinc. Attack was not even observed at rolled joints with cupronickel tubing. Proper processing of the Muntz metal alloy for tubesheet applications helps limit the susceptibility to attack. Such tubesheets are considered to have a small sensitivity to dealloying attack in power plant condensers.[4]

References

1. Sean G. Corcoran, "Effects of Metallurgical Variables on Dealloying Corrosion," *ASM Handbook*, Volume 13A, of *Corrosion: Fundamentals, Testing, and Protection*, ASM International, Metals Park, OH, 2003, pp. 287–293.
2. *The Brasses—Properties and Applications*, "Section 7: Brasses for Corrosion Resistance," Copper Development Association, www.copperinfo.co.uk, 2005.
3. Linda Garverick (ed.), "Forms of Corrosion in the Petrochemical Industry," in *Corrosion in the Petrochemical Industry*, ASM International, Metals Park, OH, 1994, p. 33.
4. Barry C. Syrett, Otakar Jonas, and Joyce M. Mancini, "Corrosion in the Condensate-Feedwater System," *ASM Handbook*, Volume 13C, *Corrosion: Environments and Industries*, ASM International, Metals Park, OH 2006, pp. 448–450.

CHAPTER 18

Intergranular Corrosion

General Description

Many materials, including metals used in most engineering applications, are made up of very small regions called grains. Grains are generally microscopic in size. If one visualizes a material's structure as being made up of atoms arranged in a particular three-dimensional lattice pattern, grains are very small regions of that lattice pattern. The lattices in the many grains within the material are randomly oriented with respect to each other. The outer edges of the grains are called grain boundaries. Intergranular corrosion refers to attack that occurs preferentially along the grain boundaries. This type of attack can penetrate deeply into the metal surface, resulting in whole grains detaching from the surface and the formation of cracklike features (Fig. 18.1).

Grain boundaries can be more susceptible to corrosion than the bulk metal for a number of reasons:

- Alloying elements that are important for corrosion resistance are depleted near the grain boundaries due to the formation of secondary phases along the boundaries.
- Collection of impurities at the grain boundaries, resulting in grain boundaries that are anodic relative to the grain interiors.
- Formation of grain boundary phases or other microstructural features that are more susceptible to corrosion than the bulk metal.
- Grain boundaries and other microstructural features, such as deformation twins, are locations of strained atom-to-atom bonds, intrinsically making these high-energy boundaries more chemically reactive and thus more prone to attack.

Intergranular attack can happen in various alloy types and environments. In cooling water applications, copper and copper alloys,

FIGURE 18.1 Uninhibited admiralty brass surface condenser tube, showing intergranular attack that resulted in detachment of whole grains, as well as the formation of a crack. Magnification: 25×. Etchant: Potassium dichromate.

stainless steels, and aluminum alloys can experience intergranular corrosion. Carbon steel can also corrode intergranularly, though attack severe enough to cause a failure is relatively rare.

Carbon Steel

Intergranular corrosion is relatively uncommon in carbon steel in cooling water environments, and it is rarely the primary cause of failure in a carbon steel component. This type of attack in carbon steels has been associated with the presence of nitrates. Concentrating mechanisms, such as ion concentration cells, can contribute to intergranular corrosion.

Copper and Copper Alloys

Of the commonly used copper and copper-based alloys, those that are known to be most susceptible to intergranular attack are Muntz metal, admiralty brass, aluminum brass, and silicon bronzes.[1] Inhibited alloys that are resistant to dealloying, such as brasses containing small concentrations of arsenic, antimony, or phosphorus, are still susceptible to intergranular corrosion. In some cases involving intergranular corrosion of uninhibited brass, metallic copper corrosion products form along the grain boundaries and annealing twins, suggesting localized dealloying along grain boundaries as the mechanism responsible for the corrosion (Fig. 18.2). Such corrosion products would not be expected in inhibited alloys since they should be very resistant to dealloying. High heat flux and the presence of deposits on the surface are often associated with intergranular corrosion.

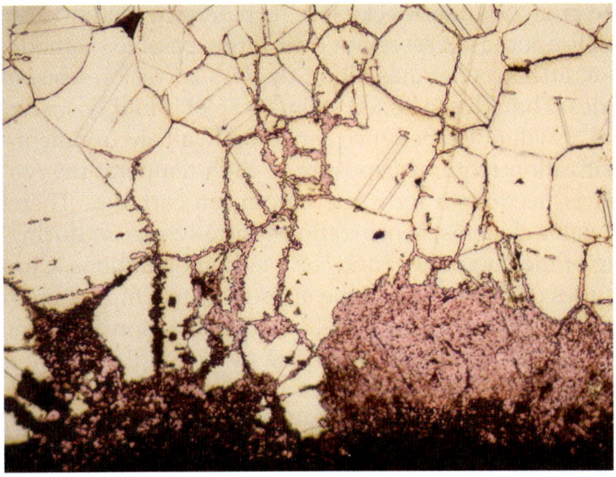

FIGURE 18.2 Intergranular corrosion in an uninhibited admiralty brass tube, showing metallic copper corrosion products penetrating preferentially along grain boundaries and annealing twins. Magnification: 200×. Etchant: Potassium dichromate.

Stainless Steels

Intergranular corrosion in stainless steels is typically associated with a microstructural condition called *sensitization*. In sensitization, chromium-rich carbides form at grain boundaries, depleting adjacent metal of chromium, an alloying element essential to corrosion resistance in stainless steels (Fig. 18.3). The chromium-depleted zone

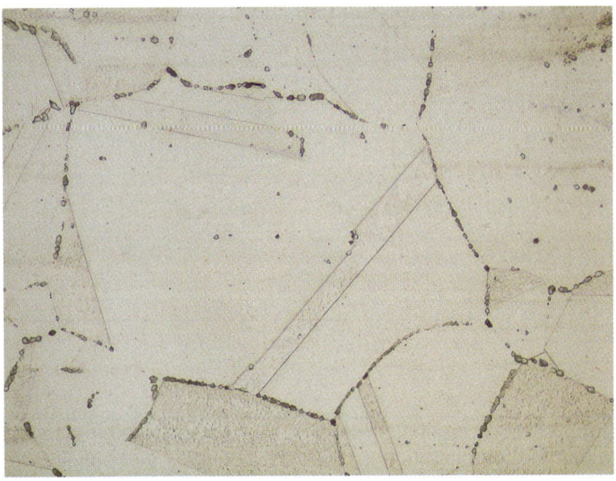

FIGURE 18.3 Austenitic stainless steel, showing small, rounded, dark-etching carbides along the grain boundaries that indicate a sensitized microstructure. Magnification: 500×. Etchant: Electrolytic oxalic acid.

along the grain boundaries can lose its passivity, thereby resulting in very small anodic areas relative to large passive, cathodic areas. Therefore, attack penetrates along the less corrosion-resistant grain boundaries. Though sensitized stainless steel is highly susceptible to corrosion, it will not experience corrosion in all environments.

Sensitization requires exposure to high temperatures, as atomic diffusion is needed to form the chromium carbides. Therefore, in cooling water systems, sensitization typically occurs during manufacturing of the component and/or welding operations. Chromium-rich carbides can form within the approximate temperature range of 900 to 1560°F (480–850°C) (Fig. 18.4). Sensitization is a function of both time and temperature, occurring most rapidly at temperatures near 1250°F (675°C). Metals that cool rapidly through this temperature range will not experience sensitization. Consequently, thin metal sections, which cool rapidly, are less susceptible to sensitization than thick sections. Limiting the carbon content of the alloy also helps to limit sensitization. Carbon contents of 0.05 percent and below are usually sufficiently low to limit carbide precipitation and resulting susceptibility to intergranular attack in austenitic stainless steels under many conditions; however, low-carbon grades typically contain at most 0.03 percent carbon. Such alloys are indicated with an "L" after the alloy number, e.g., types 304L and 316L. Austenitic stainless steels commonly used in cooling water applications generally contain a maximum of 0.08 percent carbon. Figure 18.4 illustrates that sensitization occurs much more quickly in alloys containing

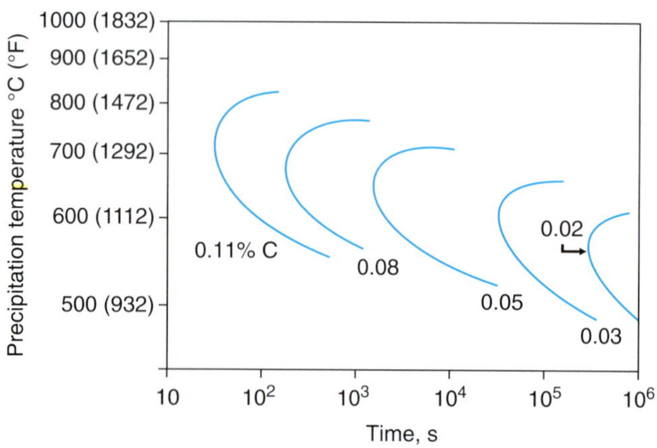

FIGURE 18.4 Effect of carbon content on $M_{23}C_6$ (metal carbide, including chromium carbide) precipitation in 18Cr-9Ni stainless steel (type 304), as detected by intergranular corrosion tests (H_2SO_4-Cu_2SO_4). Regions to the right of the curves indicate conditions that can result in sensitization.[2] *(Reprinted with permission of ASM International. All rights reserved. www.asminternational.org.)*

higher carbon concentrations than in alloys with low carbon concentrations.

Aluminum Alloys

Intergranular Corrosion

In aluminum alloys, intergranular corrosion can be promoted by secondary phases that significantly differ in nobility from the adjacent metal from which they were formed. Thus, galvanic effects cause corrosion of the more active phase (either precipitated particles or the adjacent matrix), depending on the alloy. Some magnesium- and zinc-containing precipitates in the 5XXX (aluminum-magnesium) and 7XXX (aluminum-zinc-magnesium-copper) series are anodic relative to the adjacent metal and therefore are preferentially attacked. In aluminum-copper alloys such as the 2XXX series and 7075, however, copper-containing precipitates such as $CuAl_2$ are actually more noble than the adjacent solid solution; therefore, the metal around the precipitates is preferentially attacked instead. Iron-rich precipitates can also act as cathodes relative to the more anodic matrix (see Case History 18.5).

Exfoliation Corrosion

Exfoliation in aluminum alloys is often considered to be a special type of intergranular corrosion. Exfoliation corrosion occurs within susceptible wrought microstructures, typically consisting of thin, severely elongated grains. Exfoliation is often associated with relatively high copper, zinc, or magnesium content in the alloy.

In exfoliation corrosion, attack occurs preferentially between the elongated faces of the thin, often platelike grains that result from rolling or extrusion. The corrosion products occupy a larger volume than the metal, which forces the grains apart. This results in a weak structure consisting of layers of metal and corrosion products. See Chap. 5, "Aluminum Alloys," for a more in-depth discussion of exfoliation corrosion.

Locations

Intergranular corrosion may occur in any location that is contacted by water. In carbon steels and copper-based alloys, the presence of deposits on the surface may promote concentration of aggressive species due to the formation of ion concentration cells, promoting intergranular corrosion. High heat flux may also promote corrosion; heat exchanger and condenser tubes are the most commonly affected components.

In stainless steels and aluminum alloys, susceptibility to intergranular corrosion is primarily determined by microstructural factors;

the environment does not need to be especially aggressive. Therefore, attack may occur wherever a susceptible microstructure is contacted by cooling water. However, corrosion may increase in severity if the water becomes more aggressive, either by bulk chemistry or in localized areas. Crevices, areas under deposits, or other locations that promote localized concentration of aggressive species also tend to be more susceptible to intergranular corrosion. Stainless steel heat exchanger and condenser tubes may be affected. Corrosion may also occur in welds, heat-affected zones, or anywhere else that was exposed to sensitizing conditions during manufacturing or installation. Aluminum components that most commonly experience intergranular and exfoliation attack include high-strength molds used to produce plastics and water-cooled equipment used for manufacturing of electronic components.

Critical Factors

Carbon Steel

Species such as nitrates can cause intergranular corrosion in carbon steel. One possible source of nitrates is ammonia. In highly oxygenated systems, nitrifying bacteria can convert ammonia (e.g., from a process leak) to nitrites and nitrates. This process is often associated with a drop in pH.

Copper and Copper Alloys

Deposits can result in the concentration of aggressive anions such as chloride, which can in turn promote intergranular corrosion. Intergranular corrosion has also been correlated with high heat flux in many cases.

Stainless Steels

A sensitized microstructure is typically required for intergranular attack. Sensitization can occur due to significant exposure to temperatures within the approximate temperature range of 900 to 1560°F (480–850°C), which allows for appreciable atomic diffusion. Since sensitization involves the formation of chromium-rich carbides, alloys with carbon contents above 0.05 percent are more susceptible to sensitization, while low-carbon alloys (typically below 0.03 percent carbon) are resistant to sensitization.

Typical factors that are known to promote corrosion in stainless steels, such as low pH, high chloride concentration, and high temperatures, also promote intergranular corrosion. However, the environment does not need to be particularly corrosive for intergranular attack to occur in sensitized stainless steels. Intergranular attack can occur in mild environments due to the decreased corrosion resistance of the metal adjacent to the grain boundaries.

Aluminum

A susceptible microstructure is the most important factor. If precipitates occurring along grain boundaries are sufficiently cathodic or anodic relative to the adjacent metal, the grain boundaries may be susceptible to attack. Wrought microstructures consisting of elongated grains promote exfoliation. Alloys containing significant concentrations of copper, zinc, and magnesium are particularly susceptible to intergranular-type corrosion. Slightly acidic conditions, water with appreciable conductivity, and/or galvanic coupling to a more-noble metal promote intergranular and exfoliation corrosion.

Identification

In some cases, attack at grain boundaries may be apparent during visual examination using a stereomicroscope, especially if grains are large (Fig. 18.5). Corroded areas may have a granular or sugary appearance and feel. This is due to corrosion along the boundaries of individual grains that causes them to be released from the metal surface. In the case of exfoliation corrosion in aluminum, attack can sometimes be identified visually when layers of metal and corrosion products are obviously peeling from the surface (Fig. 18.6). However, metallographic examination is typically necessary to definitively

FIGURE 18.5 Grain outlines indicating intergranular attack in an aluminum brass heat exchanger tube. Image taken with a stereomicroscope. Note the appearance of twins (straight, parallel features) within individual grains.

Figure 18.6 Corroded area along a water channel of a 7075 aluminum mold, showing sheets of metal between layers of voluminous white corrosion products, indicating exfoliation corrosion. Image taken with a stereomicroscope.

diagnose intergranular corrosion (Figs. 18.7 and 18.8). Corrosion should be apparent without any chemical etching, though etching can positively confirm the intergranular nature of the attack. In more severe cases of intergranular corrosion, whole grains can detach from a corroded surface (Fig. 18.9).

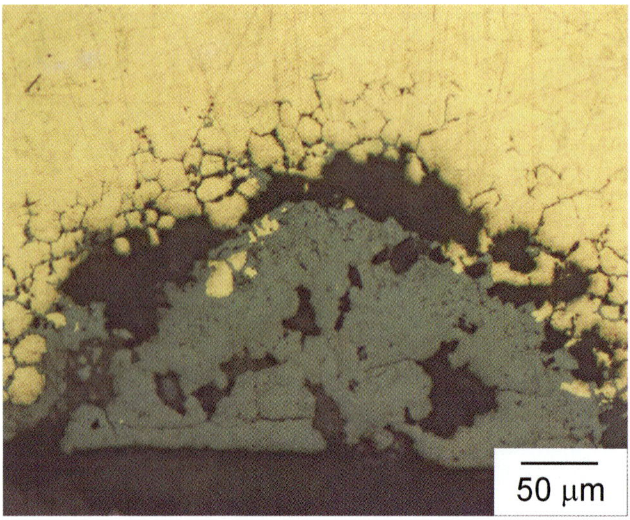

Figure 18.7 Micrograph showing intergranular attack of an uninhibited admiralty brass tube. Unetched.

FIGURE 18.8 Micrograph, showing exfoliation corrosion of an aluminum mold. Etchant: Keller's reagent.

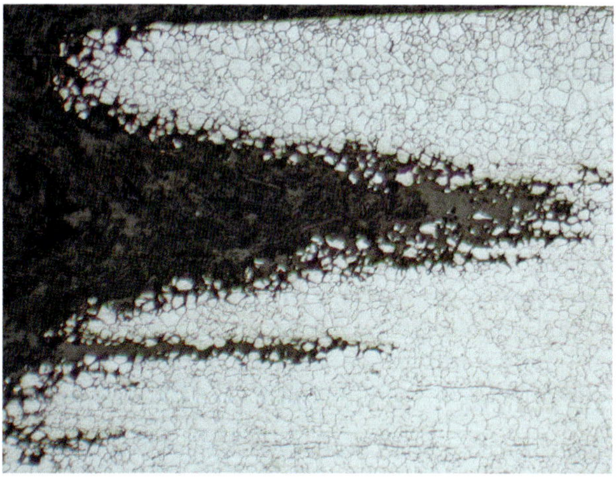

FIGURE 18.9 Whole grains separating from a corroded surface due to intergranular corrosion of a sensitized type 304H austenitic stainless steel corrosion coupon. Magnification: 100×. Etchant: Electrolytic oxalic acid.

Elimination

Carbon Steel

The source of aggressive chemical species should be identified and eliminated, if possible. For instance, if intergranular corrosion can be linked to nitrates produced by nitrifying bacteria, the source of ammonia should be located and eliminated. Bacterial growth should be controlled through the use of adequate concentrations of appropriate biocides. Since aggressive chemical species can become concentrated beneath occlusive material on the surface, clean surfaces should be maintained.

Copper and Copper Alloys

Clean surfaces should be maintained in order to reduce the likelihood of concentration of aggressive chemical species, such as sulfate and chloride, beneath occlusive material. If intergranular corrosion is a serious problem and occurs despite maintenance of clean surfaces, it may be advisable to consider alternate, more resistant metallurgies.

Stainless Steels

Since intergranular corrosion of stainless steel is typically associated with sensitization, the most effective way to avoid intergranular attack is to prevent sensitization. Some strategies are given next.

Specify Low-Carbon Grades of Stainless Steel

Since sensitization results from the formation of chromium carbides, one control method is to use stainless steels that have lower carbon concentration. Low-carbon grades of stainless steel, designated with an "L" after the alloy number, are often used for components that are to be welded. Stainless steels containing carbon concentrations of 0.03 percent or less have been shown to be very resistant to sensitization during welding. While they are not completely immune to sensitization, it is much less likely to occur than in normal grades of stainless steel. To illustrate, an austenitic stainless steel containing 0.08 percent carbon may become sensitized after a few seconds of heating within the sensitizing temperature range, while a comparable alloy with low carbon content may be heated for a few hours without becoming sensitized (Fig. 18.4).

Specify Stabilized Grades of Stainless Steel

An alternative method to prevent chromium carbide formation is to use alloying elements whose affinity for carbon is greater than that of chromium. These elements will react preferentially with the carbon, preventing chromium carbide formation and thereby leaving the chromium uniformly distributed in the metal. The carbon content of the alloy does not have to be reduced if sufficient quantities of these "stabilizing" elements are present. For example, titanium is used to produce one stabilized alloy (AISI 321), and niobium is used to provide another (AISI 347).

Alter Cooling Rate and/or Welding Methods

Sensitization is governed by both time and temperature. If a low-carbon grade is not used, preventing sensitization involves cooling quickly through the sensitization temperature range to avoid giving carbides sufficient time to form. In the case of welds, factors such as the thickness of the components to be welded, the type of welding procedure, and the time required to perform the welding will influence the likelihood of inducing potentially damaging sensitization. Experience has shown that electric arc welding, which characteristically produces intense heat over a short time, is the preferred technique for welding stainless steel that is not a low-carbon or stabilized grade.

If sensitization occurs in a stainless steel component, it may be eliminated by appropriate heat treatment of the affected metal. Heat treatment is a remedial rather than a preventive measure. Sensitized metal may be reheated to temperatures in the range of 1950 to 2050°F (1065–1120°C) for sufficient time to allow the chromium carbides to dissolve completely into solid solution in the steel. If the steel is then rapidly cooled (quenched) through the sensitization range [900–1560°F (480–850°C)], chromium carbides will not have enough time to reform. Hence, the prior sensitized microstructure is essentially thermally erased. Note that this method has definite limitations. Reheating and quenching sensitized stainless steel may not be practical in many cases, particularly for installed components that cannot be removed for controlled furnace heating and quenching operations. Also, the quenching operation can induce substantial residual stresses and warpage. Consequently, successful elimination of the problem typically involves preventive techniques rather than remedial measures.

Aluminum

Due to the nature of intergranular and exfoliation corrosion, which cause the formation of very small, tight crevices, corrosion inhibitors cannot be relied upon to prevent such attack. Therefore, strategic choice of alloy and processing techniques should be employed in order to minimize the potential for intergranular and exfoliation corrosion. Alloys containing significant amounts of copper, zinc, and magnesium tend to be particularly susceptible to exfoliation corrosion; therefore, avoidance of such alloys in water-contacting applications, if practical, can help to avoid corrosion issues.

If such alloys are desirable for other reasons, it may be possible to select heat treatments that will encourage precipitates to form more uniformly throughout the microstructure, rather than preferentially at grain boundaries. In 2XXX series alloys, for example, it has been shown that fast cooling or quenching, followed by aging or slight overaging, results in good resistance to intergranular corrosion. In 7XXX alloys, overaging is known to significantly reduce the likelihood of exfoliation corrosion. Despite a decrease in strength that accompanies overaging, the gain in corrosion resistance may make such heat treatments attractive options if exfoliation corrosion is a concern. Cladding high-strength aluminum alloys with a more corrosion-resistant outer layer is a method used to avoid attack on the alloy that is more susceptible to corrosion.

Cautions

In locations of elevated tensile stress, intergranular corrosion can occur in a cracklike fashion in a sensitized area in stainless steels. Such attack blurs the line between intergranular corrosion and stress corrosion cracking. This type of cracking is often considered to be stress corrosion cracking promoted by sensitization. However, the specific environment and circumstances of a particular case should always be taken into consideration when making a diagnosis.

Related Problems

See also Chap. 11, "Crevice and Underdeposit Corrosion (Concentration Cells)"; Chap. 16, "Galvanic Corrosion"; Chap. 17, "Dealloying"; and Chap. 22, "Stress Corrosion Cracking."

Case History 18.1

Industry:	Chemical
Specimen location:	Ammonia recovery heat exchanger
Environment:	External: Ammonia up to 145°F (63°C) Internal: Cooling water
Time in service:	3 years
Sample specifications:	Plain carbon steel tube, 15/16 in. (2.38 cm) outer diameter

A perforation occurred at the bottom of a very deep internal surface depression (Fig. 18.10). Only a thin layer of red and black oxides and/or deposits was observed in and around the depression on the internal surface. Spot tests indicated the presence of nitrates in the deposits. Numerous other pits and depressions were also observed on the surface (Fig. 18.11). Many parallel, oxide-covered, shallow grooves were oriented at approximately 45° to the longitudinal axis of the tube, suggestive of low-pH conditions (see Chap. 14, "Acid Corrosion").

FIGURE 18.10 Perforation at the base of a deep pit on the internal surface of a carbon steel heat exchanger tube.

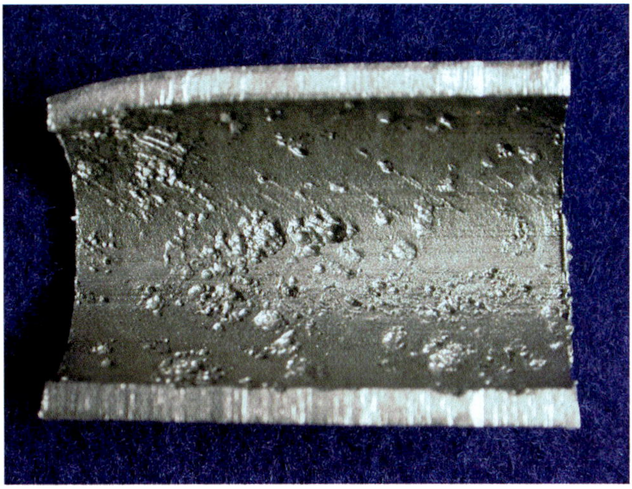

FIGURE 18.11 Internal surface of carbon steel ammonia heat exchanger tube after cleaning by glass bead blasting, showing numerous pits and depressions as well as shallow grooves oriented at approximately a 45° angle relative to the longitudinal axis of the tube.

Metallographic examinations indicated a thin layer of iron oxide–based corrosion products lining the depression containing the failure. Shallow intergranular corrosion was observed along the surfaces of the depressions, including the one that resulted in the perforation (Fig. 18.12).

The internal surface corrosion was primarily caused by a combination of oxygen corrosion and low-pH corrosion during service. Nitrates probably

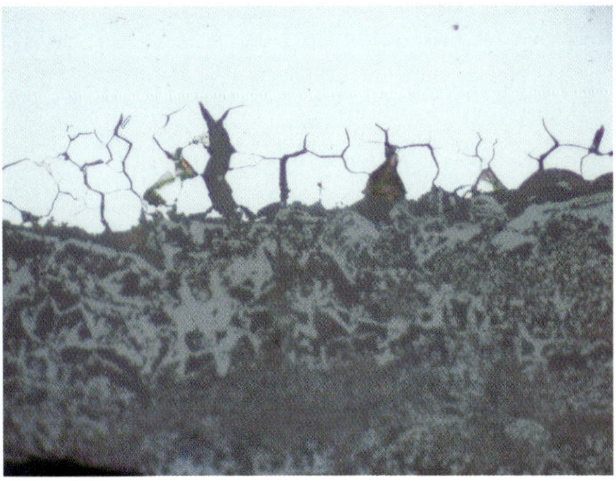

FIGURE 18.12 Intergranular corrosion along the base of a depression on the cooling water side of a carbon steel ammonia heat exchanger tube. Magnification: 500×. Unetched.

contributed to the observed intergranular corrosion. Reportedly, the cooling water system experienced leaks of both carbamate and ammonia. According to cooling water analyses, both ammonia and nitrate levels steadily increased over a six-month period. Increased nitrate levels and carbamate leaks may have corresponded to a decrease in pH and accelerated attack. Nitrifying bacteria are often responsible for increased nitrate concentrations and associated pH drops in oxygenated cooling water systems contaminated with ammonia. The microstructure of the steel included carbides along the ferrite grain boundaries, which is believed to increase the susceptibility to intergranular attack. Monitoring at exchanger inlets and outlets was recommended in order to locate the source or sources of the process leaks.

Case History 18.2

Industry:	Utility
Specimen location:	Turbine lubricating oil cooler
Specimen orientation:	Horizontal
Environment:	Internal: pH 7.9, 80°F (27°C), phosphonate, benzotriazole, tolyltriazole, phosphoric acid, polymer dispersants and scale inhibitors, sulfuric acid, bleach/bromine External: Lubricating oil, 70–80°F (21–27°C) when on standby
Time in service:	3 years
Sample specifications:	5/8 in. (1.6 cm) outer diameter, inhibited admiralty brass tubing

The section contained numerous rounded depressions on the internal surface, a few of which resulted in circular perforations (Fig. 18.13). Outlines of grains were observed along the edges of some depressions (Fig. 18.14). Some depressions were lined with green and red corrosion products (Fig. 18.15). Scanning electron

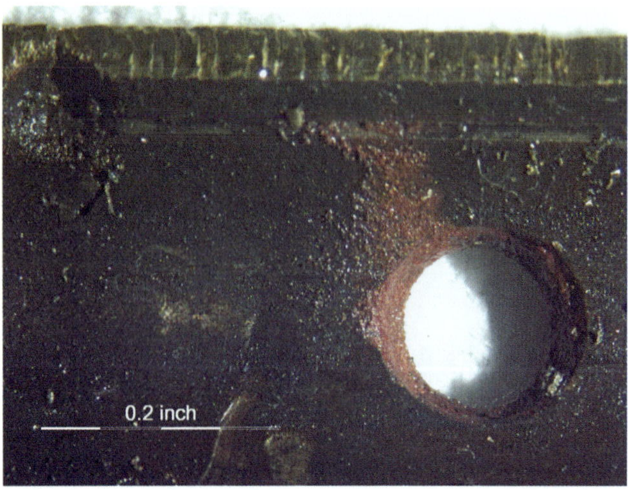

Figure 18.13 Circular perforation on the internal surface of an inhibited admiralty brass tube.

FIGURE 18.14 Edge of a depression on the cooling water side of an inhibited admiralty brass oil cooler tube, showing outlines of some grains.

FIGURE 18.15 Green and red corrosion products lining depressions on the cooling water side of an inhibited admiralty brass tube that exhibited intergranular corrosion.

microscopy with energy dispersive x-ray spectroscopy (SEM-EDS) analysis of the corrosion products indicated as much as 20 percent chlorine in the green material. Microscopic examination revealed intergranular corrosion along the edges of the perforations and depressions (Fig. 18.16). In some locations, reddish metallic copper corrosion products were observed along the edges of some grains and oxide corrosion products penetrated along the grain boundaries (Fig. 18.17).

FIGURE 18.16 Micrograph showing intergranular corrosion along the edge of a deep depression in an inhibited admiralty brass tube. Etchant: Potassium dichromate.

FIGURE 18.17 Micrograph of an area exhibiting intergranular corrosion, showing reddish metallic copper corrosion products along the edges of some grains. Unetched.

The exchanger was reportedly on standby operation for 11 months before the tube section was submitted for analysis. During the standby period, the flow rate was decreased relative to the usual flow rate. In general, low flow conditions promote corrosion. The presence of chlorine-rich corrosion products in some depressions suggested that chlorides promoted corrosion. No significant amount of deposit material was observed on the section and no cleaning was reportedly conducted before the failure occurred, suggesting that deposits did not play a role in the occurrence of intergranular corrosion. Though the tube was made from inhibited admiralty brass, the inhibitor in the alloy (arsenic, in this case) prevented dealloying from occurring but could not protect against intergranular corrosion.

Case History 18.3

Industry:	Sugar refining
Specimen location:	Surface condenser
Specimen orientation:	Horizontal
Environment:	External: Steam, condensate Internal: Cooling water
Time in service:	6 years
Sample specifications:	¾ in. (1.9 cm) outer diameter, inhibited admiralty brass tubing

A few perforations occurred within deep depressions on the internal surface of the tubing. The surface was covered with thick layers of stratified deposit material in places (Fig. 18.18). Bulk chemical analysis of the deposits by x-ray fluorescence and x-ray diffraction analyses indicated calcium carbonate as the primary constituent of the material. Microscopic examinations showed intergranular corrosion along the perforation edges and depressions (Fig. 18.7). SEM-EDS analysis indicated elevated chlorine concentrations within some depressions, relative to locations away from the depressions.

Intergranular corrosion was promoted by the thick deposit layers on the surface, which caused localized concentration of aggressive anions, such as chloride, to levels significantly higher than in the bulk water. The exchanger was reportedly hydroblasted three years before the section was received for analysis; the thick layers of deposit material covering the surface in places suggested that hydroblasting did not sufficiently remove the deposit material. Chemical analysis of the alloy indicated that it was admiralty brass inhibited with arsenic. Inhibiting elements decrease the likelihood of dealloying, but still leave the alloy susceptible to intergranular corrosion.

Figure 18.18 Internal surface of a surface condenser tube, showing thick, stratified deposit layers.

Case History 18.4

Industry:	Plastics
Specimen location:	Plastic bottle mold
Time in service:	2 months
Sample specifications:	Aluminum alloy 7075 (5.7% zinc, 2.5% magnesium, 1.6% copper, 0.19% chromium)

The mold experienced extensive corrosion along the cooling water channel (Fig. 18.19.). Fiber-like metallic protrusions as well as light-colored corrosion products and deposits were observed in the metal loss areas (Fig. 18.6). The most severe attack occurred along the sides of the channel, where the surface of the channel was approximately parallel to the direction of grain elongation (Fig. 18.20).

FIGURE 18.19 Fiberlike protrusions along a cooling water channel in an aluminum plastic bottle mold (circled).

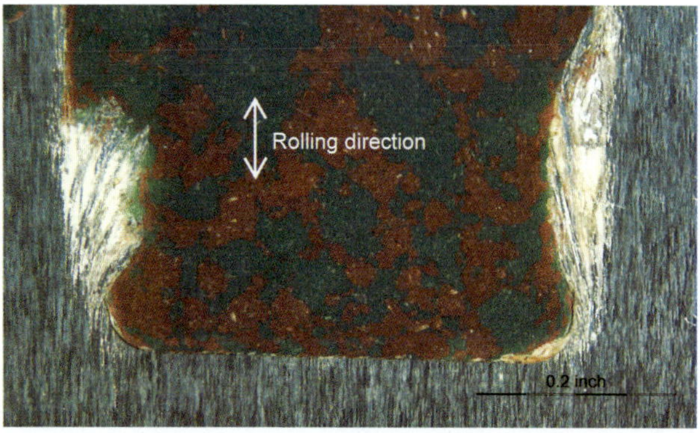

FIGURE 18.20 Cross section of the cooling water channel in an aluminum plastic bottle mold, showing elongated grains peeling away from the sides of the channel due to exfoliation corrosion. The grains are elongated along the rolling direction, indicated with an arrow.

Corrosion penetrated along the boundaries between the very elongated grains (Fig. 18.8). The direction of elongation was indicative of the direction in which rolling was conducted during manufacturing.

Alloy identification indicated a composition that was consistent with aluminum alloy 7075, a high-strength alloy that is often used in aircraft and aerospace applications. Alloy 7075 contains high concentrations of copper and zinc, which are known to promote intergranular corrosion in aluminum alloys. Due to the greatly elongated grains caused by rolling, the grains peeled away in sheets as the attack progressed, indicating the occurrence of exfoliation corrosion.

Case History 18.5

Industry:	Aluminum
Specimen location:	Caster mold
Environment:	Cooling water
Time in service:	Unknown
Sample specifications:	Aluminum casing housing a copper mold

Significant metal loss was observed in the aluminum housing of the copper mold, in a ridge that immediately surrounded the copper mold (Fig. 18.21). The most prominent metal loss occurred in teeth on the ridge (Fig. 18.22). The teeth served to direct cooling water flow; therefore, when the teeth degraded significantly, water flow was disturbed. Microscopic examinations in the metal loss areas indicated undercut pits that were surrounded by moderately deep intergranular corrosion (Fig. 18.23). SEM-EDS analysis of precipitate particles (Fig. 18.24) within intergranularly corroded areas indicated that they contained

Figure 18.21 As-received aluminum housing, in which a copper mold (not shown) was seated. The most severe metal loss occurred on the ridge surrounding the copper mold.

FIGURE 18.22 Significant metal loss in teeth on the aluminum housing.

FIGURE 18.23 Intergranular corrosion in an aluminum caster mold housing. Magnification: 100×. Unetched.

approximately 23 percent iron, 3 percent silicon and the balance mostly aluminum. In comparison, analysis of the bulk alloy indicated that it contained approximately 2 percent magnesium and 1 percent iron, with the balance aluminum. Analysis of corrosion products within corroded areas indicated as much as 9 percent sulfur and 3 percent chlorine in places. Metallic copper particles were observed on some corroded surfaces.

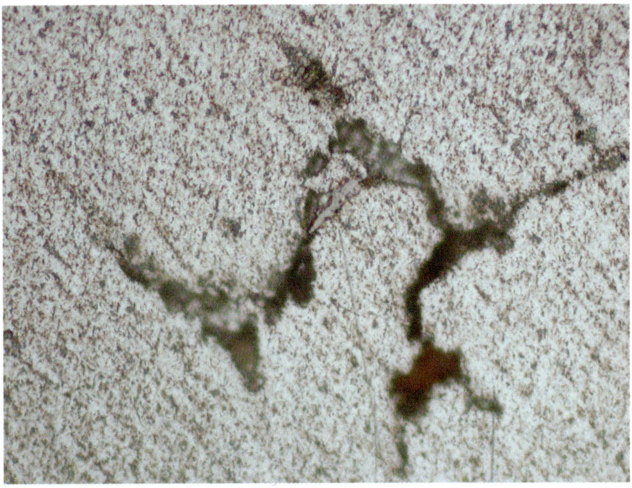

FIGURE 18.24 Intact, high-iron-content precipitate particles within intergranularly corroded areas of a cast aluminum housing. Magnification: 500×. Unetched.

The iron-rich precipitates at the grain boundaries were cathodic relative to the surrounding aluminum. Therefore, microgalvanic effects promoted intergranular corrosion significantly. The aluminum housing and copper mold were not coupled during service, ruling out macroscopic galvanic effects as the cause of the corrosion. The presence of relatively high concentrations of sulfur and chlorine in the corroded areas suggested that aggressive anions, such as sulfate and chloride, concentrated within the crevices formed due to the intergranular corrosion, further promoting attack. Corrosion inhibitors cannot be relied upon to penetrate into such tight crevices.

Case History 18.6

Industry:	Pulp and paper
Specimen location:	Paper machine suction roll
Environment:	Borehole water
Time in service:	2 months
Sample specifications:	Corrosion coupons, type 304H austenitic stainless steel

A washer-style coupon and a U-bend coupon were received for analysis (Figs. 18.25 and 18.26). Type 304H stainless steel can contain up to 0.10 percent carbon. The coupons were purposefully sensitized by heat treatment and had reportedly been exposed to borehole water from the suction roll of a paper machine for approximately two months. Orange-brown corrosion products surrounded the attachment hardware sites on both coupons.

Microscopic examinations indicated intergranular corrosion and pullout of whole grains at the attachment hardware contact sites (Fig. 18.9). Intergranular corrosion was not observed in any other location. No cracking was observed in either coupon, even at the U-bend, where tensile stresses would have been highest. SEM-EDS analysis indicated approximately 1 percent sulfur and chlorine in the corrosion products around the attachment hardware sites.

Figure 18.25 Washer-style coupon from a suction roll of a paper machine, showing orange-brown corrosion products at the attachment hardware site.

Figure 18.26 U-bend coupon from a suction roll of a paper machine, showing corrosion products around the attachment hardware site.

Corrosion at the attachment hardware sites occurred due to the concentration of aggressive anions within the crevices formed by the coupons and the attachment hardware. Sensitization of the coupons promoted the occurrence of intergranular corrosion. However, lack of attack away from the attachment hardware indicated that aggressive ion concentration within the crevice and/or beneath deposits was necessary for the intergranular corrosion to occur. This case illustrates that the susceptibility of a stainless steel sensitized to intergranular corrosion depends on the water chemistry in contact with the metal surface.

Case History 18.7

Industry:	Chemical processing
Specimen location:	Pump impeller
Environment:	Process water with sulfuric acid used to adjust pH typically between 2 and 4, up to 1000 ppm chloride, temperatures up to 170°F (77°C)
Time in service:	30 days
Sample specifications:	Cast high-alloy austenitic stainless steel

A generally corroded pump impeller from an ammonia scrubber system was received for analysis. The impeller had an alloy composition similar to CN7M, but with an appreciably lower nickel content than the chemical requirements. Alloy CN7M is considered to be a cast version of Alloy 20, which is a wrought alloy originally developed for sulfuric acid service.

The impeller surface was covered by a thin, dark corrosion product layer that contained sulfates. The surface contour was mostly irregular due to general wastage. Moderately deep, narrow grooves outlined irregular, polygonal patterns on the surface, as shown in Fig. 18.27.

The microstructure of the impeller showed appreciable dendritic segregation and many large grains, features typical of a cast component. Preferential corrosion was noted along the grain boundaries of the casting in many places, as illustrated in Fig. 18.28. In castings, segregation during solidification creates local variations in chemical composition throughout the microstructure. Some segregation along grain boundaries was noted, including elevated levels of copper. Segregation along grain boundaries allowed for intergranular attack when the metal was exposed to an aggressive environment.

FIGURE 18.27 Grooves on the impeller surface in the shape of polygons that highlight the grains of the metal.

Figure 18.28 Preferential attack along a grain boundary in the microstructure. Etchant: Electrolytic oxalic acid.

References

1. Arthur Cohen, "Corrosion of Copper and Copper Alloys," *ASM Handbook*, Volume 13B, *Corrosion: Materials*, ASM International, Materials Park, OH, 2005, pp. 125–163.
2. Charles J. Novak, "Structure and Constitution of Wrought Austenitic Stainless Steels," in *Handbook of Stainless Steels*, ed. Donald Peckner and I. M. Bernstein, McGraw-Hill, New York, 1977, pp. 4-1–4-78.

CHAPTER 19

Graphitic Corrosion

General Description

Graphitic corrosion exclusively affects cast iron components, including gray, nodular (also known as ductile) and malleable cast irons. Metal that has experienced graphitic corrosion may retain its original appearance and dimensions. Consequently, graphitic corrosion frequently escapes detection.

Although sometimes categorized as a form of dealloying, graphitic corrosion has much more in common with galvanic corrosion. To understand the basic mechanism of graphitic corrosion, some knowledge about the microstructure of cast iron is required. The microstructural feature of principal importance is the presence of flakes (gray cast iron) or nodules (nodular or malleable cast iron) of graphite embedded in a matrix of iron (Fig. 19.1). See Chap. 2, "Cast Iron," for more micrographs and an in-depth discussion of cast iron alloys. Since graphite is much more noble than the surrounding iron matrix (see Table 16.1), the graphite and iron form a microscopic galvanic couple if both are exposed to the same sufficiently conductive electrolyte (such as cooling water) (Fig. 19.2). Such a situation can occur at exposed surfaces of the casting, where both the iron-based matrix and the graphite particles contact the electrolyte. A microscopic electrochemical cell develops that is analogous to a carbon-zinc dry cell (see Fig. 16.1). The noble graphite remains intact while the iron corrodes (Figs. 19.3 and 19.4), the corrosion being driven by the potential difference generated by the galvanic couple and the electrolyte environment. Graphitic corrosion generally increases with increasing conductivity and corrosivity of the electrolyte.

It is interesting to note that the iron is not removed, but rather is transformed in situ to iron oxide. What is even more remarkable is that the volume change anticipated when iron is converted to iron oxide does not occur. Consequently, surface contours of the corroded metal remain intact, even to the point of retaining fine detail such as machining marks.

In mildly corrosive environments, such as in municipal water transmission piping, graphitic corrosion may progress slowly, requiring

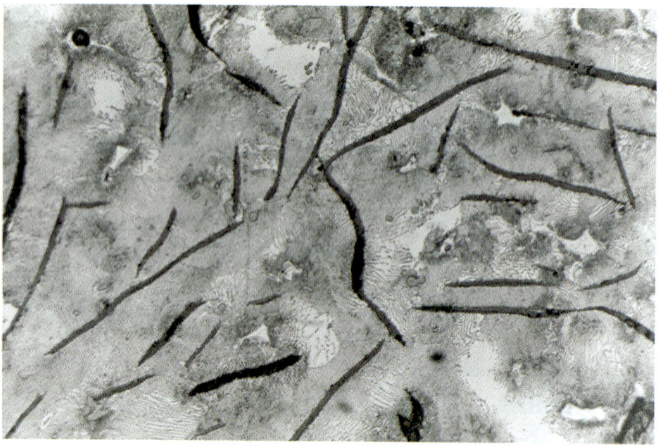

FIGURE 19.1 Typical microstructure of gray cast iron, consisting of flakes of graphite (black material) embedded in an iron-based matrix. In this case, the matrix consists of pearlite (a mixture of iron and iron carbide) Etchant: Picral.

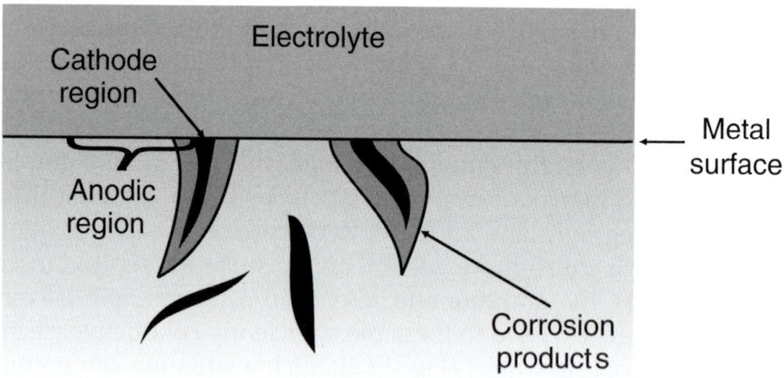

FIGURE 19.2 Schematic showing corrosion occurring around graphite flakes that are exposed at the surface of the casting. Regions of the iron matrix act as anodes, while the graphite flakes act as cathodes.

many years to effect significant damage. Thick-walled cast iron components can be found in use in Europe after more than 150 years of service. However, graphitic corrosion has caused complete penetration of thick cross sections in as little as a few months in highly corrosive environments. Although graphitic corrosion causes a substantial reduction in mechanical strength, it is well known that corroded cast iron, when sufficiently supported, may remain serviceable when internal pressure is low and shock loads are not applied.

The literature frequently states that nodular cast iron is immune to graphitization. Although it is less prone than gray cast iron to serious

FIGURE 19.3 Graphite flakes surrounded by graphitically corroded iron. Bright background is uncorroded metal. Etchant: Picral; magnification: 1000×.

FIGURE 19.4 Nodular (ductile) iron pump impeller, showing corrosion products surrounding graphite nodules. Unetched.

graphitic corrosion, the basic microstructural features required for graphitic corrosion exist in nodular irons, and cases of graphitic corrosion of this alloy have been observed by the authors.

Locations

The occurrence of graphitic corrosion is not location-specific, other than that it may occur wherever gray or nodular cast iron is exposed to sufficiently aggressive aqueous environments. This includes, and is

FIGURE 19.5 Gray cast iron heat exchanger head after partial removal of deposits and corrosion products by glass bead blasting. Graphitic corrosion completely penetrated a wall between water channels.

common to, subterranean cast iron pipe, especially in moist soil (Case History 19.1). Cast iron pump impellers and casings are also susceptible to graphitic corrosion (Case Histories 19.2–19.5). Heat exchanger shells and heads are also occasionally affected (Fig. 19.5).

Critical Factors

Two critical factors govern graphitic corrosion:

1. A susceptible cast iron microstructure containing graphite flakes or nodules
2. Exposure of the metal to an aqueous environment that is sufficiently aggressive to drive a galvanic corrosion reaction between the graphite and the iron

Though graphitic corrosion has been observed in gray, nodular, and malleable cast irons, gray cast iron is more susceptible to graphitic corrosion because of the morphology of the graphite. In gray cast iron, the graphite is in the form of flakes distributed throughout the microstructure. The surface area of graphite is much larger for the flakes than for the nodules in nodular cast irons; therefore, the driving force for graphitic corrosion is greater for gray cast iron.

Graphitic corrosion is favored by environments such as soft waters, water having a slightly acidic pH, water containing low levels (as little as 1 ppm) of hydrogen sulfide, and brackish and other high-conductivity waters. Moist soils, especially those containing sulfates, will frequently produce graphitic corrosion of unprotected gray and nodular cast iron. There is evidence that stresses may foster localized, relatively rapid graphitic corrosion. Stagnant conditions are also known to promote graphitic corrosion.

Identification

As noted earlier, complete graphitic corrosion can occur without effecting dimensional changes in the component and without altering fine surface details such as machining marks and identification numbers. This can make mere visual identification difficult. Graphitically corroded metal may, however, change color from gray to black. Affected metal is converted to a very soft material that can be dislodged with a sharp metal probe or maybe cut with a knife. Graphitically corroded metal frequently has an oily, slippery feel, especially after surface deposits are scraped away. Because of its graphite content, a chunk of corroded metal can be used for writing on paper, similar to a pencil (Fig. 19.6).

On buried pipeline, graphitic corrosion is most common on the very bottom of the pipe where it rests on the compacted bedding. This is a location that is more likely to be exposed to groundwater for an extended period of time, relative to the top or sides. When graphitic corrosion occurs under these conditions, the affected region may be a narrow zone running along the pipe bottom over some distance (Case History 19.1).

Failure of graphitically corroded cast iron will yield a brittle, thick-walled fracture. Fracture faces through the graphitically corroded region will be black or dark gray and nonmetallic in appearance (Fig. 19.7). Cross sections cut through graphitically corroded regions will readily show bright, intact metal surrounded by soft, black corrosion products (Fig. 19.8).

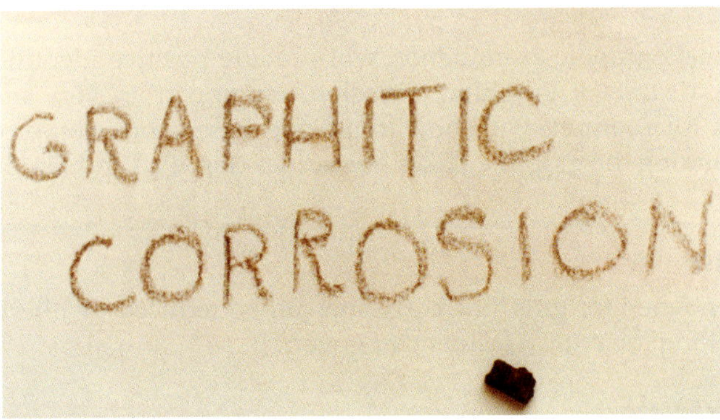

FIGURE 19.6 Graphitically corroded metal is so soft that it can be used to write on paper.

FIGURE 19.7 Brittle fracture through pipe. The gray material is graphitically corroded cast iron. Unaffected pipe wall metal is orange from normal rust.

FIGURE 19.8 Black, graphitically corroded metal at the surface surrounding bright, unaffected metal.

Metallographic examination will provide positive identification of any degree of graphitic corrosion, ranging from very mild to severe. Micrographs will show iron oxide–based corrosion products surrounding the graphite flakes or nodules (Fig. 19.3 and Fig. 19.4).

Elimination

The tendency for graphitic corrosion can be reduced by altering or controlling the critical factors that govern it.

Microstructure

Graphitic corrosion requires a microstructure that contains graphite. Avoiding a susceptible microstructure by altering material selection is a preventive rather than a remedial technique. When graphitic corrosion is anticipated, consideration should be given to specifying alternate materials. Nodular and malleable cast irons are less prone to serious graphitic corrosion than gray cast iron, but are not immune. Most corrosion-resistant cast irons containing chromium, nickel, or silicon are essentially immune to graphitic corrosion. Austenitic cast irons are also immune, as are cast steels. White cast iron, which is essentially free of graphite, is also not susceptible to graphitic corrosion; however, it is very brittle and not a suitable material for many common industrial applications.

Environment

Two approaches exist to gain control of this factor: altering the environment and isolating the metal from the environment. These are remedial measures and can be implemented during or after installation of the equipment.

1. *Alter the environment.* The key is to make alterations to the environment that will render it less aggressive, not only from the standpoint of graphitic corrosion but also with respect to other corrosion mechanisms. For example, since graphitic corrosion can be caused by exposure to slightly acidic water, a pH adjustment (neutralization) may provide sufficient protection. High-conductivity water and long periods of exposure to stagnant conditions should also be avoided if possible. Bedding and backfill for subterranean pipelines should be carefully chosen to avoid aggressive soil conditions.
2. *Isolate the metal from the environment.* If environmental alteration is impossible or not economical, it may be appropriate to separate the metal from the environment. This is commonly the economically justifiable approach and generally involves applying an appropriate protective coating to the metal. External coatings must be resistant to cold flow, especially in subterranean applications where the weight of a pipe resting on hard bedding may cause the coating to flow, opening cracks or perforations where graphitic corrosion can initiate. Internal surfaces can be lined with cement mortar. It is common practice to use cathodic protection in conjunction with coatings.

Cautions

The term *graphitic corrosion* should not be confused with the term *graphitization.* Graphitization, a distinctly different phenomenon, is the transformation of iron carbide to elemental iron and elemental carbon (graphite) that can occur in carbon-containing irons and steels at temperatures above 850°F (455°C).

It is important to note that graphitically corroded metal may be overlooked in a simple visual inspection. Gentle scraping of the surface using a sharp probe or knife should be helpful in identification, particularly in advanced cases of graphitic corrosion.

Note also that graphitic corrosion may occur preferentially in poorly accessible areas, such as the bottom of pipelines. Trouble-free service of cast iron components does not necessarily indicate that all is well, since components suffering severe graphitic corrosion may continue to operate until an inadvertent or intentional (e.g., pressure testing) shock load is applied. At this point massive, catastrophic failures can occur.

Graphitically corroded cast irons may induce galvanic corrosion of metals to which they are coupled due to the nobility of the graphite in the corrosion products covering the surface. For example, new, uncorroded cast iron or cast steel replacement pump impellers may corrode rapidly due to the galvanic couple established with a graphitically corroded cast iron pump casing. In this or similar situations, the entire affected component should be replaced. If just one part is replaced, it should be with a material that will resist galvanic corrosion, such as austenitic cast iron.

Graphitic corrosion in regions subject to high fluid velocities or cavitation may be misdiagnosed as simple erosion or cavitation. Graphitic corrosion produces relatively soft corrosion products that are more susceptible to erosion and cavitation than the original metal. Erosion will occur at significantly lower fluid velocities, and cavitation will occur at significantly lower intensities than for uncorroded metal. The continual removal of the corrosion products may obscure the evidence of active graphitic corrosion. This synergistic interaction can cause rapid wastage of components. In such cases, a metallographic examination may be required to determine if graphitic corrosion is a contributing factor.

Related Problems

See also Chap. 16, "Galvanic Corrosion"; Chap. 23, "Erosion-Corrosion"; and Chap. 24, "Cavitation Damage."

Case History 19.1

Industry:	Metals
Specimen location:	Subterranean cooling water supply line
Orientation of specimen:	Horizontal
Environment:	Internal: Cooling water External: Soil
Time in service:	25 years
Sample specifications:	10 in. (25 cm) outside diameter, gray cast iron pipe

Figure 19.9 shows a 10 ft (3 m) section of a main cooling water line that fractured along its bottom. The fracture ran the entire length of the section. The pipe wall had been converted from metallic cast iron to a soft, brittle, black corrosion

FIGURE 19.9 Section of longitudinally fractured subterranean cooling water line. Note that the fracture runs the entire length of the line.

product along a narrow zone that ran the length of the pipe. Note in Fig. 19.10 and Fig 19.7 that the corrosion product has the same contour as the original pipe. Striking the affected zone with a hammer caused the corrosion products to shatter, revealing a thin layer of intact metal beneath.

FIGURE 19.10 The black outer covering is corrosion product; the reddish-brown surface is coated with air-formed iron oxide.

Internal surfaces were coated with a thin layer of reddish iron oxides. Significant corrosion was not observed on this surface. Therefore, conditions on the external surface (soil and groundwater) were much more conducive to causing graphitic corrosion than the cooling water on the internal surface.

Microstructural examinations of the external surface revealed an interconnecting network of graphite flakes embedded in a matrix of iron oxide corrosion products.

Heavy equipment had passed over the site of failure initiation shortly before the rupture was discovered. It is probable that stresses associated with the heavy equipment initiated a fracture in the severely corroded pipe bottom.

Case History 19.2

Industry:	Metals
Specimen location:	Pump impeller
Environment:	Cooling water: 80–100°F (27–38°C), calcium 780 ppm, sodium 460 ppm, magnesium 220 ppm, chloride 1200 ppm, sulfate 300 ppm, pH 4.1, conductivity 3500 μmhos/cm
Time in service:	3 months
Sample specifications:	13 in. (33 cm) outside diameter, gray cast iron pump impeller

The impeller shown in Fig. 19.11 was removed from a cooling water return line associated with a rolling mill. Failure of these impellers was a chronic problem, requiring impeller replacement every three or four months.

Examination revealed severe metal loss on all surfaces, but it was worst on the outer edges of the vanes. Deep grooves had formed near the hub between the spiral vanes.

Most of the surface was covered with black corrosion products that were thicker in relatively low-flow areas near the hub. This layer of soft corrosion product could be shaved from corroded surfaces. Microstructural examinations revealed flakes of graphite embedded in iron oxide near the surfaces.

FIGURE 19.11 Pump impeller that suffered erosion accelerated by graphitic corrosion.

Turbulence and high fluid velocities resulting from normal pump operation accelerated metal loss by abrading the soft, graphitically corroded surface (erosion-corrosion). The relatively rapid failure of this impeller was due to the erosive effects of the high-velocity, turbulent water coupled with graphitic corrosion caused by the aggressive water chemistry (low pH, high conductivity). Erosion was aided in this case by solids suspended in the water.

Case History 19.3

Industry:	Metals
Specimen location:	Pump housing, quench-system water supply
Orientation of specimen:	Horizontal
Environment:	Untreated cooling water, pH 6
Time in service:	Unknown
Sample specifications:	Gray cast iron

Figure 19.12 shows metal loss on the throat of the pump housing. External pump housing surfaces were also affected (Fig. 19.13). Note the large tubercles (see Chap. 12, "Oxygen Corrosion"). The metal surface beneath these tubercles had sustained graphitic corrosion, in some cases to a depth of ¼ in. (0.6 cm) (Fig. 19.14).

Internal surfaces of the pump showed severe wastage (Fig. 19.15). The wasted region was free of corrosion products except for a small amount of soft, black material. Metal loss in this area was as deep as ½ in. (1.3 cm). The reddish coating partially covering the smooth area above the wasted zone in Fig. 19.15 was applied to mitigate corrosion. Where the coating was intact, corrosion did not occur. Where the coating was absent, attack was severe (Fig. 19.16).

The pump experienced graphitic corrosion. Figures 19.12, 19.14, and 19.16 illustrate typical appearances of graphitically corroded cast iron. In addition, cavitation damage (see Chap. 24, "Cavitation Damage") produced severe metal loss in specific areas (Fig. 19.15). The soft, friable corrosion products produced by graphitic corrosion are easily removed by even relatively low-intensity cavitation.

Figure 19.12 Appearance of the graphitically corroded throat of a pump housing.

FIGURE 19.13 External surface of a pump housing showing tubercles capping sites of graphitic corrosion.

FIGURE 19.14 Corroded area beneath a tubercle on the external surface. The shiny black corrosion product has been partially removed to reveal the depth of penetration.

FIGURE 19.15 Severe metal loss along an internal surface.

FIGURE 19.16 Cross section through the pump housing wall. Note the black and brown graphitic corrosion product layer near the center of the photo.

Case History 19.4

Industry:	Automotive
Specimen location:	Recirculating water pump, paint system
Orientation of specimen:	Unknown
Environment:	Treated cooling water
Time in service:	11 months
Sample specifications:	Ductile (nodular) cast iron

An impeller from a pump in the paint system was severely wasted (Fig. 19.17). All 12 pumps in the system reportedly exhibited similar wastage. The pump was idle every night and on weekends, when the impeller was exposed to stagnant conditions.

FIGURE 19.17 Ductile cast iron pump impeller that experienced severe graphitic corrosion, cavitation, and erosion damage after only 11 months in service.

The edges of the vanes were thinned to a knife blade thickness. Much of the wastage exhibited flow-related characteristics. Areas of deep metal loss occurred in repeating patterns associated with features on the impeller. Metallographic examination indicated corrosion products surrounding the graphite nodules in the microstructure (Fig. 19.4).

The severe metal loss resulted from the combined effects of graphitic corrosion, cavitation, and erosion. Extended periods of stagnant conditions during idle times significantly promoted graphitic corrosion. Analysis of the cooling water indicated high concentrations of sulfate and chloride, aggressive anions that also promoted graphitic corrosion. The soft corrosion products that primarily formed during idle periods were easily removed during operation by forces due to cavitation and erosion.

Case History 19.5

Industry:	Chemical process
Specimen location:	Well pump
Orientation of specimen:	Vertical
Environment:	Raw well water temperature 55°F (13°C), pH 7, pressure 200 psi (1.4 MPa)
Time in service:	5½ years
Sample specifications:	Gray cast iron

A well supplied cooling water to a plant. During operation of the system, a reduction in flow rate was noticed. The well was shut down and the pump inspected.

Inspection revealed severe metal loss on internal surfaces at the bases of the vanes resulting in a large, irregular perforation (Figs. 19.18–19.20). In addition, metal loss was observed in a distinct circumferential zone adjacent to internal threads (Fig. 19.19).

FIGURE 19.18 Pump casing as received. Note the large perforation near the center.

FIGURE 19.19 Longitudinally cut half-section. Note the severe wall thinning and the perforation of the wall on the left side. Compare it to the wall thickness on the right side. Also note the circumferential band of irregular metal loss just above the threads, near the bottom of the casing.

FIGURE 19.20 Close-up of the perforation in Fig. 19.18, viewed from the external surface. Graphitic corrosion of the external surface permitted erosion by turbulent water escaping through the perforation.

Metal loss in these areas produced a smooth surface, free of deposits and corrosion products. The rest of the internal surface was covered by a thin, uniform layer of soft, black corrosion products that were produced by graphitic corrosion. The graphitically corroded surfaces of the pump casing provided soft, friable corrosion products that were relatively easily dislodged by the abrasive effects of high-velocity or turbulent water (erosion-corrosion).

Failures of this type occurred every six to nine years. The well was operated intermittently, since several available wells were operated on a rotating basis. It is probable that most of the graphitic corrosion occurred during idle times due to stagnant conditions; removal of the soft corrosion products due to erosive forces occurred during operation of the pump. Changes in pump rotation were recommended to reduce long-term idle periods when stagnant conditions could cause graphitic corrosion.

CHAPTER 20

Localized Corrosion and Pitting

General Description

Localized corrosion and pitting is not considered a specific mechanism, but rather a form of corrosion on a surface. Many different mechanisms can result in pitting or localized corrosion, such as corrosion caused by exposure to a strong acid. Previous chapters have highlighted and discussed the mechanisms that often result in localized corrosion or pitting. This chapter addresses the implications associated with localized corrosion of the metal surface, discusses some factors that influence localized corrosion of different alloys, and stresses the importance of assessing localized corrosion by monitoring methods, especially corrosion coupons. In considering the forms of corrosion, it is important to understand the different types and their impact on possible failures.

General corrosion affects most or all of the surfaces of a metal component that are exposed to corrosive environments. Metal loss caused by general corrosion also tends to penetrate to a relatively uniform depth. Therefore, the expected life of a component that experiences general corrosion may be predicted fairly accurately.

Localized corrosion is defined as metal loss that occurs in isolated areas on a mostly uncorroded surface. Localized corrosion may also refer to metal loss that penetrates more deeply in isolated areas than on surrounding, more shallowly wasted surfaces (Fig. 20.1). Pitting is sometimes considered a more extreme form of localized corrosion. A pit is defined as metal loss that penetrates more deeply than the width of the opening at the surface. Pits can have many different shapes, and the morphologies can be associated with specific corrosion mechanisms in some cases.

It is vitally important to distinguish localized or pitting corrosion from general corrosion. Localized corrosion, and to a greater extent pitting, will tend to penetrate through tube and pipe walls more quickly than uniform, general metal loss. This is illustrated by the standard

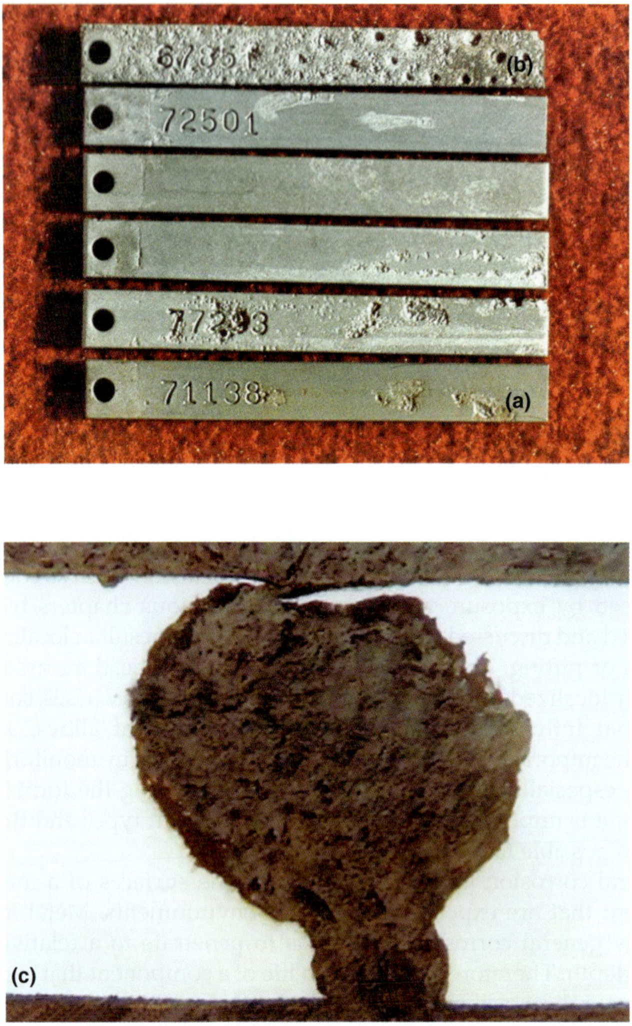

FIGURE 20.1 Examples of localized corrosion where metal loss occurs in isolated locations on a mostly uncorroded surface (*a*), or penetrates more deeply in isolated locations than on surrounding, more shallowly wasted surfaces (*b*). Pitting is a more extreme form of localized corrosion, tending to produce metal loss that penetrates more deeply than the width of the opening at the surface (*c*).

method used to calculate corrosion rates with coupons. The corrosion rate measurement is defined by the weight of metal that is removed during the exposure period. A corrosion rate calculation based upon weight loss assumes that general metal loss penetrates to a uniform depth across the entire coupon surface (Fig. 20.2). When the corrosion

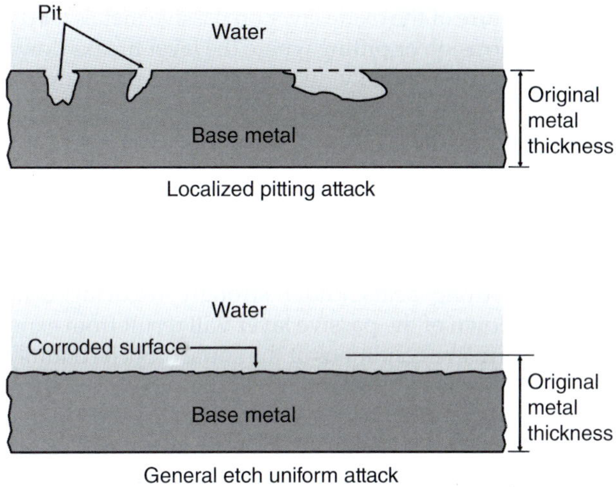

Figure 20.2 General metal loss penetrates to a uniform depth across the entire coupon surface. Localized corrosion produces deep pits that contribute low overall weight loss.

is characterized as being localized, the calculated corrosion rate by weight loss is misleading. In some cases, there may be deep pits that produce low overall weight loss, which would have a corresponding low general corrosion rate. Thus, the depth of metal loss determined by the general metal loss corrosion rate calculation would be much shallower than the depths of the actual localized corrosion or pits.

Except under extreme circumstances, there is usually an extended initiation period for pits and localized corrosion sites to develop. Initiation time may take months to years, depending upon the specific alloy and corrodent. Once pitting or localized corrosion initiates, however, it tends to penetrate at an increasing rate.[1]

There are many possible causes for localized corrosion or pitting. This chapter will touch on the various mechanisms that may cause localized corrosion in the common alloy types used in cooling water systems and will refer to other chapters within this book that will provide more in-depth information regarding those mechanisms.

Mechanism of Localized Corrosion and Pitting

Stainless Steels

In cooling water systems, corrosion-resistant metals such as stainless steels are covered with passive oxide layers that will provide excellent corrosion resistance in most environments. However, this protective oxide layer is generally in a state of constant breakdown and repair. The protective nature of the oxide layer on passive metals tends to promote localized corrosion or pitting where the layer breaks down and is not reestablished. This is mainly due to the large cathodic surface area that is covered with the passive oxide relative to the small anodic surface area where the oxide layer is compromised. The large cathode-to-anode surface area ratio drives the corrosion due to galvanic effects.

It is very rare for passive metals to experience general corrosion, which would require uniform removal of the passive layer. Such removal would require exposure to very corrosive conditions, such as exposure to strong acids under strongly reducing conditions. Localized breakdown of the passive layer will result from exposure to isolated, more corrosive conditions at the surface. Such conditions may be produced by occlusive deposits or within crevices that promote concentration cell corrosion [see Chap. 11, "Crevice and Underdeposit Corrosion (Concentration Cells)"], or by certain types of microbiologically influenced corrosion (see Chap. 13). Localized corrosion and pitting is also promoted by low or stagnant flow conditions. The formation of a deposit layer that is more noble than the stainless steel alloy, such as a manganese oxide layer, may also promote pitting of the underlying steel at locations where there are breaks in the layer. Physical disruptions in the protective oxide layer, such as at nonmetallic inclusions exposed to the surface, may also provide sites for initiation of localized corrosion or pitting.

Aside from these mechanisms, localized chemical breakdown of passive films may occur on clean surfaces away from crevices and inclusions resulting from exposure to aggressive conditions. For example, as described in Chap. 3, "Stainless Steel," pitting may occur on stainless steel alloys when surfaces are exposed to cooling water containing chlorides above certain threshold concentrations. The threshold concentration is temperature and alloy dependent. These thresholds are also influenced by other factors such as pH and oxidation reduction potential (Figs. 20.3 and 20.4). The alloy composition also appreciably influences the stability of the protective oxide layer and subsequent resistance to pitting.

Pitting may occur along welds due to heat tint oxide, which is less protective than the normal oxide that forms on stainless steels (see the "Pickling, Passivation, and Cleaning" section in Chap. 3). Microgalvanic effects can occur in weld metal due to the formation of distinct phases during solidification (see Case History 26.1, Chap. 26, "Weld Defects").

Carbon and Low-Alloy Steels

Oxide layers that form on carbon and low-alloy steels and low-chromium cast irons on surfaces that are exposed to cooling water will be much less protective than those on stainless steels and other passive metals.

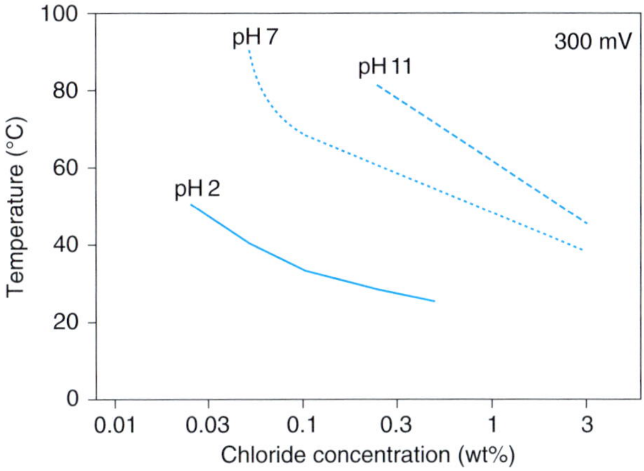

FIGURE 20.3 Curves illustrating pitting susceptibility of 316 stainless steel in water as a function of chloride concentration, pH, and temperature at one oxidation reduction potential (ORP) level.[2]

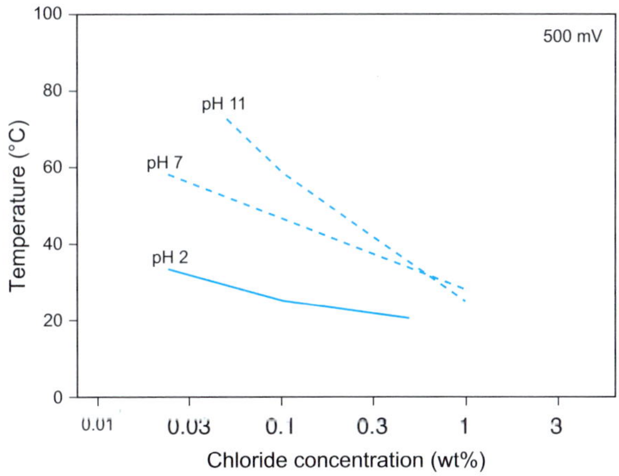

FIGURE 20.4 Curves similar to those shown in Fig. 20.3 at a different ORP level.[2]

As such, these alloys are more prone to experience general corrosion. However, such alloys may also experience localized corrosion due to nonuniform conditions that are produced on the surface. Nonuniform conditions may be produced by the formation of partially protective corrosion product and deposit layers. This may occur due to concentration cell mechanisms that are produced in localized areas (Chap. 11) or due to certain types of microbiologically influenced corrosion, as described in Chap. 13.

Nonuniform conditions may also be produced by the formation of partially protective corrosion inhibitor layers. For instance, when anodic corrosion inhibitors are used, it is vitally important to ensure that sufficient concentrations are maintained for complete protection of the metal surface. Anodic corrosion inhibitors either react with the oxide layer or form a precipitate that produces a more protective layer. The protective layer interferes with the anodic reaction of the electrochemical corrosion cell. This reduces the rate of dissolution of metal at the anode into water solution and controls the diffusion of aggressive anions to the metal surface. In order to maintain the integrity of the protective layer, anodic inhibitors must be maintained at sufficient concentration or localized corrosion will tend to occur. In addition, the type of anodic corrosion inhibitor, whether oxidizing or nonoxidizing, is also important. Oxidizing anodic inhibitors such as nitrite do not require dissolved oxygen to form more protective iron oxide layers (referred to as gamma iron oxide). Non-oxidizing anodic inhibitors such as molybdate and orthophosphate require a minimum concentration of dissolved oxygen in order to react with ferrous ions to form insoluble, impermeable precipitates at anodic sites.

Unlike with anodic inhibitors, localized corrosion or pitting does not tend to occur when cathodic corrosion inhibitors are used at insufficient concentrations. This is primarily due to the fact that such inhibitors are designed to form protective layers at larger cathodic sites.

Copper Alloys

Many cases of pitting of copper and copper alloys have been reported, although these alloys are not particularly susceptible to localized or pitting corrosion. Numerous theories have been proposed based on water chemistry conditions that are expected to promote pitting corrosion on copper alloys. However, these theories do not consistently predict whether pitting will occur. In many cases the water chemistries said to cause pitting are likely to result in the formation of concentration cells on the surface. Localized corrosion and pitting is mostly associated with concentration cell mechanisms that occur beneath occlusive deposit and corrosion product layers and possibly within crevices.

Similar to stainless steels, copper alloys are susceptible to localized corrosion at breaks in more noble manganese oxide deposit layers. Pitting may also occur on copper alloys when localized aggressive conditions are produced, such as when residual tube-forming lubricants remain on surfaces and break down to form corrosive organic acids; see the description of ant nest corrosion in Chap. 14. In addition, residual solder flux used to prepare metal surfaces for joining has been shown to cause localized corrosion, as the components in the flux are typically aggressive to copper.

Aluminum Alloys

Pitting of the more corrosion-resistant aluminum alloys, such as 1100 and 6061, is primarily caused by local destruction of protective aluminum oxide layers. This is typically due to the formation of locally acidic conditions. Perhaps the most common corrodent responsible for pitting of aluminum surfaces is chloride in water solutions. Chlorides will tend to concentrate to corrosive levels due to ion concentration cell mechanisms that are produced beneath occlusive deposit layers or between crevices. Certain types of microbiologically influenced corrosion may also cause localized corrosion or pitting on aluminum alloys, as described in Chap. 13. Localized corrosion or pitting on aluminum components may also be initiated by galvanic corrosion caused by the deposition of more noble metals onto surfaces. For instance, dissolved copper ions as low as 50 ppb in cooling water solution may plate onto aluminum surfaces to promote corrosion in localized areas. As discussed in Chap. 5, "Aluminum," the composition of aluminum alloys that contain high concentrations of certain elements such as copper or zinc makes them susceptible to localized corrosion, such as intergranular corrosion or exfoliation corrosion.

Identification

Failures due to localized corrosion or pitting often take the form of thin-edged perforations (holes). The perforations may be quite narrow, particularly in thin-walled copper alloy and stainless steel tubing.

Spots or mounds of corrosion products may be indicative of localized corrosion beneath. When corrosion product and deposit layers cover surfaces, it may be necessary to remove the material in order to confirm that localized corrosion or pitting occurred. Crevices may need to be opened to check for metal loss within the tight spaces. In copper and copper alloys, corrosion products may exhibit various colors such as blue, green, red, and/or black. In stainless steels, pits are often bare-surfaced or only lined with a thin layer of corrosion products. Chemical analysis of corrosion products may provide valuable information regarding the most likely cause of the metal loss. For example, black, sulfur-containing corrosion products, which line hemispherical pits in stainless steel, are suggestive of microbiologically influenced corrosion by sulfate-reducing bacteria. The presence of chlorides in pits indicates that concentrated chloride ions caused pitting. Unfortunately, since chloride is very water soluble, such evidence may be removed subsequent to the corrosion event, or may be rinsed away prior to section removal.

Visual or stereomicroscope examination of surfaces will clearly reveal metal loss that is deeper within localized areas than in the surrounding surfaces. The depth of localized metal loss may be measured using point micrometers, calipers, or pit depth gages.

Microscopes that are capable of measuring differences in distance between focal planes on the surface are very useful for accurate measurements along the Z axis. Such measurements are accurate only to the degree that the view provided to the microscope or the point of the micrometer will be able to reach the base of the pit or depression. Nondestructive testing such as eddy current testing and ultrasonic testing may indicate the presence of localized corrosion or pitting. However, the depth of metal loss may not be accurately measured due to interference produced by deposit layers or due to other sources of measurement error. Perhaps the most accurate measurements will be provided by examination of metallographic cross-sections through the wall of the component at appropriate locations. Such examination will capture subsurface metal loss that may penetrate deeper than what could be accessed from the surface. Unfortunately, metallographic examination requires the portion of the component to be destroyed.

Monitoring for Localized Corrosion

Analysis of corrosion coupons should include an assessment that indicates whether metal loss on the surface is general or localized. If the metal loss is due to localized corrosion, then pit depths should be measured. Microscopes with special objectives can measure pits as shallow as 0.001 in. (0.025 mm) with good accuracy.[3] When the type of corrosion on coupons is characterized as either localized or pitting, it is not appropriate to calculate a corrosion rate based upon the pit depth divided by the exposure time. Pit depth measurements should not be used to predict the service life of a component due to statistical uncertainty and because the initiation period and subsequent growth rates may vary significantly. In addition, inside a coupon rack the coupons do not experience heat transfer as in a heat exchanger, which may produce different conditions that would influence corrosion. Periodic tube sampling will provide actual, direct measurements of wall thickness loss. However, characterization of localized corrosion and pit depth measurement data from coupons is quite useful for trending purposes.

Corrosion that occurs at the location of the attachment hardware is not considered when distinguishing whether the metal loss on a coupon is general or localized. This is because the type of exposure within the crevice at the hardware attachment does not simulate actual service conditions for the system. Unfortunately, there is some weight loss associated with that crevice corrosion that cannot be avoided, which will affect the calculated general corrosion rate. However, metal loss that is localized to the attachment hardware, or is significantly deeper at that location, should be noted, as it can provide an alert that system components may be susceptible to crevice and underdeposit corrosion.

As with any coupon result, the measured corrosion rate is not the actual corrosion rate for the system. Coupon corrosion rates and the type of corrosion should only be used for trending purposes. For instance, when coupon testing indicates low general corrosion rates over an extended time period, and suddenly coupons start exhibiting localized corrosion, it may indicate that a change in conditions has occurred to promote localized corrosion or pitting. This should prompt an investigation to review system operation and determine the cause, such as a system upset or a change in water chemistry. Once the cause of localized corrosion or pitting is determined, appropriate adjustments may be made to return to a normal condition. Similar actions should be pursued if general corrosion rates increase over time. There may be some outlying results that may be difficult to explain once in a while, but for the most part coupon data tends to be useful for monitoring and control. It would also be useful to combine information from the coupon measurements with data from online corrosion measuring techniques that use corrosion probes, as well as water analyses, and other monitoring data to determine if there is a problem. See Chap. 10, "Corrosion Monitoring," for more information.

Several different electrochemical methods may be used to indicate if conditions will promote localized corrosion or pitting during a short-term test. However, no devices have been produced that can accurately and reliably measure localized corrosion rates.

Control of Localized Corrosion and Pitting

Localized corrosion and pitting may be controlled by altering the conditions that cause protective layers to break down in isolated locations on surfaces. This may be achieved by controlling deposition that can cause localized underdeposit corrosion. Localized corrosion or pitting that is produced by certain types of microbiologically influenced corrosion may be controlled through the use of proper biocontrol programs, as described in Chap. 13. In particular, deposits that are more noble than the metal, such as manganese oxide, must be substantially controlled or prevented. Crevices should also be properly sealed if possible, and low or stagnant flow conditions should be minimized or eliminated. Anodic corrosion inhibitors used for carbon and low-alloy steels and low-chromium cast irons should be used at sufficient concentrations. In the absence of deposits and crevices, stainless steels used in cooling water systems are very resistant to corrosion. However, stainless steel alloys are susceptible to pitting when aggressive anions in water solution, particularly chlorides, exceed certain concentration thresholds. Threshold chloride concentrations are strongly influenced by several factors, which include temperature, pH, and ORP, as discussed in Chap. 3 and Figs. 20.3 and 20.4.

Pitting that is associated with weld features such as heat tint oxide may be prevented by removal of the oxide through appropriate mechanical or chemical cleaning. Eliminating corrosion due to microgalvanic effects in stainless steel welds may be accomplished through proper postweld heat treatment, though that is often not possible with field-welded components.

In addition, pitting of stainless steels may be controlled by substituting susceptible alloys with more pitting-resistant alloys. For instance, 316 stainless steel, which contains a small amount of molybdenum (typically around 2 wt%), is more resistant to pitting than 304 stainless steel, which contains no appreciable concentration of molybdenum. A more in-depth discussion concerning the effects of alloying of stainless steels on pitting is presented in Chap. 3.

Case History 20.1

Industry:	Electronic component manufacturing
Specimen location:	Heat exchanger
Orientation of specimen:	Horizontal
Environment:	Internal: Once through river water External: Deionized rinse water
Time in service:	15 years
Sample specifications:	¾ in. (1.9 cm) outer diameter 304 stainless steel seam welded tube

The received section of stainless steel tubing contained multiple perforations caused by pitting that initiated on the internal surface and penetrated the tube wall. The pitting on the internal surface was localized along the seam weld of the tube (Fig. 20.5). The pitting was caused by chloride-related corrosion. High chromium concentrations in stainless steel alloys typically result in extremely thin, protective oxide layers on the alloy surface that provide passivity and corrosion resistance. Chlorides and other halides can break down the protective

FIGURE 20.5 Pitting on the internal surface of stainless steel heat exchanger tubing that was localized along the seam weld.

oxide layer and initiate pitting. Spot tests indicated the presence of chlorides at some pit sites.

Once the pits initiated, the pit environments became more corrosive over time. As corrosion proceeded, positively charged metal ions were put into solution. The positively charged ions attracted negatively charged chloride and sulfate anions, concentrating them within the pits. Hydrolysis of these anions can depress the pH locally, thus accelerating corrosion. Characteristics of the pits, such as the spherical shapes with appreciable undercutting and surfaces free of corrosion products, are typical of attack due to low pH conditions related to chloride pitting.

Chloride-related pitting of stainless steel is dependent on factors such as chloride concentration, temperatures, and pH. Elevated chloride concentrations, elevated temperatures, and lower pH levels will promote pitting. A river water analysis indicated chloride levels as high as about 600 ppm. Such elevated chloride levels at the reported operating temperatures up to 110°F (43°C) combined to produce an environment capable of initiating pitting of 304 stainless steel (see Fig. 3.13).

The pitting on the internal surface was localized along the seam weld of the tube. Welds are often more susceptible to pitting attack due to compositional differences, higher residual stresses, the presence of nonprotective oxide layers, or crevicelike defects such as incomplete fusion or surface folds. Metallographic analysis of the weld did not suggest significant variations in compositions of weld and tube metal or sensitization of the microstructure (Fig. 20.6). Therefore, stresses or surface defects at the weld were probable causes of pit initiation.

Chloride pitting can be controlled by substitution of a more resistant alloy. Austenitic stainless steels, with increased amounts of molybdenum, such as type 316 or 904L, are generally more resistant to chloride pitting at the typical operating temperatures. The use of seamless tubing would also likely be beneficial, as it would eliminate preferential sites where corrosion can occur.

FIGURE 20.6 Microstructure of the weld where pitting occurred. No evidence of significant variations in compositions of weld and tube metal or sensitization is present.

Case History 20.2

Industry:	Utility
Specimen location:	Surface condenser
Orientation of specimen:	Horizontal
Environment:	Tubeside: Open recirculating cooling water
Time in service:	18 months
Sample specifications:	1 in. (2.5 cm) outer diameter inhibited aluminum brass tubing

It was reported that the chloride concentration in the cooling water varied, and could reach as high as 2000 ppm for a period of several weeks. The condenser reportedly experienced multiple extended shutdown periods in which flow was low or stagnant. When the 18-month-old condenser was opened for inspection the first time after installation, small pits and depressions were observed on the fairly clean internal surfaces of the tubes (Fig. 20.7). Some of the pits were bare-surfaced, while others were lined with crystalline reddish and light-colored corrosion products. Scanning electron microscope (SEM) analysis of material within the pits and depressions indicated a slightly elevated concentration of chlorine.

Nondestructive testing reportedly indicated that the internal surface features penetrated through as much as 90 percent of the tube walls, although laboratory measurements of pits on the received tube section indicated a maximum penetration of 20 percent of the tube wall.

The system was treated with an azole corrosion inhibitor, which relies on a continuous supply of the chemical to the metal surface in order to maintain the corrosion-inhibiting film. The undesirably long periods of low-flow or stagnant conditions most likely resulted in an inadequate supply of the corrosion inhibitor, which allowed localized corrosion to occur. Chlorides likely contributed to the corrosion, though the excessive periods of low and no flow were the primary cause of the metal loss. The very small amount of deposit material on the internal surface suggested that deposits did not contribute significantly to the corrosion.

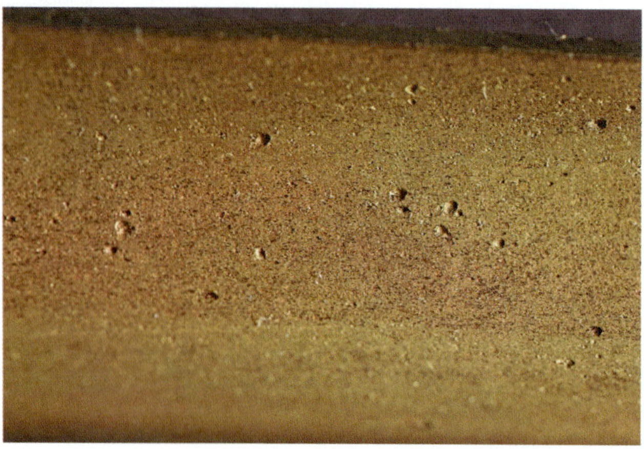

FIGURE 20.7 Small pits and depressions on the fairly clean internal surface of an aluminum brass condenser tube that was in service for 18 months.

References

1. Mars G. Fontana and Norbert D. Greene, *Corrosion Engineering*, 2d ed., McGraw-Hill, New York, 1978, p. 49.
2. Pahkala, A., Suutala, N., Karjalainen, P., "Effect of environmental parameters of the pitting resistance of stainless steels," Applications of Stainless Steel Conference Proceedings, Norberg, H., Bjorklund, J., eds., Stockholm, June 9–11, 1992, v1, pp. 468–476.
3. In use at the coupon laboratory of Nalco, an Ecolab Company.

SECTION 3

Cracking

CHAPTER 21

Corrosion Fatigue

General Description

Corrosion fatigue, first mentioned in the literature in 1917, refers to the generation of cracks resulting from the combined effects of cyclic tensile stresses and corrosion (Fig. 21.1). The role of the environment sets corrosion fatigue apart from pure mechanical fatigue. In corrosion fatigue, damage produced by the combined effects of the cyclic tensile stresses and corrosion is significantly greater than the total damage produced by cyclic stresses and corrosion separately. In purely mechanical fatigue, chemical reaction of the material with the environment does not play a significant role in the cracking.

Corrosion fatigue cracks typically develop according to the following sequence:

- During the first cycle of sufficient tensile stress, brittle corrosion products/oxides that have formed on the metal surface fracture, opening a microscopic crack in the brittle material that exposes the underlying metal surface.
- The exposed metal at the tip of the crack reacts with the environment, generally by oxidation, forming a microscopic notch in the metal surface.
- During the ensuing cycles of sufficient tensile stress, the corrosion products or oxides tend to preferentially fracture at the notch, causing the crack to deepen.

As this cyclic process continues, a wedge-shaped crack propagates through the metal.

No metal commonly used in cooling water systems is immune to corrosion fatigue. Some reduction of the metal's resistance to cracking under applied cyclic stresses is observed if the metal is corroded, even mildly, by the environment in which the stressing occurs. Corrosion fatigue produces fine to broad fissures with little or no branching. They may be filled with dense corrosion products. The fissures may occur singly, but they commonly appear as families of parallel

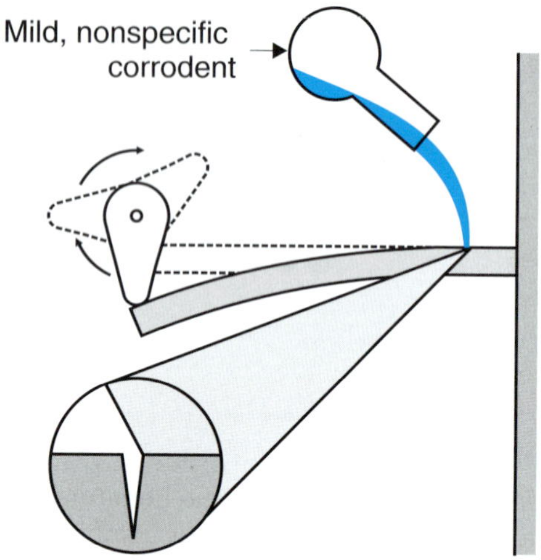

FIGURE 21.1 Schematic representation of a corrosion fatigue crack propagating under the influence of a cyclic bending stress and exposure to a mild, nonspecific corrodent.

FIGURE 21.2 A family of short, transverse corrosion fatigue cracks originating on the external surface of a tube.

fissures or cracks (Fig. 21.2). They are frequently associated with pits, grooves, or other penetrations into the metal surface that provide locations of stress concentration that promote crack initiation. Like other forms of environmentally induced cracking, corrosion fatigue produces thick-walled fractures, even in ductile metals (Fig. 21.3).

Locations

Common sites for corrosion fatigue are locations where components are constrained, such as tubesheets, where cyclic stresses may be intensified (see Case History 21.1). Corrosion fatigue cracking at mid-span regions due to flow-induced vibration in heat exchanger tubes has been reported. Corrosion fatigue crack initiation at pit sites is

FIGURE 21.3 Blunt fracture edges typical of corrosion fatigue cracks.

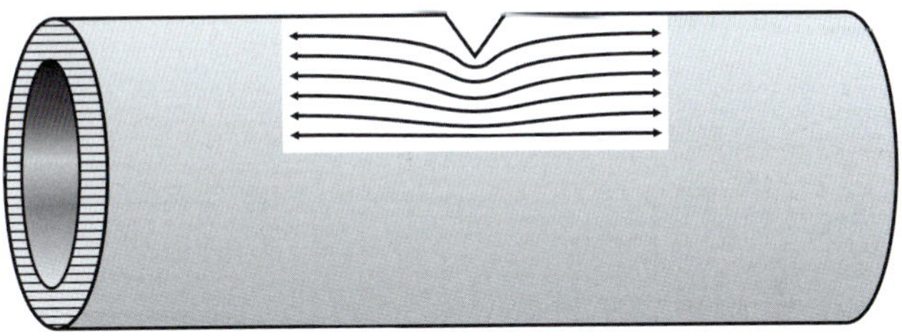

FIGURE 21.4 Schematic representation of localized increase in stress at a surface discontinuity.

common, as applied cyclic tensile stresses are concentrated at the bases of the pits. Anywhere that surface irregularities or discontinuities exist can also be preferential sites for corrosion fatigue cracking due to the stress concentration caused by such features (Fig. 21.4). However, corrosion fatigue cracks do not require surface irregularities or discontinuities to initiate. Locations of elevated residual tensile stresses, such as near welds or at bends, can also be preferred sites for corrosion fatigue cracking.

Critical Factors

Numerous factors can have a potentially significant effect on corrosion fatigue cracking. Most of these relate to stress and the corrosiveness of the environment. The effect of these factors varies with the metals and environments involved.

Stress

One of the two most important stress factors affecting corrosion fatigue is the magnitude of the tensile stress. The overall stress level is generally much lower than the yield strength of the metal. However, in general, higher tensile stresses increase crack growth rate and the number of cracks initiated.

In addition, the greater the amplitude of the tensile stress, the fewer the cycles necessary to produce a failure. In some cases, a fatigue limit exists such that a metal can endure an infinite number of stress cycles without failure as long as the amplitude of the stress is below a certain level (Fig. 21.5). Carbon steel exhibits a fatigue limit in air, whereas nonferrous metals generally do not. However, it is important to note that a fatigue limit is generally characteristic of pure fatigue and does not typically exist for corrosion fatigue. Corrosion reduces the magnitude of stress, time, and/or number of stress cycles needed to cause a cracking failure.

Elevated static residual tensile stresses, such as from welding or forming operations like bending, are additive to cyclic tensile

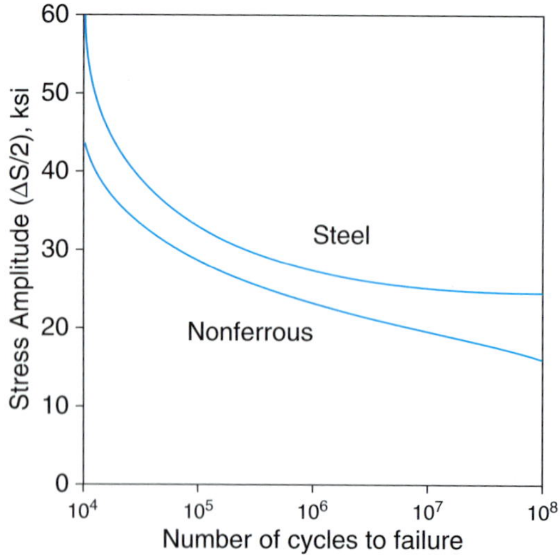

FIGURE 21.5 Schematic plot of stress amplitude versus number of cycles to failure, showing that the curve for carbon steel asymptotically approaches a fatigue limit, or a stress below which fatigue failure does not occur. Most other materials do not exhibit a fatigue limit. (*Reprinted with permission of ASM International. All rights reserved. www.asminternational.org.*)

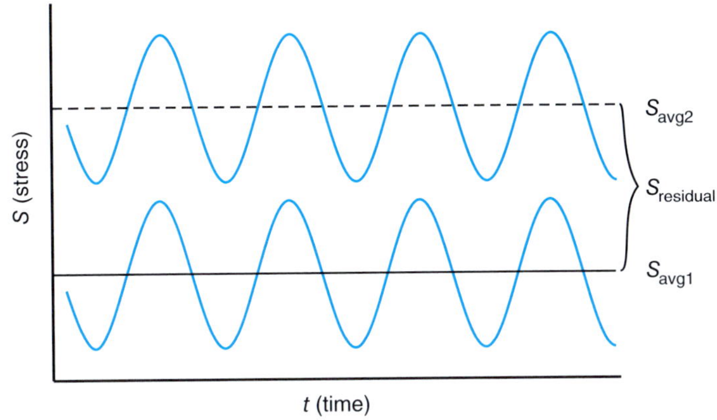

FIGURE 21.6 Schematic plot showing the additive nature of static residual tensile stress. The static residual tensile stress ($S_{residual}$) raises the original average tensile stress (S_{avg1}) to S_{avg2}.

stresses (Fig. 21.6). Such static tensile stresses can cause corrosion fatigue to occur at lower cyclic stress amplitudes than it otherwise would. Static tensile stress also increases the severity of corrosion fatigue cracking, thus decreasing the number of cycles necessary to cause failure.

The other main stress factor affecting corrosion fatigue is the frequency of the cyclic stress. Because (a) corrosion is an essential component of the failure mechanism and (b) corrosion processes typically require time for the interaction between the metal and the environment, the corrosion fatigue life of a metal depends on the frequency of the cyclic stress. Relatively low stress frequencies permit adequate time for corrosion to occur; however, high stress frequencies may not allow sufficient time for the corrosion processes necessary for corrosion fatigue. As stress frequency increases, the failure mode gradually shifts from corrosion fatigue to pure fatigue at very high frequencies.

Corrosion

The environment does not need to be especially corrosive for corrosion fatigue to occur. Corrosion fatigue can occur under relatively mild conditions as long as the cyclic tensile stresses are sufficiently high. The rate of crack growth generally increases as the corrosiveness of the environment increases. Factors that influence the corrosivity of the environment vary with the metal under consideration.

It is common to find corrosion fatigue cracks initiating from sites where corrosion occurred, as discussed previously. Locally elevated stresses are sufficiently high to rupture protective corrosion product layers. In uncorroded areas, the stress level is lower and may be insufficient to cause damage to the corrosion product layer. It is also quite possible that the local environment at a pit site is more corrosive due to the concentration of aggressive chemical species there and, therefore, more prone to produce corrosion fatigue in the presence of cyclic stress.

Identification

Corrosion fatigue typically produces relatively straight, unbranched cracks and fissures (Fig. 21.7). It is common for multiple parallel cracks to form. The cracks and fissures are often tight, especially if the cyclic tensile stresses responsible for the cracking are of high frequency. The tightness of the cracks may make them difficult to detect visually. Metallographic analysis reveals that the cracks propagate transgranularly; that is, they travel through grains, rather than along grain boundaries. They are commonly found at the bottom of localized corrosion sites such as pits and depressions, although they can form in the absence of pits. The cracks are oriented perpendicularly to the direction of the cyclic tensile stress responsible for the cracking. The fractures resulting from corrosion fatigue have a thick-walled, brittle appearance, even in ductile metals. Corrosion products commonly cover the fracture faces but may not be present on the surface of the component. In some cases, fissures are filled or lined with corrosion products that clearly contain a centerline crack (Fig. 21.8).

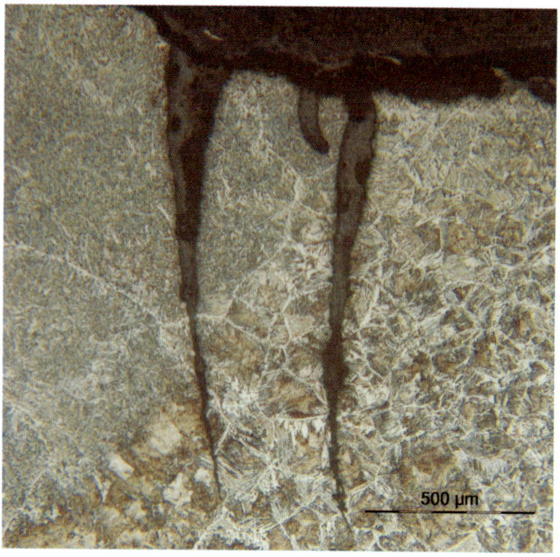

FIGURE 21.7 Metallographic cross section, showing generally straight, transgranular, wedge-shaped corrosion fatigue fissures at a weld joining two segments of carbon steel piping with different outer diameters. Etchant: Nital.

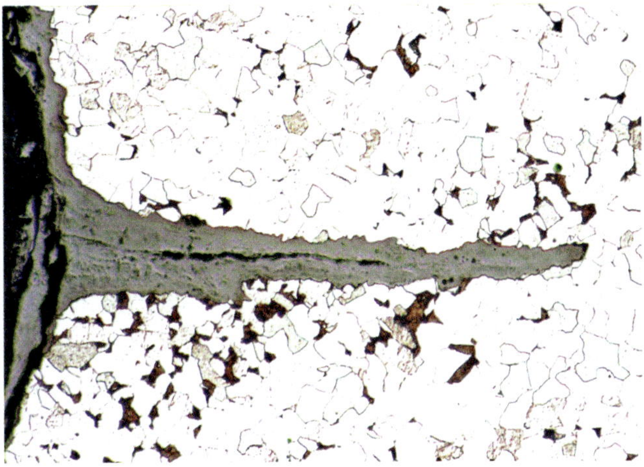

FIGURE 21.8 Metallographic cross section showing a wedge-shaped fissure filled with oxide corrosion products containing a centerline crack. Etchant: Nital.

Corrosion fatigue cracks can be detected by nondestructive testing techniques such as magnetic particle inspection, radiography, ultrasonic testing, and dye penetrants. However, positive identification requires metallographic examination. In shell-and-tube heat exchangers, corrosion fatigue may occur in many tubes simultaneously if they are subjected to a similar environment and cyclic tensile stress magnitude. Nondestructive testing of tubes at locations similar to those in which cracks are observed can be useful to identify other cracked tubes.

Elimination

Mitigation or elimination of corrosion fatigue cracks involves gaining control of the critical factors that govern the mechanism.

Stress

Reducing or eliminating cyclic stresses will minimize or prevent corrosion fatigue. Reduction or elimination of cyclic stresses frequently requires equipment design modifications. This requires attention to external and internal sources of cyclic stress. External sources include vibrations from machinery and water hammer. Internal sources of cyclic stress may include thermal expansion and contraction due to substantial and rapid temperature variations and flow-induced vibration. In heat exchangers, cracking from flow-induced vibration is often restricted to peripheral rows of tubes. Reduction of flow-induced vibration requires correct spacing of tube supports and/or a change in the placement and design of baffles. A change of tube metallurgy without an appropriate alteration of tube supports and/or baffling may cause flow-induced vibration problems where none existed previously. Complete failure of tube supports due to corrosion or other issues can result in vibrations that would not occur under normal operation.

Corrosion

Techniques for controlling the corrosion factor can be categorized into three basic approaches, as outlined in the following list.

1. *Alter the environment to render it less corrosive.* This approach may be as simple as maintaining clean, smooth metal surfaces without corrosion product layers or surface irregularities. It is well known that the chemistry of the environment beneath deposits and corrosion products can become substantially different than that of the bulk environment. This difference can lead to localized underdeposit corrosion [see Chap. 11, "Crevice and Underdeposit Corrosion (Concentration Cells)"]. The pit and depression sites may then act as preferential locations of corrosion fatigue when cyclic stresses are applied. The specific steps taken to reduce corrosivity vary with the metal under consideration. In general, appropriate adjustments to the water chemistry and chemical treatment program should be evaluated. Appropriate use of corrosion inhibitors may reduce or eliminate localized corrosion, which can produce initiation sites for corrosion fatigue cracks. The effectiveness of inhibitors depends upon their application to clean metal surfaces.
2. *Choose a more corrosion-resistant metallurgy.* Since a metal's susceptibility to corrosion fatigue depends in part on its corrosion resistance in a particular environment, improving corrosion resistance also improves resistance to corrosion fatigue. However, care must be taken that change to a different metal does not result in trading corrosion fatigue cracking for some other form of environmentally sensitive cracking, such as stress corrosion cracking.
3. *Separate the metal from the environment with a physical barrier.* Galvanized coating layers, such as zinc or zinc and aluminum, prevent corrosion of steel by providing a physical barrier as well as by acting as a sacrificial anodic material in the event of a break in the coating. Inert coating materials can be applied to physically protect metal from the environment. Note that simply reducing corrosion may not be sufficient to eliminate corrosion fatigue cracking. Reducing or eliminating cyclic tensile stresses is the most effective way to eliminate corrosion fatigue cracking.

Cautions

Corrosion fatigue cracks can be difficult to see with the naked eye because they are often tight. Nondestructive testing techniques should be used in suspect areas. Evidence of gross corrosion does not necessarily accompany corrosion fatigue cracking. Deposits and/or corrosion products may cover crack sites, rendering them undetectable until the material is removed. Simple visual examinations of crack appearance may be insufficient to distinguish between the various types of failure mechanisms that produce brittle, thick-walled fractures, such as corrosion fatigue and stress corrosion cracking. In most cases a formal metallographic examination is required.

Related Problems

See Chap. 22, "Stress Corrosion Cracking."

Case History 21.1

Industry:	Utility
Specimen location:	Surface condenser tube, inlet end
Orientation of specimen:	Horizontal
Environment:	Internal: Circulating water pH 7.8–8.2 External: Steam and condensate treated with an oxygen scavenger and ammonia, pH 7.8–9.2, temperature ambient to 150°F (66°C)
Time in service:	25 years
Sample specifications:	1 in. (25 mm) outer diameter, 30 ft (9 m) long, inhibited admiralty brass

Figure 21.9 shows the appearance of numerous failures of tubing that occurred near the water box inlet end. Cracking of this type was a recurring problem in this condenser. Approximately 9 percent of the tubes in the condenser were plugged.

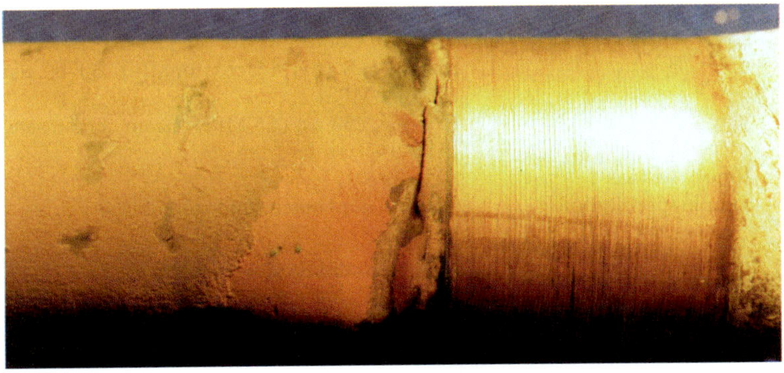

Figure 21.9 Circumferential corrosion fatigue crack adjacent to location of tubesheet.

FIGURE 21.10 Straight, unbranched crack on the surface (magnification 33×).

The condenser was in cyclic service, although the failures occurred while the boiler itself was out of service.

The failures occurred due to circumferential cracks adjacent to the rolled section of tube where it contacted the tubesheet. (Slight distortion of the tube occurred during its removal from the tubesheet.) Figure 21.10 shows the straight, unbranched character of the crack.

Microstructural examinations of tube wall cross sections taken from the cracked region revealed straight, unbranched cracks running directly across the metal grains (transgranular), typical of corrosion fatigue cracks. The cracks originated on the external surface, although evidence of very shallow corrosion fatigue fissures was also observed on the internal surface. Corrosion was superficial.

The cyclic stresses responsible for this failure were apparently bending stresses associated with thermal expansion and contraction due to the cyclic nature of operation.

Case History 21.2

Industry:	Refinery
Specimen location:	Surface condenser: Bed air blower cat cracker regeneration system
Orientation of specimen:	Horizontal
Environment:	Internal: Cooling water 90–120° F (32–49°C), pH 6.6–6.8 (pH 4 occasionally), 30–60 psi (0.2–0.4 MPa) External: 650 psi (4.5 MPa), steam condensate, pH 8–9
Time in service:	10 years
Sample specifications:	¾ in. (19 mm) outer diameter, admiralty brass

Transverse cracks shown in Fig. 21.11 were found in a total of 15 tubes from random areas of the condenser. The equipment was in continuous service for two years. Cracks were not observed previously.

Visual examinations revealed that the cracks originated on both the internal and external surfaces, although there were more on the internal surface. Thin coatings of deposits partially covered both the internal and external surfaces.

FIGURE 21.11 Transverse cracks on the internal surface.

FIGURE 21.12 Short, transgranular fissures that originate from depressions on the external surface. Magnification, 100×. Etchant: NH_4OH-H_2O_2.

Microscopic examinations revealed that the cracks were unbranched and transgranular, typical of corrosion fatigue cracks. The examinations also revealed that the cracks tended to initiate at sites of intergranular corrosion or at depression sites, both of which acted as stress concentrators (Fig. 21.12). Intergranular corrosion and pitting of brass are frequently associated with deposits covering the metal surface (see Chap. 18, "Intergranular Corrosion").

The orientation of the cracks revealed that they resulted from cyclic bending stresses or cyclic axial stresses. Depressions and intergranular corrosion intensified the applied stresses and significantly promoted corrosion fatigue cracking.

Case History 21.3

Industry:	Utility
Specimen location:	Service water drain line
Orientation of specimen:	Unknown
Environment:	Internal surface: River water treated with orthophosphate and scale/corrosion inhibitor External surface: Ambient conditions
Time in service:	Unknown
Sample specifications:	1 in. (2.5 cm) outer diameter (OD) piping welded to 1½ in. (3.8 cm) OD piping; carbon steel

FIGURE 21.13 Tight, straight crack that extends around the majority of the smaller-diameter piping. The crack occurred near a weld between two piping segments of different diameters.

The section contained a tight transverse crack around the majority of the circumference of the smaller-diameter piping near the welded joint (Fig. 21.13). Metallographic examination indicated that the through-wall crack occurred just outside the heat-affected zone (HAZ) (Fig. 21.14). Other fissures observed within the HAZ appeared to originate from both the external and internal surfaces

FIGURE 21.14 Cross section, showing the through-wall crack just outside the heat-affected zone (arrow). Shallow corrosion fatigue fissures also originated from the external surface within the fusion zone (circled). Etchant: Nital.

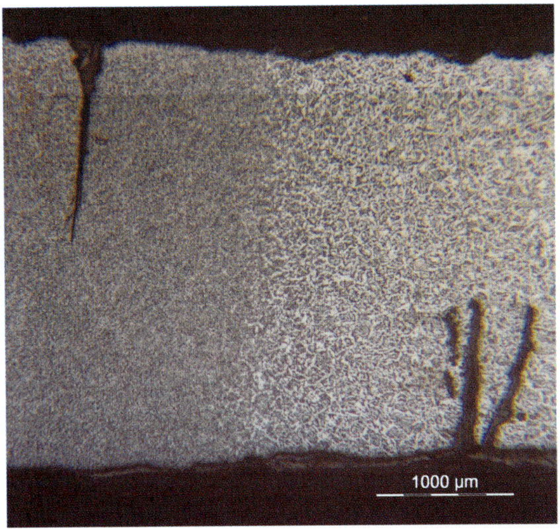

FIGURE 21.15 A location close to the through-wall crack, showing fissures originating from both the external and internal surfaces. Etchant: Nital.

(Fig. 21.15). Fissures apparently originating from the external surface were also observed within the fusion zone of the weld (Fig. 21.7). The through-wall crack and fissures penetrated transgranularly. The crack and fissures were mainly straight, except for some fine, minor branching in places. Tubercles on the internal surface were caused by oxygen corrosion.

The cracks and fissures were caused by corrosion fatigue cracking. Intermittent operation and/or flow-related vibrations probably caused the cyclic tensile stresses. Elevated residual tensile stresses due to welding increased the overall stress level and promoted corrosion fatigue cracking. The crack that ultimately caused the failure initiated on the waterside. Oxygen corrosion on the internal surface contributed to corrosion fatigue fissures on that surface, though it was clear that the cyclic tensile stresses were much more instrumental in causing the failure than waterside corrosion.

Case History 21.4

Industry:	Food
Specimen location:	Lubricating oil heat exchanger
Orientation of specimen:	Unknown
Environment:	Internal surface: Lubricating oil External surface: Cooling water
Time in service:	Unknown
Sample specifications:	¾ in. (1.9 cm) outer diameter (OD) piping U-bend; admiralty brass

A deep, transverse, wedge-shaped fissure penetrated from the external surface at a baffle contact point (Fig. 21.16). Green corrosion products were also observed surrounding this location. The fissure penetrated transgranularly, characteristic of corrosion fatigue (Fig. 21.17). No fissures were observed in

FIGURE 21.16 External surface of a brass heat exchanger tube, showing a fissure along a baffle contact point. Green corrosion products are also observed in the surrounding area.

FIGURE 21.17 Deep fissure on the external surface. Shallow intergranular corrosion was observed in all examined locations, but was significantly deeper at baffle contact points. Magnification: 25×. Etchant: Potassium dichromate.

any other examined locations. Baffles are locations of rigid restraint, and are common preferred areas for corrosion fatigue cracking to occur. Possible causes of the cyclic tensile stresses responsible for the cracking included vibrations or thermal expansion and contraction due to temperature variations.

Intergranular corrosion was observed in all examined locations on the external surface, and was significantly deeper at the baffle contact points. The metal loss caused by intergranular corrosion likely promoted crack initiation, as it served as a location of stress concentration. Though corrosion contributed to the propagation of the fissure, the tightness of the fissure suggested that the cyclic tensile stresses were the primary contributor to the formation of the fissure, and that corrosion was secondary.

Case History 21.5

Industry:	Chemical
Specimen location:	Cooling tower basin, joint between wall and floor
Orientation of specimen:	Horizontal and vertical
Environment:	Internal surface: Cooling water, 70–80°F (21–27°C) External surface: Ambient
Time in service:	3 years
Sample specifications:	Welded plate, type 304 austenitic stainless steel

A brittle-type fracture occurred along a fillet weld joining the wall and floor of a cooling tower basin (Fig. 21.18). Figure 21.19 shows the bottom of the cooling tower and the weld repair that was made after the failure occurred. The approximate location of the received section is shown schematically. The fracture surface exhibited ratchet marks, which form at locations where multiple cracks propagating on different planes in the metal join together. In addition, radial features extending from a point on the internal surface suggested that cracking originated from that surface (Fig. 21.20). Metallographic examination showed that the crack occurred at the edge of the fusion zone of the fillet weld and propagated along a straight, unbranched path through the plate (Fig. 21.21).

The weld was a natural site of stress concentration. Residual tensile stresses due to welding probably also promoted the cracking. The cyclic tensile stresses

Figure 21.18 As-received section of welded type 304 stainless steel plate from the bottom of a cooling tower basin, showing the fracture along the fillet weld.

FIGURE 21.19 Approximate location of the received section from the cooling tower basin, showing the weld repair.

FIGURE 21.20 Fracture surface, showing radial features extending from the apparent crack origin, adjacent to the internal surface (arrow).

were probably caused by vibrations. No cracking or corrosion was observed away from the failure, indicating that this failure was primarily caused by the combination of cyclic and static tensile stresses. Since corrosion was not a major contributor to the failure in this case, it was probably closer to pure fatigue than corrosion fatigue. Improved support structures were installed to prevent a recurrence of cracking.

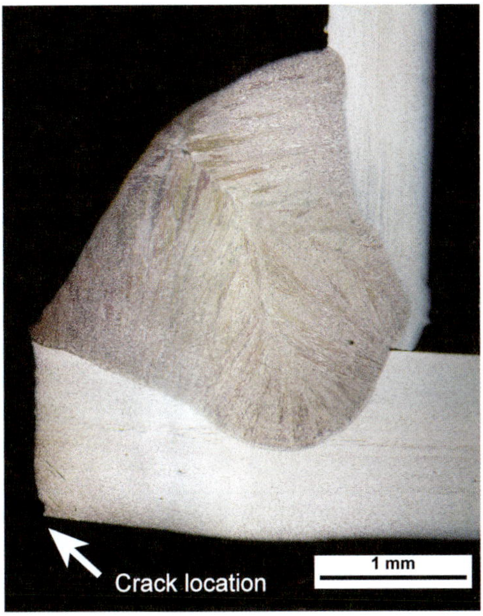

FIGURE 21.21 Micrograph, showing the crack location at the edge of the fillet weld. Etchant: Electrolytic oxalic acid.

Case History 21.6

Industry:	Metals
Specimen location:	Cooling system pan for extruded brass alloys
Orientation of specimen:	Horizontal
Environment:	Sprayed cooling water
Time in service:	7–8 months
Sample specifications:	Radial section from an annular 1 in. (2.5 cm) thick carbon steel plate with drainage holes

The section was removed from a steel pan that transferred extruded brass alloys to a conveyor at elevated temperatures, reportedly 1350–1400°F (732–760°C). The extruded material was air cooled to about 950°F (510°C) for about 15 minutes, then subjected to cooling water sprays. The pan typically experienced 3 to 4 runs each day, five times per week.

Many tight cracks and fissures extended between the drainage holes, as well as radially away from the edges of the holes (Fig. 21.22). Close examination of the brittle fracture surface revealed thumbnail-shaped cracks adjacent to the holes and shear lips along the plate edges (Fig. 21.23). Metallographic examination indicated that the fissures were wedge-shaped, transgranular, and filled with oxide-based corrosion products that were cracked along the centerline (Fig. 21.8).

Corrosion fatigue cracking was caused by thermal expansion and contraction due to the repeated contact with very hot material, followed by cooling with water. Although some very shallow corrosion fatigue fissures were located on the top surface of the pan, the cracks through the pan initiated at the center drain hole, where the cyclic tensile stresses concentrated. The stresses were

Figure 21.22 Plate section from a pan that transferred high-temperature brass alloys through cooling water sprays, showing tight cracks and fissures extending between and away from drainage holes.

Figure 21.23 Crack face, showing thumbnail-shaped cracks on either side of a drainage hole. Shear lips are visible toward the bottom of the photo.

concentrated around the drain holes in the pan, and were probably greatest between the two closely spaced holes. Design changes to reduce the stresses around the drain holes in the pan, including shape, size, and placement alterations, were suggested.

Case History 21.7

Industry:	Plastics
Specimen location:	Molds for plastic components
Orientation of specimen:	Unknown
Environment:	Internal surface: Cooling water External surface: Plastic component
Time in service:	Unknown
Sample specifications:	Long, narrow, tubular mold components; chromium-plated tool steel

Tight longitudinal cracks occurred in several plastics mold components, causing cooling water leaks that resulted in frequent shutdowns for repairs. In some locations, light mechanical pressure during metallographic sectioning caused pieces of the sections to break off (Fig. 21.24). Cross sections through the tubular mold components showed multiple wedge-shaped, longitudinal cracks and fissures that initiated on the internal surface (Fig. 21.25). The cooling water surfaces were

FIGURE 21.24 Longitudinal cracks joined together to cause breakage of pieces due to light mechanical force during sectioning. Iron oxide corrosion products are visible on the internal surface.

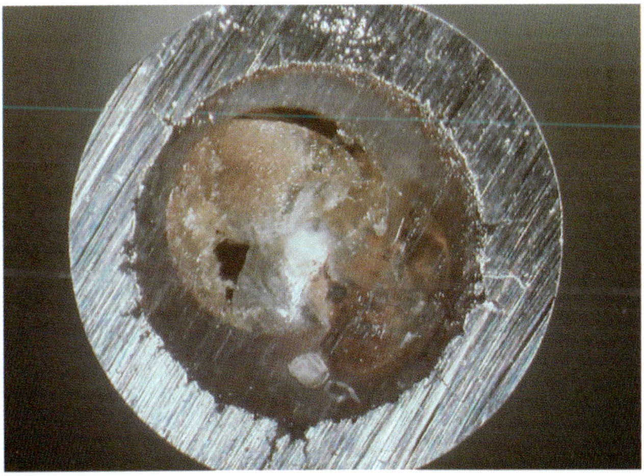

FIGURE 21.25 Cross section of a tubular component, showing multiple wedge-shaped, longitudinal fissures extending from the internal surface. This particular component contained plastic in the cooling water channel.

FIGURE 21.26 General metal loss and several wedge-shaped, longitudinal fissures penetrating from the internal surface. The fissures are lined with iron oxide–based corrosion products. Magnification: 25×. Unetched.

generally covered with brown and black iron oxide–based corrosion products, including tubercles. Metallographic examinations indicated that the fissures were lined with iron oxide–based corrosion products (Fig. 21.26). Though the components were reportedly chromium-plated throughout, a continuous plating layer was not observable visually on the cooling water side. However, microscopic particles of remnant chromium from the coating were observed in the corrosion products on the waterside (Fig. 21.27).

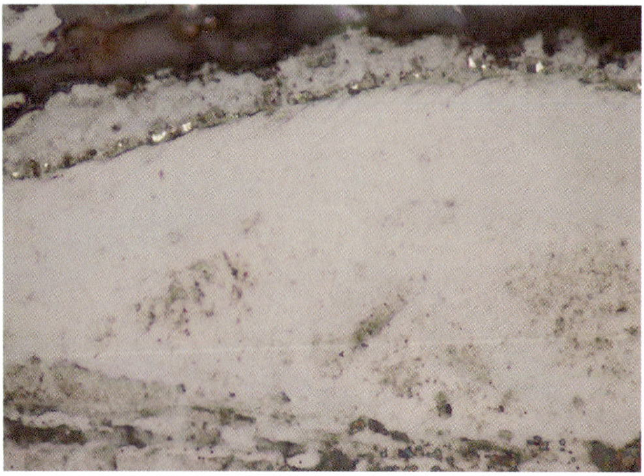

FIGURE 21.27 Dense iron oxide corrosion products on the internal surface, showing remnant particles of the chromium coating. Magnification: 200× Unetched.

Corrosion fatigue cracking was caused by the cyclic stresses supplied by thermal expansion and contraction of the metal during mold heating and cooling cycles. Due to the requirements of the molding process, the rapid temperature changes that caused the cracking would not be avoidable. It is possible that deposit and corrosion product layers contributed to cracking to some degree by insulating the surfaces and by restricting cooling water flow. Under such conditions, the metal temperature would become higher during heating cycles, causing it to thermally expand more significantly. The larger temperature change with the heating and cooling cycles would result in higher cyclic tensile stress levels. Oxygen corrosion contributed to cracking, but it was considered to be a secondary factor, as evidenced by several very tight fissures that formed. Some cracks may have initiated at the bases of pits formed due to corrosion, but others formed where smooth, general metal loss occurred. Therefore, the pits were not necessary for deep cracks to form.

Galvanic effects probably enhanced the severity of corrosion on the internal surface. Initial cracks or other defects in the chromium plating layer caused an unfavorable, high cathode-to-anode surface area ratio between the plating (cathode) and the exposed steel (anode) of the tubular mold. Due to shielding of the surface by the plating layer, corrosion inhibitors were not able to protect the underlying steel adequately. Because of all of these circumstances, corrosion may have been worse during the initial stages than if the surface were not covered with a chromium plating layer. If at all possible, components exposed to a potentially corrosive electrolyte (such as water) should not be plated with more noble metals.

CHAPTER 22

Stress Corrosion Cracking

General Description

Stress corrosion cracking (SCC) has been defined as cracking of a susceptible alloy under the combined action of corrosion and static tensile stress (Fig. 22.1). The stress and corrosion components interact synergistically to produce cracks, which initiate on the surface exposed to the corrodent and propagate in response to the stress state.

Possibly the earliest documented case of SCC is a well-known story. In India during the early 20th century, the British Army stored brass ammunition cartridges in horse barns during the rainy season, exposing the cartridges to ammonia vapors. The combination of the ammonia and high residual tensile stresses from cold forming of the cartridge cases resulted in what was then referred to as "season cracking." Storing the cartridges elsewhere alleviated the problem.

Stress corrosion cracks are always perpendicular to the principal tensile stress. Longitudinal or transverse crack orientations in tubes are common (Figs. 22.2 and 22.3). Occasionally, both longitudinal and transverse cracks are present on the same tube (Fig. 22.4). Varied orientations are often associated with complex stress states. Less frequently, SCC is a secondary result of another primary corrosion mode. In such cases, the cracking, rather than the primary corrosion, may be the ultimate cause of failure (Fig. 22.5). Not all corrosive environments will cause stress corrosion cracking.

Few, if any, failure mechanisms have received as much attention as SCC. Yet despite an enormous research effort over many years, an acceptable, generalized theory that satisfactorily explains all elements of the phenomenon has not been produced. The process of SCC is complex, and a review of the various theories regarding cracking mechanisms is beyond the scope of this book. However, the mechanisms generally involve some type of metal dissolution (such as dealloying, oxidation, etc.) or hydrogen evolution and resulting embrittlement at the crack tip. Considering the variety of combinations of alloy types and corrosive species that can result in SCC, it is likely that different mechanisms apply

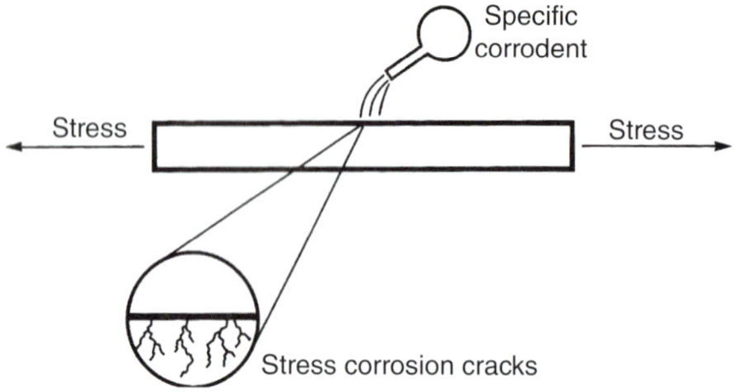

Figure 22.1 Stress corrosion cracks result from the synergistic interaction of tensile stress and specific corrodents.

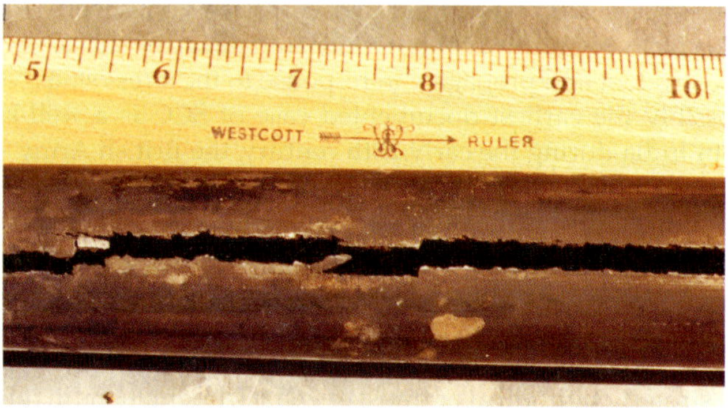

Figure 22.2 Longitudinal stress corrosion cracks in a heat exchanger tube; the broad gap between the crack faces reveals that high-level residual hoop (circumferential) stresses from the tube-forming operation provided the stress component required for SCC.

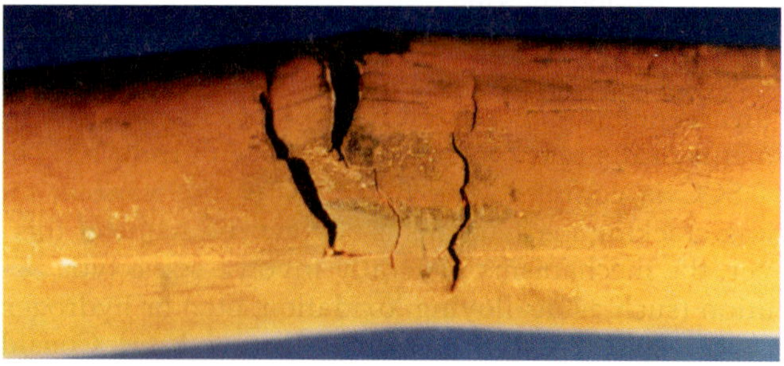

Figure 22.3 Transverse stress corrosion cracks in a condenser tube; the presence of the cracks along just one side indicates that a bending moment provided the stress.

FIGURE 22.4 Both longitudinal and transverse stress corrosion cracks on a brass heat exchanger tube that had been exposed to ammonia. Note the branching of the cracks.

FIGURE 22.5 Stress corrosion cracks associated with corrosion. The stress corrosion cracks were the ultimate cause of the failure.

in different situations. Nevertheless, the basic characteristics of SCC are well-known, and a wealth of practical experience permits at least a comfortable working knowledge of the phenomenon.

It is important to realize that the conditions causing SCC may occur not only during normal operation of equipment but also during startup, shutdown, idle periods, or system upsets. Stresses and environmental conditions under these circumstances can be quite different than those encountered during normal operation.

Locations

Due to the wide variety of environments to which cooling water components are exposed on the cooling water and process sides, SCC can occur in many different locations throughout the system. However, some factors that make a location particularly susceptible to SCC include:

- High residual stresses, such as at or near welds or rolled-in tube ends.
- High service stresses, such as points of physical constraint.

- The potential for concentrating corrosive agents due to the formation of concentration cells (such as within crevices) or evaporation [see Chap. 11, "Crevice and Underdeposit Corrosion (Concentration Cells)."]
- Condensation, whether during service or idle periods, that can dissolve gaseous corrodents and result in SCC. A thin film of condensed moisture on the metal surface may have a much higher corrodent concentration than any bulk fluid that has collected.

Critical Factors

There are two main requirements for SCC:

1. Sufficient static tensile stress
2. One or more corrodents

Sufficient Static Tensile Stress

The minimum threshold stress that is required to cause SCC is difficult to define since it depends on a number of factors such as alloy composition, concentration of corrodent, and temperature. For instance, when corrodent concentration is appreciably high, the stresses required to produce cracking can be significantly lower than when the corrodent concentration is low. However, for each combination of environment and alloy system, there appears to be a threshold stress below which SCC will not occur. Threshold stresses can vary from 10 to 70 percent of yield strength depending on the alloy, the environment, and the temperature.[1] In some cases, stresses near the yield strength of the metal are necessary.

Stresses may be applied and/or residual. Examples of applied stresses are those arising from thermal expansion and contraction, pressurization, and service loads. Applied stresses may be continuous or intermittent. Examples of residual stresses are those arising from welding, fabrication, and installation. The importance of residual stresses in SCC should not be underestimated. Residual stresses may approach the magnitude of the yield stress. Design stresses are typically well below the yield stress. In general, the propensity for and severity of cracking increases as stress levels rise. However, SCC can occur within the range of typical service stresses. Figure 22.6 illustrates how increasing tensile stresses decrease the amount of time necessary for SCC to occur.

Physical discontinuities in a metal surface can concentrate applied tensile stresses, effectively increasing the nominal stress at the discontinuity (see Fig. 21.5 in Chap. 21, "Corrosion Fatigue"). Stress-concentrating discontinuities can arise from three sources:

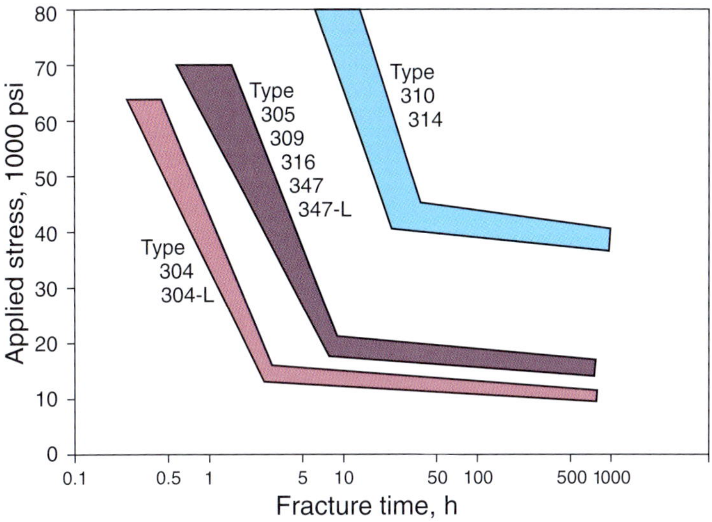

FIGURE 22.6 Threshold stresses of various austenitic stainless steel alloys in boiling 42 percent magnesium chloride solution.[2] *(Courtesy of McGraw-Hill.)*

1. Design: Notches, holes, keyways, welds, sharp corners, etc.
2. Fabrication and installation: Mechanical damage, arc strikes, weld-related cracking, rough grinding or machining, etc.
3. Localized deterioration: Corrosion (especially pitting and intergranular attack), erosion, cavitation, mechanical wear, etc.

Since susceptibility to SCC commonly increases as stress level increases, stress corrosion cracks tend to occur more readily at discontinuities, whereas smooth areas may not be affected. However, this is not a hard and fast rule. It is not uncommon for stress corrosion cracks to form on smooth surfaces.

One or More Corrodents

Essentially all industrial metals are susceptible to SCC if exposed to a number of corrodents. However, one of the unusual and interesting features of SCC is the relative specificity of the corrodents. Specific corrodents are known to cause SCC for certain alloy systems, provided that the static tensile stress level, temperature, and other necessary conditions are sufficient for SCC. It is important to note that not all chemical species that cause corrosion of an alloy can also cause SCC. In addition, environments that produce SCC are not necessarily corrosive to the metals in the unstressed state. General or localized metal loss is not necessary for stress corrosion cracking to occur, and often does not accompany it. Note that oxygen and at least a small

Alloy Type	Corrodent
Austenitic stainless steel	Chlorides Hot concentrated caustic Hydrogen sulfide
Carbon and low-alloy steels	Hot concentrated caustic Concentrated nitrate solutions Anhydrous ammonia Carbonate and bicarbonate solutions
Copper-based alloys	Ammonia (vapors and solutions) Amines Sulfur dioxide Nitrates, nitrites
Titanium	Ethanol Methanol Seawater (certain alloys) Hydrochloric acid (10% solution)
Aluminum alloys*	Halides (especially chloride)

*More dependent upon alloy composition and microstructure than other alloy types

Table 22.1 Selected Stress Corrosion Cracking Agents

amount of moisture (in most cases liquid water) are necessary for SCC to occur.

Common corrodents known to cause SCC in alloys often used in cooling water systems are listed in Table 22.1. This is not an exhaustive list. Over time, the list of corrodents known to cause SCC in industrially relevant alloys continues to grow as research continues.

Corrodents often cause SCC in some alloy systems but not others. For example, austenitic stainless steels (300 series) are susceptible to cracking in chloride solutions but are unaffected by ammonia. Brasses, on the other hand, will crack due to exposure to ammonia in the presence of moisture, but not chlorides. The corrodent need not be present at high concentrations. In the case of chloride SCC of austenitic stainless steels, cracking has occurred at corrodent levels of just a few parts per million (ppm).

Of the metals commonly used in cooling water systems, austenitic stainless steels and copper-based alloys (particularly brasses) are most frequently affected. In cooling water environments, SCC of austenitic stainless steels is most often associated with chlorides. Susceptibility of stainless steel alloys to chloride SCC is very dependent on alloy composition, especially the nickel content (Fig. 22.7). Higher nickel content austenitic stainless steel alloys are more resistant to chloride SCC. However, such alloys are not commonly used for many reasons, particularly higher cost (see Chap. 3, "Stainless Steel"). Other stainless steel alloys are more resistant to chloride SCC than the more

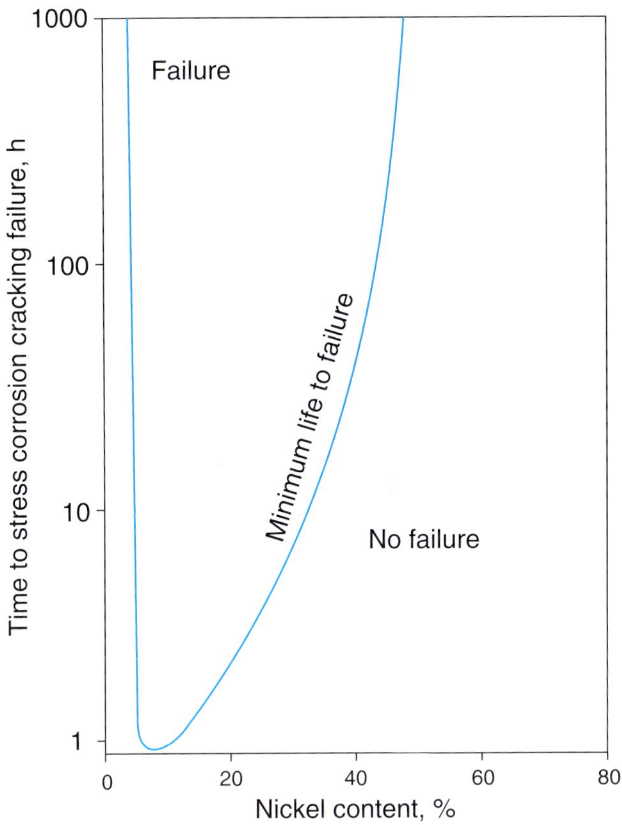

FIGURE 22.7 Stress corrosion cracking of iron-nickel wires in boiling 42 percent magnesium chloride solution. Higher nickel contents generally result in greater resistance to chloride SCC.[3] (*Courtesy of McGraw-Hill.*)

commonly used 304 and 316 austenitic grades. Figure 22.8 shows relative resistances to stress corrosion cracking for a few stainless steel alloys. Types 2304 and 2205 are duplex stainless steels. Note the increased likelihood of SCC as chloride concentration increases.

In copper-based alloys, ammonia is the most well-known corrodent associated with SCC. Other chemical species that may produce ammonia, such as some ammonium salts that dissociate, can also cause SCC. Decomposition products of nitrogen-containing material (such as biological materials) can include ammonia. Ammonia can also be produced by some kinds of bacteria, such as denitrifying bacteria, which utilize nitrate and nitrite and produce nitrogen gas and ammonia.

Pure copper and cupronickel alloys are very resistant to SCC in ammoniacal atmospheres. In brasses, susceptibility to SCC generally increases with increasing zinc content up to a point. Figure 22.9

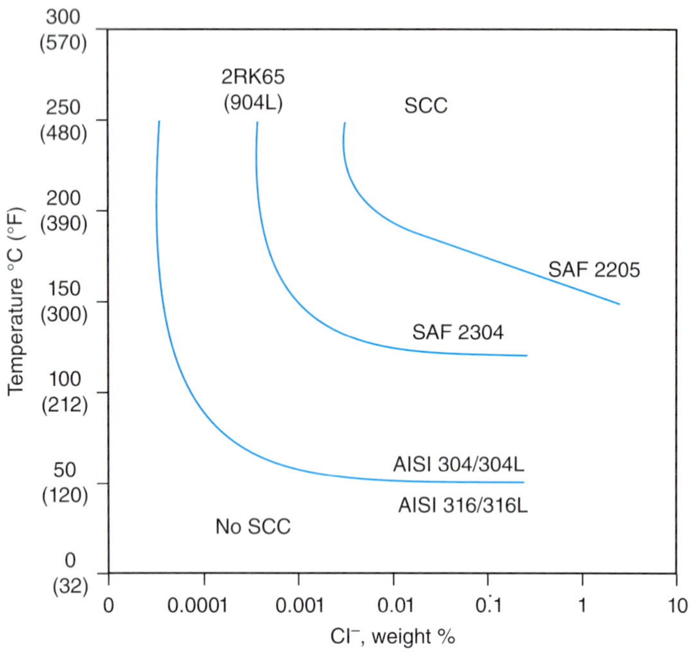

FIGURE 22.8 Stress corrosion cracking resistances for various austenitic and duplex stainless steel alloys in oxygen-bearing chloride solutions of neutral pH. (*Courtesy of AB Sandvik Materials Technology.*)

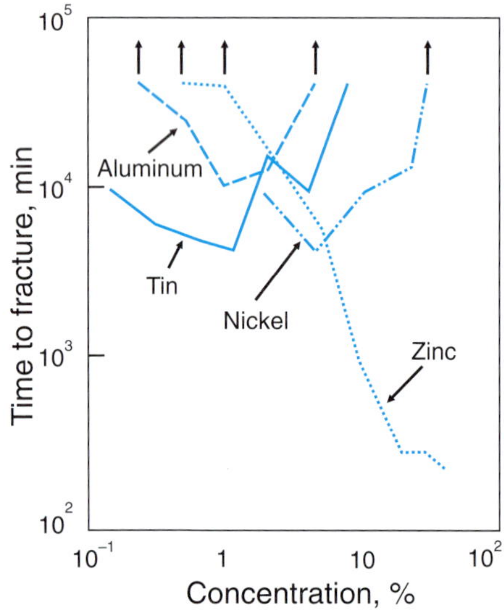

FIGURE 22.9 Effects of various concentrations of zinc, nickel, tin, and aluminum on time to fracture of copper due to SCC in a moist ammoniacal atmosphere.[4] (*Reprinted with permission of ASM International. All rights reserved. www.asminternational.org.*)

illustrates how higher zinc concentrations correlate to shorter times to fracture in a moist ammoniacal atmosphere. Data for nickel, tin, and aluminum are also given in the graph. For those alloying elements, times to fracture exhibit approximate minimum values at certain concentrations.

SCC in aluminum alloys is much more dependent on alloy composition and microstructure than other alloy types, and less dependent upon specific corrodents. Therefore, the line between intergranular corrosion and SCC is quite blurry for aluminum alloys. Water is the key chemical component necessary to produce SCC in aluminum alloys. Halides, particularly chloride, have been found to accelerate SCC.

Other factors that may also influence susceptibility to SCC include temperature, pH, corrodent concentration, time, presence of oxidizing agents, alloy composition, microstructure, and strength. In-depth discussion of all these items is beyond the scope of this book; however, a few of the factors are briefly discussed next.

Temperature

Like many mechanisms involving corrosion processes, elevated temperatures promote SCC in many situations; however, high temperatures are certainly not required for all types of SCC. For example, in the case of copper alloys, ammonia SCC can occur at room temperature, as in the case of the cracked ammunition cartridges described in the beginning of this chapter. In most cases, there are no well-defined minimum temperatures for SCC to occur. However, for chloride SCC of austenitic stainless steels, it is well documented that SCC is rarely observed at temperatures below approximately 120 to 130°F (49–60°C) (Fig. 22.8). Some types of SCC are known to occur only at temperatures significantly higher than those typically found in industrial cooling water systems. For example, caustic SCC of carbon and low-alloy steels requires both a high concentration of caustic (usually more than 5 percent) and temperatures in excess of 200 to 300°F (93–150°C). In stainless steels, temperatures approaching the boiling point of water [212°F (100°C)] are regarded as dangerous when caustic concentrations are high due to the risk of SCC. At very low caustic concentrations, however, much higher temperatures can be tolerated.

Corrodent Concentration

As might be expected, higher corrodent concentrations generally promote SCC. Figure 22.10 shows the results of an experiment in which a porous, insulating material was used to wick a sodium chloride solution to stressed, heated specimens of type 304 stainless steel. The heat provided a strong concentrating mechanism. Higher chloride concentrations greatly increased susceptibility to SCC, as evidenced by the much shorter time to failure when compared with the more dilute solutions.

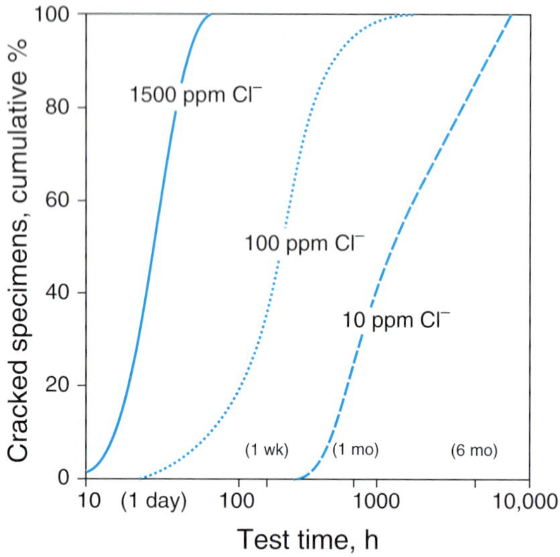

FIGURE 22.10 Effect of chloride concentration on susceptibility of type 304 austenitic stainless steel to chloride SCC using a wick test at 100°C.[5] *(Reprinted with permission of ASM International. All rights reserved. www.asminternational.org.)*

Note that low corrodent concentrations in the bulk fluid cannot be relied upon to prevent SCC from occurring if there is the possibility of localized concentration. Various mechanisms can cause corrodents to become concentrated locally relative to the bulk fluid. The locally higher corrodent concentration may result in corresponding localized SCC. Concentration mechanisms include:

- *Concentration cells.* Permeable deposits, crevices, pits, preexisting cracks, and other conditions that result in physical shielding can lead to concentration of a corrodent in the stagnant solution, which can be 10 to 100 times or more greater than that measured in a bulk fluid (see Case History 22.3 and Chap. 11).
- *Evaporation.* Evaporative concentration can produce concentrations of 100,000 times or more in certain circumstances. Heat transfer surfaces, liquid and vapor interfaces, and regions where wetting and drying conditions occur are areas subject to evaporative concentration (see Case Histories 22.3, 22.4, and 22.6).
- *Vapor spaces.* In addition to the possibility of wetting and drying conditions, vapor spaces may allow gaseous corrodents, such as ammonia and sulfur dioxide, to concentrate to high levels in thin films of condensate (see Case Histories 22.1 and 22.2).

The difficulty in accurately estimating the degree of local concentration remains one of the principal reasons susceptibility to SCC in a specific environment or circumstance is difficult to predict. Measurement of nominal stresses or levels of corrodent in the bulk environment can be quite misleading as predictors of SCC susceptibility.

pH

In the case of chloride SCC of austenitic stainless steels, lower pH promotes cracking at lower temperatures and chloride concentrations (Fig. 22.11).

In the case of brass, studies have shown a complex relationship between pH and time to cracking in an ammoniacal copper sulfate solution (Fig. 22.12). Cracking occurs most quickly at near-neutral pH

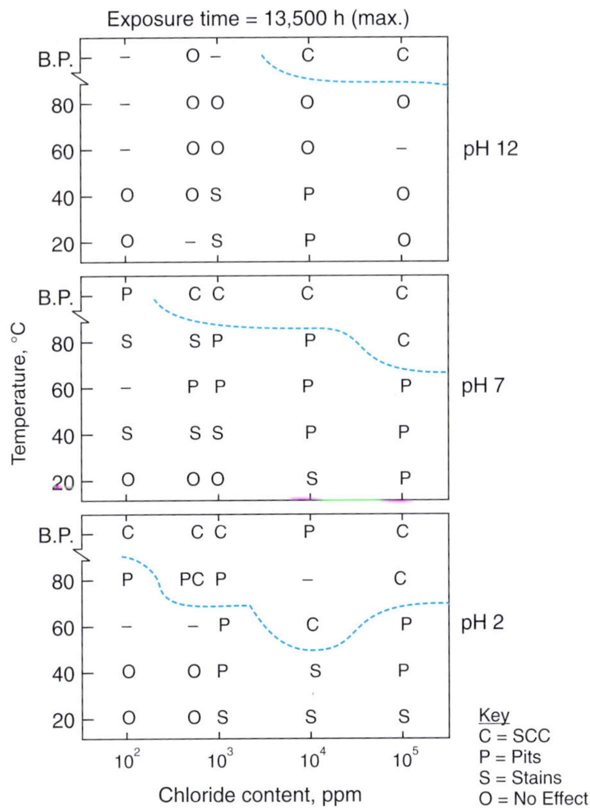

FIGURE 22.11 Effects of chloride content, pH, and temperature on SCC of type 304 austenitic stainless steel in sodium chloride solutions.[6] *(Reprinted with permission of ASM International. All rights reserved. www.asminternational.org.)*

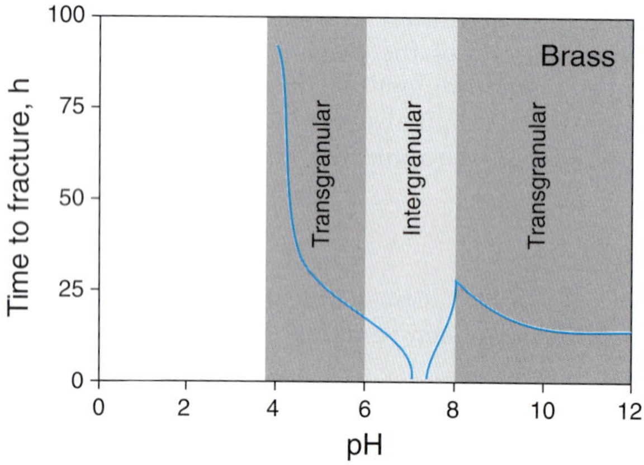

FIGURE 22.12 Effect of pH on time to fracture for brass in an ammoniacal copper sulfate solution at room temperature.[7] (*Reprinted with permission of ASM International. All rights reserved. www.asminternational.org.*)

levels and tends to occur intergranularly. Transgranular cracking dominates in more alkaline and acidic solutions. At pH values below 4, brass is quite resistant to SCC.

Microstructure

Features within the microstructure of a metal may make it more susceptible to SCC. One prominent example is sensitization of stainless steel, which is the precipitation of chromium carbides at grain boundaries. This precipitation effectively robs the adjacent metal of chromium, which is crucial for the corrosion resistance of stainless steel. Intergranular attack that occurs preferentially in a location of elevated tensile stress blurs the line between intergranular corrosion and intergranular stress corrosion cracking (also see Chap. 18, "Intergranular Corrosion"). Grain boundary precipitates in aluminum alloys can also promote intergranular corrosion and, in the presence of sufficient tensile stress, intergranular stress corrosion cracking.

In some cases, segregation of certain elements (impurities) at the grain boundaries can promote intergranular corrosion or stress corrosion cracking without the precipitation of distinct secondary phases. Phosphorus segregation is thought to promote intergranular SCC in carbon and low-alloy steels, though relatively harsh conditions are still needed to cause SCC such that it should not occur in typical cooling water environments. Segregation of elements such as phosphorus, silicon, and sulfur has been observed in austenitic stainless steels, but it is not certain whether such elements play a definite role in SCC. In aluminum, segregation of magnesium has been associated with intergranular SCC.

Identification

Stress corrosion cracks tend to be fine and branched. Unaided visual identification prior to failure can be difficult due to the typical tightness of stress corrosion cracks. Fractures that occur due to SCC are brittle and are typically thick-walled regardless of the ductility of the metal. Evidence of metal loss on surfaces and along cracks, and accumulations of corrosion products, are sometimes not observed. Stains in the cracked region or localized metal loss may be apparent in some cases. Stress corrosion cracks tend to originate at physical discontinuities, such as pits, notches, and corners. Areas that may possess high residual stresses, such as welds or arc strikes, are especially susceptible. A low-power hand lens will greatly aid in identifying cracked locations. Cracking may be detectable using nondestructive testing methods such as dye penetrant, magnetic particle, ultrasonic, eddy current, radiographic, or acoustic emission techniques. Stress corrosion cracking failures generally result in leaks from crack sites.

In some cases, cracks propagate along grain boundaries, which can be more susceptible to corrosion than the material within the bulk of the grains. Such cracking is referred to as intergranular (see Chap. 18). In other cases, cracks propagate transgranularly, or across grains. Metallographic examination can provide positive identification of stress corrosion cracks, as well as identify the nature of the cracking: intergranular, transgranular, or mixed mode (both intergranular and transgranular) (Figs. 22.13–22.15). Such characteristics may provide some information about the corrodent. In austenitic stainless steels such as types 304 and 316, lateral corrosion along coring or ferrite stringers is

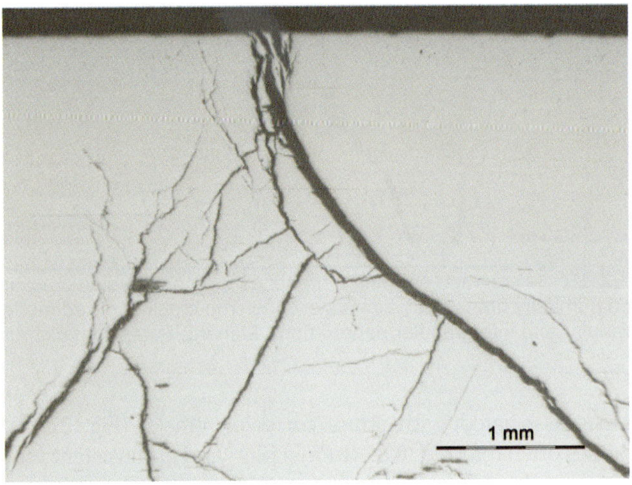

FIGURE 22.13 Micrograph showing highly branched chloride stress corrosion cracks in austenitic stainless steel. Unetched.

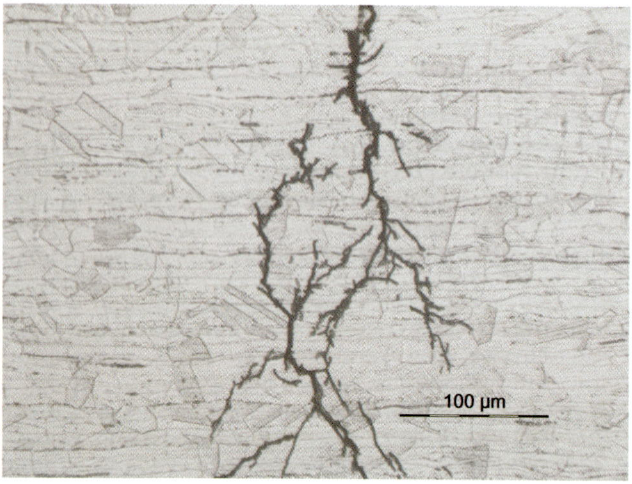

FIGURE 22.14 Micrograph showing transgranular penetration of chloride stress corrosion cracks in austenitic stainless steel. Etchant: Electrolytic oxalic acid.

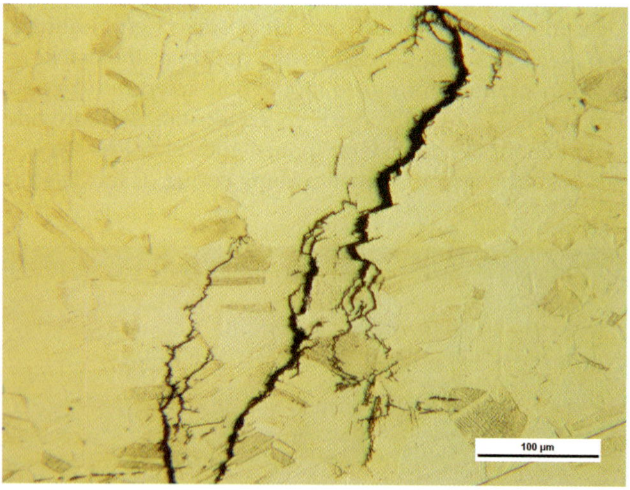

FIGURE 22.15 Micrograph showing a branched stress corrosion crack in an inhibited admiralty brass heat exchanger tube. The crack is mixed mode (both transgranular and intergranular penetration). Etchant: Potassium dichromate.

sometimes associated with stress corrosion cracks (Fig. 22.16). Such metal loss is often referred to as "tunnel pits" and is consistent with chloride corrosion. Metallographic cross sections can also reveal the path of crack propagation and hence, the surface on which the cracks originated, if such information is not evident by visual inspection.

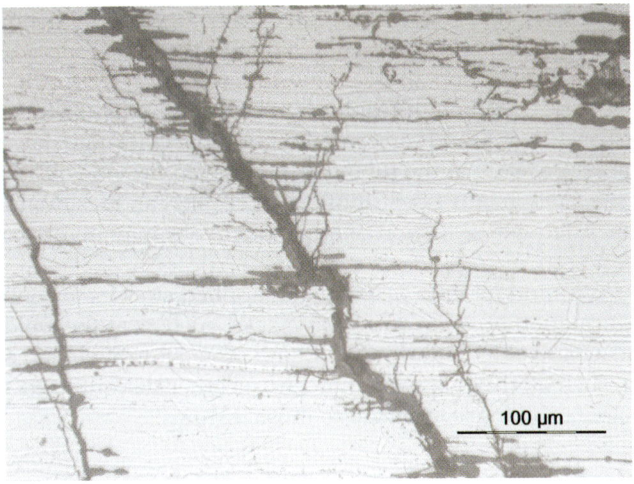

Figure 22.16 Micrograph showing lateral corrosion (tunnel pits) associated with chloride stress corrosion cracks in an austenitic stainless steel tube. The corrosion penetrates along coring and ferrite stringers. Etchant: Electrolytic oxalic acid.

Elimination

The two principal factors governing SCC are tensile stresses and exposure to a specific corrodent. These two factors interact synergistically to produce cracking. Removal or sufficient reduction of one factor can potentially prevent cracking. If such control measures cannot be implemented, or are ineffective, then the use of alternate alloys or materials should be considered.

Tensile Stress

Residual

Stress-relief annealing of components and assemblies following cold working or welding operations reduces residual tensile stresses. Note, however, that annealing has no effect on applied stresses. Annealing may be feasible for smaller components, but it is unrealistic for very large components or field-welded assemblies. Some components, such as those fabricated from thin stainless steel, may experience warpage due to heat treatment. Residual stresses due to welding are unavoidable, but the use of good welding technique can prevent them from becoming excessive.

Applied

Since cracks propagate perpendicularly to the tensile stress, the orientation of the cracking can provide information on the origin of the stress responsible for the damage. For instance, if transverse cracks occur in a tube, the stresses may have been supplied by bending or thermal expansion and contraction. Longitudinal cracks in a tube or pipe suggest that hoop stresses due to internal pressurization contributed to SCC, although residual stresses from manufacturing can also be considerable and can cause longitudinal cracking. The magnitude of the pressure in the tubing or piping can provide information about the likely relative contributions of applied and residual tensile stresses. Adequate reduction of applied stresses may involve equipment redesign.

It is common for both residual and applied stresses to contribute to the formation of stress corrosion cracks, such as at the welds of a load-bearing welded attachment. Eliminating sites of stress concentration, such as sharp corners, notches, or other discontinuities may also help if cracking is localized to those features.

Corrodent

If the responsible specific corrodent cannot be entirely removed from the environment, it may be beneficial to reduce its concentration, since susceptibility to SCC is frequently concentration-dependent. However, if this method of elimination is chosen, care must be taken to avoid conditions that will allow corrodents to concentrate locally, such as those that promote evaporation or localized boiling, alternate wetting and drying conditions, crevices, or deposits. Note that there are usually no well-defined guidelines for maximum corrodent concentrations for avoiding or eliminating SCC. For instance, chloride SCC of 304 and 316 stainless steel has been known to occur in the presence of just a few parts per million of chloride under certain conditions. Localized concentration of corrodents, such as in crevices or under deposits, can produce conditions that are conducive to SCC even if the corrodent concentration in the bulk fluid is low.

Changing the Material

Since metals are typically sensitive to specific SCC agents, successful elimination of SCC problems can be effected by changing the metallurgy of the affected component to a material that will not crack in the existing environment. This method takes advantage of the specificity of the environment and alloy system relationship. However, it is important to identify the specific corrodent responsible for cracking since some corrodents affect more than one alloy system. Choosing an alternate metallurgy may be the only viable option if sufficient stress reduction and/or elimination of the corrodent is not feasible or practical.

In the case of austenitic stainless steels, possible substitutes include superaustenitic stainless steels, ferritic stainless steels, and duplex stainless steels (Fig. 22.8). Duplex stainless steels contain both austenite and ferrite in their microstructures. Even if microscopic stress corrosion cracks occur in the austenite phase regions, the ferrite regions prevent the cracks from growing beyond a microscopic size due to the ferrite phase's high resistance to SCC. For copper-based alloys, cupronickel alloys and copper tend to be much more resistant to SCC than brasses. Unfortunately, an alloy upgrade to improve resistance to SCC often comes with a significant increase in cost.

Cautions

Stress corrosion cracks tend to be fine, tight, and easily overlooked until failure occurs. Various nondestructive techniques are available to aid in the discovery of cracks, such as dye penetrant, eddy current, ultrasonic, and radiographic techniques.

It is possible to confuse SCC with other brittle cracking phenomena, such as corrosion fatigue cracking. Confirmation of SCC typically requires a metallographic examination. On thin-walled components, the surface from which the cracking originates may not be apparent through visual examination. In these cases, a formal metallographic examination may be required to assure positive identification of the surface from which the cracks originate.

Due to the complexity of the SCC phenomenon, preventive techniques have been expressed in general rather than specific terms. It should be recognized that exceptions to these general guidelines may exist and that cracking could occur in some specific circumstance even if the guidelines are followed.

Related Problems

See Chap. 11, "Crevice and Underdeposit Corrosion (Concentration Cells)"; Chap. 18, "Intergranular Corrosion"; and Chap. 21, "Corrosion Fatigue."

Case History 22.1

Industry:	Chemical process
Specimen location:	Surface condenser tube
Specimen orientation:	Horizontal
Environment:	Internal: Treated cooling water, chromate 10–13 ppm External: Steam and condensate at a partial vacuum of 4–10 in. (10–25 cm) of mercury
Time in service:	17 years
Sample specifications:	¾ in. (19 mm) outer diameter, admiralty brass

A routine inspection of the tube bundle during a plant outage revealed fine cracks (Fig. 22.17). Scattered longitudinal cracks were observed along the lengths of most tubes. The external surface was covered with a thin film of black copper

FIGURE 22.17 A network of longitudinal cracks along the external surface of an admiralty brass condenser tube. (Magnification: 7.5×.)

oxide and deposits. The bundle had been exposed to ammonia levels that produced 0.5 ppm of ammonia in the accumulated condensate.

Metallographic analysis revealed branched, transgranular cracks that originated on the external surface.

The longitudinal orientation of the cracks revealed that hoop (circumferential) stresses caused by residual stresses from tube manufacturing and/or internal pressurization provided the necessary stresses. Ammonia was the specific corrodent responsible for SCC.

Note that the 0.5 ppm of ammonia measured in the accumulated condensate may not have accurately reflected the potentially higher level of ammonia dissolved in a thin film of moisture that formed on the external surfaces of the tubes in the region of initial condensation. This film of water could have been saturated with ammonia, increasing the potential for SCC.

Case History 22.2

Industry:	Chemical process
Specimen location:	Tube from a heat exchanger—instrument air cooler
Specimen orientation:	Horizontal
Environment:	Internal: Air entering the exchanger at 115°F (45°C) External: Treated cooling water at 50°F (10°C)
Time in service:	4 months
Sample specifications:	5/8 in. (16 mm) outer diameter, admiralty brass

Transverse cracks, shown in Fig. 22.18, occurred in many tubes of this new exchanger. All cracks occurred at the air inlet end of the cooler (Fig. 22.19) and had the brittle appearance typical of stress corrosion cracks.

Close examinations of the internal surface under a low-power stereomicroscope revealed, in addition to the open cracks illustrated, numerous tight secondary cracks. Tube surfaces away from the cracks were smooth, uncorroded, and free of deposits.

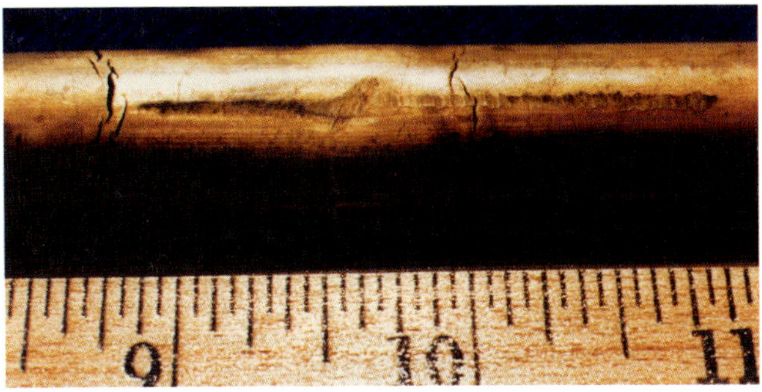

FIGURE 22.18 Discontinuous, transverse cracks in an admiralty brass tube from an air cooler.

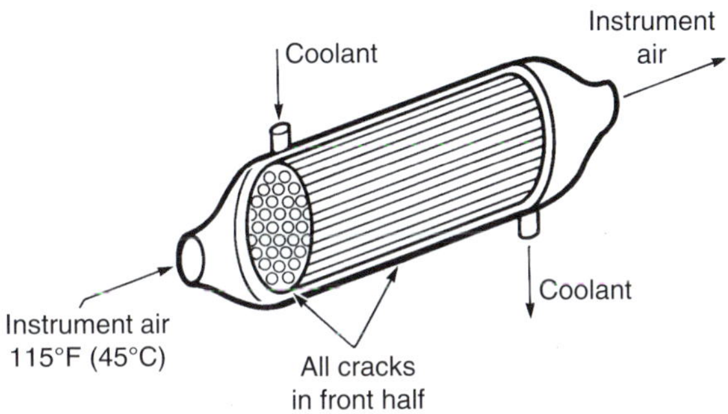

FIGURE 22.19 Schematic of air cooler, showing location of cracks at the air inlet end.

Microstructural examinations revealed that the tube was annealed brass. All cracks originated on the internal surface at pit sites. Cracks were fine, branched, and ran through the metal grains (transgranular). Crack density over the specimens examined averaged 40 cracks per linear inch (16 cracks/cm).

The transverse orientation of the cracks, coupled with their predominance along one side, revealed that bending of the tube induced the responsible tensile stresses. The rapidity of crack propagation and the high crack density indicated that the stress level was high.

The confinement of the cracks to a specific area of the cooler suggested that condensate from atmospheric moisture initially formed in the area and dissolved a corrodent from the atmosphere such as ammonia, sulfur dioxide, or nitrogen oxides. Since the previous cooler had been in service for 20 years, it was believed that the rapid failure of the new exchanger was due principally to very high bending stresses, which may have been induced during construction of the cooler.

Case History 22.3

Industry:	Chemical process
Specimen location:	Heat exchanger tube
Specimen orientation:	Horizontal
Environment:	Internal: A mixture of steam, condensate, hydrogen, nitrogen, and carbon dioxide; outlet temperature 165°F (75°C), inlet temperature 265°F (130°C), condensate pH 10 External: Treated cooling water, outlet temperature 105°F (40°C), inlet temperature 85°F (30°C), pH 7.5, chloride 150–200 ppm, phosphate 10–12 ppm, zinc 2–4 ppm
Time in service:	15 years
Sample specifications:	1 in. (25.4 mm) outer diameter, 304 stainless steel

The transverse fissures illustrated in Figs. 22.20 and 22.21 are similar to failures that had occurred in many tubes. All failures occurred near the tubesheet at the hot end of the exchanger (Fig. 22.22).

Figure 22.20 shows the proximity of the cracks to the rolled end of the tube. The cracks occurred from just one side. The original crack was tight, but distortion of the tube during its removal from the exchanger expanded the cracks, giving them a fissure-like appearance in the photographs.

FIGURE 22.20 Short, transverse fissure adjacent to the rolled end of a type 304 stainless steel heat exchanger tube.

FIGURE 22.21 Fissures originating on the external surface of a type 304 stainless steel heat exchanger tube. (Magnification: 7.5×.)

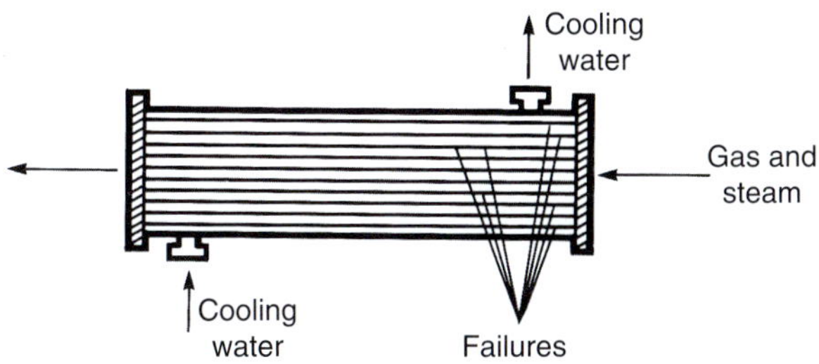

FIGURE 22.22 Schematic of heat exchanger, showing location of cracks.

Microstructural examinations revealed that the cracks originated on the external surface. The cracks were branched and ran across the metal grains (transgranular).

The transverse orientation of the cracks along just one side of the tube revealed that bending provided the stresses for cracking. The tubesheet acted as a constraint to the bending, intensifying stresses in the tube wall adjacent to the sheet.

Although 200 ppm of chloride is sufficient to induce SCC in a highly stressed metal, it is possible that chlorides were concentrated in the cracked region, perhaps by localized boiling or by concentration beneath deposits. It is worth noting in this regard that all of the cracking occurred at the hot end of the exchanger, most likely because the metal temperature in that region was above 120°F (49°C). Note that metal temperatures can be appreciably higher than the outlet bulk water temperature, depending upon heat transfer from the process water side to the cooling water side.

Case History 22.4

Industry:	Chemical process
Specimen location:	Heat exchanger tube from a carbon dioxide compressor intercooler
Specimen orientation:	Horizontal
Environment:	Internal: Carbon dioxide, temperature 280–163°F (138–73°C), pressure 1735 psi (12 MPa) External: Cooling water treated with phosphate, zinc, and a biodispersant, temperature 104°F (40°C), pressure 64 psi (441 kPa), pH 8–9, chloride 200 ppm
Time in service:	2 years
Sample specifications:	9/16 in. (1.4 cm) outside diameter, 304 stainless steel tube

A total of 13 tubes in the exchanger had suffered cracking (Fig. 22.23). The cracks were predominantly longitudinally oriented, although some portions were transverse in orientation.

Metallographic examination revealed that the cracks originated on the external surface (Fig. 22.24). The cracks were highly branched and transgranular, typical of chloride stress corrosion cracking of austenitic stainless steels. The thick-walled fracture faces were also typical of stress corrosion cracking.

Figure 22.23 Cracks on the external surface of a type 304 stainless steel heat exchanger tube. The cracks are primarily longitudinal, but they are transverse in orientation in some portions.

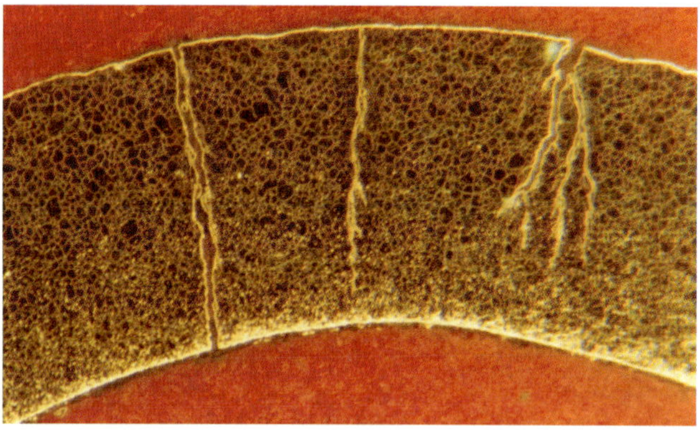

Figure 22.24 Stereomicroscope photo of a mounted cross section, showing cracks originating on the external surface. Note the branching. (Magnification: 33×.)

The longitudinal orientation of the cracks revealed that the stresses responsible for the cracking were hoop stresses (circumferentially oriented) resulting from the internal pressure. The probable specific cracking agent was chloride, which may have concentrated at possible boiling sites on the external surface. Note that internal and external operating conditions and environments, although physically separated by the tube wall, each contributed a necessary condition (chloride and sufficient temperature) to induce stress corrosion cracking.

Recommended remedial measures included reduction or elimination of chlorides or replacement of the 304 stainless steel with an alloy that is more resistant to chloride stress corrosion cracking.

Case History 22.5

Industry:	Chemical process—plastics
Specimen location:	Coolant piping connections in extrusion barrel cooling jacket
Specimen orientation:	Vertical
Environment:	Internal: Softened cooling water, temperature 250°F (120°C), pressure 25 psi (172 kPa), pH 7.8–8.5 External: Ambient
Time in service:	4 months
Sample specifications:	½ in. (1.3 cm) outer diameter, 304 stainless steel tube

Figure 22.25 depicts the arrangement of connections in the extrusion barrel cooling jacket. Cracking of these stainless steel connections (Figs. 22.26 and 22.27) were occurring as frequently as every four to six weeks.

Close visual inspection of the blunt, fractured ends revealed small secondary cracks along the internal surface. Internal surfaces were covered with a layer of deposit material that produced a highly alkaline aqueous solution (Fig. 22.27).

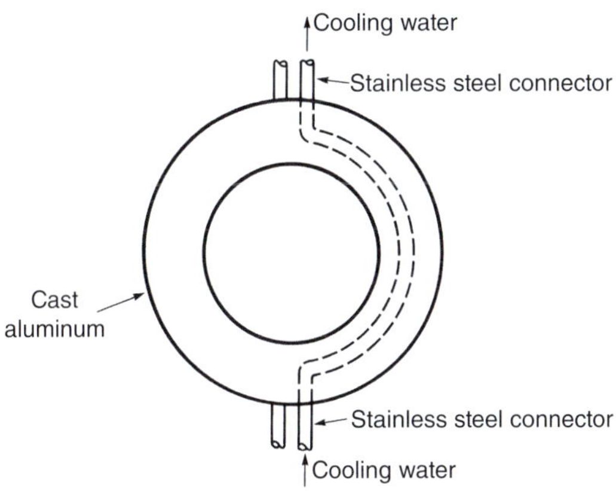

FIGURE 22.25 End view of plastics extrusion barrel cooling jacket.

FIGURE 22.26 Stainless steel connectors, showing circumferential fractures.

FIGURE 22.27 Cracked connectors; note brittle, thick-walled character of the cracked faces and the accumulation of alkaline deposits on the internal surface of the connector on the left.

Metallographic examinations revealed highly branched, predominantly transgranular cracks originating on the internal surface.

The transverse (circumferential) crack path revealed that the stresses responsible for SCC were axially oriented; that is, the tube was pulled at its ends. Residual tube-forming stresses may also have contributed in this case. The specific cracking agent was probably caustic, which was apparently concentrated by evaporation when water flashed to steam in the failure locations.

If stresses could not be sufficiently reduced, a metal possessing greater resistance to caustic SCC was recommended. Alteration to the environment, such as eliminating alkalinity or increasing system pressure to prevent flashing (thereby minimizing the potential for concentration of caustic or other corrodents), may have also prevented failure by this mechanism; however, due to the high temperature and lack of a definite maximum chloride concentration to prevent SCC, chloride SCC would still have been a concern.

Case History 22.6

Industry:	Chemical process—plastics
Specimen location:	Orthodichlorobenzene (ODCB) column condenser tube
Specimen orientation:	Vertical
Environment:	Internal: Process stream, ODCB 99.4%, $COCl_2$ 0.5%, HCl 0.1%; inlet temperature 220°F (105°C); outlet temperature 175°F (80°C) External: Treated cooling water inlet temperature 80°F (27°C), outlet temperature 130–212°F (55–100°C)
Time in service:	3 months
Sample specifications:	1½ in. (3.8 cm) outer diameter, 304 stainless steel tube

Numerous tubes in this new heat exchanger had suffered transverse cracks (Fig. 22.28). All cracking occurred at the top of this vertical condenser, within 1 in. (2.5 cm) of the rolled areas at the top tubesheet, in the vapor space (Fig. 22.29).

FIGURE 22.28 Circumferential cracks along one side of the heat exchanger tube. Note proximity of cracks to the plane of the tubesheet, as indicated by the tubesheet grooves (left). Expansion of the cracks may have occurred during removal of the tube from the exchanger.

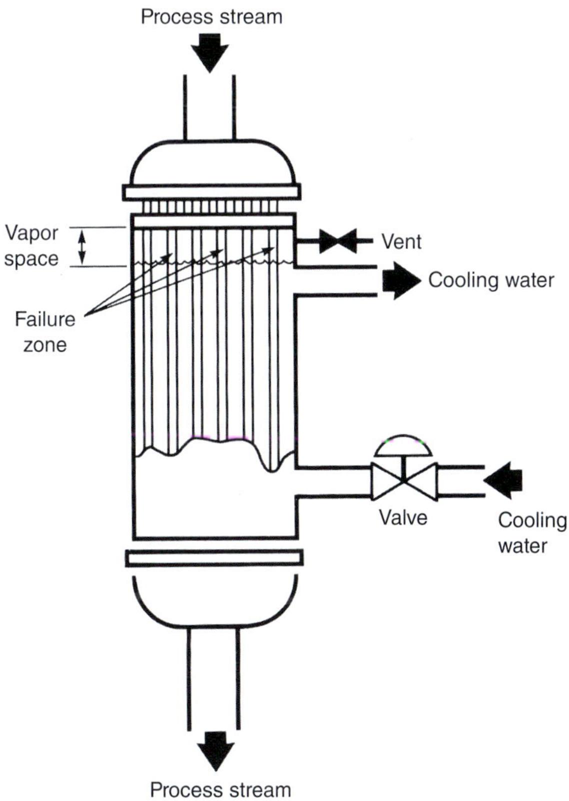

FIGURE 22.29 Schematic representation of vertical heat exchanger showing location of cracks in the vapor space.

Metallographic examinations revealed branched, transgranular cracks originating on the external surface (cooling water side). Analysis of material covering the crack surfaces revealed the presence of chlorine.

The visual and microscopic appearance of the cracks, coupled with the presence of chlorine-containing corrosion products on the cracked surfaces, indicated that the failure occurred due to chloride SCC. The transverse orientation of the cracks and their location on just one side of the tubes near their entrance into the tubesheet revealed that the stresses responsible for the cracking were bending stresses concentrated near the plane of constraint (tubesheet).

Several factors appeared to have contributed to the delivery of chloride to the stressed area. First, early in the short life of the exchanger, the cooling water was contaminated with hydrochloric acid, resulting in a depression of pH to 4. Second, the valve regulating cooling water flow into the condenser was almost closed (open 20 percent), resulting in low flow of cooling water. Consequently, at times the cooling water would exit the condenser at the boiling point. The vent provided to discharge normal quantities of vapors from the vapor space was apparently incapable of accommodating the large quantities of vapor produced by the boiling. Water droplets containing chloride were transported to the failure location. Chloride concentrated on the tube surfaces due to evaporation.

Case History 22.7

Industry: Food

Specimen location: Cooling water return piping from an evaporator, corn syrup refinery

Specimen orientation: Vertical

Environment: Alternating steam and cooling water

Time in service: 2 years

Sample specifications: 3½ in. (8.9 cm) outer diameter, type 316L austenitic stainless steel

The section consisted of two straight segments of piping joined together by short segments welded to form an S shape (Fig. 22.30). Many tight, branched cracks

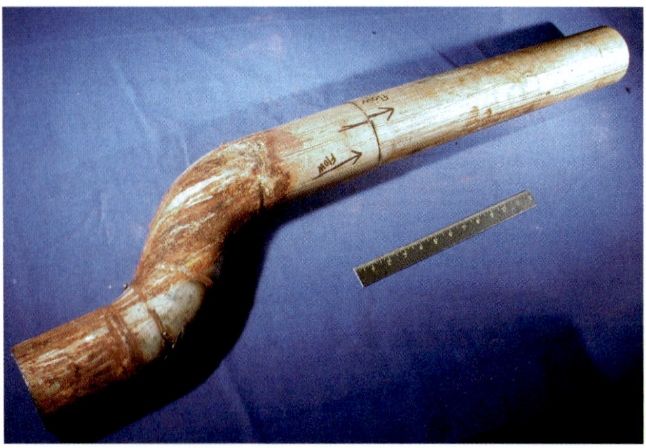

Figure 22.30 Type 316L stainless steel piping section that includes several segments welded together to form an S shape.

FIGURE 22.31 External surface after cleaning by glass bead blasting, showing tight, branched cracks along and slightly away from a circumferential weld.

were evident close to and slightly away from the circumferential welds (Fig. 22.31). Chemical spot tests of material on both the internal and external surfaces indicated the presence of chlorides. Scanning electron microscopy/energy dispersive spectroscopy (SEM/EDS) analysis of material on both surfaces identified chlorine, with as much as 10 percent detected in deposits on the internal surface, along the cracks. Metallographic examination revealed many branched, transgranular cracks that primarily originated from the internal surface (Fig. 22.32). Corrosion occurred laterally (known as tunnel pits), along coring (segregation of alloying elements) and ferrite stringers (Fig. 22.16). Corrosion along the ferrite stringers

FIGURE 22.32 Metallographic cross section, showing many highly branched cracks at and near a circumferential weld (weld metal can be seen on the left side). The cracks primarily originate from the internal surface. Etchant: Electrolytic oxalic acid.

occurred due to microgalvanic effects between the ferrite and austenite (see Chap. 16, "Galvanic Corrosion").

The characteristics of the cracks were typical of chloride stress corrosion cracking. The general proximity of the cracks to the welds suggested that residual stresses from the welding process contributed significantly to the tensile stresses required for cracking. Residual tensile stresses from pipe manufacturing and applied stresses due to bending may have also contributed.

The internal surface of the piping was alternately exposed to steam and cooling water for undisclosed periods of time, with temperatures reaching up to 220°F (104°C) when steam flowed through the piping. The chloride content of the cooling water averaged about 480 ppm, and the internal surface of the pipe was never cleaned. The alternating steam and cooling water flow allowed for concentration of the chloride in the cooling water and most likely resulted in cooling water temperatures exceeding 120 to 130°F (49–54°C) at times. Alternate metallurgies, such as duplex stainless steels or suitable nickel-based alloys, were recommended in order to alleviate chloride SCC in this application.

Case History 22.8

Industry:	Building
Specimen location:	Plate-and-frame heat exchanger
Specimen orientation:	Vertical
Environment:	Side 1: Closed-loop water Side 2: Domestic hot water, 140°F (60°C)
Time in service:	3 years
Sample specifications:	Two-ply welded corrugated plate, type 304 austenitic stainless steel

Two corrugated sheets were welded together along the inlets and outlets of the plate. Many tight, branched cracks occurred between the inlet and outlet holes (Fig. 22.33). Most of the cracking occurred outside of the water-contacting area of the plate. Cracks were observed on both plies of the plate, though they were more numerous on the domestic hot water side. Numerous tight cracks

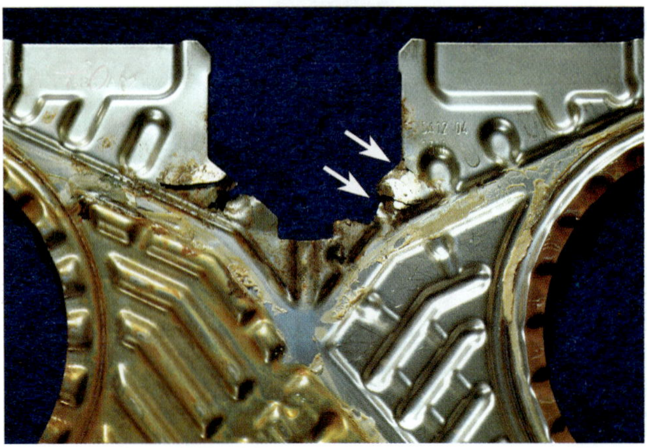

Figure 22.33 Domestic hot water side, showing tight, branched cracks between inlet and outlet holes.

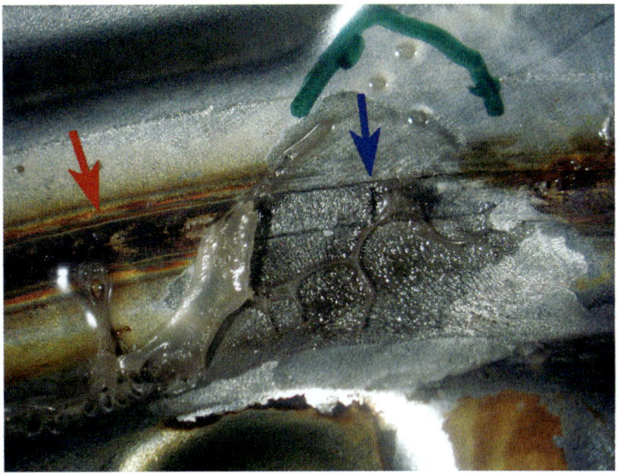

FIGURE 22.34 Weld joining two plies of the plate (red arrow). Note also the darkened appearance of the weld due to heat tint oxide. Transverse cracks occurred across the welds (blue arrow). Jagged metal loss occurred near some cracks.

occurred transversely to the welds joining the plies (Fig. 22.34). Metal loss areas with jagged interior surface contours were observed near some cracks. Heat tint oxide was observed along the welds (see "Pickling, Passivation, and Cleaning" section in Chap. 3 for more information on heat tint oxide).

Metallographic cross sections across the cracks indicated that they primarily originated from the hot water side and had many tunnel pits along their lengths (Fig. 22.35). The cracks penetrated transgranularly (Fig. 22.14). Cross sections

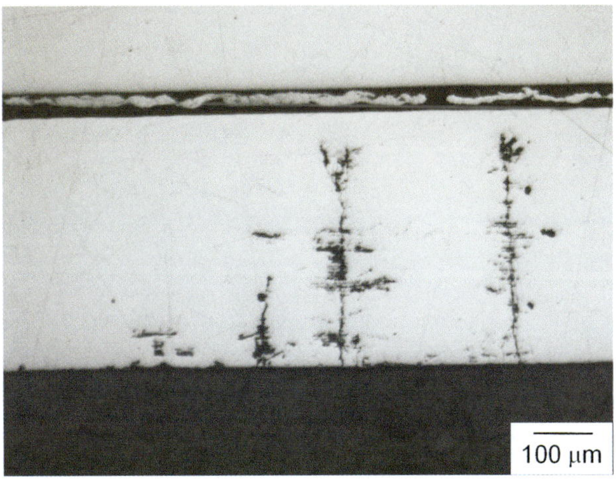

FIGURE 22.35 Micrograph showing tight, branched cracks originating from the domestic water side. Many tunnel pits extend from the cracks. Unetched.

through the metal loss areas indicated that they were deep and undercut. SEM-EDS analysis indicated low concentrations (1 percent or less) of chlorine in the cracked areas.

The temperature of the domestic hot water was sufficiently high for chloride stress corrosion cracking to occur. Residual stresses due to welding and forming of the corrugated sheet provided the tensile stresses necessary for cracking to occur. Once cracks penetrated one ply, water leaked into the space between the plies and caused further cracking in areas that should not have contacted water. Heat tint oxide most likely promoted metal loss along the welds. A single annealed plate was recommended instead of a two-ply plate to eliminate residual stresses due to welding and decreased corrosion resistance due to heat tint oxide.

References

1. Mars G. Fontana and Norbert D. Greene, *Corrosion Engineering*, 2d ed., McGraw-Hill, New York, 1978, p. 96.
2. Fontana and Greene, *Corrosion Engineering*, p. 96.
3. Fontana and Greene, *Corrosion Engineering*, p. 104.
4. W. R. Warke, "Stress-Corrosion Cracking," *ASM Handbook*, Volume 11, *Failure Analysis and Prevention*, ASM International, Materials Park, OH, 2002.
5. Russel H. Jones (ed.), *Stress-Corrosion Cracking of Stainless Steels*, ASM International, Materials Park, OH, 1992, p. 105.
6. Jones, *Stress-Corrosion Cracking of Stainless Steels*, p. 106
7. Warke, "Stress-Corrosion Cracking."

SECTION 4

Mechanical Damage

CHAPTER 23

Erosion-Corrosion

General Description

Erosion-corrosion can be defined as the accelerated degradation of a material resulting from the joint action of erosion and corrosion when the material is exposed to a rapidly moving and/or turbulent fluid. Metal can be removed in the form of solid particles of corrosion product or, in the case of severe erosion-corrosion, as dissolved ions.

Ideally, metal surfaces in a well-designed, well-operated cooling water system will establish an equilibrium with the environment by forming a layer of protective corrosion products. This covering effectively isolates the metal from the environment, thereby limiting further corrosion. Any mechanical and/or chemical conditions that affect the ability of the metal to form and maintain this protective layer can lead to metal loss. Erosion-corrosion is a classic example of such a combination of mechanical and chemical conditions. A typical sequence of events is listed here. Steps 2 and 3 may occur almost simultaneously.

1. A layer of corrosion products already exists, or forms on an unprotected metal surface.
2. The layer is mechanically removed by the abrasive effects of a high-velocity, impinging, or turbulent fluid.
3. Corrosion products are re-formed on the metal surface.

Metals that depend on a relatively thick, protective layer of corrosion products for corrosion resistance are frequently subject to erosion-corrosion. One example is the mechanically weak layer typically formed on cast iron alloys. This is due to the poor adherence of these layers relative to the thin films formed by the classical passive metals, such as stainless steel and titanium. Both stainless steel and titanium are very resistant to erosion-corrosion in most cooling water environments. However, most other metals that are commonly used in cooling water systems are susceptible to erosion-corrosion. Soft metals, such as copper, some copper-based alloys, and aluminum alloys are especially susceptible.

Locations

Favored locations for erosion-corrosion are areas exposed to high flow velocities, fluid impingement, and turbulence. Tees, bends, elbows (Fig. 23.1), pumps, valves (Fig. 23.2), and inlet ends of heat exchangers (Figs. 23.3 and23.4) are commonly affected. Turbulence may be created downstream of crevices, ledges (Fig. 23.5), welds, abrupt cross-section changes (Fig. 23.6), and other features that disrupt flow. Flow may be disrupted downstream from localized thick deposit or corrosion product layers. Debris lodged in tubes or water channels will also create downstream turbulence.

Erosion-corrosion problems on the external surfaces of tubes are frequently associated with impingement of high-velocity gases,

FIGURE 23.1 Erosion-corrosion at an elbow in a brass piping assembly. Note also the horseshoe-shaped depressions and comet tails aligned with flow direction in the straight section.

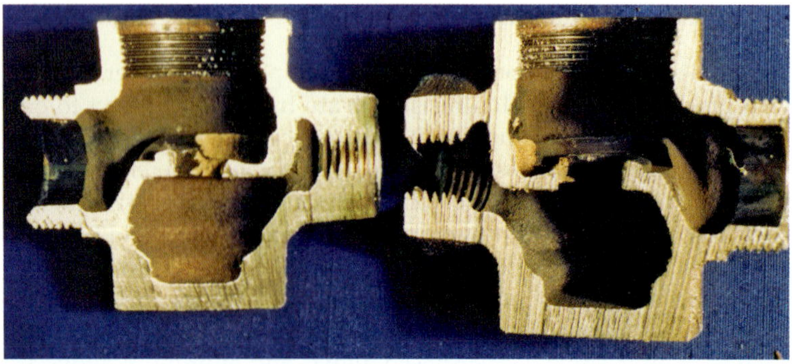

FIGURE 23.2 Split valve body, showing metal loss at the valve seat.

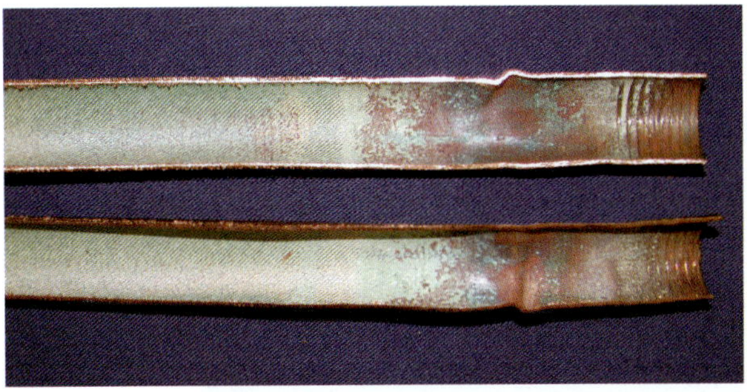

FIGURE 23.3 Internal surface of enhanced copper chiller tubing, which sustained the most severe metal loss near the inlet end (right side of photo). Note the smooth surface appearance and the general lack of deposits close to the inlet end, though some material may have spalled due to deformation during tube removal.

FIGURE 23.4 Tubesheet of a two-pass heat exchanger, showing erosion damage of the copper tubes at the inlet of the first pass (right side of photo), but not at the outlet of the second pass (left side of photo).

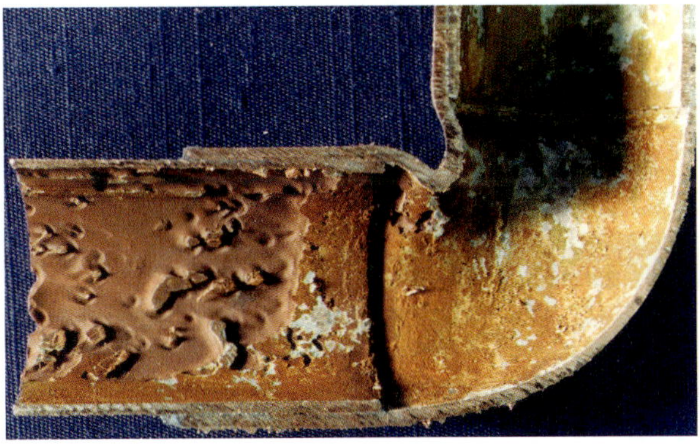

FIGURE 23.5 Turbulence created by the ledge at the tube end caused erosion-corrosion immediately downstream.

FIGURE 23.6 Severe erosion-corrosion damage in a copper pipe soldered to a reducing fitting. The fitting provided both a decrease in cross-sectional area and a ledge that disrupted flow. A perforation occurred within a depression (circled). Deposits on the eroded surface indicated that erosion-corrosion was not actively occurring shortly before the section was removed from service.

such as steam, that contain entrained water droplets. In surface condensers, this typically involves peripheral tubes at the inlet nozzle to the shell side (Fig. 23.7). Baffle and tube interfaces may also be affected.

Critical Factors

Velocity

Erosion-corrosion is a fairly complex failure mode influenced by both environmental factors and metal characteristics. Perhaps the most important environmental factor is velocity. A threshold velocity is

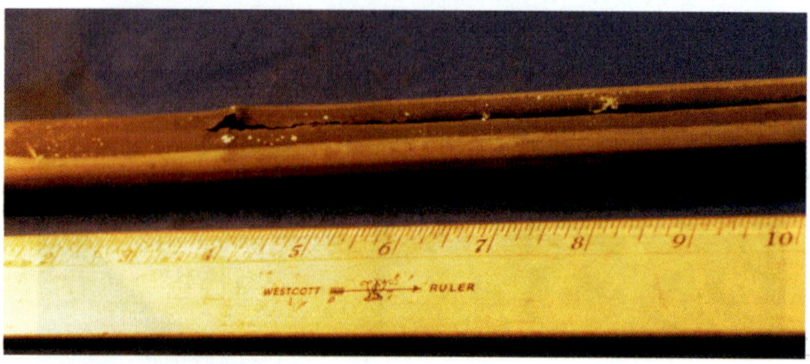

FIGURE 23.7 Erosion-corrosion damage on the surface of a brass tube facing the steam inlet nozzle in a surface condenser.

Alloy	Alloy Type	Maximum Design Velocity	
		Clean Seawater	Good Freshwater
C12200	Copper	3 ft/s (0.9 m/s)	4 ft/s (1.2 m/s)
C23000	Red brass	4 ft/s (1.2 m/s)	5 ft/s (1.5 m/s)
C44300	Admiralty brass	6 ft/s (1.8 m/s)	9 ft/s (2.7 m/s)
C60800	Aluminum bronze	8 ft/s (2.4 m/s)	11 ft/s (3.4 m/s)
C68700	Aluminum brass	9 ft/s (2.7 m/s)	11 ft/s (3.4 m/s)
C70600	Cupronickel (90-10)	8 ft/s (2.4 m/s)	11 ft/s (3.4 m/s)
C71500	Cupronickel (70-30)	15 ft/s (4.6 m/s)	15+ ft/s (4.6+ m/s)

Table 23.1 Maximum Recommended Fluid Velocities for Select Copper Alloys[1]

often observed, below which metal loss is negligible and above which the metal loss rate increases appreciably as velocity increases. The threshold velocity varies with the metal, the chemical makeup of the environment, and other factors such as temperature. Of the alloy types that are commonly used for piping and tubing in cooling water systems, copper alloys are the most susceptible to erosion-corrosion. Recommended maximum design fluid velocities for a few common copper alloys are given in Table 23.1. These values should only be used as guidelines, since factors such as localized turbulence and water chemistry can still cause erosion-corrosion even when flow velocities are below the recommended maximum values.

Plugging tubes in a heat exchanger without decreasing the flow of water to the exchanger increases the flow velocity in the remaining unplugged tubes. A significant number of plugged tubes can translate into a significant flow velocity increase in the unplugged tubes; therefore, it may be advisable to consider decreasing water flow appropriately to avoid erosion-corrosion in such situations. However, such a decrease in water flow may not be desirable because it would reduce cooling of the process side.

Turbulence

Turbulence can result in erosion-corrosion in localized areas, while other areas away from the region of disturbed flow remain unattacked. Impediments to water flow resulting from inadequate equipment design or lodgment of foreign objects in the flow path can have a dramatic effect on the erosion-corrosion process. Much of this influence is linked to the creation of turbulence and the simple increase in fluid velocity past obstructions. The importance of these factors is quickly recognized when the phenomenon of threshold velocity is considered.

For example, unsatisfactory equipment design has caused inlet end erosion due to improper distribution of water in the water box of a heat exchanger. Poor design and construction practices have caused erosion-corrosion of heat exchanger tubesheets due to unsatisfactory fluid distribution or leakage past partition plates. Deep channels ("worm holes") have been observed in tube surfaces due to leaking tube-to-tube sheet joints. Water channels should have sufficient cross-sectional area to provide the required cooling without excessive water flow velocities. In addition, connections between water channels of different components must be properly aligned to avoid undesirable turbulence.

A common cause of erosion is partial obstruction of flow paths by foreign bodies. At the inlet end of a heat exchanger, for example, debris such as rocks, sticks, cooling tower fill, and wood chips may lodge in tube ends or be held against the tubes by water flow. The nominal velocity of the water past the obstruction increases according to the degree of obstruction. It can be shown that if two-thirds of the tube opening is blocked, the water velocity past the obstacle can increase by 2½ times.

Some debris may enter the tube and then become lodged. This condition may be more likely to occur on internally rifled tubing, as the rifling ridges may provide ledges for debris to catch. Turbulence created at the site of the debris often produces crescent- or irregularly shaped erosion-corrosion patterns (Figs. 23.8–23.10).

Material

Different materials offer varying degrees of erosion resistance. Figure 23.11 shows approximate normalized relative erosion resistances of various alloy types. The rankings are based on the rule of

FIGURE 23.8 Erosion site resulting from lodgement of a foreign body inside a tube. (Magnification: 8×.)

FIGURE 23.9 Erosion site resulting from lodgement of a foreign body inside a tube. (Magnification: 8×.)

FIGURE 23.10 Erosion site resulting from lodgement of a foreign body inside a tube. (Magnification: 8×.)

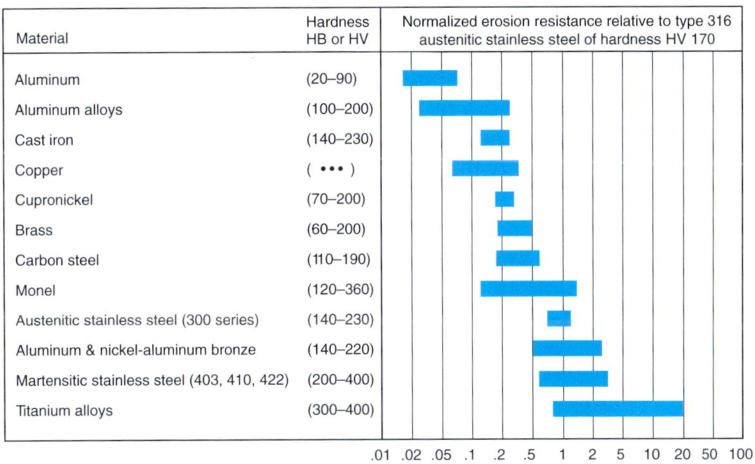

FIGURE 23.11 Approximate normalized erosion resistances of various metals and alloys, relative to type 316 austenitic stainless steel with hardness of HV 170.[2] (*Reprinted with permission of ASM International. All rights reserved. www.asminternational.org.*)

thumb that alloys with higher hardness tend to be more erosion-resistant than softer alloys. However, note that this rule may only apply when comparing similar alloys; it may not be applicable when comparing different alloy types. Therefore, such erosion resistance values should only be used as rough guides, as actual relative erosion resistances may differ. For example, experience has shown that cupronickel is generally more erosion-resistant than copper and brass, though it is placed below both copper and brass in the ranking. Also, this ranking does not take into account the effect of the environment, such as the corrosiveness of the water.

When corrosion plays a significant role in erosion-corrosion, both the characteristics of the corrosion products and the characteristics of the metal itself are important, since removal of the corrosion products is a significant part of the erosion-corrosion process. Logically, any corrosion products on the surface must be removed before erosion of the metal itself can occur. In erosion-corrosion cases, the resistance of the material to corrosion in the environment may be just as important as its erosion resistance. In some situations, an alloy that is relatively corrosion-resistant in a particular environment, but not especially erosion-resistant, may perform better than an alloy that is technically more erosion-resistant, but not as corrosion-resistant in that environment.

Particulate

Although erosion-corrosion can and does occur in the absence of entrained particulate, it can be promoted and/or accelerated by a dispersion of particulate matter (sand, silt, corrosion products, deposits) entrained in the fluid. The character of the particulate, and even the fluid itself, substantially influences the effect. Eight major characteristics are influential: particle shape, particle size, particle density, particle hardness, particle size distribution, angle of impact, impact velocity, and fluid viscosity. For example, angular particles are generally more erosive than rounded particles. At a given velocity, larger particles have more kinetic energy than smaller particles, so larger particles tend to be more erosive. Particles also tend to be more erosive with increasing density and hardness. Entrained gas bubbles can also accelerate erosion-corrosion in a manner similar to particulate material.

Fluid Characteristics

In addition to fluid velocity, other characteristics of the eroding fluid can exert a marked influence on the erosion-corrosion process. Among the important factors are the following:

1. *Corrosivity.* In general, the more corrosive the fluid, the greater the rate of metal loss. For example, seawater is generally more corrosive than freshwater; therefore, materials may be less resistant to erosion-corrosion in seawater than in

freshwater, which is illustrated by Table 23.1. It is interesting to note that the fluid velocity can affect the apparent corrosiveness of a fluid. For instance, for carbon steels exposed to solutions containing dissolved oxygen, carbon dioxide, or hydrogen sulfide, higher velocities increase the supply rate of these dissolved gases to the metal surface, thereby increasing the corrosion rate.

2. *Temperature.* In general, the higher the temperature, the higher the corrosion rate.
3. *pH.* The effect of pH on erosion-corrosion varies with the metal and environment combination. See Chaps. 1 through 5 for in-depth discussions of the effects of water chemistry on various metals and alloy types commonly used in cooling water systems.

Identification

The classic signature of erosion-corrosion is the formation of horseshoe-shaped depressions or features such as comet-shaped depressions with tails, and grooves that are oriented along the direction of fluid flow (Figs. 23.12–23.13). Sand dune–like contours can develop on generally corroded surfaces (Fig. 23.14). Erosion-corrosion may also

FIGURE 23.12 Horseshoe-shaped depressions on the internal surface of a brass heat exchanger tube.

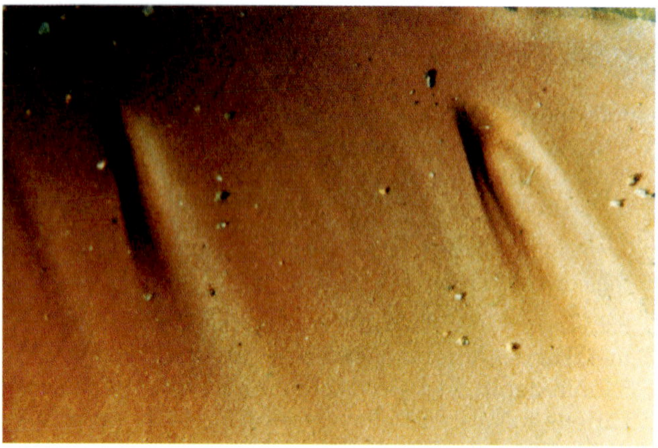

FIGURE 23.13 Comet-tail erosion patterns on copper. (Magnification: 15×.)

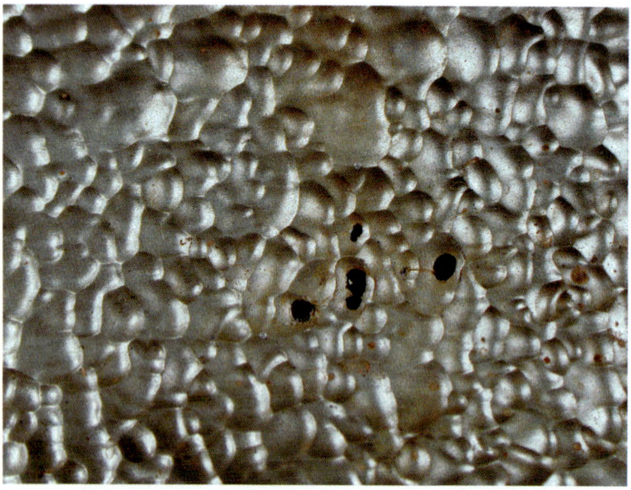

FIGURE 23.14 Sand dune–like appearance of the internal surface of a carbon steel cooling water return pipe. Note the shiny metallic appearance of the surface, indicating that erosion-corrosion was actively occurring shortly before removal from the system.

produce irregular islands of unattacked metal (Fig. 23.15). Localized metal loss tends to occur immediately downstream of features that disrupt flow, such as welds, ledges, or debris adhering to the surface. Occasionally, erosion-corrosion will produce smooth, almost featureless, surface contours (Fig. 23.16), although even in such cases, flow-related features are often observed around the perimeter of the affected region. General metal loss tends to occur when fluid velocities are very high.

FIGURE 23.15 General metal loss surrounding irregular islands of intact metal on the internal surface of air conditioner tubes.

FIGURE 23.16 Relatively smooth, lightly rolling contour of the internal surface at the inlet end of a surface condenser tube.

Affected areas are often essentially free of deposits and corrosion products, particularly if erosion-corrosion was actively occurring shortly before examination. Freshly eroded metal surfaces may have a metallic luster. However, deposits and/or corrosion products may be found nearby, sometimes on islands of unattacked metal within eroded areas (Fig. 23.17). Eroded areas may be covered with deposits and corrosion products if erosion-corrosion was not actively occurring shortly before examination.

Impingement of liquid and/or gas on a metal surface often produces wastage that is rough and jagged. A common example of impingement attack on tubes containing cooling water is on the steam side in a surface condenser, opposite the inlet nozzle (Fig. 23.7 and Fig. 23.18).

Visual examination of affected surfaces is often sufficient to determine that the cause of attack is erosion-corrosion. However, metallographic examinations can often provide information regarding other contributing factors.

FIGURE 23.17 Deposits covering intact metal surfaces at sites of horseshoe-shaped erosion-corrosion depressions. (Magnification: 15×.)

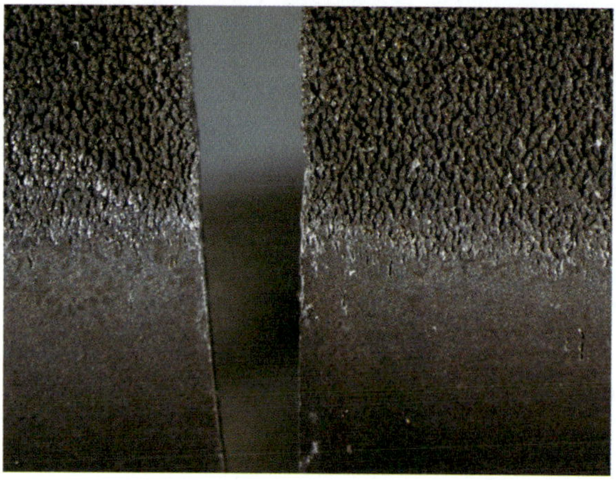

FIGURE 23.18 External surface of a titanium condenser tube, showing a sandy appearance in an area of severe metal loss due to steam impingement. A straight fatigue crack ultimately caused the failure.

Metallographic examinations in eroded areas will indicate bare surfaces (often with no visible oxide layer) if erosion occurred shortly before the sample was removed from service. Depressions may be undercut on their upstream sides (Fig. 23.19). Rifling ridges may be worn down, producing wavelike profiles (Fig. 23.20). If entrained particulate matter contributed to erosion-corrosion, plastic deformation may be evident in a shallow zone along the eroded surface (Fig. 23.21).

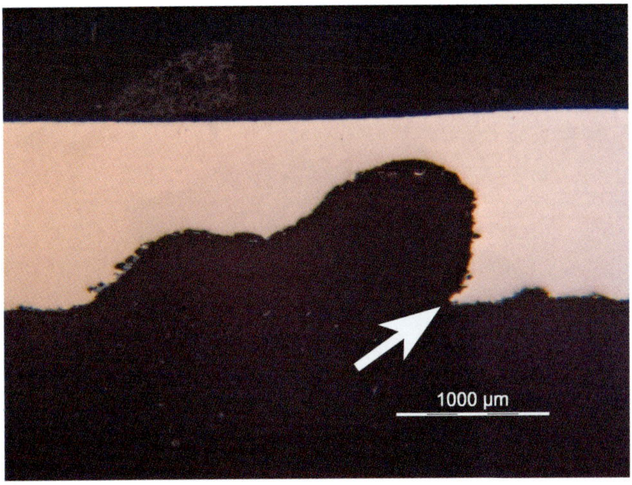

Figure 23.19 Depression caused by erosion-corrosion in a copper pipe. The depression is undercut on the upstream side (arrow). Unetched.

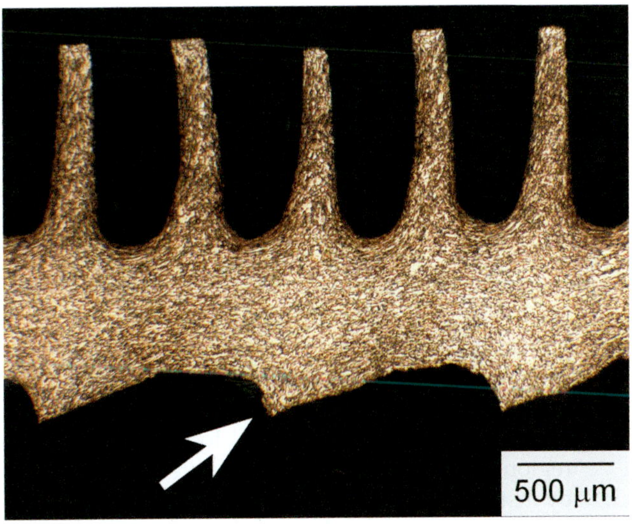

Figure 23.20 Longitudinal cross section through enhanced copper tubing, showing a wavelike profile of the rifling ridges on the internal surface due to metal loss from erosion-corrosion (arrow). Unattacked ridges would have a more rectangular profile. Etchant: Potassium dichromate.

Nondestructive testing methods, such as eddy current testing and ultrasonic testing, may be used to identify locations of thinned pipe or tube walls, though visual inspection is necessary to positively identify erosion-corrosion as the cause of the wastage.

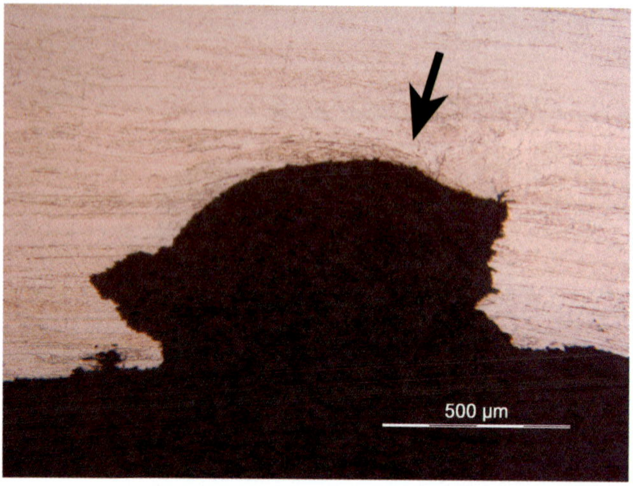

FIGURE 23.21 Cross section through a depression caused by erosion-corrosion in copper piping, showing shallow plastic deformation that suggests impact by entrained hard particles (arrow). Etchant: Potassium dichromate.

Elimination

Due to the many factors that can contribute to the erosion-corrosion process, several approaches can mitigate or eliminate the problem.

1. Because erosion-corrosion is directly linked to velocity and turbulence, reducing both velocity and turbulence are important first steps in elimination. Reducing velocity and turbulence often requires design changes such as the following:
   - Increasing tube or nozzle diameters.
   - Increasing bend radii.
   - Eliminating or diminishing features that disrupt water flow.
   - If erosion-corrosion damage is localized to a water inlet region, altering the distribution pattern of incoming water by installing diffusers in the inlet region.
   - For inlet or outlet end erosion-corrosion of tubes in heat exchangers, sleeves, inserts, or ferrules may be installed into the tube ends. These should be a minimum of 5 in. (12.7 cm) long. The ferrules may be made of nonmetallic or erosion-resistant metals, such as stainless steel, if galvanically compatible. The end of the ferrule should be feathered to prevent turbulence.

   Deterioration of tubes near the inlet nozzle on the shell (steam) side of surface condensers due to impingement of water and steam mixtures can be alleviated through the use of appropriately placed and sized baffles or impact plates, or by applying clip-on impingement shields to the tubes (see Case History 23.8 and Case History 23.9).

2. Because alterations to equipment design can be cumbersome and expensive, a more practical approach may be to change the metallurgy of affected components. Metals used in typical cooling water environments vary in their resistance to erosion-corrosion (see Fig. 23.11).

   The resistance of a metal to erosion-corrosion significantly depends on the tenacity of the corrosion product layer it forms in the environment to which it is exposed. Zinc (brasses), aluminum (aluminum brass), and nickel (cupronickel) alloyed with copper increase the layer's tenacity. An addition of ½ to 1¼ percent iron to cupronickel can greatly increase its erosion-corrosion resistance for the same reason. Similarly, chromium additions to iron-based alloys and molybdenum added to austenitic stainless steels will increase resistance to erosion-corrosion.

   In some situations, it may be possible to only change the metallurgy of affected components. For example, impingement attack on peripheral tubes at the inlet nozzle of a condenser can be alleviated in some situations simply by specifying more erosion-resistant metals, such as 70:30 cupronickel, stainless steel, or titanium, for the affected tubes. Note that, in cases of particularly severe impingement, even metals considered to be very erosion-resistant can still experience metal loss (see Case History 23.9). As with any change in metallurgy, caution should be taken not to create any potentially harmful galvanic couples. Also see the note under "Cautions" regarding tube vibration.

3. Entrainment of particulate matter, such as sand and silt, in cooling water can contribute significantly to erosion-corrosion. Water analysis may indicate if there are appreciable amounts of suspended solids in the water. In these cases, it is important to eliminate or reduce the amount of particulate by settling or filtration. It may also be necessary to reduce or eliminate entrained gas bubbles since they can accelerate erosion-corrosion in a similar manner to particulate matter.

4. Significant localized erosion-corrosion is also commonly associated with lodgment of large foreign objects (sticks, stones, shell fragments) on the face of a heat exchanger tubesheet or within tubes. As a source of erosion-corrosion, this condition has probably been underemphasized. Reduction or elimination requires frequent cleaning of screens and filters and periodic cleaning of the exchanger.

5. Corrosivity of the water has a direct influence on erosion-corrosion susceptibility. Opportunities for reduction of corrosivity are often limited, but the following are offered as guidelines:
   - Eliminate or reduce the amount of potentially aggressive chemical species in the water, such as chloride, sulfate, and sulfide.
   - Decrease water temperature.
   - Appropriately adjust pH.
   - Appropriately adjust hardness or alkalinity to avoid soft water conditions.
6. Use of corrosion inhibitors. Because corrosion is a fundamental part of the erosion-corrosion process, inhibitors that will reduce corrosion under conditions of high fluid velocity, or help maintain the protective corrosion product layers, have been a cost-effective method of limiting erosion-corrosion in some situations. For example, injection of ferrous sulfate either intermittently or continuously has been successful in inhibiting erosion-corrosion of some copper-based alloys in brackish or seawater cooling systems. However, it should be noted that corrosion inhibitors may be stripped from the surface by the high-velocity fluid flow.
7. Cathodic protection can also be useful for limiting the occurrence of corrosion that may contribute to erosion-corrosion, although in heat exchangers, its ability to protect tube interiors is generally limited to the first 4 to 6 in. (10–15 cm) of tube length. Such systems, however, must be properly designed, operated, and maintained to be effective.

Cautions

A classic feature of erosion-corrosion is the directional character of the metal loss. The metal loss will be oriented along the direction of fluid flow or according to turbulence patterns. However, other corrosion modes may produce directionality in metal loss and could be confused with erosion-corrosion. For example, attack on steel by an acid may produce directional metal loss along mandrel marks and along machined surfaces. A corrosive fluid dripping or running down in response to gravity may also produce flow-oriented metal loss.

When combating erosion-corrosion by changing tube metallurgy, caution must be exercised to ensure that no unwanted galvanic couples are created. In heat exchangers, baffle spacing should be appropriate for the mechanical characteristics of the tube metallurgy so that vibration-associated cracking problems are not introduced.

Corrosion tests of metals under static conditions reveal nothing relating to erosion-corrosion susceptibilities. It is entirely possible that a metal with sufficient corrosion resistance under static conditions will fail in service, when excessive fluid velocity and/or turbulence produces erosion-corrosion. Similarly, it has been observed that galvanic corrosion between coupled, dissimilar metals may be accelerated or even initiated under flow conditions when little or no galvanic corrosion is observed under static conditions (see Chap. 16, "Galvanic Corrosion").

Case History 23.1

Industry:	Utility
Specimen location:	Surface condenser
Orientation of specimen:	Horizontal
Environment:	Internal: River water treated with biocides, 35–150°F (2–46°C), pH 8.2
	External: Steam and condensate, pH 8.5–9.0
Time in service:	25 years
Sample specifications:	7/8 in. (22 mm) outer diameter, admiralty brass

Internal surface metal loss of the types illustrated in Fig. 23.22 and Fig. 23.23 affected approximately 45 tubes over a four-month period. It was noted that metal loss along the bottom half of the tubes was more severe. Significant numbers of failures of this type had not been noted before, although it was known that entrainment of silt in the cooling water occurred seasonally.

In this case, reduced river water levels resulting from a drought caused an increase in the amount of sand and silt entrained in the water, thereby increasing

FIGURE 23.22 Metal loss from erosion-corrosion in its early stages. Note the elongated depressions aligned with water flow.

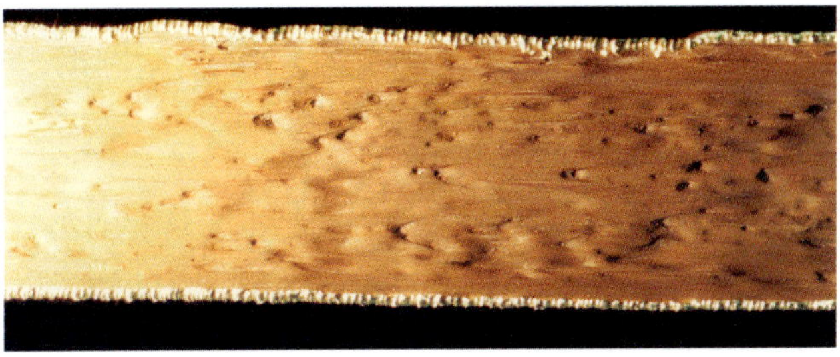

FIGURE 23.23 Erosion-corrosion in its mature form. Note the horseshoe-shaped depressions aligned with flow direction.

its erosive character. This also accounted for the greater metal loss along the bottom half of the tubes, where heavy particulates would have tended to hit the surface with increased frequency.

Note the incomplete attack on the surface in Fig. 23.22. This illustrates erosion-corrosion in its early stages. Compare this to the general metal loss illustrated in Fig. 23.23. Note the superimposed horseshoe-shaped depressions aligned with water flow, which illustrates erosion-corrosion in its later stages.

Case History 23.2

Industry:	Utility
Specimen location:	Surface condenser
Orientation of specimen:	Horizontal
Environment:	Internal: Brackish water treated with tolyltriazole and a chlorine-based biocide External: Steam and condensate
Time in service:	2 years
Sample specifications:	1 in. (25 mm) outer diameter, 90:10 cupronickel tubes

One received set of tubes had been cleaned before submission for examination. The sections contained randomly distributed pinhole perforations at the bottoms of large pits (Fig. 23.24). Associated with these sites was an irregular, arrowhead-shaped region of shallow corrosion that was covered with corrosion products consisting of sparkling cuprous oxide crystals. These regions were immediately upstream of the pit sites. Eddy current tests performed earlier in the year had not detected pitting at that time.

At this point in the investigation, the relationship between the pits and the arrowhead-shaped regions of corrosion was uncertain. Several possible causes for the pitting were considered, such as biological and/or microbiological activity, and debris (concrete chips, etc.) lodged in the tubes, but each was tentatively dismissed as improbable since none of the proposed mechanisms adequately accounted for all observations.

Figure 23.24 Perforation downstream of an arrowhead-shaped area of corrosion. (Magnification: 7.5×.)

FIGURE 23.25 Elongated mounds of fibrous debris attached to the internal surface. (Magnification: 7.5×.)

A few months later, a second set of tube sections was submitted for examination. These tubes had not been cleaned. Close examination revealed arrowhead-shaped mounds of fibrous debris lodged on the tube wall (Fig. 23.25). Dislodgement of these mounds revealed an arrowhead-shaped region of shallow corrosion containing sparkling crystals of cuprous oxide, essentially identical to those observed in the first batch of samples.

The fibrous strands were examined and identified as seed hairs from grass. Subsequently, fibrous material was collected from the cooling tower basin and seed pods from grass growing in a bog located adjacent to and upwind of the cooling towers. This material was also examined and found to be identical to the seed hairs that composed the mounds adhering to the tube wall.

At this point, the sequence of events eventually leading to the tube wall perforations could be established:

1. Seed hairs from the grass formed, matured, and were carried by the wind into the basin of the cooling tower.
2. They collected in the cooling tower basin and were eventually carried into the cooling water system, where their fibrous character permitted them to adhere to rough spots on the tube walls.
3. These spots then became localized collection sites for seed hairs, silt, and corrosion products, forming a mound that was shaped into an arrowhead by water flow.
4. The mounds shielded the tube wall, causing underdeposit corrosion [see Chap. 11, "Crevice and Underdeposit Corrosion (Concentration Cells)"].
5. When a critical mound size was attained, turbulence was created immediately downstream of the mounds, resulting in erosion and eventual perforation of the tube wall.

The eddy current tests performed earlier detected no pits because pit formation depended on the contamination of the system by seed hairs, which occurred after the eddy current testing.

This analysis underscores the importance of examining failed components before they are cleaned or altered in any way. It also demonstrates the potential complexity of failure analysis and that it is imperative to discard explanations that do not adequately account for all relevant observations. Also important to note are good background information on the system and the potential involvement of environmental factors, such as the seasonal development of seed hairs in a field of grass near the cooling tower, that eventually contributed to perforation of condenser tubes.

Case History 23.3

Industry:	Utility
Specimen location:	Condenser tube
Orientation of specimen:	Horizontal
Environment:	Internal: Treated cooling water adjusted with sulfuric acid for pH control and sodium hypochlorite added as a biocide; pressure 50 psi (345 kPa), temperature 100–120°F (38–49°C), water velocity 7 ft/s (2.1 m/s), pH 8.0–8.4, sulfate 500–1000 ppm, chloride 100–450 ppm, total hardness 500 ppm External: Steam and condensate
Time in service:	4 years
Sample specifications:	1 in. (25 mm) outer diameter, 304 stainless steel tubes

Severe, highly localized internal surface metal loss (Fig. 23.26) perforated a total of 100 tubes over a four-year period. Visual inspections disclosed numerous additional damaged but, as yet, unperforated tubes. The metal loss was confined to the first 8 in. (20 cm) of the inlet ends of the tubes.

The areas of metal loss had a bright metallic luster and irregular shapes (Figs. 23.27 and 23.28). Microstructural examinations of the metal loss regions did not reveal any plastic deformation along the internal surface.

The rust-colored concrete chips shown in Fig. 23.29 were removed from tube ends in the condenser. Inspection of the water box revealed large quantities of debris adhering to the tubesheet. The gouging was caused by the lodgment of this hard debris at the inlet ends of the tubes. Intense turbulence downstream from the lodged debris was sufficient to cause highly localized erosion, even in a typically very erosion-resistant material such as stainless steel. Rubbing of the hard debris at the contact points may have also contributed to the metal loss.

Figure 23.26 Severe localized metal loss on the internal surface of a stainless steel condenser tube.

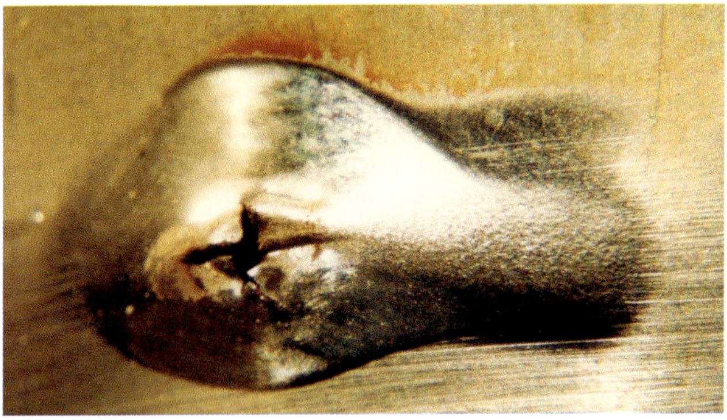

FIGURE 23.27 Close-up of a perforation shown in Fig. 23.26. (Magnification: 7.5×.)

FIGURE 23.28 Elliptical metal loss area on the internal surface, ½ in. from the inlet end. The long axis parallels the longitudinal tube axis. (Magnification: 7.5×.)

FIGURE 23.29 Concrete chips, covered with orange iron oxide, that were removed from the tube ends.

Case History 23.4

Industry:	Pulp and paper
Specimen location:	Lubricating oil cooler
Orientation of specimen:	Horizontal
Environment:	Internal: Mill water, chlorinated (0.2 ppm residual), 120–160°F (49–71°C), pH 7.1–7.3, conductivity 100 μmhos/cm External: Synthetic lubricating oil, 175°F (79°C)
Time in service:	3 months to 3 years
Sample specifications:	3/8 in. (10 mm) outer diameter, brass

The mill experienced 30 to 40 heat exchanger failures over the previous nine months. Service life of the brass tubes in these exchangers varied from three months to three years.

Close examinations revealed that metal loss was confined to the internal surface, within ½ in. (13 mm) of the inlet ends of the tubes (Fig. 23.30). Note the two perforations near the bottom of the image. Generally, the metal loss was confined to the region within the tubesheet, but occasionally perforations occurred just beyond the tubesheet, resulting in leakage of cooling water.

Turbulence at the inlet end caused erosion-corrosion. Problems of this type can be remedied by using appropriate inserts or ferrules.

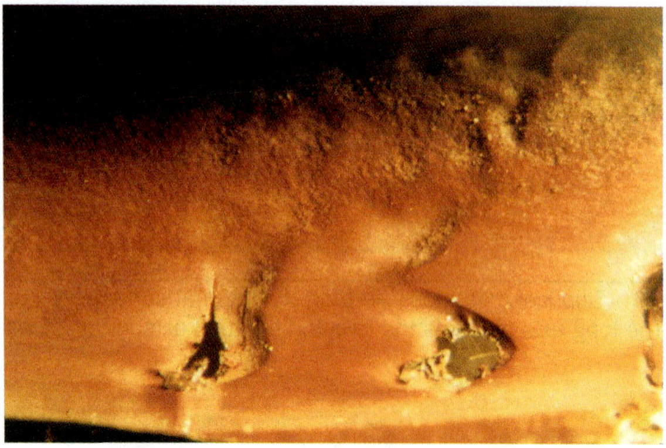

FIGURE 23.30 Erosion depressions and perforations at the inlet end of a brass tube.

Case History 23.5

Industry:	Hospital
Specimen location:	Chilled water control valve
Environment:	Internal: Chilled water External: Atmosphere
Time in service:	Unknown
Sample specifications:	Cast leaded brass

The cross-sectional schematic shown in Fig. 23.31 illustrates the spatial relationship between the eroded components shown in Figs. 23.2, 23.32, and 23.33.

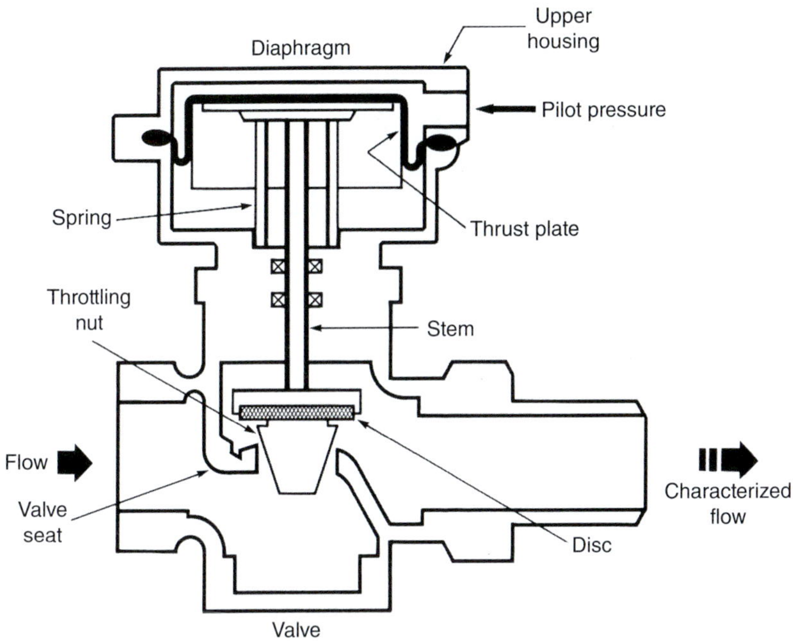

FIGURE 23.31 Cross-sectional schematic of chilled water control valve. (*Reprinted with permission from Landis & Gyr Powers, Inc., Buffalo Grove, Illinois.*)

FIGURE 23.32 Erosion-corrosion of a valve seat (bright metal in center of photograph).

FIGURE 23.33 Metal loss by erosion-corrosion at edges of a throttling nut. (Magnification: 7×.)

Highly localized metal loss at the seat of the globe valve is apparent in Fig. 23.2. Figure 23.32 shows the same component, close up. Wasted surfaces had a bright, metallic luster and were free of corrosion products or deposits. Metal loss along the edge of the throttling nut was also apparent (Fig. 23.33). Erosion-corrosion of these components was caused by high-velocity turbulent flow through a narrow opening that resulted from incomplete opening of the valve. In this case, erosion was the dominant factor in the metal loss. Corrosion was a minor contributing factor. Since globe-type valves such as this one are typically recommended for throttling flow, it was recommended that a more erosion-resistant material be considered for this application.

Case History 23.6

Industry:	Utility
Specimen location:	Service water pump impeller
Orientation:	Vertical
Environment:	Once-through cooling water
Time in service:	5 years
Sample specifications:	9½ in. (24 cm) diameter, silicon brass

The leading edges on most vanes were thinned down to knife edges (Fig. 23.34). Portions of the leading edges of some vanes were repaired. Most of the vane surfaces and the hub surfaces on the inlet side had a generally smoothened contour. Horseshoe-shaped depressions were evident in the metal loss areas. The vane surfaces generally had a shiny, metallic appearance, free of deposit or oxide layers. The impeller had a reddish appearance in places, particularly on the sides. Metallographic examinations indicated a dendritic microstructure, typical of cast brass. Porous copper corrosion products in the interdendritic areas indicated that shallow dealloying occurred in the deeply corroded areas, as well as at locations away from the severe attack (see Chap. 17, "Dealloying").

The damage was caused by erosion-corrosion, with dealloying as the primary corrosion mechanism. The porous copper corrosion products that result from dealloying are mechanically weak and are easily removed by rapid and turbulent fluid flow. Since rapid and turbulent fluid flow is intrinsic to the operation of a pump, it was recommended that more erosion-resistant alloys, such as aluminum bronze, cupronickel, and stainless steel, be investigated as alternative metallurgies.

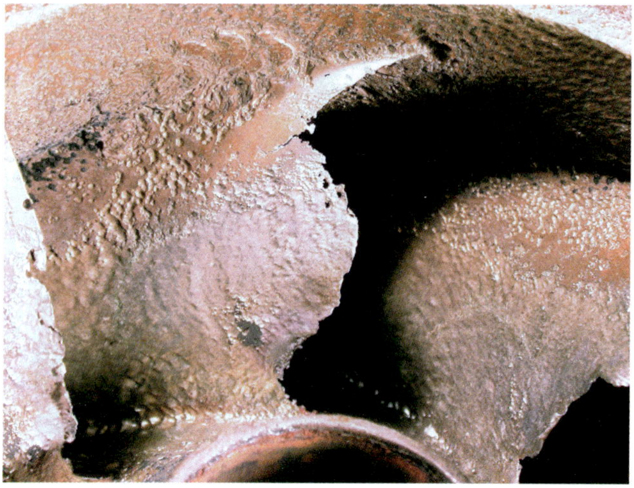

FIGURE 23.34 Severe flow-related wastage on the vanes of a silicon brass pump impeller.

Case History 23.7

Industry:	Utility
Specimen location:	Chiller condenser
Orientation:	Horizontal
Environment:	Internal: Cooling water External: Refrigerant 134
Time in service:	4 years
Sample specifications:	¾ in. (1.9 cm) diameter enhanced 90/10 cupronickel tubing

A small, thin-edged perforation occurred near the outlet end of the tube (Fig. 23.35). The perforation occurred in one of two deep, irregularly shaped

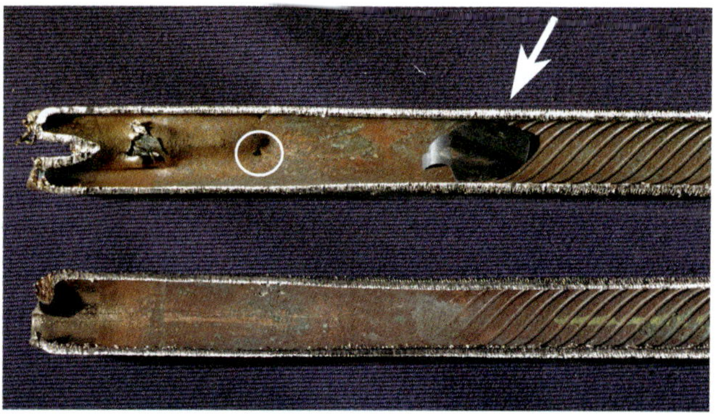

FIGURE 23.35 Internal surface of a section of enhanced cupronickel chiller tubing, showing a small perforation (circled) downstream from a large chip of black plastic (arrow).

FIGURE 23.36 Close-up view of the perforation shown in Fig. 23.35, which occurred in an area of very deep internal surface metal loss.

areas of internal surface metal loss (Fig. 23.36). The wasted surfaces were smoothened and generally free from deposit or corrosion product layers. Metallographic examinations did not indicate any evidence of plastic deformation in the wasted areas. The wastage occurred downstream from a large black plastic chip, which was not easily dislodged from the tube. The plastic chip had previously been observed during borescope examinations. Infrared spectral analysis of the black plastic chip indicated that it consisted of polyvinyl chloride (PVC). Away from the deeply wasted areas, a few very small corrosion product mounds were scattered across the internal surface.

The metal loss and perforation were caused by erosion-corrosion. The average flow velocity through the tubes was reportedly 6.2 ft/s (1.9 m/s). This value was below the maximum recommended velocity to avoid erosion in 90/10 cupronickel (8 ft/s, or 2.4 m/s; see Table 23.1). However, the lodged debris disrupted water flow, creating significant turbulence that resulted in localized metal loss. The plastic debris may have originated from cooling tower fill.

According to a water analysis from the cooling tower that supplied the unit, the cooling water contained a very high concentration of chloride (9500 ppm) and an appreciable concentration of sulfate (1700 ppm). Such anions are known to reduce the resistance of cupronickel to erosion-corrosion. Corrosion product mounds away from the erosion damage suggested that the water was somewhat corrosive toward the cupronickel. However, erosion caused by foreign debris lodged in the tube was the primary cause of the metal loss.

Case History 23.8

Industry:	Utility
Specimen location:	Condenser tube, top row immediately adjacent to steam inlet
Orientation of specimen:	Horizontal
Environment:	Internal: Untreated river water External: Wet steam, pH 9.0–9.6
Time in service:	17 years
Sample specifications:	½ in. (13 mm) outer diameter, admiralty brass

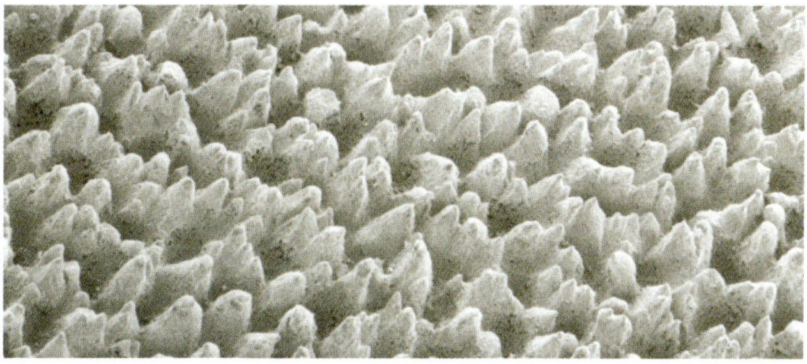

FIGURE 23.37 Scanning electron micrograph showing cones of brass created by impingement of high-velocity wet steam. (Magnification: 50×.)

The longitudinally oriented split apparent in Fig. 23.7 was located along the top of the tube facing the steam inlet nozzle. This was one of several tubes in this area having similar longitudinal splits. Leakage of river water from the tubes resulted in boiler feedwater contamination, which was a major factor in subsequent tube failures in the boiler.

The split was centered in a distinct band of severe wall thinning from the external surface. Close examinations of this surface under a low-power microscope revealed a multitude of small, well-formed brass cones pointing straight up from the surface (Fig. 23.37). The external surface on the bottom of the tube retained its smooth, original surface.

The metal loss and resulting split were caused by erosion from impingement of high-velocity steam in which droplets of water were entrained. The conical features in the affected region were typical of damage by erosion due to impingement.

Although the coolant (river water) was at relatively low pressure, measurements revealed a residual hoop stress in the tube of approximately 9000 psi (62 MPa). The longitudinal rupture occurred from the combination of applied and residual stresses after erosion had sufficiently reduced the wall thickness.

Several suggestions were provided for eliminating the erosion damage:

1. Improve moisture separation from the steam before it enters the condenser.
2. Specify metals that are more resistant to erosion, such as cupronickel, nickel-copper alloys, and stainless steels. Only affected tubes need to be replaced by these metals.
3. Install a deflector shield to protect the area experiencing erosion. Note that shields could be clipped to the tubes themselves.

Case History 23.9

Industry:	Utility
Specimen location:	Surface condenser
Orientation:	Horizontal
Environment:	Internal: Cooling water External: Steam
Time in service:	Unknown
Sample specifications:	1 in. (2.5 cm) diameter titanium tubing

A longitudinal band of metal loss occurred on the external surface of a titanium condenser tube section. Where the metal loss occurred, many sharp, flow-aligned cones gave the surface a sandy appearance (Fig. 23.18). A single straight, unbranched, transverse crack extended from the wasted area into the unwasted portion of the tube. Near the crack edge, numerous other fissures originated from between cones.

The wastage was caused by erosion due to impingement of steam that contained entrained water droplets. The transverse crack was caused by fatigue due to cyclic bending loads and vibration caused by the impinging steam. Thinning of the tube wall due to erosion weakened the tube wall and facilitated the cracking. Also, the locations between the cones provided sites of stress concentration that promoted the formation of fatigue fissures.

It was recommended that steam separation equipment be examined to ensure proper separation of water from the steam. Periodically replaceable protective shielding for the tubing was also suggested.

References

1. Richard A. White, *Materials Selection for Petroleum Refineries and Gathering Facilities*, NACE International, Houston, TX, 1998.
2. F. J. Heymann, "Liquid Impingement Erosion," in *ASM Handbook*, Volume 18, *Friction, Lubrication and Wear Technology*, 10th ed. ASM International, Materials Park, OH, 1992.

CHAPTER 24

Cavitation Damage

General Description

The first recorded observation of cavitation was made in 1754 by Euler. Cavitation is the near-instantaneous formation and collapse of vapor bubbles in a liquid subject to rapid, intense, localized pressure changes. The deterioration of a material resulting from its exposure to a cavitating fluid is referred to as cavitation damage. Cavitation is sometimes considered to be a form of erosion, as both types of wastage are caused by mechanical damage inflicted by the fluid. However, the mechanisms and wastage morphologies of cavitation and erosion are quite different, though both types of damage are sometimes found together on components such as pumps due to the turbulent and high-velocity flow during operation.

Figure 24.1 is a simplified representation of the cavitation process. Figure 24.1A represents a vessel containing a liquid. The vessel is closed by an airtight plunger. When the plunger is withdrawn (B), a partial vacuum is created above the liquid, causing vapor bubbles to form and grow within the liquid. In essence, the liquid boils without a temperature increase. If the plunger is then driven toward the surface of the liquid (C), the pressure increases and the bubbles collapse (implode). These three steps occur in a matter of milliseconds in a cavitating liquid.

Cavitation damage results from hydrodynamic forces created by the collapsing vapor bubbles. Implosion of a vapor bubble creates a microjet of water that is ejected from the collapsing bubble at velocities that may range from 330 to 1650 ft/s (about 100–500 m/s).[1] Normally, the energy of this microjet is quickly absorbed by the surrounding fluid. However, if the bubble collapse occurs adjacent to a metal surface, the microjet may strike the surface, dislodging corrosion products (Fig. 24.2) and/or locally deforming the metal itself. Single impacts result in little, if any, metal damage. However, repeated impacts cause continual dislodging and re-formation of corrosion products on the metal surface, which would otherwise be protective in a well-managed cooling water environment. Eventually, this process results in significant metal loss. When the metal itself is directly affected by cavitation (i.e., when there is not enough time between impacts to re-form the corrosion product layer), continuous impacts cause removal of chunks of metal.

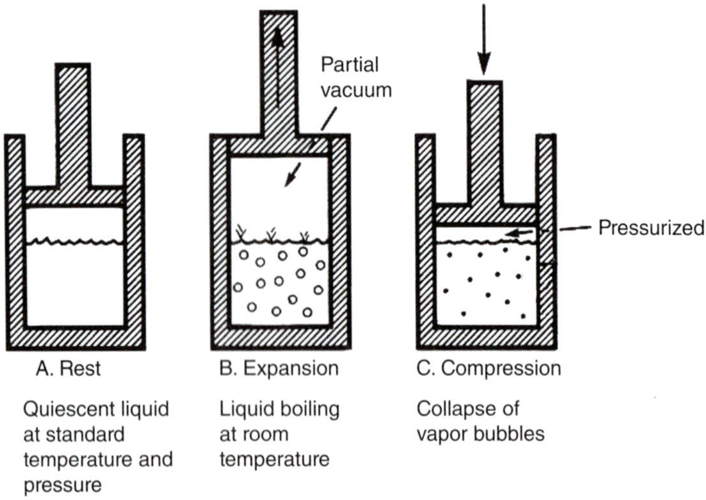

FIGURE 24.1 Schematic representation of cavitation—a cross section through a vessel and plunger enclosing a fluid. (A) Plunger stationary, liquid at standard temperature and pressure; (B) plunger withdrawn, liquid boils at room temperature; (C) plunger advanced, bubbles collapse.

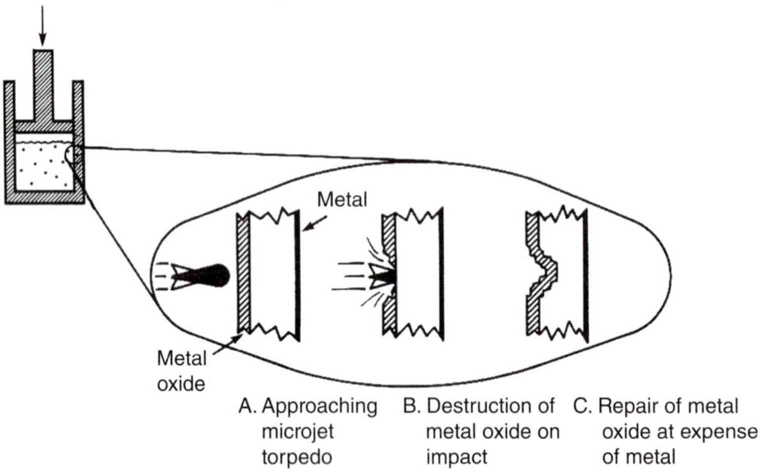

FIGURE 24.2 Disintegration of protective corrosion product by impacting microjet torpedo.

Metal damage due to cavitation can be both rapid (Figs. 24.3–24.6) and severe. However, cavitation damage can have a time dependency. An incubation period may be observed, during which little or no damage is apparent. Following the incubation period, the rate of metal loss may increase sharply. After reaching a maximum value, the rate of damage may then either decline to a lower steady-state value or fluctuate unpredictably.

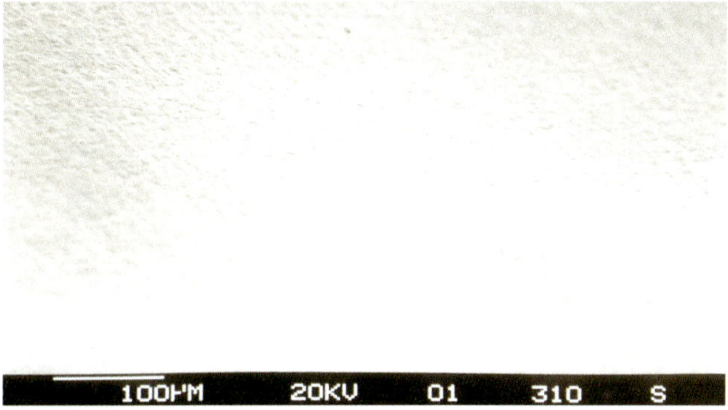

Figure 24.3 A scanning electron micrograph used to show progressive deformation in aluminum foil exposed to a cavitating fluid for successively longer periods. This image shows the surface prior to exposure. (Magnification: 350×.)

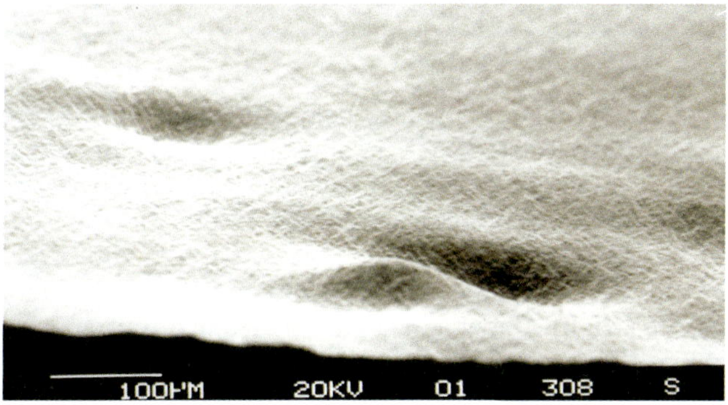

Figure 24.4 The surface of Fig. 24.3 exposed for 5 seconds.

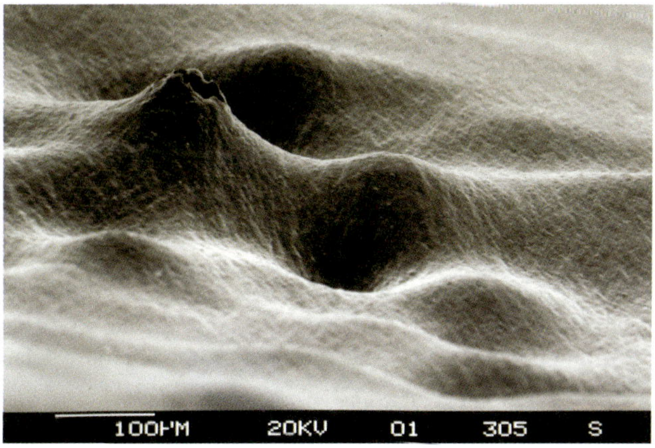

Figure 24.5 The surface of Fig. 24.3 exposed for 10 seconds.

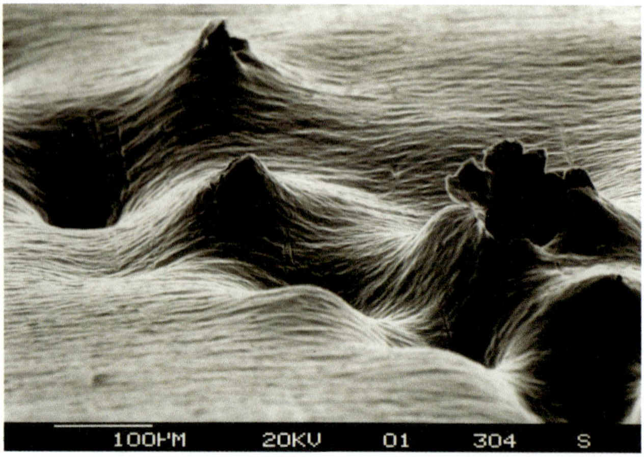

Figure 24.6 The surface of Fig. 24.3 exposed for 20 seconds.

Locations

In general, cavitation damage may be possible wherever an unstable state of fluid flow exists or where substantial pressure changes are encountered. Susceptible locations include those downstream from sharp discontinuities, surface irregularities, or other features on metal surfaces that result in turbulence. Areas where flow direction of the fluid is suddenly altered (Fig. 24.7), and regions where the cross-sectional areas of the flow passages are changed, are also particularly susceptible to cavitation damage.

Damage will be confined to the bubble-collapse region, usually immediately downstream of the low-pressure zone. Components exposed to high-velocity or turbulent flow, such as pump impellers, pump housings, valves, and tee fittings, are often subject to cavitation. The suction side of pumps (Case History 24.3) and the discharge side of regulating valves (Fig 24.8 and Case History 24.4) are frequently affected. Tube ends, tubesheets, and shell outlets in heat exchanger equipment have been affected, as have cylinder liners in diesel engines (Case History 24.1).

Critical Factors

The essential condition governing cavitation is unstable fluid flow. Unstable fluid flow is affected by the properties of the fluid itself, the physical characteristics of the surfaces exposed to the liquid, and equipment operating practices. These factors are discussed in more detail next.

Properties of the Fluid

Temperature, pressure, chemical composition of the fluid, and air bubble content can affect the tendency of the fluid to cavitate. For example,

FIGURE 24.7 Cavitation damage at bottom of a tee at the impact site of in-flowing water.

FIGURE 24.8 Cavitation damage on the discharge side of a regulating valve.

minute air bubbles or suspended solid particles in the fluid can act as nucleation sites for cavitation bubbles, thereby increasing the tendency of the fluid to cavitate. Increasing pressure decreases susceptibility to cavitation; decreasing pressure increases susceptibility to cavitation.

Corrosion can significantly increase the rate of metal loss due to cavitation, though it is certainly not required for cavitation to occur. Cavitation damage in the complete absence of corrosion has been demonstrated (e.g., roughening of polished glass in cavitating distilled water). However, in cooling water environments it is highly probable that corrosion is a contributing factor.

Physical Characteristics of Exposed Surfaces

Cavitation is favored by:

- Rough surfaces, which provide more nucleation sites for bubbles than smooth surfaces
- Sharp surface discontinuities
- Rapid changes in flow direction
- Abrupt changes in cross-sectional area

In addition, metals that form a tenacious, protective oxide layer on the surface, like stainless steel and titanium, are generally more resistant to cavitation damage than those that do not.

Equipment Operating Practices

Circumstances that may induce cavitation include vibration, excessively high flow rates, insufficient head pressure to the suction side of pumps, and operation of valves in a partially open or closed position.

In cooling water systems, pumps are probably the components that most commonly sustain cavitation damage. There are five main reasons for cavitation in pumps, as summarized in the following list. More in-depth discussions of the causes of cavitation may be found in the literature.[2]

1. Vaporization: Caused by factors on the suction side, such as throttling of valves or clogged filters. This results in excessively low pressure, which causes localized boiling and subsequent cavitation to occur. This is the most common type of cavitation in pumps.
2. Internal recirculation: Restricted flow on the discharge side causes liquid to recirculate within the pump. This causes the liquid to overheat and/or pass through narrow areas at high speed. Possible reasons for restricted discharge-side flow include a closed valve or a clogged filter.
3. Vane passing syndrome: Blade tips of the impeller pass too close to the cutwater of the pump casing. This may be caused by an oversized impeller or improper repair or rebuild of the impeller and/or housing to incorrect specifications.

4. Air aspiration: Caused by air drawn into the pump. Air may enter through a variety of locations, including pump shaft packing, joints, or valves in the suction piping.
5. Turbulence: Results from the formation of vortexes in the suction flow due to inadequacies in the design of the suction piping system.

Identification

In some cases, a cavitating pump may produce noise similar to stones hitting a metal surface. However, normal pump noise may drown out such sounds. Visually, cavitation typically produces sharp, jagged, spongelike metal loss, even in ductile materials. The affected region is free of deposits and accumulated corrosion products if cavitation has been recent.

Typically, affected surfaces are highly localized to specific regions, although if cavitation is severe and widespread, the area affected may be extensive (Fig. 24.9). On symmetrical components with repeating elements, such as the vanes of pump impellers, the pattern of damage may repeat itself at identical locations on each element (Fig. 24.10). The pattern of areas with cavitation damage on an impeller may provide information about the specific reason for the cavitation (Fig. 24.11).

Damage due to cavitation and strong mineral acids can have similar surface appearances. However, cavitation results in highly specific and

FIGURE 24.9 Cast iron pump impeller with severe cavitation damage.

FIGURE 24.10 Silicon red brass pump impeller, showing a repeating pattern of metal loss at the edges of the vanes.

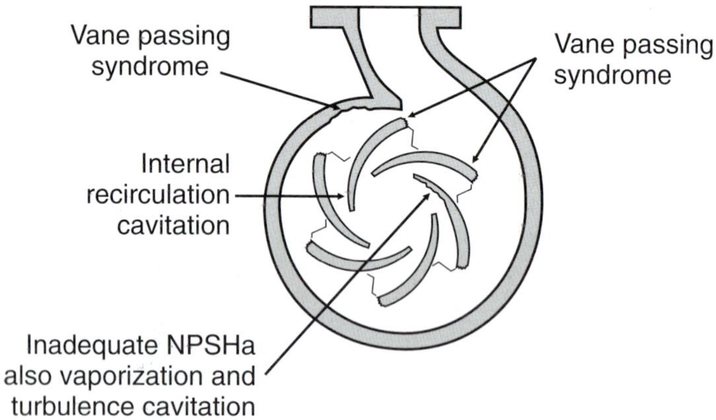

FIGURE 24.11 The pattern of cavitation damage in a centrifugal pump can provide information on the specific reason for the cavitation. Some of the common locations of cavitation damage and their typical causes are indicated.[3] (*Courtesy of The Bachus Company Ltd.*)

localized areas of attack, whereas strong acid corrosion typically produces widespread general metal loss (see Chap. 14, "Acid Corrosion").

Examination of the microstructure of the cavitated surface will sometimes reveal evidence of deformation such as twins (Neumann bands) in carbon steel or cast iron (Fig. 24.12). Deformation twins are indicative of a high rate of plastic deformation, such as that which

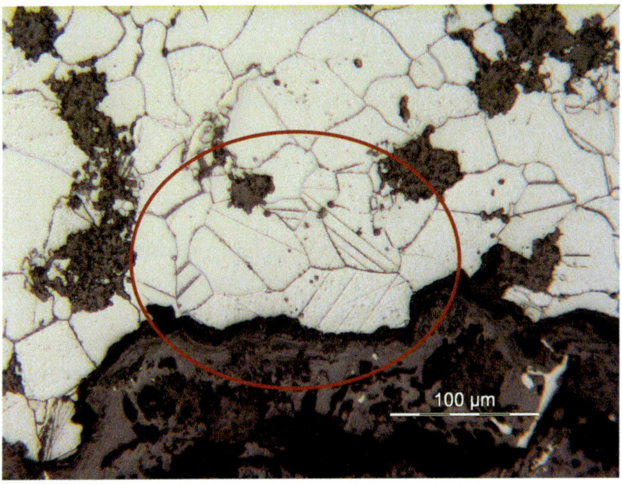

FIGURE 24.12 Micrograph showing deformation twins (Neumann bands) in a ductile cast iron pipe elbow that exhibited cavitation damage. Etchant: Nital.

occurs due to shock loading. The twins appear in micrographs as well-defined parallel lines that terminate at grain boundaries. Other metals may exhibit evidence of cold work.

Elimination

Several approaches, which are listed here, are available to alleviate or eliminate cavitation damage problems. The best remedial action will depend on the factors influencing cavitation.

- Redesign of equipment
- Alteration of operating procedures
- Change of materials
- Use of protective coatings
- Alteration of environment

Redesign of Equipment

Of the various approaches to reducing cavitation damage, redesign of equipment is probably the most successful. The following changes have proven useful:

- Eliminate abrupt changes in cross-sectional area within tubes, pipes, orifices, and flow channels.
- Increase radii of curvature of bends in the fluid flow path.

- Remove surface discontinuities.
- Smoothen surface contours to reduce bubble nucleation sites.
- Ensure that pump components are adequately sized and aligned, especially when performing repairs or replacements.

Alteration of Operating Procedures

In principle, altering operating procedures involves minimizing localized pressure changes. In practice, this involves such things as maintaining adequate feed to the suction side of a pump impeller (maintaining adequate net positive suction head), reducing flow velocity through a heat exchanger, or eliminating or reducing the magnitude and frequency of vibration in piping or a cylinder liner. Increasing system pressure can often reduce pressure changes. Filters upstream of cavitating pumps should be checked and cleaned because something as simple as a clogged filter can restrict flow enough to cause cavitation.

Change of Materials

If redesign of the equipment or changing operating procedures cannot eliminate cavitation damage, it may be necessary to choose an alternate material. Historically, numerous attempts have been made to correlate resistance to cavitation damage with certain material properties, such as hardness. It has been observed that cavitation resistance relates in a general way to a metal's hardness and ductility. Therefore, alloys such as 18-8 stainless steels are often recommended when alloys such as carbon steel, cast iron, or copper alloys experience cavitation damage. However, cavitation is unfortunately too complex for such a simple relationship between hardness and cavitation resistance to be universal. Resistance to cavitation damage is significantly influenced by factors such as the corrosive nature of the cavitating fluid and the intensity of the cavitation. Hence, no specific material properties may be used to reliably rank metals according to cavitation resistance in all environments. As always, care should be taken to avoid any undesirable galvanic couples when changing alloys.

Weld overlays of stainless steel or cobalt-based wear-resistant and hard facing alloys, such as Stellite, may salvage damaged equipment. In addition, weld overlays incorporated into susceptible zones of new equipment may provide cost-effective resistance to cavitation damage.

Use of Coatings

Resilient materials such as rubber, other elastomers, and some plastics may be useful in certain applications, especially under conditions of low cavitation intensities. Such materials reduce bubble nucleation sites and absorb the implosion energy of the bubbles. However, disbondment of the coating can occur at the metal-coating interface at high cavitation intensities, even if the exposure is brief.

Alteration of Environment

If the cavitation intensity is low and corrosion is a significant accelerating factor, appropriate corrosion inhibitors can be useful. Notable successes have been achieved with corrosion inhibitors that are used in diesel engine coolants to control cavitation damage on cylinder liners.

Cautions

It is possible to confuse corrosion by a strong mineral acid with damage by cavitation. Both can produce jagged, undercut, spongelike surface contours. The two can be distinguished, however, by noting that attack by a strong acid generally affects all exposed surfaces and is therefore not as location-specific. Cavitation is localized to regions experiencing rapid pressure changes and is therefore quite location-specific.

Related Problems

See also Chap. 14, "Acid Corrosion"; Chap. 23, "Erosion-Corrosion."

Case History 24.1

Industry:	Automotive
Specimen location:	Diesel engine cylinder liner
Orientation of specimen:	Vertical
Environment:	Internal: Hot combustion gases, moving piston External: Cooling water containing ethylene glycol
Time in service:	Unknown
Sample specifications:	Gray cast iron

Figure 24.13 illustrates the typical appearance of cavitation damage along opposite sides of a diesel engine cylinder liner. The vertical pit alignment, the presence of the pits at opposite sides of the liner, and the comparatively smaller population of pits along one side are typical of cavitation damage of cylinder liners. Note also the jagged, spongelike, highly localized metal loss shown in Fig. 24.14. This is characteristic of cavitation damage in general.

Cavitation occurred on the cooling water side due to vibrations generated during the operation of the engine. Possible sources of such vibration included piston slap, loose bearings, and a crankshaft or piston that was out of balance.

In order to minimize cavitation, it was necessary to reduce or eliminate the vibration. Addition of corrosion inhibitors to the coolant were suggested, as they have been successful in controlling cavitation damage in some cases.

Figure 24.13 Typical vertical alignment of cavities resulting from cavitation damage in a diesel engine cylinder liner.

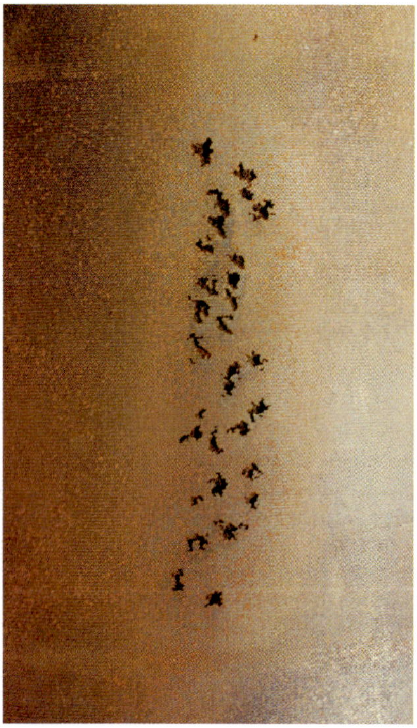

Figure 24.14 Jagged, cavernous pits typical of cavitation damage.

Case History 24.2

Industry:	Automotive
Specimen location:	Diesel engine spacer
Orientation of specimen:	Vertical
Environment:	Cooling water containing ethylene glycol
Time in service:	200 hours
Sample specifications:	Cast aluminum

Figure 24.15 shows the throat of the spacer. Damage was in the form of pits that were localized to both sides of a 2¼ in. (5.7 cm) long vertical opening.

The pits were clustered into groups near the center and along the edges of the bore (Fig. 24.16). The interiors of the pits were jagged and visually free of deposits or corrosion products. The cavitation damage on the spacer surface was due to vibrations from operation of the engine.

Metallographic examinations revealed a thin coating of corrosion products and no evidence of cold working of the surface metal, suggesting that cavitation affected the corrosion product layer rather than the metal directly. Therefore, it was thought that the use of an appropriate corrosion inhibitor could reduce cavitation damage.

FIGURE 24.15 Zones of cavitation damage segregated to both sides of a vertical opening in an aluminum diesel engine spacer.

FIGURE 24.16 Aluminum diesel engine spacer, showing a spongelike surface contour typical of cavitation damage. (Magnification: 8×.)

Case History 24.3

Industry:	Metals
Specimen location:	Pump impeller and diffuser
Orientation of specimen:	Vertical
Environment:	Untreated cooling water-oil emulsion, pH 4.3–5.1, 135–145°F (57–62°C)
Time in service:	7 years
Sample specifications:	Ferritic gray cast iron

Figure 24.17 illustrates severe damage to a cast iron pump impeller due to cavitation. The characteristically jagged, spongelike metal loss almost completely consumed the vanes of the impeller. The diffuser exhibited severe cavitation damage as well (Fig. 24.18). Some flow-related metal loss was observed on the

FIGURE 24.17 Cast iron pump impeller, showing severe cavitation damage that virtually consumed the vanes.

FIGURE 24.18 Back of the cast iron diffuser, showing severe metal loss.

diffuser vanes, indicating that erosion-corrosion also occurred, though cavitation caused the bulk of the damage.

Microstructural examinations revealed shallow graphitic corrosion (see Chap. 19) both in and away from the wasted areas. The evidence of graphitic corrosion indicated one of the following:

1. Cavitation was occurring intermittently.
2. Cavitation was more or less continuous, but the cavitation intensity was low enough that only the mechanically weak layer of corrosion products was dislodged.

The pump supplied a water-oil emulsion for cooling rolls during hot and cold rolling of aluminum. The emulsion had a relatively low pH, which most likely promoted graphitic corrosion. The graphitic corrosion was not excessive, as evidenced by the shallow corrosion away from the wasted areas. However, graphitic corrosion did accelerate metal loss due to cavitation and erosion-corrosion.

Recommended solutions were to ensure sufficient coolant supply to the pump and to check for any clogged filters or other sources of flow obstruction, particularly on the suction side. Changing the pump metallurgy to a more corrosion- and cavitation-resistant material, such as stainless steel, was also a potential solution if operational changes could not alleviate the cavitation problem.

Case History 24.4

Industry:	Steel
Specimen location:	Main supply line to spray cooling system
Orientation of specimen:	Horizontal
Environment:	Internal: Untreated lake water, intermittently chlorinated External: Air
Time in service:	6 years
Sample specifications:	Cast carbon steel

The damage illustrated in Fig. 24.19 occurred in an expansion flange located immediately downstream of a valve. The valve was operated in a partially closed position.

Jagged, spongelike metal loss occurred downstream of a circumferential weld bead joining the flange to the expansion pipe section (Fig. 24.20). An area of similar metal loss was observed further downstream, in the region of diameter increase (Fig. 24.21). Metal loss was most severe slightly downstream from the weld bead, which protruded from the inner diameter and likely caused turbulence where weld repairs from the external surface closed a large perforation.

Cavitation damage was caused by severe turbulent flow past a partially closed valve, coupled with flow disruption across the intruding weld bead as well as localized pressure reduction in the region of the diameter increase.

Potential remedial measures included:

- Operating the valve in the fully open position
- Specifying a more cavitation-resistant material such as stainless steel, or carbon steel overlaid with a cavitation-resistant material

FIGURE 24.19 Damage to the internal surface of a supply line.

FIGURE 24.20 Jagged metal loss downstream of protruding weld bead.

FIGURE 24.21 Metal loss downstream of flange wall expansion.

Case History 24.5

Industry:	Utility
Specimen location:	Cooling water supply pump
Orientation of specimen:	Vertical
Environment:	Cooling water
Time in service:	2 years
Sample specifications:	Silicon red brass pump impeller (82% copper, 14.0% zinc, 4.0% silicon)

The impeller exhibited localized areas of rough metal loss at the edges of the vanes (Fig. 24.10 and Fig. 24.22). Overlapping, comet-shaped depressions were observed along the edges of the shrouds. In most locations, the impeller was reddish or copper-colored. However, the surface was brass-colored in the center of deeply wasted areas.

The repeating pattern of rough metal loss was indicative of cavitation damage. However, the comet-shaped depressions on the shrouds were characteristic of erosion-corrosion damage. Microscopic examinations confirmed that dealloying occurred, as suspected due to the copper color of the surface in most places (see Case History 17.7 in Chap. 17, "Dealloying"). The soft, mechanically weak metallic copper corrosion products that resulted from dealloying accelerated the metal loss caused by cavitation and erosion-corrosion. A harder material, such as aluminum bronze, was suggested to reduce cavitation damage.

FIGURE 24.22 Close-up view of rough metal loss at the edges of the vanes. The surface of the impeller generally had a reddish color, though the areas of deepest wastage were brass-colored at the center.

References

1. Frederick G. Hammitt and Frank J. Heymann, "Liquid-Erosion Failures," *ASM Handbook,* Volume 11, *Failure Analysis and Prevention*, ASM International, Metals Park, OH, 1986.
2. L. Bachus and A. Custodio, *Centrifugal Pumps and Everything You Need to Know About Pumps*, The Bachus Company Inc., 2003.
3. Bachus and Custodio, *Centrifugal Pumps and Everything You Need to Know About Pumps.*

SECTION 5

Material and Design Issues

The true influence of flaws and defects on component failure is commonly misunderstood. This misunderstanding often arises from one of two misconceptions. The first misconception can be clarified by simple definitions of a flaw and a defect.

Flaw: A flaw can be defined as an imperfection in a material that does not affect its usefulness or serviceability. A component may have flaws and still retain its usefulness. This fact is recognized by most material codes that permit, but limit, the size and extent of flaws. This is particularly true of welds, which commonly contain harmless flaws. In addition, metal typically contains a small population of microscopic, nonmetallic inclusions that do not adversely affect mechanical properties or significantly contribute to corrosion.

Defect: A defect is an imperfection in a component, resulting from the manner in which it was manufactured, shipped, handled, or installed. A defect may contribute materially to failure or limited serviceability. Failures in cooling water systems resulting directly or indirectly from defects account for less than 1 percent of all failures.

The second misconception involves the perception of what constitutes a defect. A defect is not simply a visually observable imperfection or internal imperfection that can be identified by various monitoring techniques. Defects, from a failure-analysis standpoint, may also result from microstructural features that are produced during production, heat treatment, and forming of the alloy. Certain microstructures may lead to cracking or highly localized corrosion. These considerations will be covered in Chap. 25, "Material Defects," and Chap. 26, "Weld Defects."

CHAPTER 25

Material Defects

General Description

The percentage of failures that occur solely from material defects is quite low, less than 1 percent of all failures. It is more common for defects to combine with specific environmental factors to produce failures, such as cracking or localized corrosion.

Nearly every manufacturing, shipping, and installation process is a potential source of a defect. In recognition of this, manufacturers mount major efforts to minimize or eliminate defects. Although an exhaustive list of defects is beyond the scope of this book, some common defects found in components used in cooling water systems will be discussed.

Defects in finished components can sometimes be traced as far back as the initial casting process for the alloy due to the formation of voids or inclusions. Subsequent mechanical forming operations conducted on the original casting will deform, flatten, and elongate the original defect. Defects in seamless tubes can originate in extrusion or drawing processes.

Defects in tubes, pipes, and other formed components may appear on the surfaces or they may be hidden subsurface within the wall. The defects often cannot be detected by visual inspection. Some common defects are listed here and are described in more detail in the section on "Identification."

- *Gouges and grooves.* Gouges and grooves are caused by mechanical removal of metal during fabrication and/or installation. Gouges tend to be localized, while grooves tend to be elongated. Depending upon the cause of metal deformation, the gouge or groove may be relatively deep. Sharp grooves can act as stress concentrators, which can lead to various kinds of cracking.
- *Laps.* Laps consist of deformed, thin flaps of metal that form from the surface of the component and fold over onto the surfaces of adjacent areas. The laps, which are not fused to the

surface, may form during mechanical rolling, extrusion, drawing, die threading, or tapping processes. Enhanced tubes that are used in chillers are particularly susceptible to the production of lap defects by the mechanical processes of forming external surface fins and internal surface rifling. Metallographic examination will reveal if the folded metal formed from the component itself as a lap, and was not formed due to transfer of another foreign metal onto the surface.

- *Seams.* As a defect, a seam is distinct from pipe or tube weld seam defects that result from a welding process (discussed in Chap. 26, "Weld Defects"). Seams can be found in nonwelded (seamless) tubes. They can originate from large pores or nonmetallic inclusions in the original cast alloy. They are larger than laps, are produced by a forming process, and remain unfused. At times, they will appear in a spiral pattern in tubing and piping. Seams can be very tight and appear as hairlines on the surface. If a seam is deep, it can cause failure when the component is pressurized. Metallographic examination will reveal if the folded metal formed from the component as a lap and was not formed due to transfer of another foreign metal onto the surface. In addition, metallographic examination will definitely distinguish a seam defect from a weld seam defect.
- *Dents, crimps, corrugations, and flattening.* These appear as simple deformation of the normal geometry of the component. Corrugations may be produced in thin-walled tubes at bends.
- *Laminations.* A lamination is a separation within the wall of a component caused by the presence of highly elongated nonmetallic inclusions in the alloy. Laminations generally are aligned parallel to the surfaces.
- *Microstructural irregularities.* Aside from nonmetallic inclusions, microstructural irregularities in the original as-manufactured component may promote corrosion. Such irregularities may be caused by segregation of alloying elements. One example is stainless steel that is sensitized during manufacturing. Another is the presence and distribution of significant amounts of undesirable secondary phases in the microstructure. Also, substantial plastic deformation produced during forming operations may impart residual stresses that are high enough to promote both corrosion and cracking mechanisms. Microstructural defects such as sensitized welds in stainless steels, laminations, substantial plastic deformation, and internal defects may only be identified by metallographic examination. Examples of preferential corrosion along sensitized grain boundaries and inclusions are presented in Chap. 3, "Stainless Steel," and Chap. 26, "Weld Defects."

- *Casting Defects.* Pores or tunnels in castings may result from entrapped gases or shrinkage voids that form during solidification. Nonmetallic inclusions such as slag or other contaminants may also serve as defects. Hot cracks, also known as hot tears, may form during cooling of the casting under applied or residual stress. Visual examination of surface defects or cross-sectioning through internal defects reveals pores or tunnels. When viewed under a low-power microscope, the surfaces of shrinkage void defects may be irregularly shaped and rough-surfaced and may have distinct solidification features (e.g., dendrites). Porosity formed by trapped gases will tend to have smooth and rounded features.

Surface defects, if sufficiently severe, may cause failures by themselves. More commonly, they act as triggering mechanisms for other failure modes. For example, open laps or seams form crevices that may develop into concentration cells for aggressive anions that produce corrosion.

Forming Defects

Gouges, grooves, laps, and other surface defects produced by fabrication processes may act as stress-concentration sites leading to fatigue, corrosion fatigue, or stress corrosion cracking. Some manufacturing activities such as drawing, straightening, heat treating, welding, bending, flaring, and even rough handling can produce invisible defects, such as harmful levels of residual stress. This can lead to failure by cracking mechanisms and/or corrosion.

Grooves, laps, and dents may be produced during the formation of threads on piping and couplings. If these defects produce gaps between the mating threads in the joints where they are installed, water leakage may occur. Water leakage may in turn facilitate corrosion on thread surfaces within the joint. Severe separation between threads caused by thread-forming defects or subsequent corrosion damage may cause water to leak out through the joint. Thread sealants may help to seal the gaps formed by thread forming defects to prevent water leakage. However, in some cases the defects are severe enough to allow water to leak even when thread sealants are used. Dents and corrugations in tubing can induce erosion failures, especially in soft metals such as copper and brass.

Casting Defects

Some components, particularly in valves and pumps, are cast into their final shape. During casting, voids, porosity, and other defects may be produced on the surfaces or internally. The voids are a casting defect resulting from insufficient feed of molten metal during the casting process. Shrinkage voids often occur due to uneven cooling of different sections of a casting. During solidification of the molten metal, shrinkage cavities are created due to the volume change that occurs when the metal solidifies. Additional liquid metal must be supplied to fill the cavities as the casting solidifies, or shrinkage voids may be produced. Porosity in a cast component may result from entrapped gases. Since gases tend to float, pores are typically found near surfaces that face upward during casting. Hot tears, also known as hot cracks, may form due to contraction of the component during cooling.

Internal defects may not promote further damage unless their size, location, and shape significantly reduce structural integrity. In such cases, mechanical cracking failures such as fatigue or tensile overload may occur. Casting defects exposed to surfaces may promote concentration cell corrosion or microbiologically influenced corrosion. Due to irregularities in surface contours at casting defects, flow may be disrupted to produce erosion-corrosion.

Locations

In general, manufacturing defects can exist in any component at any location. However, defects resulting from fabrication or installation activities tend to occur in specific locations, such as:

- Tube ends where they are rolled into tubesheets
- Bends
- Welds
- Threaded joints
- Cast components

Critical Factors

The critical factors governing defect-related failures are the severity of the defect and the environmental conditions with which the defect may interact to produce further damage and failure. A defect, if sufficiently severe, may cause failure without contributions from environmental conditions. However, defect-environment interactions are more common.

Equipment can operate satisfactorily with components containing significant defects simply because there are no environmental conditions with which the defect may interact to produce failure. This explains why equipment may operate satisfactorily for years and then fail shortly after some seemingly innocuous change is made in the operating environment. For example, a dent or corrugation in a copper heat exchanger tube may remain harmless at normally acceptable flow velocities, but localized downstream turbulence may be sufficient to promote erosion at increased flow velocities. Residual stress in a brass tube at a bend or at the rolled end of the tube may provide years of trouble-free service until a surface is exposed to water containing high concentrations of dissolved ammonia. Following such exposure, failure may occur by stress corrosion cracking shortly thereafter. Laps, seams, or sensitization in a stainless steel component may remain unaffected until a change in operation produces conditions that cause crevice corrosion or localized corrosion. Hence, defect-related failures tend to be opportunistic.

Identification

The locations of most defects can be detected using one or more appropriate nondestructive testing techniques such as eddy current and ultrasonic testing, or simple dye penetrant and magnetic particle testing for surface defects. In the absence of routine nondestructive testing, identification of surface defects in installed equipment may be limited to visual inspection. In some cases, the type of surface defect may not be definitely identified using visual inspection alone. Destructive metallographic examination will positively identify the causes of all defects.

Defects that can be detected visually, without additional metallographic examination, include:

- Gouges and grooves
- Dents, crimps, corrugations, and flattening

Some surface defects that may be identified by visual examination but require metallographic examination for positive identification include:

- Laps
- Seams

Some defects that cannot be definitely identified by examination of surfaces, and therefore will require destructive metallographic examination, include:

- Microstructural irregularities and laminations
- Casting defects

Elimination

Elimination of defect-related failures has a limited chance of success once equipment is installed. The key to successful elimination is to prevent defective components from being installed in the first place. This is accomplished by following stringent quality control practices.

Stringent quality control practices require adequate inspections at all stages, including manufacturing, shipping, handling, and installation. In addition to visual inspections, several nondestructive testing techniques are available. However, it should be recognized that no individual or series of nondestructive testing techniques is capable of detecting all defects.

The available nondestructive testing techniques in common use for cooling water systems are ultrasonic tests, eddy-current tests, radiographic (x-ray) inspection, magnetic particle and dye penetrant testing, pressurized gas testing, hydrostatic tests, and visual examinations. All of these techniques must be executed by technically qualified personnel.

Each test has specific limitations and may miss defects of various sizes and orientations. The specific limitations for a particular technique must be understood for correct interpretation of results.

Water used in any testing procedure should be free of corrodents and of substances whose decomposition products may be corrosive. Metal surfaces should be dried after testing and kept dry prior to service.

Cautions

Distinctions need to be drawn between simple material flaws and service-threatening defects. Rejection of components that are merely flawed can be economically unjustifiable. Distinguishing between flaws and defects generally requires the judgment of an experienced inspector.

Related Problems

See also Chap. 26, "Weld Defects."

Case History 25.1

Industry:	Plastics
Specimen location:	Air cooler heat exchanger
Environment:	External: Ambient conditions Internal: Chilled water
Time in service:	6 years
Sample specifications:	½ in. (1.3 cm) outside diameter, copper tube containing U-bend

A U-bend section contained a perforation along the intrados (Fig. 25.1). Corrugations formed at the bend intrados during original tube-bending operations. Smooth-surfaced depressions formed on the internal surface at the peaks of the corrugations in the tube wall (Fig. 25. 2). The perforation occurred at the base of the deepest depression (Fig. 25.3). All of the depressions formed due to erosion. Erosion resulted from the turbulent flow of water that was produced by the disruption of flow across the surface at the corrugations in the bend. The use of tubing containing such corrugations should be avoided.

FIGURE 25.1 U-bend section containing a perforation along the intrados. Note the corrugations that were produced during tube bending.

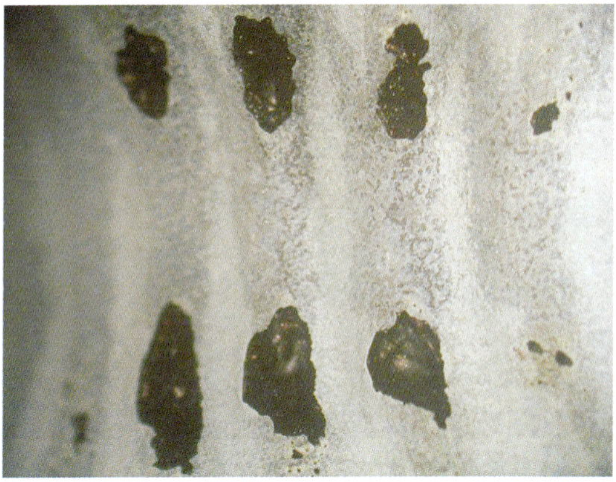

FIGURE 25.2 Smooth-surfaced depressions on the internal surface at the peaks of the corrugations in the tube wall.

FIGURE 25.3 Perforation at the base of the deepest depression.

Case History 25.2

Industry:	Building cooling system
Specimen location:	Chilled water loop
Environment:	External: Ambient
	Internal: Chilled water
Time in service:	3 years
Sample specifications:	Steel piping, externally threaded, 2½ in. (6.4 cm) outer diameter

Water leaked from two threaded joints in the chilled water loop after about three years of service. High-magnification views of the threads at one joint revealed forming defects as smeared and folded-over steel and cracklike features that produced uneven surface contours (Fig. 25.4). The appearance of the defects on the threads was consistent with a machining defect referred to as "chatter". Portions of the threads that were covered with thread sealing material were also uneven (Fig. 25.5). Transverse cross sections were taken through the threads

FIGURE 25.4 High-magnification views revealed defects produced by machining the external threads. The defects consisted of smeared and folded-over steel and cracklike features that produced uneven surface contours.

FIGURE 25.5 Uneven threads covered with thread sealing material.

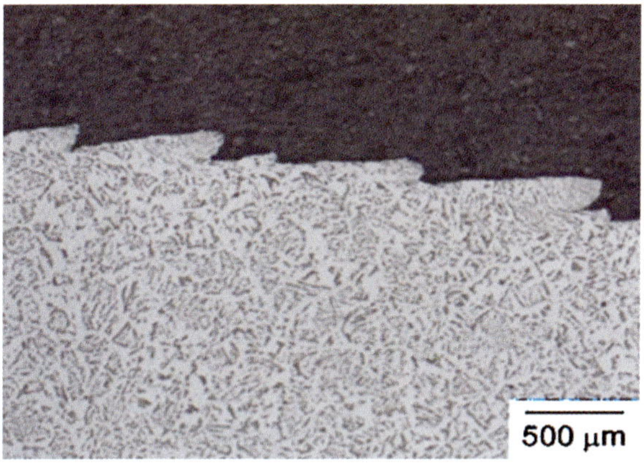

FIGURE 25.6 Metallographic cross sections through the threads. Plastically deformed, folded-over flaps of pipe metal (laps) produced thread surfaces with wavy, uneven contours.

for metallographic examination (Fig. 25.6). Plastically deformed, folded-over flaps of pipe metal (laps) produced thread surfaces with wavy, uneven contours. Water leakage through the joint was caused by the thread defects as well as ineffective sealing of the threads by sealing material.

Water leaked past the threads for another steel pipe nipple in the chilled water loop. Gouges formed along one side of the threads within the joint (Fig. 25.7). Figure 25.8 shows a metallographic cross section through the gouged thread face in the joint that experienced water leakage. A narrow gap formed between these gouged threads and the mating cast iron thread surfaces.

In both cases, the gaps between the engaged threads allowed water to progressively penetrate through, to subsequently allow shallow corrosion to occur

FIGURE 25.7 Gouges formed along one side of the threads within the joint.

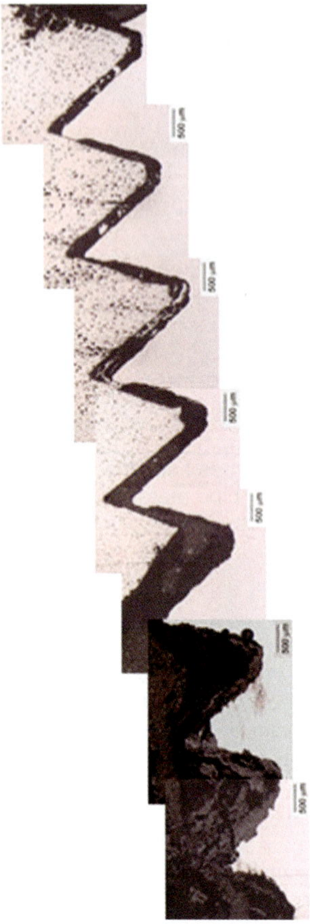

FIGURE 25.8 Metallographic cross section through the gouged thread faces in the joint that experienced water leakage. A narrow gap formed between the gouged threads on the external surface of the pipe and the mating thread surfaces of the cast iron fitting.

on the thread surfaces along the way. Eventually, the combined effects of corrosion and water seepage allowed water to penetrate completely through the joints.

Case History 25.3

Industry:	Building cooling system
Specimen location:	Chiller evaporator
Orientation of specimen:	Horizontal
Environment:	Internal: Chilled water: 44°F (7°C) External: Refrigerant 54°F (12°C)
Time in service:	10 years
Sample specifications:	¾ in. (1.9 cm) outer diameter, externally finned, internally rifled copper tube

Eddy current testing revealed excessive wall thickness loss in several enhanced copper evaporator tubes. Examination of one of the tubes revealed corrosion products that formed in scattered locations between the rifling ridges on the internal surface (Fig. 25.9). Metallographic analysis revealed narrow laps in the rifled regions (Fig. 25.10). The laps, which were surrounded by deformed metal, formed during tube manufacture. The tight crevices that formed at the sites of the laps allowed concentration cells to develop. Chlorides concentrated within some of the crevices to produce corrosive, acidic environments that caused shallow metal loss (Fig. 25.11).

Figure 25.9 Corrosion products formed in scattered locations along the rifling ridges on the internal surface of an enhanced copper chiller tube.

Figure 25.10 Narrow laps in the rifled regions.

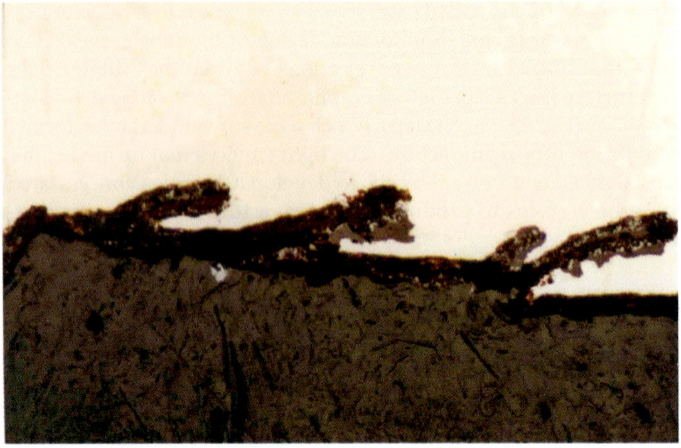

Figure 25.11 Corrosion at crevices formed by laps.

Case History 25.4

Industry:	Steel
Specimen location:	Stack plate cooling section, blast furnace
Orientation of specimen:	Slanted
Environment:	Internal: Cooling water treated with scale inhibitor External: Refractory
Time in service:	1 year
Sample specifications:	Cast copper

Figure 25.12 shows the end profile of a sectioned stack plate with deep, irregularly shaped casting voids at the intersection of the walls. Sectioning through these void zones revealed deep internal tunnel porosity (Fig. 25.13). Microscopic examination revealed that the porous areas showed distinct solidification features (dendrites), which indicated that they formed due to shrinkage during solidification (Fig. 25.14).

Figure 25.12 Cross section through a stack plate showing casting voids located at the junction of two walls (black spots at center of photograph).

Figure 25.13 Sectioning through the void zones shown in Fig. 25.12 revealed deep internal tunnel porosity.

Figure 25.14 When viewed under a low-power microscope, the contours of porous areas showed distinct solidification features.

Case History 25.5

Industry:	Commercial building
Specimen location:	Fire protection system piping
Environment	External: Ambient
	Internal: Untreated water for fire protection
Time in service:	10 years
Sample specifications:	Steel piping containing Victaulic coupling grooves at the ends, 3½ in. (8.9 cm) outer diameter

A section of piping contained a small perforation at a corner of the coupling groove (Fig. 25.15). The forming operation that produced the groove reduced the pipe wall thickness significantly at the groove. Metallographic examination also revealed that the grains were deformed significantly at the rolled groove (Fig. 25.16). Dimensional measurements of the coupling groove suggested that the groove was deeper than the typical standard roll groove specifications. In addition to the associated reduced tube wall thickness, this region was also a preferred location for corrosion. Due to residual stresses, deeper depressions formed at the groove than elsewhere on the internal surface.

Figure 25.15 Perforation at a corner of the coupling groove in a Victaulic coupled pipe.

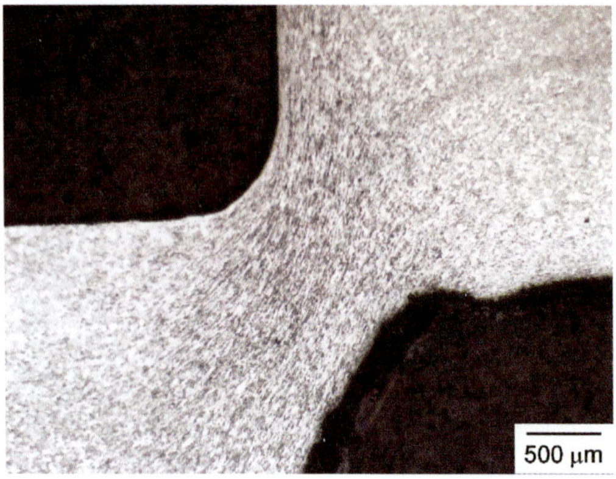

Figure 25.16 The forming operation reduced the pipe wall thickness significantly at the groove. Metallographic examination revealed that the grains were deformed significantly at the groove and experienced preferential metal loss.

CHAPTER 26

Weld Defects

Welding is defined as joining two or more pieces of material by applying heat, pressure, or both, with or without filler material, to produce a localized union through fusion or recrystallization across the interface.[1]

A metallic weldment consists of the base metal or metals to be joined, a fusion zone, and surrounding heat-affected zones (HAZ) (Figs. 26.1 and 26.2). The fusion zone may be produced by melting an appropriate filler metal or by melting the base metal itself. The fusion zone should be free of significant porosity and nonmetallic inclusions, and it should form smoothly flowing surface contours with the components being joined. The weldment should not contain excessive residual welding stresses within the weld metal or in surrounding heat-affected zones.

Sound, fully satisfactory welds are the rule rather than the exception. As a joining technique, welding remains unrivaled as the method of choice for critical applications where the joint requires superior strength and integrity. An efficient weld joint should be as strong as the base metal that is joined. Carefully prepared welding procedures, faithfully followed by competent welders, consistently produce trouble-free welds. "Trouble-free" does not imply flawlessness, however. This is recognized by all major welding codes, which permit, within strict limitations, certain types of weld flaws.

The subject of weld defects is quite extensive, and complete coverage is well beyond the scope of this book. Therefore, this chapter will focus on specific types of weld defects which are of general concern in cooling water systems. Most of the discussion will involve steels, primarily plain carbon steel and stainless steels, which are commonly used materials in cooling water systems. The types of welds that are most commonly employed in cooling water systems are autogenous (do not use filler metal) resistance welds for tube and pipe seams, and arc welds that use filler metal to join piping, tubing, and various components.

A *weld defect* may be defined as a metallurgical or structural interruption in the weldment that significantly degrades the properties of the weld with respect to its intended use. Weld defects are not necessarily location specific. Any weld is a potential site of a defect.

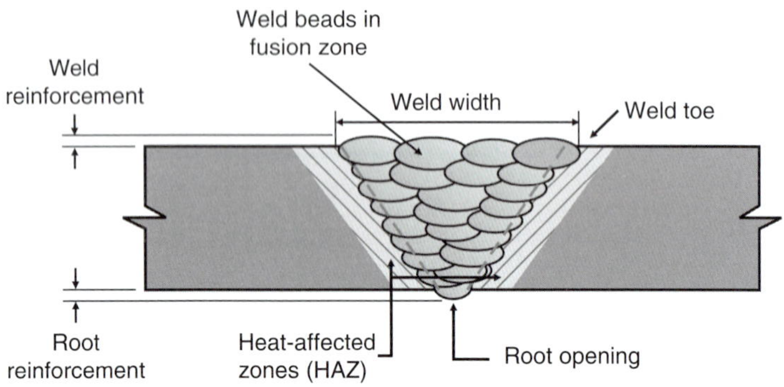

FIGURE 26.1 Schematic of a multipass weld produced with filler metal.

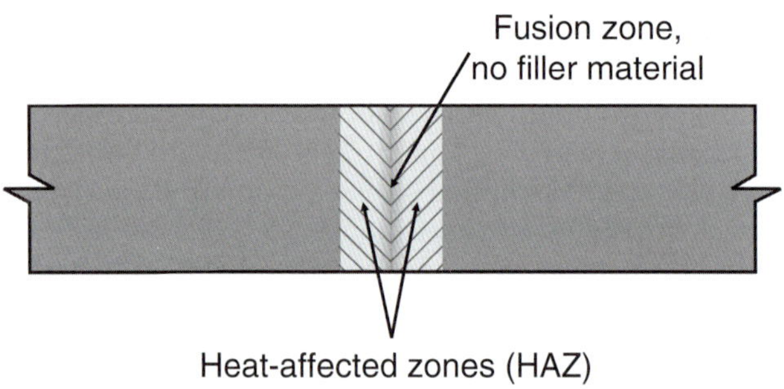

FIGURE 26.2 Schematic of autogenous weld produced without filler metal.

Various types of weld defects that are encountered in cooling water systems will be described in detail in the following sections.

Specific Weld Defects

Incomplete Fusion

Incomplete fusion, as the term implies, refers to incomplete melting and attachment of the mating components within a weld joint. Incomplete fusion may occur in multipass arc welds between individual weld beads, between the base metal and weld metal, or at any point in the welding groove (Fig. 26.3).

When incomplete fusion defects emerge at surfaces, they can be observed visually if accessible, or by boroscope inspection on the tubeside. An example of an incomplete fusion defect is shown in Fig. 26.4.

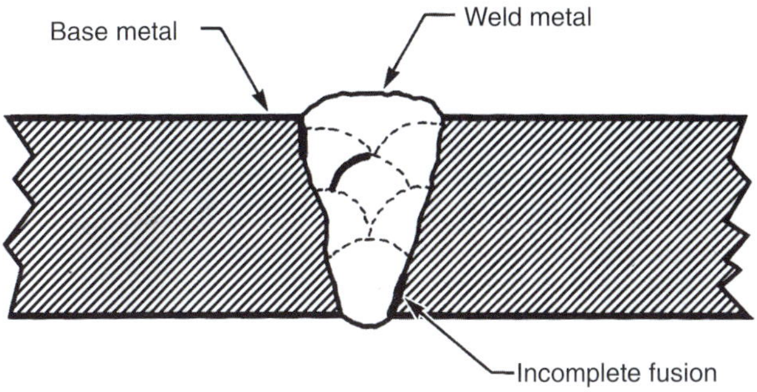

FIGURE 26.3 Schematic showing common sites of incomplete fusion.

FIGURE 26.4 Example of incomplete fusion along edge (top) of weld bead.

Dye penetrant and magnetic particle inspection may be used to aid in the identification of weld defects that are exposed to surfaces. Ultrasonic, radiographic, and eddy-current testing can detect incomplete fusion defects both at surfaces and within the weld. However, positive identification requires metallographic analysis. Incomplete fusion defects may occur in welds that do not use filler metal, such as electric resistance welds that are used to manufacture seams in tubing and piping. In such cases, incomplete fusion is characterized by a straight, V-shaped continuous or discontinuous groove along the seam (Figs. 26.5 and 26.6). This is referred to as a weld seam defect.

Due to stress concentrating effects, incomplete fusion at surfaces is more critical than internal incomplete fusion. Concentration of applied stresses at such defects can lead to failure by mechanical fatigue, corrosion fatigue cracking, and stress corrosion cracking. Crevices that are exposed to the waterside surfaces are susceptible to crevice corrosion

FIGURE 26.5 V-shaped continuous groove formed by an incomplete fusion defect at an electric resistance weld used to produce a seam during original pipe manufacturing.

FIGURE 26.6 Metallographic cross section through a pipe weld seam with an incomplete fusion defect. Note the flat surfaces of the V-shaped groove.

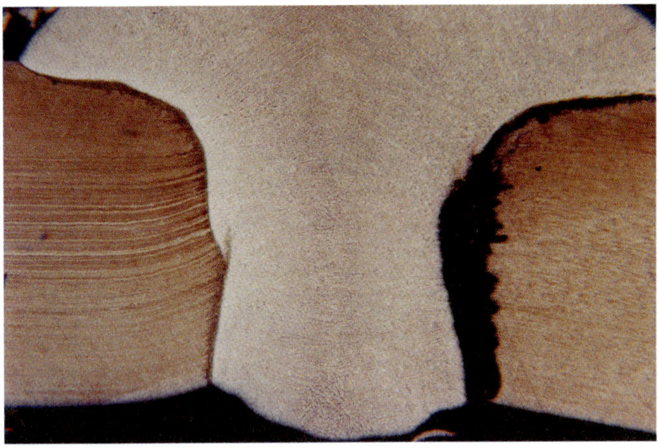

FIGURE 26.7 Cross section of stainless steel weld showing crevice corrosion along a site of incomplete fusion.

and microbiologically influenced corrosion. Figure 26.7 is a metallographic cross section which shows where crevice corrosion occurred at an incomplete fusion defect.

Elimination

The basic cause of incomplete fusion is failure to elevate the temperature of the base metal or of the previously deposited weld metal to the melting point. Incomplete fusion can be eliminated by ensuring proper welding travel speed, proper electrode size, sufficient current, and proper electrode manipulation. Incomplete fusion can also be caused by poor joint preparation, such as failure to remove flux, oxides, and nonmetallic materials adhering to the surfaces to be welded.

Inadequate Joint Penetration

In some cases weld filler metal incompletely penetrates through the thickness of the joint (Fig. 26.8). This usually applies to the initial weld pass, or to passes made from one or both sides of the joint. When inadequate joint penetration defects emerge at surfaces, they can be observed visually if accessible, or by boroscope inspection on the tubeside (Fig. 26.9). Dye penetrant and magnetic particle inspection may be used to aid in the identification of weld defects that are exposed to surfaces. Internal and surface inadequate joint penetration defects may also be detected by ultrasonic, radiographic, and eddy-current testing. However, positive identification of the defect as an inadequate joint penetration defect requires metallographic analysis (Fig. 26.10).

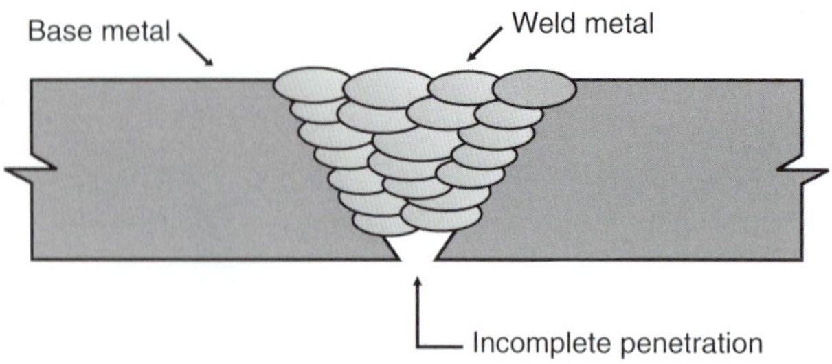

FIGURE 26.8 Schematic showing incomplete penetration defect. Note the presence of a fissure extending into the fusion zone of the weld from the defect.

FIGURE 26.9 Inadequate joint penetration defect.

Components used in cooling water systems require full-penetration welds for proper structural integrity. In addition, failures by mechanical fatigue, corrosion fatigue, stress corrosion cracking, and various corrosion mechanisms, particularly crevice corrosion and microbiologically influenced corrosion, have been associated with incomplete penetration defects.

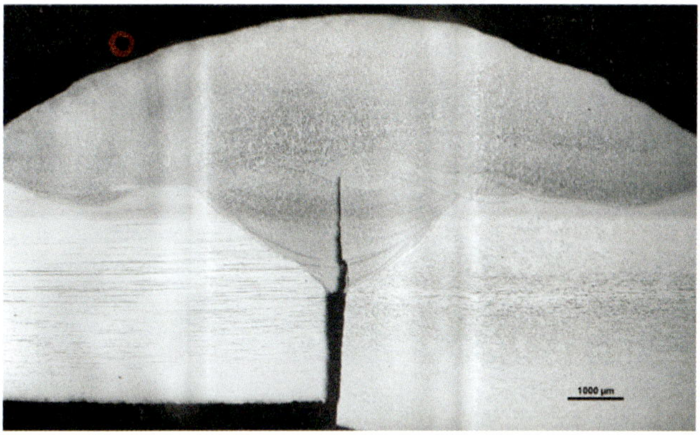

FIGURE 26.10 Metallographic cross section of inadequate joint penetration defect.

Elimination

Inadequate joint penetration is generally caused by insufficient heat to attain fusion temperatures along the entire weldment. This may result from unsatisfactory groove design for the weld, the use of an electrode that is too large, excessive weld travel rate, or insufficient welding current.

Excess Penetration

An excess penetration defect occurs when filler metal that is used during arc welding disrupts the weld bead at the root of the weld. Excess penetration defects in cooling water equipment may affect welded tube-to-tubesheet joints and butt welds in piping. Some excess penetration defects are called "burnthrough." Excess penetration can exist as thick buildups of weld metal on the back side of the weld (Fig. 26.11). Excess penetration may also cause concavity of the weld metal on the back side of the weld, sometimes referred to as "sink." Excess penetration may be identified through visual inspection and/or nondestructive testing such as radiography and ultrasonic testing. Metallographic examination is not required, but it may aid in the examination.

Excessive weld metal buildup is undesirable because this material can disrupt water flow, possibly causing localized deposition and subsequent underdeposit corrosion. Localized turbulence downstream of excessive weld metal protruding from the surface can promote erosion, particularly for softer metals such as copper alloys or aluminum.

Concavity produced by excess penetration can provide sites of stress concentration that may promote the initiation of fatigue or corrosion fatigue cracks.

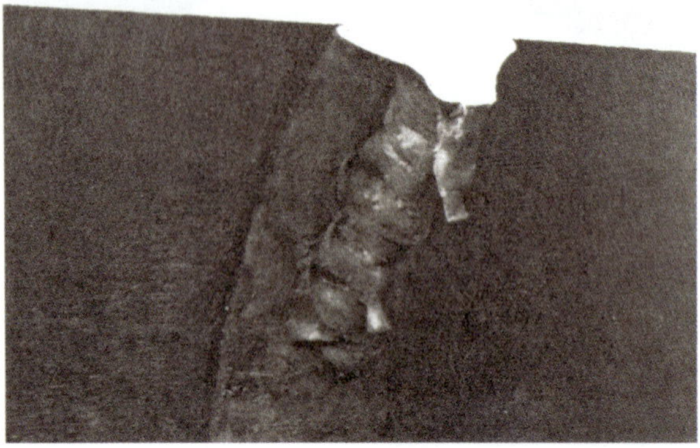

FIGURE 26.11 Excess penetration defect that caused thick buildups of weld metal on the back side of the weld. (*Reprinted with permission from Helmut Thielsch, Defects and Failures in Pressure Vessels and Piping, New York, Van Nostrand Reinhold, 1965.*)

Elimination

Excess penetration is frequently caused by improper welding techniques, excessive welding heat relative to the thickness of the tube or pipe, poor joint preparation, and poor joint alignment.

Undercutting

The term *undercutting* refers to the creation of a continuous or intermittent groove melted into the base metal at either the interface of the weld metal and the surface of the base metal (toe of the weld) or at the bottom or base (root) of the weld (Fig. 26.12). Undercutting at exposed surfaces may be identified visually. Nondestructive testing may indicate

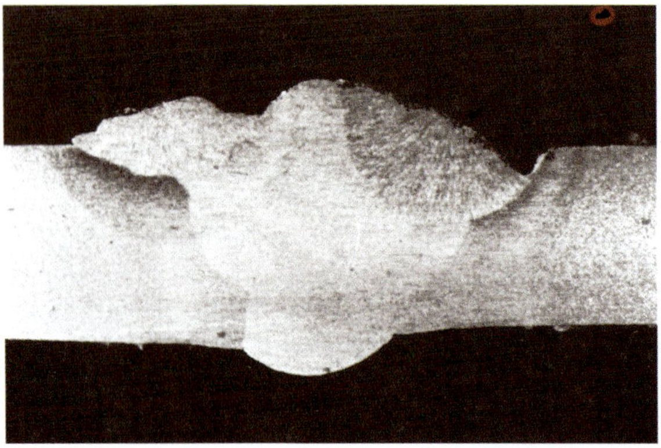

FIGURE 26.12 Undercutting at the interface of the weld metal and surface. (*Reprinted with permission from Helmut Thielsch, Defects and Failures in Pressure Vessels and Piping, New York, Van Nostrand Reinhold, 1965.*)

the possibility of appreciable undercutting. Metallographic analysis is required to positively diagnose subsurface undercutting. Depending upon depth and sharpness, undercutting may provide points of stress concentration that can promote cracking by either fatigue or corrosion fatigue mechanisms.

Elimination

Undercutting is generally caused by using excessive welding currents for a particular electrode, or maintaining an arc that is too long. Serious undercutting may be repaired either by grinding or by depositing additional weld metal.

Porosity

Porosity defects occur in welds that use filler metal. Porosity is most often caused by the entrapment of gas bubbles in the weld metal. This results either from decreased solubility of a gas as the molten weld metal cools, or from chemical reactions that occur within the weld metal. Porosity is rarely exposed to the surfaces of the weld. Internal porosity may also be produced by shrinkage of the weld metal during solidification (also known as shrinkage cavities), depending upon the size of the weld and other factors. The distribution of pores within the weldment can be classified as uniformly scattered porosity, cluster porosity, or linear porosity. Linear porosity, where pores are aligned along a plane in the weld, can appreciably reduce the mechanical strength of the weld. Porosity at and near surfaces will also have a significant effect on the integrity of the weld (Fig. 26.13). Pores that emerge at the surface may be revealed by visual examination, and discovery may be aided by magnetic particle or dye-penetrant inspections. Internal defects may be revealed by nondestructive testing techniques, including radiography, ultrasonic, and eddy-current testing.

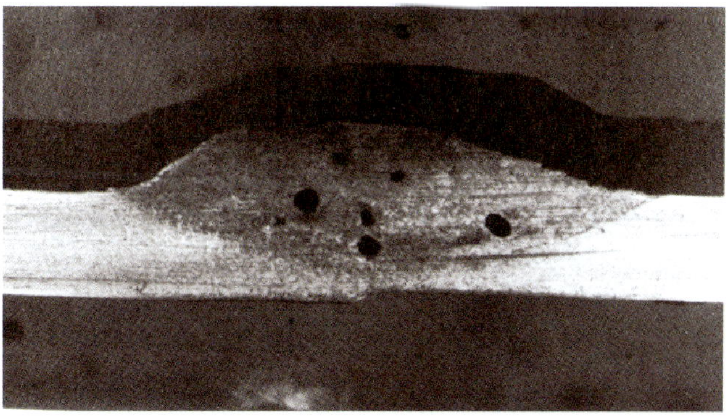

Figure 26.13 Porosity in weld bead. (*Reprinted with permission from Helmut Thielsch, Defects and Failures in Pressure Vessels and Piping, New York, Van Nostrand Reinhold, 1965.*)

Porosity can only be positively identified by conducting metallographic analysis.

Elimination

Porosity can be limited by using clean, dry materials to limit chemical reactions that can produce gases. Maintaining proper weld current and arc length will also tend to reduce porosity.

Slag Inclusions

Slag inclusions are nonmetallic solids trapped in the weld deposit or between the weld metal and base metal. Slag forms by a high-temperature chemical reaction, which may occur during a welding process that employs filler metal and fluxing agents. Inclusions form due to poor welding practice and/or contamination of the weldment by metal oxides and deposits. These inclusions may be present as isolated particles, clusters of particles, or as continuous or interrupted bands (Fig. 26.14). Slag entrapment can also occur if weld metal temperature is too low or if solidification is too rapid. Slag inclusions will not be visually identifiable except if the slag particles are large and emerge at the weldment surfaces. Nondestructive radiography, eddy-current, and ultrasonic testing techniques can reveal slag inclusions of appreciable size. However, slag inclusions must be positively identified through metallographic analysis.

Service failures are generally associated with surface-lying slag inclusions or large internal inclusions (Fig. 26.15). Large internal inclusions can significantly reduce the cross-sectional area of the wall, decreasing the mechanical strength of the weld.

Elimination

The number and size of slag inclusions can be minimized by maintaining weld metal at low viscosity, preventing rapid solidification, and maintaining sufficiently high weld metal temperatures. Since slag is less dense than the weld metal, it will float to the surface if

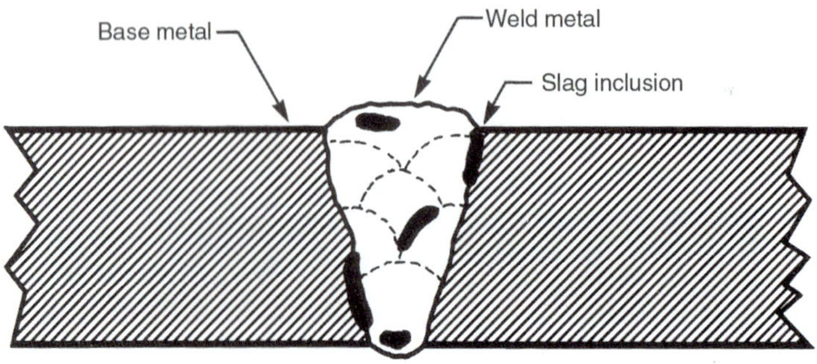

FIGURE 26.14 Schematic of common slag inclusion sites in a weld.

FIGURE 26.15 Metallographic cross sections through weld showing internal slag and porosity defects.

unhindered by rapid solidification. Therefore, preheating the components to be welded or high weld heat input may prevent slag entrapment. Weld metal contamination must also be avoided.

Galvanic Corrosion

Galvanic corrosion refers to the preferential corrosion of the less noble metal of two dissimilar metals that are in electrical contact in the presence of a conductive fluid (see Chap. 16, "Galvanic Corrosion"). The corrosion potential difference, the magnitude of which depends on the metal-pair combination and the nature of the fluid, drives a corrosion reaction that causes the less noble metal to corrode. The galvanic series for various metals in seawater is shown in Chap. 16, Table 16.1. Galvanic potentials may vary with temperature, time, flow velocity, and composition of the fluid. Galvanic corrosion of any type is most severe in immediate proximity to the junction of the coupled metals. It is well known that galvanic corrosion may occur when two dissimilar metals are welded together or when the weld metal itself is sufficiently less noble than the metals that are joined.

However, the corrosion of the weld metal, as presented here, is a special case of general galvanic corrosion. Galvanic corrosion at welds may occur on a macroscopic scale, where all of the weld metal corrodes, or on a microscopic scale due to the presence of dissimilar phases within the weld metal. The primary identifying feature is confinement of metal loss to the weld bead (Fig. 26.16). Macroscopic metal loss may occur if the weld metal is less noble than the base metal. The large cathodic surface area of the base metal relative to the anodic surface area ratio of the weld metal further promotes galvanic corrosion of the weld.

Figure 26.16 Metallographic cross section showing corrosion confined to weld bead.

For some alloys, corrosion results from the manner in which some weld metals solidify. Weld metal may experience alloy segregation as the molten metal solidifies. This microsegregation may result in marked differences in chemical composition within the weld, including the formation of a second phase. The difference in nobility between the phases or metals in the weldment can result in galvanic corrosion of the less noble metal or phase. The less noble weld material will corrode away, leaving behind the skeletal remnants of the more noble metal (Figs. 26.17 and 26.18).

Susceptibility to corrosion is greatly increased if second phases form interconnected or continuous networks. In industrial applications, this type of microgalvanic corrosion is often associated with stainless steels. Occasionally, galvanic corrosion of weld beads produces large cavities covered by a thin outer skin of weld metal. Even close examinations of such sites under a low-power microscope may fail to reveal the cavities (Fig. 26.19). However, probing with a pick may expose the underlying cavities (Fig. 26.20). It is common for such corrosion to be detected by fluid leakage. Such defects may be detected by nondestructive radiography and ultrasonic testing techniques. However, positive identification of microgalvanic corrosion of welds requires metallographic examination (Fig. 26.21). In advanced stages, base metal immediately adjacent to the weld bead may be affected.

FIGURE 26.17 Metallographic cross section showing preferential corrosion of the less noble of two phases in a weld bead at a weld joining austenitic stainless steel alloys.

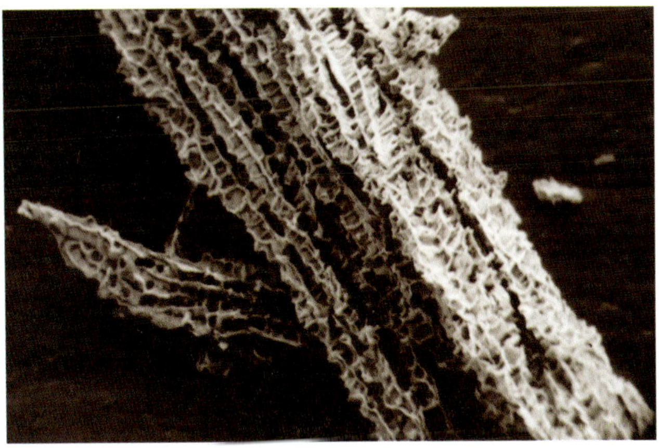

FIGURE 26.18 Scanning electron microscope view of skeletal remnants of noble phase.

Elimination

Galvanic corrosion of welds requires the presence of substantial compositional differences within the weld metal and exposure of such a weld to a sufficiently aggressive environment. If the aggressiveness of the environment cannot be sufficiently reduced, significant compositional differences within the weld metal must be avoided if it does not negatively affect the integrity of the weld. This requires following proper welding techniques, protecting the molten weld metal

FIGURE 26.19 Location of weld bead where radiography indicated a large pit.

FIGURE 26.20 Same site as Fig. 26.19 after probing with a sharp pick.

from interaction with the environment, and choosing the correct weld filler-metal alloy.

Galvanic Corrosion Cautions.

Certain types of stainless steel weld beads are metallurgically designed to form two compositionally distinct phases to reduce the effects of weld metal solidification shrinkage and to control hot cracking. Such welds must be used prudently when exposure to aggressive environments is anticipated. Also, high weld heat input has been linked to microstructural galvanic corrosion of welds (see Case History 26.1).

Sensitization (Weld Decay)

A well-known metallurgical weld defect in stainless steel alloys is termed *sensitization* or *weld decay*. Sensitization occurs in narrow heat-affected zones within the base metal that is located immediately adjacent to the

FIGURE 26.21 Metallographic cross section of a weld revealing a cavity covered by a thin skin of weld metal.

fusion zone of the weld. Sensitization, in a welding context, refers to the formation of chromium carbides along the grain boundaries of the metals being joined. These carbides form within a specific temperature range during the welding process. Stainless steel is susceptible to sensitization when it is heated to the range of 900 to 1560°F (480–850°C). Since any welding operation involving stainless steel will produce such temperatures in the metals being joined, it would appear that all welded stainless steel would sensitize. However, sensitization is a function of both time and temperature, occurring most rapidly at temperatures near 1250°F (675°C). Metals that cool rapidly through this temperature range will not experience sensitization. Consequently, thin metal sections, which cool rapidly, are less susceptible to sensitization than thick sections.

The formation of chromium carbides causes a severe depletion of dissolved chromium in the metal in a narrow zone surrounding each affected metal grain (Fig. 26.22). Since the corrosion resistance of stainless steel is directly linked to the concentration of chromium in the alloy, the loss of chromium adjacent to grain boundaries within narrow zones renders them susceptible to corrosion in a sufficiently

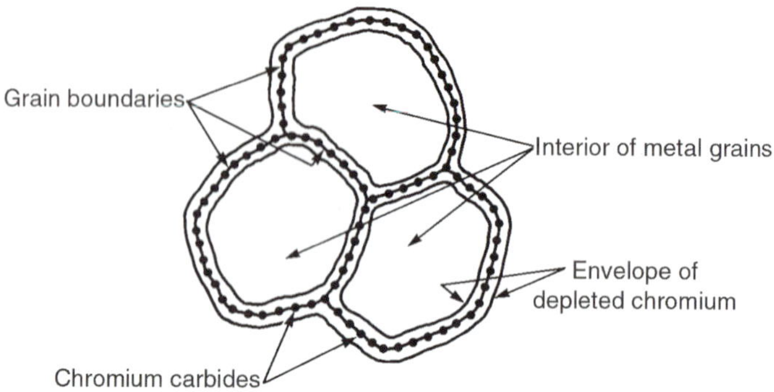

Figure 26.22 Schematic representation of sensitized grain boundaries in stainless steel.

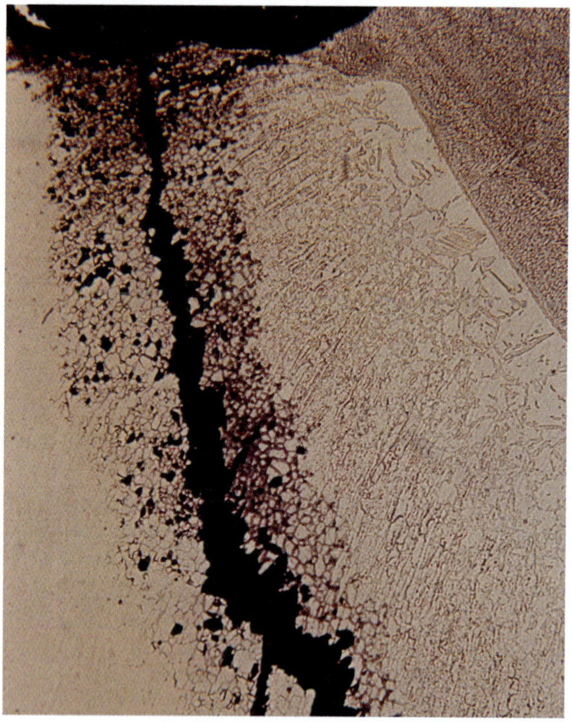

Figure 26.23 Preferential corrosion penetrating along sensitized grain boundaries in heat-affected zone of weld with ferritic stainless steel as the base metal.

aggressive environment (Fig. 26.23). Because metal loss may occur within narrow sensitized heat-affected zones, the damage is sometimes called "knife-line attack." Though sensitized stainless steel is highly susceptible to corrosion, it will not experience corrosion in all environments.

Sensitization cannot be identified by simple visual examination. Typically, the weld and metal adjacent to the heat-affected zones will be unattacked. The corroded area may have a granular or sugary appearance and feel. This is due to corrosion along the boundaries of individual grains that causes them to be released from the metal surface. Affected metal may lose its metallic ring. Positive identification of sensitization requires metallographic examination.

Elimination

Sensitization may be avoided by specifying proper base metal alloys and welding techniques as follows:

- Specify low-carbon grades of stainless steel.
- Specify stabilized grades of stainless steel.
- Alter welding methods.

Specify low-carbon grades of stainless steel. Since sensitization results from the formation of chromium carbides, one control method is to use stainless steels that have lower carbon concentration. Stainless steels containing carbon concentration of 0.03 percent or less have been shown to be resistant to sensitization during welding. The low-carbon grades of 304 and 316 stainless steel are designated 304L and 316L, respectively.

Specify stabilized grades of stainless steel. An alternative method to prevent chromium carbide formation is to use alloying elements whose affinity for carbon is greater than that of chromium. These elements will react preferentially with the carbon, preventing chromium carbide formation and thereby leaving the chromium uniformly distributed in the metal. The carbon content of the alloy does not have to be reduced if sufficient quantities of these "stabilizing" elements are present. For example, titanium is used to produce one stabilized alloy (AISI 321) and niobium is used to provide another (AISI 347).

Alter welding methods. Sensitization is governed by both time and temperature. Consequently, factors such as the thickness of the components to be welded, the type of welding procedure, and the time required to perform the welding will influence the likelihood of inducing potentially damaging sensitization. Experience has shown that electric arc welding, which characteristically produces intense heat over a short time, is the preferred technique for welding stainless steel that is not a low-carbon or stabilized grade.

If sensitization occurs in a stainless steel component, it may be eliminated by appropriate heat treatment of the affected metal. Heat treatment is a remedial rather than a preventive measure. Sensitized metal may be reheated to temperatures in the range of 1950 to 2050°F (1065–1120°C) for sufficient time to allow the chromium carbides to redissolve completely into solid solution in the steel. If the steel is then rapidly cooled (quenched) through the sensitization range of 900 to 1560°F (480–850°C), chromium carbides will not have enough time to re-form. Hence the prior sensitized microstructure is essentially thermally erased. Note that this method has definite limitations. Reheating and quenching sensitized stainless steel may not be practical in many cases, particularly for installed components that cannot be removed for controlled furnace heating and quenching operations. In particular, the quenching operation can induce substantial residual stresses and warpage. Consequently, successful elimination of the problem typically involves preventive techniques rather than remedial measures.

Sensitization (Weld Decay) Cautions

Note that low-carbon or stabilized grades of stainless steel do not possess intrinsically greater corrosion resistance than their unadjusted counterparts. Their sole value in typical cooling water systems results from their resistance to sensitization that can result when the metals are welded or are otherwise heated to temperatures within the sensitizing range. It is therefore not necessary or justifiable to specify low-carbon or stabilized grades of stainless steel for typical cooling water system components that will not be welded or heated.

Dissimilar Metal Welds

In some designs, substantially different alloy compositions, such as low-alloy steels and stainless steels, are joined by welding. Due to different thermal expansion coefficients and thermal conductivities of the different alloys, significant stresses may be imparted to the weld joint during the welding process. Under extreme conditions, fatigue or corrosion fatigue cracks may form due to the application of thermally induced stresses. There is much debate about metallurgical influences on cracking. During welding of low-alloy steel to stainless steel, carbon diffusion results in localized decarburization of the low-alloy steel, and corresponding localized carburization of the stainless steel near the interface. The decarburization in the low-alloy steel has been suspected to initiate cracks. Cracks form within the decarburized base metal in close proximity to the weld metal due to the loss of strength. Nondestructive testing may be conducted at dissimilar metal welds to determine if cracking occurred. Metallographic examination is required to determine the exact cause of cracking.

Elimination

When welding low-alloy steels to stainless steels, austenitic stainless steel weld metal may be replaced with appropriate nickel alloys, which are similar in thermal expansion coefficient to low-alloy steel, and have lower carbon solubility.

Cracking

As discussed earlier, cracking may occur indirectly by fatigue or corrosion fatigue mechanisms due to the application of cyclic tensile stresses at sites of various weld defects. However, cracks may form at welds or within heat-affected zones during or soon after the welding process. Cracks may be oriented in the transverse or longitudinal direction, and they may also travel at irregular angles. Cracks may open to surfaces, and can be wide, but frequently are tight. Such cracks are typically thick-walled and exhibit very little, if any, plastic deformation.

Cracks may form due to the welding process through two primary mechanisms, known as hot cracking and cold cracking. The specific causes for such cracking are described next.

> **Hot cracking** of carbon and low-alloy steel weld metal occurs immediately upon solidification. Such cracks are primarily and most commonly caused by the presence of excessive amounts of low-melting sulfides and phosphides in the weld metal. These low-melting materials reduce the strength and ductility of the weld. Under stresses either applied to the joint or caused by contraction during cooling, cracks or tears may form in the weld.
>
> Hot cracking may also occur in austenitic stainless steels. Austenitic stainless steel weld beads should contain between 5 and 10 percent ferrite in order to control hot cracking. This is normally accomplished by using proper filler metal composition. It is common for the filler metal to contain higher concentrations of ferrite-forming alloying elements such as chromium and molybdenum, and lower concentrations of austenite-stabilizing elements such as nickel, than are present in the base metal. For the same reasons, columbium-bearing and higher nickel content austenitic stainless steel alloys are more susceptible to hot cracking.
>
> **Cold cracking** of carbon and low-alloy steels occurs after the weld has cooled. Such cracks may form days after the welding procedure, and they will rarely form in the weld metal. They are often located within heat-affected zones. Thick components are more susceptible to cold cracking. Cold cracking is usually associated with infiltration of hydrogen into the metal. The hydrogen is most often provided by excessive moisture in the shielding gas or electrode. Oil, grease, and rust are other potential sources of hydrogen. In addition, cold cracking typically occurs in hardened steels. Low-alloy and high-carbon steels are more likely to become hardened within the weld and heat-affected zones than low-carbon (mild) steels.

In general, hot and cold cracks at surfaces may be detected visually, with the aid of dye penetrant or magnetic particle inspection.

Nondestructive testing may identify possible sites of cracking. Internal and surface cracks may be detected by radiography or ultrasonic testing techniques. However, positive identification requires metallographic analysis.

Elimination

Hot cracking may be prevented by using steel that has low phosphorus and sulfur content (<0.05 percent). Heat may be reduced during welding to shorten solidification time that will also reduce the amount of segregation. Joint design should be optimized. In multipass welding, care must be exercised to repair cracks that formed in the initial passes before subsequent passes are applied. Precautions must also be taken to avoid creating stress-concentrating notches in welds and to keep residual welding stresses to a minimum.

Cold cracking can be controlled by reducing the amounts of moisture or other hydrogen-bearing contaminants in the electrode, in the shielding gas, and on component surfaces. Preheating or postheating will control or prevent cold cracking by allowing the hydrogen to diffuse out of the metal.

Laminations

Laminations are nonmetallic inclusions (impurities) that tend to be oriented parallel to the surfaces of wrought plate, piping, and tubing. Inclusions generally originate in the original cast alloy and are expanded into thin sheets during subsequent hot-working processes. Laminations can cause defects or cracks in welded components (Fig. 26.24). Welding imposes stresses that can cause the lamination to separate. In addition, the heat of welding may cause hot cracking to occur at sulfide inclusions in steels. Cracks formed along inclusions may then propagate into the weld itself. Typically, such defects occur when laminations are large or when the inclusion population is high. In either case, the quality of the alloy would be considered to be poor. It is possible for laminations to be present in isolated portions of a tube, pipe, or plate. Therefore, locations of defects of this type are essentially unpredictable.

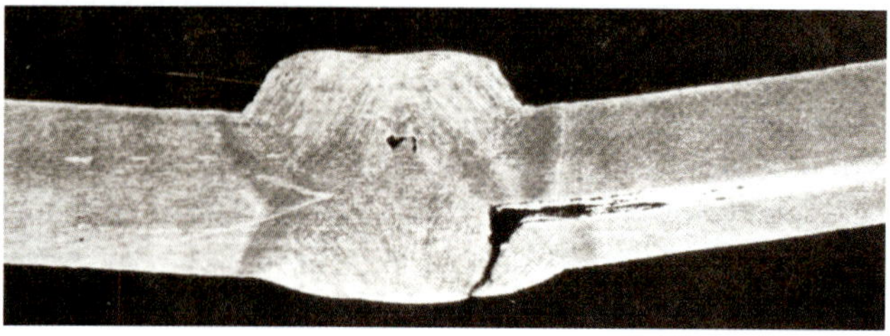

Figure 26.24 Cracking in weld deposit caused by lamination in steel base metal. (*Reprinted with permission from Helmut Thielsch,* Defects and Failures in Pressure Vessels and Piping, *New York, Van Nostrand Reinhold, 1965.*)

Visually, cracks produced by welding metal containing laminations cause jagged, stepwise tears in the base metal. The crack may propagate into the weld itself. Visual evidence of laminations is not definite. Positive identification requires metallographic examination.

Elimination

Since laminations result from steelmaking and steel-forming processes, little can be done to eliminate defects once they have survived quality control inspections. If laminations are suspected, ultrasonic testing or radiography may identify them. They may also be observed visually at cut ends of plate, pipe, or tubes if the cut intersects the lamination. However, positive identification requires metallographic examination.

Welding Stresses

Although not commonly listed as a weld defect, high residual welding stress nevertheless provides an essential contributing factor to stress-corrosion and corrosion fatigue cracking, corrosion, and other failures. It differs in an important respect from other types of weld defects in that residual stresses cannot be visually identified or revealed by conventional nondestructive testing techniques. Residual stress is defined as a stress in the metal that is not supplied by externally applied forces. As such, residual stresses are imparted to the metal by various forming operations and welding. Residual stresses at welds are produced primarily by thermal expansion and contraction between the hot weld metal and the much cooler base metal. Weld metal shrinkage during solidification may also provide residual stresses. Residual stress levels may reach the yield strength of the metal itself.

Corrosion and cracking will tend to occur more readily in locations of high residual stress, often developing within the weld or adjacent base metal. High residual welding stresses in or immediately adjacent to the weld may be the primary contributor to stress corrosion cracking. Applied stresses from external loading of these components can add to the residual weld stresses, producing even higher overall stress levels. This can increase the susceptibility to stress corrosion cracking and can affect the orientation and location of crack paths. See Chap. 21, "Corrosion Fatigue," and Chap. 22, "Stress Corrosion Cracking," for more in-depth discussions concerning the effects of residual stress.

Unfortunately, conventional nondestructive testing cannot detect high residual stresses. Hardness testing may be conducted to indicate if residual stresses are elevated to the point that they may promote cracking or corrosion. Preferential corrosion or localized cracking may highlight locations with high residual stress.

Elimination

Residual stresses caused by large temperature differences between the weld bead and adjacent metal can be reduced by preheating the metals to be welded, especially if the sections are thick. Proper welding techniques, especially the use of appropriate weld filler metals to minimize weld metal shrinkage, can significantly reduce residual stresses. Stress-relief annealing heat treatments may be used following welding to reduce residual stresses. However, when stress-relief annealing stainless steel components, care must be taken to avoid slow cooling through the sensitization range.

Poor Weld Joint Design

There are five basic types of weld joints that use filler metal (Fig. 26.25). These types of joints are:

1. Butt joint
2. T-joint
3. Corner joint
4. Lap joint
5. Edge joint

Misalignment of the base metal components at weld joints that employ filler metal is commonly called poor fit-up or mismatch. When the base metal components are misaligned, they may not be bonded properly by filler metal (Fig. 26.26). The types of joints that are most commonly used, and will tend to be affected by poor fit-up in cooling water systems, include butt, T, and corner joints.

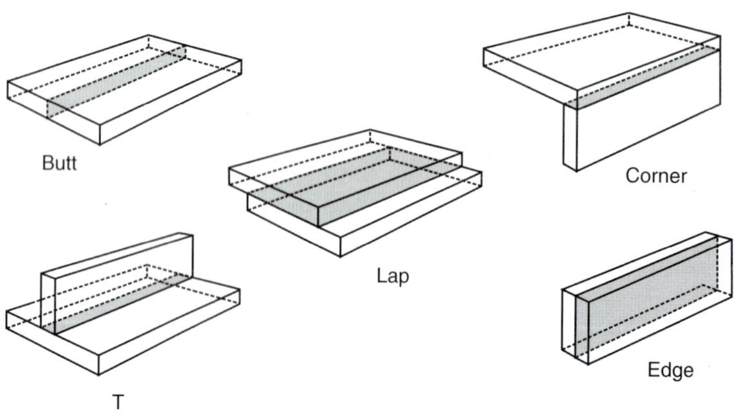

FIGURE 26.25 Basic weld joint designs.

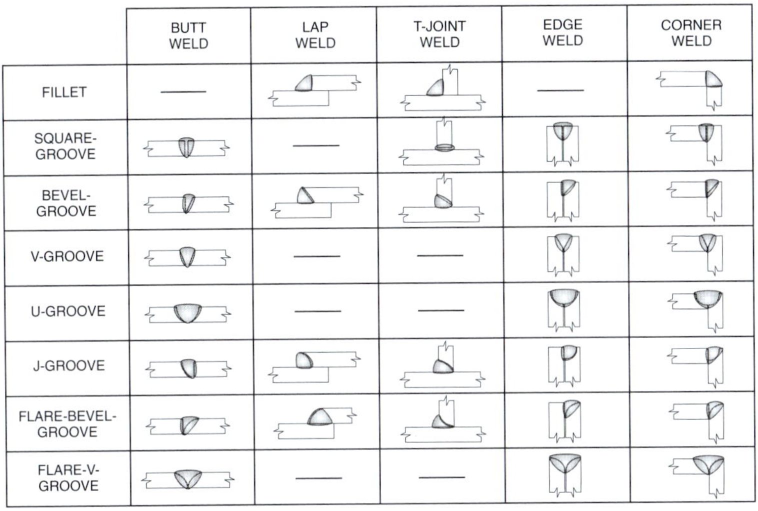

FIGURE 26.25 *(Continued)*

FIGURE 26.26 Significant misalignment of the base metal at a butt weld. An example of poor fit-up.

A proper weld joint requires that the base metal components within such joints have a proper gap width. For some welds, the gap is tight, while others require a small weld root opening. This will allow filler metal to penetrate properly through the gap, but will not allow excessive penetration or burnthrough. This may require the formation of proper grooves at the mating edges of the components to be welded. Depending upon the thickness of the components, the recommended groove depth is typically greater than about 1/8 in. (0.318 cm). At least a portion of the root faces at the edges of the base metal components to be welded must be directly aligned. Butt welds should be directly in-line, while corner and T-joints should be perpendicular. When welding thin components of less than 0.105 in. (0.267 cm), corner joints must overlap (flush corner joint), but thicker components may use a full open corner joint that is stronger.

When the base metal components are poorly fit up, one or more defects may be produced. These may include incomplete or excessive weld metal penetration defects, high residual stresses, and poor structural integrity. Poor fit-up may promote the formation and growth of fatigue, corrosion fatigue, or stress corrosion cracks. Under extreme conditions, where joint strength is insufficient, a tensile overload failure may occur. Poor fit-up of welds may be detected visually, or through the use of suitable nondestructive testing techniques such as radiographic and ultrasonic testing. However, positive identification requires metallographic analysis.

Elimination

Poor fit-up may be prevented through the use of grinding, mechanical distortion, or the use of proper clamps or fixtures, if practical. This may be done prior to welding to properly align the components and to provide the proper gap width between them.

Additional Considerations

Weld spatter, arc strikes, slag, and oxides that cover surfaces following welding may promote corrosion or cracking on certain alloys. For instance as discussed in Chap. 3, "Stainless Steel," thermally formed oxides may be produced during welding on stainless steels, which is commonly known as heat tint (Fig. 3.18). Heat tint should be controlled adequately through proper use of inert shielding gases. If such measures are not taken, or upset conditions occur, an excessive heat tint layer may form that should be removed to provide optimum corrosion resistance. Thermally formed oxides on stainless steels are not as protective as oxide layers that are produced normally when exposed to air or oxygenated water or during passivation treatments. Dark heat tint oxides should be removed using a suitable pickling and passivation treatment to allow the normal passive film to form on the surface.

Welding debris such as weld spatter, shavings, filings, grinding chips, and even welding tools have found their way into tubes and piping. If this debris is not removed, it can cause partial blockage of cooling water flow that can result in product defects and/or underdeposit corrosion. Increased turbulence around such material may promote erosion of softer materials. Substantial defects on the internal surfaces of piping and tubing and deep within the weld metal are usually inaccessible for repair and would require removal and rewelding.

The importance of a competent, experienced weld inspector in evaluating weld defects cannot be overstated. Competent weld inspection requires, in part, distinguishing visually unappealing but serviceable welds from visually satisfying but defective welds. Such judgment requires knowledge of and experience in the techniques of welding, weld repair, and nondestructive testing, as well as service requirements, metallurgy, and defect interpretation. The mere presence of pores, slag inclusions, sensitization, and other defects in a weldment does not necessarily constitute sufficient grounds for rejecting the weld. A perfect weld is not a weld without flaws but a weld that will satisfactorily perform its intended function.

Case History 26.1

Industry:	Utility
Specimen location:	Distribution headers, recirculating cooling system
Orientation of specimen:	Horizontal, vertical
Environment:	Internal: Cooling water treated with corrosion inhibitors and sodium hypochlorite biocide, 75°F (24°C), 50 psi (345 kPa), pH 7–8 External: Ambient, atmospheric
Time in service:	2 years
Sample specifications:	14 in. (35.5 cm) diameter, 316L stainless steel pipe; 308L weld filler metal

Operation of the cooling water system was intermittent, resulting in long periods (up to 30 days at a time) of no-flow conditions. After 2½ years, leaks were found at welded pipe junctions. Radiographic examinations revealed numerous additional deep corrosion sites at welds that had not yet begun to leak.

Several of the welded junctions were removed from the system for metallographic examination (Fig. 26.27). The internal surface was covered with reddish and tan deposits and corrosion products. The underlying metal surface itself retained a bright, metallic luster.

Figure 26.27 Internal surface of 14-in.-diameter stainless steel pipe. The circumferential weld is in the center of the specimen.

Close examination of the weld under a low-power stereoscopic microscope revealed small openings (Fig. 26.19). Probing these sites with a pin revealed a large pit that had been covered by a thin skin of weld metal. These sites contained fibrous, metallic remnants (Fig. 26.20). Examination under a scanning electron microscope further revealed the fibrous character of the material, as seen in Fig. 26.18, and also the convoluted shapes of the individual fibers (Fig. 26.28). The shape of the fibers was consistent with a dendritic structure that was produced during welding. Energy-dispersive spectrographic analysis of this material revealed the compositions in Table 26.1. Note that the fibers contain appreciably higher concentrations of chromium, and lower concentrations of nickel, typical of the ferrite phase in the weld.

Fig. 26.21 shows a transverse cross section through a pitted weld. Note the cavernous pit below a small opening at the surface.

The only significant corrosion observed in the entire system was confined to the weld beads along the internal surface. Corrosion occurred due to a microstructural galvanic couple formed between two distinct phases in the weld-bead microstructure (Figs. 26.29 and 26.30). The less noble phase corroded away, leaving behind the skeletal remnants of the more-noble phase.

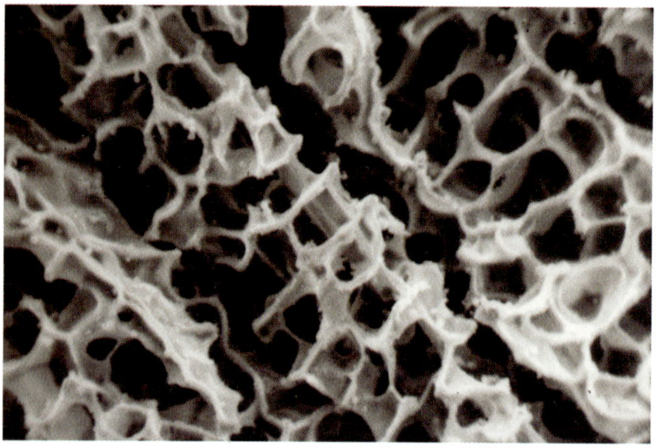

Figure 26.28 Scanning electron microscope view of convoluted contours of individual fibers.

Element	Fibers (%)	Weld Metal (%)
Iron	57	66
Chromium	33	22
Silicon	3	0.5
Nickel	2.5	10
Molybdenum	2	1.5

Table 26.1 Chemical Compositions of Fibers and Weld Metal

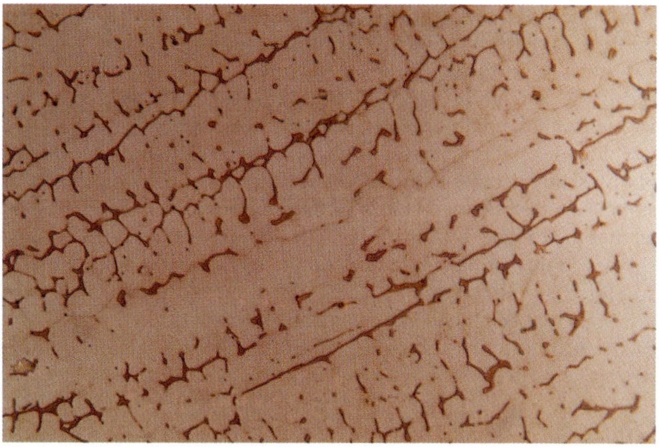

Figure 26.29 Phase 1: (noble) dark script. Phase 2: light background. (Etchant: Nitric acid.)

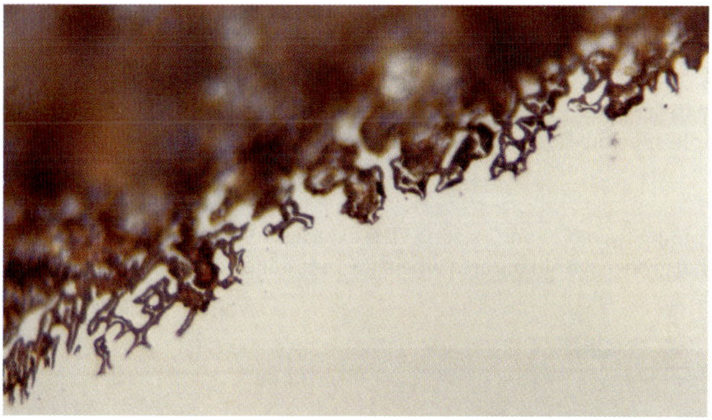

Figure 26.30 Preferential corrosion of Phase 2; Phase 1 remains unaffected. (Unetched.)

The triggering mechanism for the corrosion process was localized depassivation of the weld metal surface. Depassivation (loss of the thin film of chromium oxides that protect stainless steels) can be caused by deposits or by microbial masses that cover the surface [see Chap. 11, "Crevice and Underdeposit Corrosion (Concentration Cells)" and Chap. 13, "Biologically Influenced Corrosion"]. Once depassivation occurred, the critical features in this case were the continuity, size, and orientation of the noble phase. The massive, uninterrupted network of the second phase (Figs. 26.18 and 26.28), coupled with its exposed orientation along the surface of the weld bead, made this particular region of the weld bead susceptible to self-sustained galvanic corrosion once depassivation occurred. Research has demonstrated that the microstructural features of a large, continuous, surface-lying second phase can be produced by excessively high heat input during the welding process. Corrosion of this type could be avoided by reducing heat input during welding.

It is worth noting that corrosion failures essentially identical to the one presented here have been frequently attributed in the literature to microbiologically influenced corrosion. Whereas microbial masses may participate in the triggering mechanism by inducing localized depassivation, depassivation can occur under many environmental circumstances that have nothing to do with microbiological activity. In the laboratory, corrosion of this type has been induced simply by scratching (thereby depassivating) the surface of a weld having a susceptible microstructure. What is perhaps more crucial in mitigating this problem is that the diagnosis of microbiologically influenced corrosion ignores the critical importance of the second-phase microstructure as an essential galvanic driving force for the corrosion process. Once corrosion at such a site begins, it can continue in a self-sustained manner (autocatalytically), without microbial intervention, by spontaneous concentration of chloride ions and hydrogen ions by hydrolysis, a mechanism reported in the corrosion literature (see Chap. 11).

Case History 26.2

Industry:	Steel
Specimen location:	Emergency cooling water supply line to a continuous caster mold
Orientation of specimen:	Vertical
Environment:	Internal: Softened cooling water, 70°F (21°C), treated with corrosion inhibitors and nonoxidizing biocide External: Ambient air
Time in service:	5 years
Sample specifications:	Galvanized steel pipe 6¾ in. (17.1 cm) outer diameter, ½ in. (1.3 cm)

The piping leaked at one location. The external surface exhibited a straight, longitudinal groove within the area where the perforation and leak occurred (Fig. 26.31).

FIGURE 26.31 The external surface exhibited a straight, longitudinal groove within the area where the perforation and leak occurred.

FIGURE 26.32 Larger iron oxide tubercles formed along the seam than in other locations on the internal surface of the pipe.

The internal surface was covered with moderately thick iron oxide layers and tubercles. Larger iron oxide tubercles formed along the weld seam than in other locations (Fig. 26.32). Beneath the hard outer shells of the tubercles, components of the corrosion products and deposits were acidic in a distilled water solution. No significant amounts of chloride were present. However, moderate concentrations of sulfate and very high concentrations of sulfide were present.

Removing the tubercles revealed internal surfaces that experienced fairly shallow, general metal loss in most locations except along the weld seam. Along the weld seam a deep, straight groove was present that contained a perforation at one location (Fig. 26.33).

FIGURE 26.33 Along the seam on the internal surface, a deep, straight groove was present.

FIGURE 26.34 Metallographic cross section at the seam near the failure revealed deep, sharp-edged, V-shaped grooves at both the internal and external surfaces. (Etchant: Nital.)

Metallographic cross sections at the weld seam near the failure revealed deep, sharp-edged, V-shaped grooves at both internal and external surfaces (Figs. 26.34 and 26.35). The V-shaped grooves formed due to weld seams that were incompletely fused during original pipe manufacturing. These grooves were deepened by corrosion. Moderately deep metal loss occurred on the surfaces of the grooves. An incompletely fused seam would have promoted the formation of both differential oxygen concentration cells and ion concentration cells. Aggressive anions, particularly sulfate and sulfide, diffused and concentrated beneath layers and within crevices to levels that were several times that in the bulk water to produce corrosive, acidic environments. The presence of sulfide suggested that sulfate-reducing bacteria may have also contributed to corrosion.

FIGURE 26.35 Higher magnification metallographic cross section at the seam near the failure revealed a deep, sharp-edged, V-shaped groove at the internal surface.

Perforations occurred when the combined effects of the incomplete fusion defects of the weld seam and corrosion thinned the pipe wall.

Case History 26.3

Industry:	Chemical process
Specimen location:	Heat exchanger
Orientation of specimen:	Horizontal
Environment:	Internal: Cooling water—phosphate, zinc, and polymer treatment, 70°F (21°C) External: Gaseous environment containing ammonia, 250°F (121°C)
Time in service:	10 years
Sample specifications:	¾ in. (1.9 cm) outside diameter, carbon steel

Figure 26.36 illustrates one of many similar pinhole perforations found in this and other tubes within the exchanger. Figure 26.37 shows the typical appearance of the internal surface immediately after removing the tube from the exchanger. Laboratory acid cleaning of the internal surface revealed a defective weld seam in the tube (Fig. 26.38). All of the perforations occurred along the weld seam.

FIGURE 26.36 Pinhole perforation in a steel heat exchanger tube, viewed from the external surface.

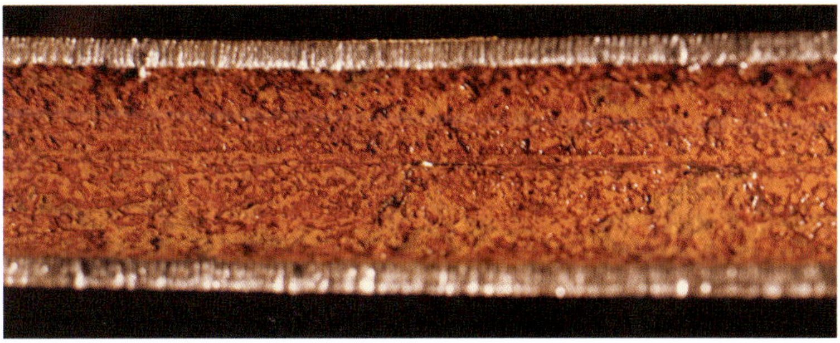

FIGURE 26.37 Typical appearance of the internal surface. Note the small pit centered on the longitudinal seam.

Figure 26.38 Close-up of the internal surface following acid cleaning. Note the perforation in the tube wall along the weld seam.

Microstructural examinations revealed V-shaped grooves along the tube seam, some extending into as much as 50 percent of the tube wall thickness. The characteristics of the grooves indicate that they formed due to incomplete fusion of the weld seam. The incompletely fused seam provided a crevice in which concentration cells developed (see Chap. 11). The resulting localized corrosion deepened the grooves until perforations occurred.

Longitudinal alignment of perforations in a seam-welded tube is a strong indication that the seam contains incomplete fusion defects.

Case History 26.4

Industry:	Chemical process
Specimen location:	Water storage tank
Environment:	Internal: Warm softened water External: Ambient air
Time in service:	4 years
Sample specifications:	Longitudinally seam-welded carbon steel cylindrical tank wall. 4 ft (122 cm) in diameter, 10 ft (305 cm) long, with flat, circular carbon steel end caps welded to ends of cylinder

A cylindrical water storage tank was installed on a metal frame several feet above the ground. The cap on one end of the tank ruptured along an attachment weld, spilling its contents. The force of the rupture caused the tank to fall to the ground, denting the tank wall in localized areas on both ends. The weld metal for the failed end cap incompletely penetrated the gap between the cap and the end of the cylindrical tank wall. Visual and metallographic examination revealed that the cap was attached to the cylindrical vessel by a thin weld bead that covered the external surface of the joint, with little to no joint penetration (Figs. 26.39 and 26.40). Significant porosity was also present within portions of the weld (Fig. 26.41).

The failure resulted from corrosion fatigue cracking that penetrated through the weld bead. The thin weld bead did not provide sufficient joint strength. The porosity within the weld beads further reduced weld strength. Cracking was promoted by stress concentration that occurred at the tight crevice formed by

FIGURE 26.39 Rupture edge at the end of the cylindrical vessel. The edge was attached to the cap by a thin weld bead that covered the external surface of the joint. The original cut end is unaltered.

FIGURE 26.40 Metallographic cross section revealed that the end of the cylindrical vessel was covered by a thin weld bead on the external surface of the joint, with little to no joint penetration. (Etchant: Nital.)

FIGURE 26.41 Significant porosity within portions of the cap weld.

the lack of weld metal penetration into the joint. The crack penetrated within a localized portion of the weld and ran along the rest of the weld around the entire circumference of the cap. A previous leak was repaired on the external surface. The repair weld also did not penetrate completely through the joint (Fig. 26.42). In addition, the seam weld for the cylindrical portion of the tank also consisted

FIGURE 26.42 Previous leak was repaired on the external surface which did not penetrate completely through the joint.

FIGURE 26.43 Seam weld for the cylindrical portion of the tank consisted of an arc weld with weld metal that did not completely penetrate through the joint. (Etchant: Nital.)

of an arc weld with weld metal that did not completely penetrate through the joint (Fig. 26.43). The joint opened wide when the tank wall was dented following rupture of the end cap. Proper nondestructive testing such as radiographic or ultrasonic testing would have identified the lack of penetration defects at the welds prior to installation of the tank.

Case History 26.5

Industry:	Chemical process
Specimen location:	Demineralized water supply line
Orientation of specimen:	Horizontal
Time in service:	10 years
Environment:	Internal: 150°F (66°C) Untreated demineralized water External: Ambient air
Sample specifications:	16 in. (40.6 cm) outside diameter 316L stainless steel piping

A tight, straight crack occurred along the centerline of the weld where the ends of the mating pipe segments were misaligned due to poor fit-up. Misalignment of the joint also promoted the formation of an incomplete weld metal penetration defect (Fig. 26.44). Metallographic examination revealed that weld metal penetrated approximately midway through the pipe wall at some locations (Fig. 26.45). Remnants of a beveled groove on the external surface were also revealed. Where the piping was properly aligned at the circumferential, multipass weld joint, a sound weld was produced with proper fusion and full penetration of the filler metal, and no cracking occurred.

The tight transgranular crack penetrated from the base of the incomplete penetration defect, through the weld metal to the external surface due to fatigue cracking (Fig. 26.46). The locations of incomplete penetration served as stress concentration sites that promoted cracking. The piping segments were significantly out of alignment along the through-crack, indicating that residual stresses

FIGURE 26.44 Misalignment of the joint promoted the formation of an incomplete weld metal penetration defect.

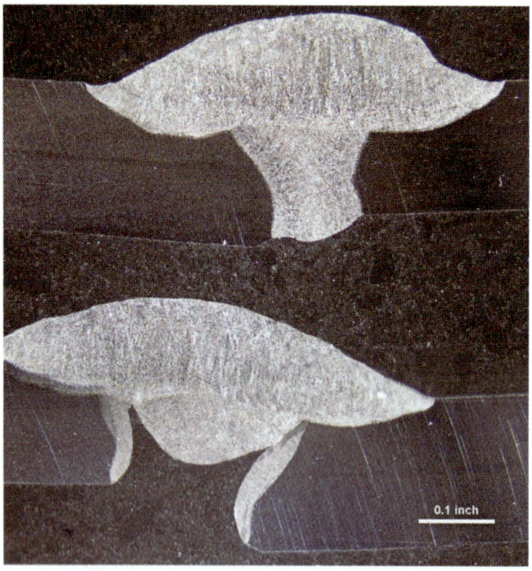

FIGURE 26.45 Metallographic cross section revealed that weld metal penetrated approximately midway through the pipe wall. (Etchant: Electrolytic oxalic acid.)

on the joint were very high in the location of the failure. Furthermore, welds are typical locations of elevated residual tensile stresses due to the welding process itself. The misalignment did not appear to follow a regular pattern, suggesting that one or both of the piping sections may have been slightly out of round.

FIGURE 26.46 Tight, straight, transgranular fissure penetrated through the weld metal toward the external surface at the incomplete penetration defect. (Etchant: Electrolytic oxalic acid.)

The shapes of the ends of piping sections must match properly at butt welds. If the mating ends of the pipes are distorted, they may be mechanically ground or deformed properly, or fixtures may be used to match the ends prior to welding.

Reference

1. *ASM Metals Handbook*, 8th ed., Volume 1, p. 40.

CHAPTER 27

Equipment Design and Operation

This chapter addresses important factors to consider when determining the root cause for a failure, corrosion, cracking, or other damage to components in cooling water systems. The factors are typically characterized as mechanical, operational, and chemical (MOC) influences. This chapter discusses the various important, specific MOC influences in cooling water systems.

Corrosion and failure mechanisms in cooling water systems are affected by all three of the following factors to varying degrees:

1. Mechanical—The mechanical and physical properties of the materials of construction that are used in the system are important. Of equal importance is the design of the components as they are constructed within the cooling water system or within a particular component such as a heat exchanger. The corrosion and cracking resistance of the material must be suitable for exposure to both the specific cooling water chemistry and the process fluids.
2. Operational—The operating condition is an important contributing factor, particularly when there are significant and rapid variations in certain parameters. Important operating parameters include flow rates, temperature, intermittent or continuous service, idle time conditions, and layup practice.
3. Chemical—The allowable water chemistry and chemical treatment that may be used in the water depend upon the type of system. Contamination of the cooling water, or changes to the water chemistry due to variations in operating practice, may affect corrosion or cracking of system components.

Mechanical Design Effects

Effects of Material Properties on Corrosion and Failure Mechanisms

The materials of construction used in cooling water systems should be chosen based upon a few key factors. Prior experience or field testing data of alloys in various environments in literature references should be considered before choosing an alloy. Corrosion testing in the actual environment may also be conducted to determine suitability. Of significant importance are the mechanical properties such as strength, hardness, ductility, and fabricability. Physical properties such as thermal conductivity and thermal expansion coefficient also affect serviceability. The resistance of the material to corrosion and stress corrosion cracking on both the cooling water and process sides should also be assessed. The cost of the alloy relative to the benefits achieved is perhaps the primary deciding factor for material choice.

In most cases the material is properly vetted and experiences suitable service life under normal conditions. Poor or insufficient information obtained about the operating conditions during the design phase can lead to the selection of materials not suitable for service, resulting in premature failures. In some cases it is presumed that an appropriate material is used, but unknown abnormalities in the manufacturing processes, fabrication, or alloy composition of the component compromise its mechanical properties, cracking resistance, or corrosion resistance. Some examples of alloy composition problems include preferential corrosion of the zinc-rich phase in two-phase, high zinc content brasses (see Chap. 17, "Dealloying" and Case History 27.1), exfoliation corrosion of certain aluminum alloys (see Chap. 5, "Aluminum"), and preferential corrosion along residual ferrite phase in austenitic stainless steels (see Chap. 3, "Stainless Steel").

Component Construction

Enhanced Tubing

Certain heat exchanger tubes are formed with spiraled rifling on the internal surface (see Fig. 11.50). The intent of the rifled grooves is to promote swirling and turbulent flow of cooling water through the tube. It is also common for internally rifled tubes to contain fins on the external surface. The combination of internal rifling and external fins is designed to maximize efficiency of heat transfer. For this reason, such tubing is commonly known as "enhanced tubing." Perhaps the most common use of enhanced tubing is for copper and copper alloy tubing that is used in chillers in heating, ventilation, and air conditioning (HVAC) systems.

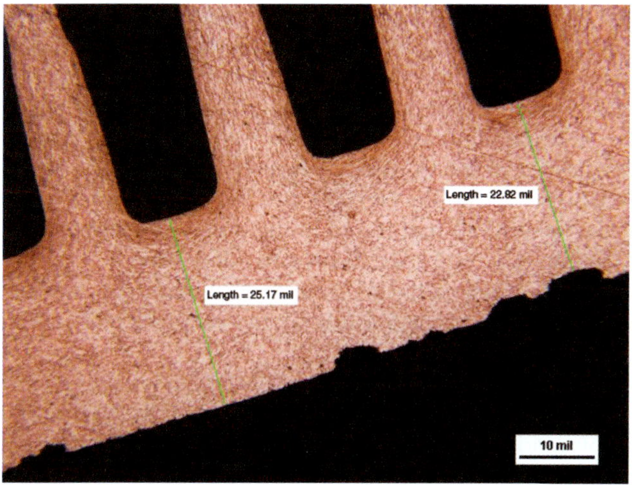

FIGURE 27.1 Nominal wall thickness for enhanced copper tube between fins is about 0.026 in. (0.066 cm).

When enhanced tubing is formed, the wall between external surface fins and internal surface rifling may be quite thin. Typical nominal wall thickness for enhanced copper tubing ranges from about 0.019 to 0.035 in. (0.048–0.089 cm) (Fig. 27.1). Such thin tube walls have low tolerance for metal loss and cracking, so fairly shallow pits and fissures, or even forming defects, will penetrate to a fairly high percentage of the original wall thickness. Eddy-current testing is a nondestructive testing technique which is commonly used to measure tube wall thickness in heat exchangers. Eddy-current testing conducted on thin-walled enhanced tubing may report significant wall thickness loss percentages with relatively shallow pits, cracks, or forming defects on tube internal or external surfaces. In addition, the internal surface rifling grooves will tend to entrap deposits more readily than smooth-surfaced tubing (see Fig. 11.49). Intended flow between the grooves will be impeded, decreasing the heat transfer and promoting under-deposit corrosion. In order for enhanced tubing to maintain optimal performance, proper periodic mechanical or chemical cleaning should be conducted. Eddy-current test results are affected to some degree by internal surface deposits. Due to the increased surface area of internally rifled tubing, higher concentrations of copper corrosion inhibitors are recommended than for smooth-walled tubing.

Plate-and-Frame Heat Exchangers

Crevices form at the many contact points in a plate-and-frame heat exchanger. These are typically at the peaks of the corrugations of adjacent plates within an assembled plate stack (Fig. 11.20). This is inherent to the design of the exchanger, which relies upon narrow flow passages to achieve high-efficiency heat transfer. However, the crevices that form between the plate-to-plate contact points promote localized corrosion due to the concentration of aggressive anions, such as chloride, sulfate, and sulfide [see Chap. 11, "Crevice and Underdeposit Corrosion (Concentration Cells)"]. The aggressive anions are drawn into the crevices, producing concentrations which are much higher than that found in the bulk water, resulting in corrosive, acidic environments. Such corrosion may be controlled to some degree by proper alloy selection based upon water chemistry, and by maintaining low aggressive anion concentrations and conductivity in the water.

In order to assure high-efficiency heat transfer, the narrow flow passages must not be occluded by deposits. In order to achieve proper water flow and to prevent additional underdeposit corrosion, it is recommended that higher-purity cooling waters be used for plate-and-frame heat exchangers. Unfortunately, some units that use poor-quality cooling water containing excessive suspended and/or dissolved solids are often susceptible to deposition. Deposition substantially reduces heat transfer and may initiate underdeposit corrosion.

Cooling Water Flow Channel Design

The design of cooling water flow passages may affect the flow rate, the direction of water flow, and fluid pressure in certain locations. Erosion or erosion-corrosion due to impingement or induced turbulence, particularly for susceptible materials such as copper and aluminum, may occur when sudden changes in tube or pipe diameter or flow direction occur. When water flows through a tube, pipe, or other component to another component that has a much larger cross-sectional area, there may be a significant change in flow velocity and pressure. A significant pressure drop may occur at the inlet, which may cause cavitation. Such a design may also cause the flow velocity through the larger cross-sectional area component to decrease significantly and promote settling of suspended solids and corrosion. Small-diameter orifices located downstream may be significantly occluded or blocked by accumulated deposits (see Case History 27.2).

Shellside Cooling in Shell-and-Tube Heat Exchangers

Cooling water may flow through the tubes (tubeside) or across the external surfaces (shellside) in shell-and-tube heat exchangers. Tubeside cooling water flow is preferred. However, certain conditions, such as high pressure of process fluids, may require cooling water to flow across the shellside. Flow across the shellsides of tubes is much less uniform than through the tubes (Fig. 27.2). Flow is highly variable, particularly at and around baffles and tubesheets, where stagnant flow conditions can form. Suspended solids tend to be trapped

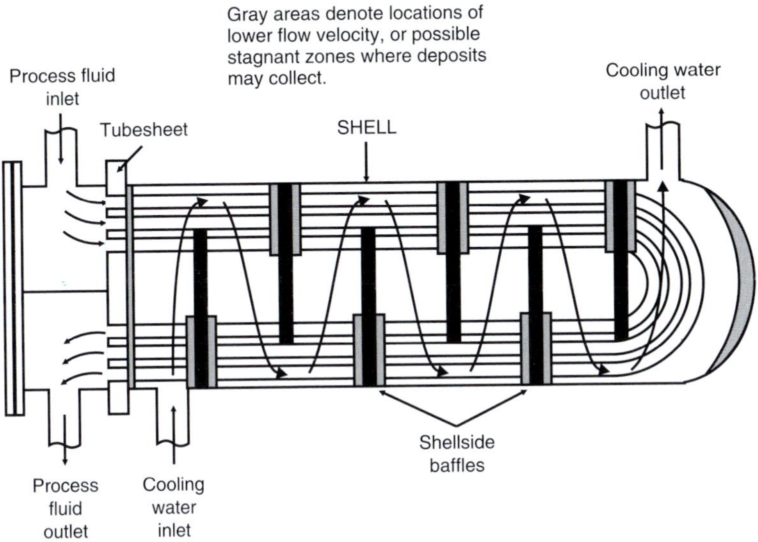

FIGURE 27.2 Flow across the shellsides of tubes is highly variable, particularly at and around baffles and tubesheets, where stagnant zones can form, and they may entrap suspended solids.

along baffles and tubesheets to promote underdeposit corrosion, and they serve as a barrier between the metal surface, the inhibitors, and biocides. Tightly spaced tubes also tend to entrap suspended materials. It is not uncommon to experience deposition and corrosion on shellsides, even when bulk flow rates seem to be sufficient, that is, at or above the recommended 3 ft/s (0.9 m/s).

Design and Fabrication Issues

Components within a cooling water system should be constructed to avoid excessive applied static stresses and cyclic tensile stresses to control corrosion and cracking. Components should be properly supported to avoid excessive deflection. For instance, it is necessary to install more tube supports in a shell-and-tube exchanger that uses titanium tubing than in one that uses stainless steel heat exchanger tubing, in order to control fatigue cracking. However, excessive rigidity may also restrain components and locally concentrate applied stresses. Sharp corners, notches, dents, or other sources of insufficient design thickness will also provide sites where applied stresses will concentrate. Significant stresses may develop at the contact points for dissimilar metals that have substantially different thermal expansion coefficients. For instance, when carbon steel and austenitic stainless steels contact or are welded together, and they are heated or cooled rapidly, the difference in expansion or contraction may impart substantial stresses at the interface. Wide and rapid, cyclic temperature variations may cause fatigue or corrosion fatigue cracks to form and grow at the interface.

Electric Current Effects

Dielectric Couplings

As described in Chap. 16, "Galvanic Corrosion," when two dissimilar metals that have a significant difference in galvanic potential directly contact, the less noble metal is susceptible to corrosion, while the more noble metal does not corrode. If such dissimilar alloys must be used, it is recommended that they be separated by a suitable electrically insulating material. Ideally, one of the components would contain a dielectric coupling, which is typically composed of a thick, impermeable plastic material that covers the joint. Such a coupling will prevent metal-to-metal contact. Other insulating materials may be used in threaded joints such as thread sealing tape or gel compound. However, such thread sealing materials are typically quite thin, and may easily become damaged or degrade to allow metal-to-metal contact (Fig. 27.3).

Cathodic Protection

It is common practice for dissimilar metals to be coupled in cooling water systems in applications such as shell-and-tube heat exchangers. Copper alloy tubing is often mechanically rolled into carbon steel tubesheets. In order to control galvanic corrosion of the carbon steel tubesheet, it is also common practice to install protective coatings on

FIGURE 27.3 Thin, shredded, and degraded thread sealing material may allow metal-to-metal contact.

FIGURE 27.4 Sacrificial anode attached directly to a steel heat exchanger headbox surface using a threaded connection.

the waterside of the tubesheet. If the coating fails, galvanic corrosion may penetrate quite deeply since there is a large cathode (more noble copper alloy) to anode (small exposed areas on carbon steel tubesheet) surface area ratio. Therefore, it is recommended that a suitable cathodic protection system be installed.

Coated carbon steel components are often protected using sacrificial anodes. Sacrificial anodes are made of suitable alloys that are appreciably less noble than the components that they are protecting in the environment. The anodes are attached directly to the component to be protected, ensuring metal-to-metal contact, typically using threaded connections (Fig. 27.4). For carbon steel, sacrificial anodes are commonly composed of zinc, magnesium, or aluminum alloys. The dissimilar metals form a galvanic couple. Therefore, the anode material will corrode sacrificially to the tubesheet metal. It is important to periodically inspect the condition of sacrificial anodes. When the anode is consumed to a certain degree, it must be replaced. In addition, if the exposed surfaces of the anode are covered with deposit and corrosion product layers, then the efficiency of the cathodic protection will be reduced. Impressed current cathodic protection systems may also be used, but they are much more complex and much less commonly used in cooling water systems than sacrificial anode systems.

Stray Current Corrosion

Stray current corrosion can occur when a current that passes through a metal component leaks (strays) from the component into an electrolyte. The electrolyte may be cooling water, but it also may be wet soil surrounding buried piping. Where the current leaves the metal, in order to balance the charge of the leaked current, metal ions go into

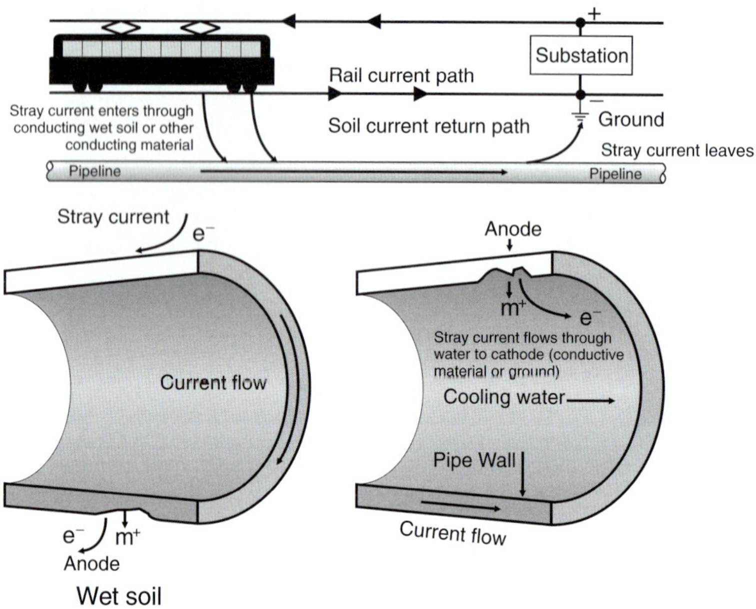

FIGURE 27.5 Schematic drawings showing stray current corrosion mechanism.

solution in the electrolyte, resulting in metal loss (Fig. 27.5). In many cases, claims for stray current corrosion are not well founded. However, when the metallic component carries significant current, and the water has suitable conductivity, the conditions will certainly be conducive to stray current corrosion.

Stray current corrosion may occur in components that carry currents, whether by design or inadvertently. Perhaps the most common and well-known example of a component that may accidentally carry electric currents and experience stray current corrosion is buried piping that is installed near buried power cables that leak current. Examples of water-cooled systems that, by design, carry electric currents include electric arc furnaces and magnetic stirring systems for steel continuous casting as well as for electronic component manufacturing. In these systems, common locations for currents to stray into the cooling water are at junctions of current-carrying metals and nonconducting materials such as rubber hoses. Typically, the current leaves the metal into the cooling water near the interface of the rubber hose, which is where the metal loss occurs. Localization and proximity of typically severe metal loss at locations of preferred current leakage are perhaps the most distinguishing characteristics of stray current corrosion. The morphology of metal loss and corrosion products that form are not distinctly associated with stray current corrosion.

For systems that are susceptible to stray current corrosion, it is common practice to mandate a very low conductivity limit for the cooling water to control current leakage into the water. Corrosion control is difficult in low-conductivity systems for metals such as carbon steel, copper alloys, and aluminum. Corrosion inhibitors must either impart low conductivity to the water, must be used at low concentrations, or both. Corrosion inhibitors will be overwhelmed by the stray current corrosion process. It is recommended that more resistant materials such as stainless steel be used, along with the use of low-conductivity water. Note that metal loss due to stray current corrosion will not be reflected by coupon corrosion rate measurements, since properly installed coupons are isolated from the system.

Atmospheric Corrosion

Atmospheric corrosion as discussed in this section is similar to conditions that may be encountered by structures that are exposed outdoors. Outdoor structures are exposed to rainwater, salt spray, and alternate wetting and drying conditions. Uncoated carbon and low-alloy steel components are particularly susceptible to atmospheric oxygen corrosion due to exposure to water that is oxygenated by the surrounding air. However, in cooling water systems, instead of rainwater and salt spray from seawater, cooling water is exposed to component surfaces. Certain systems employ direct contact cooling of components or products by directly spraying water across their surfaces. An example is secondary (spray water) cooling of continuous cast steel. Directly sprayed surfaces are not consistently submerged in water. When surfaces are cooled in this manner, the cooling water doesn't flow across all system components in a uniform manner as when water flows through a pipe or tube. Cooling water contact may be highly variable, causing some surfaces to experience alternate wetting and drying conditions. Thus, surfaces are not exposed to a consistent flow of properly treated cooling water that contains controlled concentrations of corrosion inhibitors, biocides, and dispersants, so they are therefore susceptible to corrosion.

Between wetting cycles, residual cooling water will evaporate, causing dissolved ions to concentrate. Certain anions such as chloride and sulfate may concentrate to corrosive levels due to evaporation. When halogens are used as biocides, they may evolve from the water solution when it is sprayed and/or heated. When halogen concentrations are high due to increased demand, the evolved halogen gases may produce very corrosive, oxidizing conditions in thin moisture films that form on surfaces.

Corrosion at Air/Water Interfaces (Waterlines)

Corrosive conditions may be produced in air spaces located directly above waterlines due to concentration of aggressive ions and evolved

FIGURE 27.6 Corrosion along air/water interface on carbon steel component.

halogens (Figs. 3.12 and 27.6). At the meniscus or waterline, or along the borders of water droplets that cling to surfaces, differential oxygen concentration cells may form. At such locations, the dissolved oxygen concentration will be much higher at the interface than in the bulk water. The surfaces at these locations will be cathodic to the adjacent areas that are exposed to lower dissolved oxygen concentration. The areas contacting water with lower dissolved oxygen concentration will be anodic and will experience localized corrosion, typically in the form of grooves. Aggressive anions may also concentrate at the interface.

Control of Atmospheric and Waterline Corrosion

Under atmospheric and waterline conditions, corrosion cannot be controlled as effectively as in a pipe or tube with properly treated water flowing through it at adequate flow velocity. Metal loss on carbon steel cooling water system components is similar to atmospheric oxygen corrosion that occurs on building structures. Atmospheric corrosion on large structural components may be controlled most effectively through the use of proper coating materials such as epoxies that are resistant to the environment. Coating materials may degrade over time, particularly when exposed to harsh conditions. Therefore, coatings require periodic inspection and maintenance to ensure continued effective protection of the underlying metal. Cooling tower sheet metal and evaporative condenser tubes, which are exposed to cascading water and intermittent water spray, are highly susceptible to atmospheric corrosion. Thus it is common practice to use sheet metal and tubing that is made of materials that are resistant to atmospheric corrosion. A common material of construction for such sheeting and tubing is galvanized steel (Fig. 27.7). Galvanized steel is commonly

FIGURE 27.7 Galvanized steel evaporative condenser tubing exposed to water sprays for cooling.

employed for building materials that are used outdoors such as fencing, utility poles, and highway guardrails. Since galvanized steel is used for evaporative condensers due to atmospheric exposure, it is very important for the cooling water chemistry to be controlled to prevent the formation of "white rust." Recommended water chemistry to control white rust is presented in Chap. 15.

Case History 27.3 describes atmospheric and other corrosion problems that are encountered in continuous caster spray chambers where the cast steel strand is directly cooled by water sprays.

Operational Issues

Water Flow

One of the most important parameters that affect corrosion, deposition, corrosion inhibition, and microbiological control is cooling water flow. Excessive flow velocity, impingement, or excessively turbulent flow of water may cause erosion or erosion-corrosion of softer metals such as copper or aluminum alloys. Cast irons, carbon, low-alloy, or stainless steels are much less susceptible to erosion. Flow disruptions resulting from surface irregularities or improperly aligned or poorly designed flow passages are possible causes for impingement or turbulence and the resulting erosion or cavitation.

When flow rates are low or stagnant, suspended solids tend to settle along the bottom sides of horizontally oriented tubing and piping and in shielded locations to promote underdeposit corrosion and microbiologically influenced corrosion by anaerobic bacteria. In addition, when flow rates are low or stagnant, corrosion inhibitors and biocides are consumed and are not adequately replenished as they would be when flow rates are optimal. Ideally, a flow rate of at least 3 ft/s (0.9 m/s) should be achieved. Certain exchangers, particularly those that are oversized, may vary or throttle cooling water flow to control or prevent overcooling of a process. In some cases, in order to avoid excessive cooling rates for the process, cooling water flow rates may be reduced substantially, possibly to near stagnancy. Stagnant flow may be defined as a flow velocity of 1 ft/s (0.3 m/s) or lower. In some cases, stagnant flow and subsequent corrosion and failures occur due to obstructions caused by debris that collects at tube inlets (Figs. 11.51–11.53).

Idle Time Corrosion

Care is taken to ensure that water chemistry, chemical treatment, and water flow are optimized during service to control corrosion, deposition, and microbiological growth. However, during extended idle periods when the system does not operate, it is uncommon for proper flow of treated water to be maintained due to the high pumping cost. Instead, it is more common for water flow to be valved off, leaving the system full of stagnant water, or for the system to be drained.

If the system is not drained, but is placed into wet layup for an extended period of time, perhaps weeks to months, suspended material will settle, and the corrosion inhibitors and biocides that were sufficient during normal operation will eventually be exhausted. Oxygen corrosion, underdeposit corrosion, and microbiologically influenced corrosion will ensue. If the system is drained, residual water will evaporate from surfaces over time and aqueous corrosion will cease. However, care must be taken with the design of components to ensure that pools of stagnant water do not remain for extended periods of time (Fig. 27.8). In such locations, the same problems will occur as in systems that are laid up wet with poor layup practice.

In order to control corrosion during extended idle periods, it is recommended that proper layup practice be instituted. Proper wet layup practice will typically involve the use of corrosion inhibitors and biocides that are appropriate for the system and are used at higher concentrations than for normal service. Periodic sampling may be necessary to ensure that optimum water chemistry and treatment residuals are maintained. Periodic recirculation of the treated water is also recommended.

Special care must be exercised during wet layup of equipment exposed to cold climates. When water stored in piping or tubing is allowed to freeze, expanding ice can impart stresses that are sufficient to deform the wall or even cause it to rupture. Such damage may be avoided by using suitable fluids that are resistant to freezing such as glycol/water mixtures, or by draining the system and using dry layup practice.

FIGURE 27.8 Idle time corrosion at location where a stagnant pool of water collected at the bottom of a drained cooling water channel.

In systems that are laid up dry, but retain standing pools of water, efforts should be made to remove the residual water. Horizontal piping may be slanted, and drain valves may be installed to improve drainage. Hot, dry air may be blown through the system, or pans filled with desiccants may be installed strategically to remove the water. When a system is laid up dry, the water in the deposit and corrosion product layers will evaporate, producing brittle materials that will tend to flake off when flow is restored. Hydrated deposit and corrosion product layers in water-filled piping tend to be much more adherent.

In some cases corrosion control during dry layup may be improved by applying soluble oil treatment just prior to shutdown. Following drainage, a thin layer of water-repellent oil will remain on surfaces. This layer prevents the surfaces from being wetted, so corrosion will not occur. Such treatments are well suited to systems that are frequently drained and filled. However, if soluble oil treatment is used, it should be removed prior to startup.

Hydrotesting

Hydrotesting is conducted in cooling water systems at high pressure prior to startup to ensure component and system integrity. Corrosion can be significant if the water used for hydrotesting was allowed to remain stagnant in the system for a long period of time. Failures may not immediately develop during hydrotesting, but corrosion that initiated during that period may continue during service. Hydrostatic testing of equipment and piping should follow proper procedures. The procedures typically require the use of treated water of suitable quality. In addition, the equipment should be drained if it is not going to be placed into service soon after testing to avoid prolonged stagnant periods.

Startup After Layup

When systems that are not laid up properly are restarted after a long-term layup, it is common for substantial amounts of corrosion products and deposits to be swept downstream. This material will become suspended in the water. When cooling water contains a large amount of suspended solids, particularly iron oxide, corrosion inhibitors will attach to them. This reduces the amount of corrosion inhibitor that is needed to protect the surfaces. Consequently, corrosion rates may increase, generating more corrosion products, which may cause a snowball effect. On top of that, the suspended material will drop out in low-flow areas, promoting underdeposit corrosion, and serving as a barrier that prevents inhibitors and biocides from reaching the metal surface to provide protection.

Free Cooling

HVAC chillers typically operate using two separate cooling water loops, one for the chilled water that provides cooling for the building, and another for cooling refrigerants such as Freon for centrifugal chillers and lithium bromide for absorption chillers. When the outside air temperature is low enough, typically lower than 48°F (8.9°C), savings may be achieved by using the open recirculating cooling water from the cooling tower that is normally used to cool the condenser. The colder water can be used to supply what is called "free cooling" for the chilled water system, saving energy costs to run the chiller.

When free cooling was used in the past, it was common practice to directly divert the open recirculating condenser water into the closed recirculating chilled water system. In this way, water from each system became contaminated by the other. The chilled water system tends to suffer more severe consequences, since open recirculating cooling water tends to be dirtier than the chilled water. This promotes deposition and microbiological growth within the chilled water system. Also, the corrosion inhibitors that are used in one system are introduced to the other. The closed-loop treatment chemicals, which are properly suited for the closed loop, are at least partially lost to the open recirculating cooling water. Certain corrosion inhibitors that are used in a closed loop, such as nitrite, are not suitable for use in an open recirculating system, since they may promote the growth of denitrifying bacteria. In addition, the inhibitors used in open recirculating cooling water systems are not as effective in closed-loop systems, in part due to the combined cooling water chemistry. Also, closed-loop corrosion inhibitor programs are much more effective for controlling corrosion under low or stagnant flow conditions than typical open recirculating cooling water programs. More modern free cooling system designs employ a heat exchanger that separates the chilled water from the open recirculating cooling water, thus preventing cross-contamination.

Alternating Steam and Cooling Water

Certain systems are designed to alternately heat and then cool a process through alternate use of steam and cooling water. Ideally, the system should be designed using parallel, separate heating and cooling coils. However, certain systems feature alternating steam and cooling water injection into the same coil or fluid chamber. Such practice may be highly problematic. During normal cooling service, the materials of construction will be exposed to oxygenated waters. Corrosion of susceptible materials such as carbon steels and copper alloys may be controlled by corrosion inhibitors that are used in the cooling water. However, when the cooling water is subsequently displaced by steam, water evaporation causes aggressive anions to concentrate to much higher levels than in the bulk cooling water. In addition, dissolved solids will precipitate as liquid water evaporates. Also, heat from the steam exposure will degrade the protective inhibitor layers, creating a need for replenishment. All of these conditions will promote corrosion. Even materials such as 304 or 316 stainless steel, which will resist corrosion by oxygenated water, will be susceptible to pitting as chloride concentrates at higher temperatures. These stainless steels are also highly susceptible to chloride stress corrosion cracking when temperatures increase above about 120°F (49.4°C) (see Case History 22.7). In addition, rapid changes in temperature due to alternating steam and cooling water exposure will cause the metal to expand and contract during each cycle. Depending upon the thermal expansion coefficient of the metals or contacting metals, fatigue or corrosion fatigue cracking may occur.

Skin Temperature

Deposition and corrosion are significantly affected by the skin temperature, or metal temperature at the waterside surface. The skin temperature is controlled by cooling water flow velocity, cooling water temperature, process fluid temperature, heat flux, thermal conductivity of the metal, and thickness of the wall of the component. Deposition of inversely soluble salts and corrosion increases as a function of increasing skin temperatures. In addition, certain corrosion inhibitors are less effective above certain threshold temperatures.

Severe Temperature Variations

Basic oxygen furnace (BOF) hoods, electric arc furnace hoods, hot blast valves, continuous caster rolls, and hot mill rolls may experience severe temperature variations during service or due to intermittent service. Wide and rapid temperature variations apply cyclic tensile stresses of substantial magnitude that promote corrosion fatigue cracking. Corrosion fatigue cracking may be controlled most effectively by minimizing the magnitude and frequency of the applied stresses. Corrosion fatigue may occur even when corrosion rates are very low, simply due to the cyclic mechanical fracture of the protective oxides at the tips of the advancing fissures. However, some benefit may be obtained in slowing the rate of corrosion fatigue crack growth by controlling corrosion, particularly if corrosion rates are high. It should be emphasized that control of corrosion may slow down the rate of corrosion fatigue crack growth, but it will not prevent crack formation or propagation.

Oxygen corrosion on carbon steel components that are cooled by cooling water produces iron oxide layers on the surfaces. Porous, voluminous iron oxide corrosion products and deposits will reduce heat transfer to increase metal temperatures during higher temperature service. The increase in metal temperature will increase temperature variations between high- and low-temperature service. Consequently, the components experience increased expansion and contraction, which accelerate corrosion fatigue crack growth.

Nondestructive testing techniques such as ultrasonic testing may be conducted during maintenance procedures. Such testing can detect excessively deep cracks and can monitor for the progression of crack growth over time.

Chemical Issues (Water Chemistry and Treatment)

Once-Through, Open Recirculating, and Closed Recirculating Systems

Corrosion control in once-through cooling systems is limited by the concentration of corrosion inhibitors and biocides that may be added to the cooling water. The concentration of certain corrosion inhibitors will be limited by environmental discharge restrictions for water that is returned to a lake, river, or cooling pond. In addition, it is common for once-through cooling water systems to use very high volumes of water. Therefore, economics may limit the amount of inhibitor and biocide that can be added to the water. Lower inhibitor and biocide concentrations may allow corrosion rates to increase relative to what may be achieved in recirculating systems, where higher treatment levels are typically allowed. The source of cooling water used in once-through systems may also significantly affect corrosion and deposition. For instance, high-chloride-content seawater and brackish water will be significantly more corrosive than freshwater. In addition, seasonal and even daily changes in cooling lakes, ponds, or rivers can affect deposition and corrosion. The amount of suspended silt particles in surface waters that can affect deposition and corrosion may vary over time.

Unlike once-through systems, cooling water used in recirculating systems is not typically discharged directly into a lake, river, or pond. Therefore, corrosion inhibitors and biocides in recirculating systems may be used at higher and more effective concentrations than in once-through systems. In addition, cooling water chemistry may be controlled through the use of acids and alkalinity supplements, blowdown, and filtration as necessary. Therefore, corrosion rates in recirculating systems may typically be controlled to lower levels than in once-through systems.

Closed recirculating systems commonly use higher-purity feedwater than open recirculating systems, and they typically experience low water losses, so water chemistry may be controlled more effectively. Without appreciable water losses, inhibitor and biocide concentration may be maintained at higher levels, allowing excellent control of corrosion and microbiological growth. However, when substantial leaks occur in closed cooling water systems, which then require significant makeup, corrosion control becomes more challenging due to the introduction of increased amounts of dissolved gases, dissolved solids, and other contaminants. In addition, increased makeup requirements will increase chemical treatment costs. Severe leakage may require dosing chemical treatment at lower, less desirable concentrations in order for the program to be economical.

Water chemistry in open recirculating cooling systems is more difficult to control than in closed recirculating systems due to water losses caused by evaporation, drift, leakage, and environmental exposure to dusts and gases that are scrubbed from the air through cooling towers. These conditions require the addition of makeup water, blowdown, sidestream filtration, and chemical treatment in order to achieve proper balance of water chemistry. Corrosion rates are affected by several factors, including conductivity, dissolved ion and gas concentrations, holding time index, temperature, and other factors.[1]

Water Chemistry and Treatment Issues

Ground and surface water chemistries in various geographic locations will vary significantly. In some cases the source water is satisfactory for use as makeup water in specific types of cooling water systems. As discussed previously, corrosion and deposition are affected by several different water chemistry factors. In general, some of the more significant water chemistry factors that affect corrosivity include pH, alkalinity, conductivity, concentration of aggressive anions such as chloride, and halogen levels. Certain dissolved gases may significantly contribute to corrosion or stress corrosion cracking of certain alloys. For instance, ammonia in water solution may significantly affect copper alloys (see Chap. 15, "Alkaline Corrosion," and Chap. 22, "Stress Corrosion Cracking"). Corrosion may be controlled through the use of proper water pretreatment and chemical treatment.

In some systems, recycled water is used for purposes of water conservation and cost control. The use of recycled waters or those that have high dissolved ion concentration, high conductivity, and run at high cycles of concentration tend to promote deposition and corrosion. Higher organic, phosphate, and ammonia concentrations may promote microbiological growth and corrosion as well. Such waters, or cooling waters that are exposed to in-leaked organic, ammoniated, and certain other process contaminants, will increase halogen demand, which may also contribute to corrosion.

In some cases it will be necessary to use pretreatment systems to remove suspended and dissolved ions in the water to make it suitable for use in certain cooling water systems, particularly closed recirculating systems. Pretreatment may be necessary to control scaling, fouling, corrosion, and stress corrosion cracking of certain alloys, depending upon temperatures and other factors.

Selection of appropriate treatment chemicals is, of course, a crucial part of maintaining a cooling water system that is as trouble-free as possible. Due to the variety of cooling water systems and water sources, treatment chemical selection is a complex topic that is beyond the scope of this book. Proper implementation of a chemical treatment program is just as important as chemical selection. Overfeed of certain treatment chemicals can increase the potential for corrosion. For example, it is generally recommended that the free chlorine residual (which includes chlorine in incoming water, as well as that added to the system) in a system should not exceed 0.5 ppm for extended periods of time to avoid excessive corrosion of carbon steel components. Consistently high free chlorine residuals (>1.0 –1.5 ppm) are likely to cause corrosion of stainless steel, copper alloys, and carbon steel in cooling water systems. High free chlorine residuals can also adversely affect other treatment chemicals. Underfeed of corrosion inhibitors will, of course, promote corrosion.[2]

Low-Conductivity Cooling Water Systems

In some systems where water-cooled components carry high electrical currents, it is a requirement for the cooling water conductivity to be very low. This is necessary in order to control the leakage of current into the water, which will reduce the component's efficiency. In addition, leakage of current into the water may cause a phenomenon called stray current corrosion. Stray current corrosion was described in a previous section in this chapter. Low-conductivity water will contain very low concentrations of dissolved ions, including calcium and other dissolved mineral salts as well as anions and cations that may produce significant conductivity at even low concentrations, such as sodium and chloride. Typically, low-conductivity water may be supplied as distilled or demineralized water. Such waters will be very corrosive to carbon and low-alloy steels, low-chromium cast irons, and copper alloys due to the lack of calcium hardness and alkalinity that would help to produce protective layers on the surface. Corrosion of such alloys may be controlled through the use of suitable corrosion inhibitors. However, due to the conductivity restriction required by the process, the amount of standard cooling water inhibitors that may be used is limited, and may provide insufficient protection. Increased corrosion will supply metal ions to the water to increase conductivity, which in turn may promote further corrosion. Special low-conductivity corrosion inhibitors and nonoxidizing biocides may be required for such systems (Case History 27.6).

Nonchemical Devices

The use of proper chemical treatment to control corrosion, scaling, and biofouling in cooling water systems has proved successful for many years. In an effort to save chemical treatment costs and to reduce the concentration of chemicals discharged into waste streams, various types of mechanical devices have also been promoted for use in cooling water systems for many years. These devices, which do not supply chemicals to the water, are claimed to control corrosion, scaling, and bacterial growth. They may be magnetic, electromagnetic, or electrostatic, or they may be frequency generators. Theories used to explain how the devices are purported to control scaling and corrosion are suspect. Several studies have been conducted over the years at various university, industrial, and institutional labs which prove that such devices do not control scaling or corrosion.[3] In many cases, chemical treatment that was used in cooling water systems that had no problems with scaling or corrosion was discontinued, and nonchemical devices were installed. Subsequent service using only the nonchemical devices produced deposits, and corrosion rates increased. This increases cleaning frequency and reduces component service life dramatically in many cases. Figure 27.9 shows a tube section from a high-temperature water system which used softened water, dispersants, and corrosion inhibitors for many years. A nonchemical device was installed on the system, and the use of softened water and other chemical treatments was discontinued. Over a period of five weeks following the change, thick, highly stratified layers of

FIGURE 27.9 Tube section from a high-temperature water system which used softened water, dispersants, and corrosion inhibitors for many years. A nonchemical device was installed on the system, and the use of softened water and other chemical treatments was discontinued. Over a period of five weeks following the change, thick, highly stratified layers of calcium carbonate–rich deposit layers formed on the internal surface of the tube. These layers reduced heat transfer so substantially that they had to be chemically removed to allow the system to return to normal operation.

calcium carbonate–rich deposit layers formed on the internal surface of the tube. These layers reduced heat transfer so substantially that they had to be chemically removed to allow the system to return to normal operation.

Case History 27.1

Industry:	Electronics
Specimen location:	Process chilled water loop for sputter coating
Specimen:	Connecting fittings
Orientation of specimen:	Horizontal
Environment:	Internal: Closed recirculating cooling water External: Ambient air
Time in service:	4 years
Sample specifications:	5/16 in. (0.79 cm) brass fittings

Brass fittings experienced premature cracking failures. The brass fittings were electroplated throughout with a layer of nickel (Fig. 27.10). The average microhardness measurement for the brass fittings was about 75 Rockwell B. Chemical analysis revealed that the fitting alloy had an average composition of 59.7 percent copper, 37.2 percent zinc, 2.9 percent lead, and 0.2 percent tin, typical of a free-machining brass alloy. Metallographic analysis revealed that the brittle nickel layer covering unused fittings was very thin and contained numerous cracks (Fig. 27.11). The brass fittings failed due to the formation of cracks that penetrated

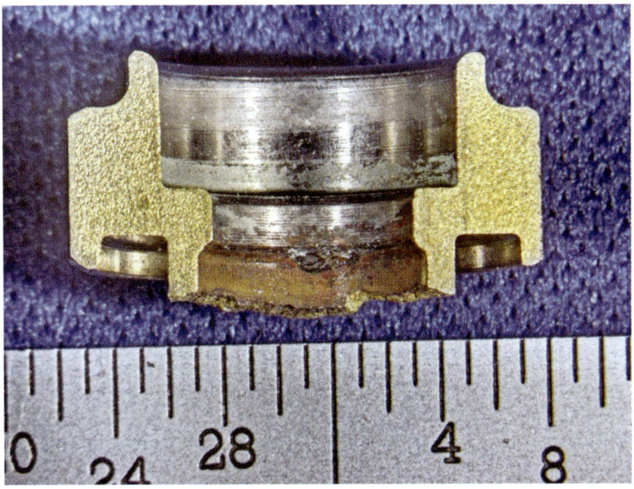

Figure 27.10 Brass fitting plated with a layer of nickel that was partially lost during service.

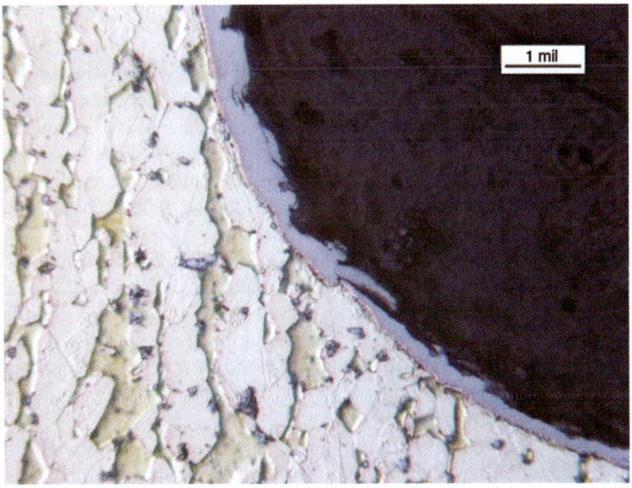

Figure 27.11 Metallographic cross section reveals cracks in the thin, brittle nickel plating.

from the internal surface where the original thin and brittle nickel plating layer was removed. At cracks or defects in the plating layer, the underlying brass was exposed. Shallow corrosion that occurred on the exposed brass undermined the nickel layer, allowing it to detach from the brass substrate and spall under applied stresses and under forces applied by cooling water flow.

The fissures that penetrated through the walls of the brass fittings formed only within isolated locations due to fatigue cracking resulting from the application of

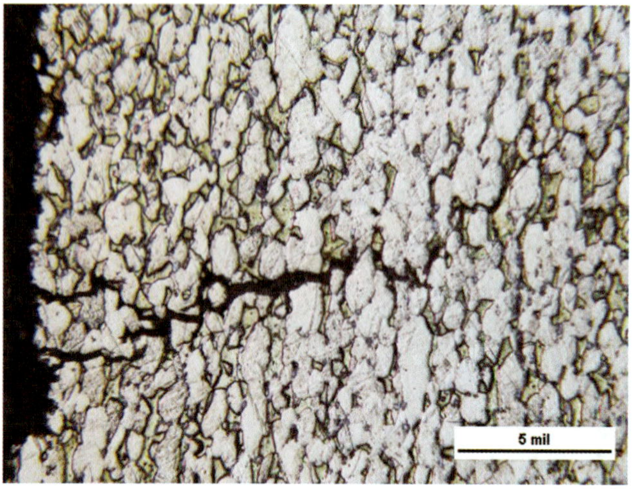

FIGURE 27.12 Fissure that penetrated through the wall of the brass fitting due to fatigue cracking. Etchant: Potassium dichromate.

cyclic tensile stresses (Fig. 27.12). Fatigue cracking may be controlled by minimizing the magnitude and frequency of the cyclic stresses. The stresses may have been supplied by several possible sources during installation and operation. Static, impact, and vibrational stresses applied to the fitting may have provided some stresses that contributed to fatigue cracking. In addition, the unit was cycled on and off frequently during normal service, which resulted in pressure variations, thus imparting cyclic tensile stresses. Temperature variations contributed cyclic stresses to some degree due to thermal expansion and contraction of the fitting. The cracking resulted from the combination of cyclic and other applied stresses, and residual stresses from forming and installation of the fittings. Weak materials are more susceptible to fatigue cracking than strong materials. The fittings were made from soft, weak brass containing substantial amounts of even softer lead precipitate (Fig. 27.13). The lead was added to the alloy to improve machinability, which also reduced the strength of the material. The alloy is much more susceptible to fatigue than many other materials.

In addition, the fittings experienced shallow waterside corrosion that penetrated preferentially along the beta phase in the microstructure. Beta phase tends to form when zinc concentration exceeds about 32 percent. The beta phase in the alloy is anodic to the surrounding alpha phase. Therefore, it will tend to corrode preferentially due to a microgalvanic corrosion mechanism (Chap. 17). However, the influence of corrosion on the growth of the fatigue fissures and cracks was insignificant. Crack propagation also penetrated preferentially along the uncorroded beta phase in many locations.

If a brass alloy is desired, it is recommended that a proper lead-free, single-phase, lower zinc content admiralty brass alloy be used. For greater strength, a copper alloy such as cupronickel was recommended. The thin nickel plating did not provide adequate resistance to corrosion or cracking. Therefore the use of such plating on the waterside was considered to be unnecessary.

FIGURE 27.13 Microstructure consisting of two-phase brass containing substantial amounts of lead precipitate. Etchant: Potassium dichromate.

Case History 27.2

Industry:	Steel
Specimen location:	Continuous slab caster
Specimen:	Bearing housing
Orientation of specimen:	Horizontal
Environment:	Internal: Closed recirculating cooling water External: Hot cast steel strand, spray chamber environment
Time in service:	2 years
Sample specifications:	Carbon steel housing with 304 stainless steel shield on side facing strand.

A continuous caster roll seized in a bearing during service. Investigation revealed that the grease within the bearing degraded. Bearings for continuous caster rolls must be adequately cooled in order to ensure that the lubricating grease does not degrade. Seizing of the roll in the bearing will cause the roll to overheat or wear unevenly and will affect conveyance of the strand. Therefore, corrosion and deposition must be controlled in the cooling water supply and return lines and throughout the cooling water channels inside the bearing housing.

The cooling water channel at the top of the housing that was in closest proximity to the strand was sawcut, which revealed that the cross section of the channel was completely occluded by corrosion product and deposit layers (Fig. 27.14). The mating outer wall of the housing and a sawcut portion of the opposite wall within the channel were pulled apart for examination (Fig. 27.15). The top outer wall of the bearing housing was made of stainless steel, and was essentially unattacked. However, the opposite wall of the cooling water chamber, which was made of carbon steel, was covered with thick, black iron oxide tubercles and corrosion products that were overlaid with layers of deposit. These tubercles, which formed due to corrosion on the surface beneath, grew to the point that they contacted the opposite walls of the cooling water chamber and blocked water flow.

FIGURE 27.14 Sawcut cooling water channel at the top of a bearing housing, which was in closest proximity to the strand, revealed that the cross section of the channel was completely occluded by corrosion product and deposit layers.

FIGURE 27.15 Carbon steel cooling water chamber surface was covered with thick iron oxide tubercles and corrosion products that were overlaid at outlet end by collected, settled deposits.

The deposits and corrosion products were removed to reveal surfaces that experienced general metal loss. Metal loss on the carbon steel surfaces within the cooling water channel was primarily caused by oxygen corrosion.

The cooling water inlet port had a much smaller cross-sectional area than the cross-sectional area of the cooling water chamber within the bearing housing (Fig. 27.16). With this increase in cross-sectional area, flow velocity within the cooling water channel decreased substantially. Corrosion inhibitors and biocides were much less effective as flow velocity was low or stagnant, and thick

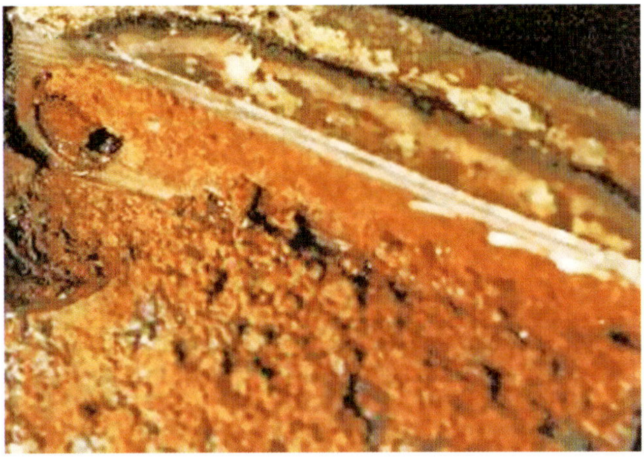

FIGURE 27.16 Cooling water inlet and outlet ports had a much smaller cross-sectional area than the cross-sectional area of the cooling water chamber within the bearing housing.

layers of corrosion products and deposits accumulated on surfaces. Cleaning was recommended for piping, narrow orifices, and water channels that were covered with thick corrosion product and deposit layers to restore sufficient water flow. In addition, cooling water flow rates needed to be increased through the system to control settling of deposits and corrosion. Changing the design of the inlet and outlet ports and the cooling water channel was suggested to improve flow.

Case History 27.3

Industry:	Steel
Specimen location:	Continuous caster
Specimen:	Spray chamber components and structure
Environment:	Hot cast steel strand, spray cooling water, steam, condensate, and mold powder by products
Sample specifications:	Carbon steel structural beams and segment housings, austenitic stainless steel spray water supply piping, clad rolls

Secondary or spray cooling of continuously cast steel is an example of direct contact cooling. As the steel strand leaves the end of the primary cooling system in the mold, it further solidifies with secondary cooling (Fig. 27.17). This is accomplished through direct contact of the strand surfaces with water or air/water mixtures that are sprayed through nozzles. The nozzles are installed at the outlet of the mold and between the rolls that convey the strand down the caster. The spray water that is directed onto the strand will rebound and overspray, and will cascade down the structural components within the spray chamber (Figs. 27.18 and 27.19). The heat of the strand causes the spray water to boil, producing steam that can flow throughout the spray chamber and condense on cooler surfaces (Fig. 27.20).

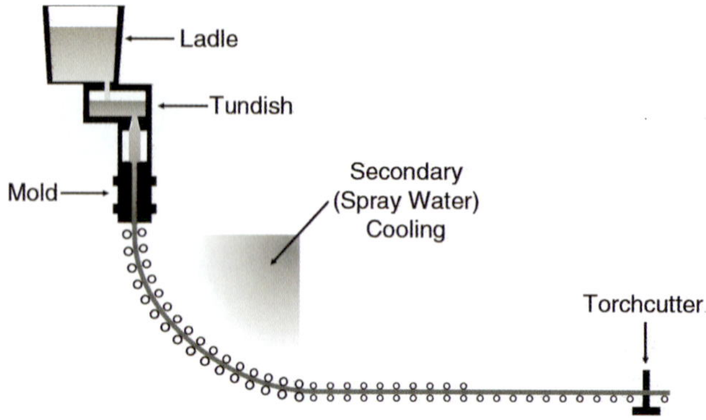

Figure 27.17 Schematic showing secondary cooling of continuous cast steel strand by direct contact spray water cooling.

Figure 27.18 Spray water flowing from nozzles that are installed between rolls that convey the strand.

Figure 27.19 Cascading, nonuniform spray water flow contacting spray chamber components.

Figure 27.20 Steam produced when spray water boils off hot steel strand.

The structural components within the spray chamber of a steel continuous caster are primarily made of carbon steel and low-alloy steels. The metal loss that occurs on these components is primarily caused by atmospheric oxygen corrosion. This corrosion results from exposure of the surfaces to oxygenated water. The water becomes oxygenated by the surrounding air. Corrosion rates due to atmospheric oxygen corrosion within the horizontal zones of caster spray chambers may vary depending upon design, operation, and water chemistry. The volume of the enclosure (if it is enclosed) and the efficiency of venting of water vapor and aggressive agents influences the amount of corrosion that will occur. Atmospheric corrosion rates on exposed carbon steel structures in a caster spray chamber due to exposure to mildly corrosive, oxygenated water may vary from about 10 to 30 mpy (0.25–0.75 mmpy), based upon corrosion coupon studies. These corrosion rates are similar to atmospheric oxygen corrosion of exposed steel on outdoor structures that are constantly exposed to misting conditions.

However, some casters use spray water containing high chloride concentrations (>300 ppm). In some cases, spray waters have a high halogen demand due to heavy organic fouling from greases, oils, and hydraulic fluids. Very corrosive conditions will be produced by the concentration of aggressive anions due to evaporation, and due to ion concentration cell mechanisms beneath deposit and corrosion product layers. Under such conditions, uncoated carbon steel structural components may experience much higher corrosion rates than in casters that use mildly corrosive waters. Corrosion rates may exceed 100 mpy (2.54 mmpy). In some casters subjected to highly corrosive conditions, structural beams required replacement within two years of service. It is common to find carbon steel components covered with thick, highly stratified layers of iron oxide, some of which peels away from the surfaces (Fig. 27.21). The stratification indicates that corrosion occurred during many separate events. When the cooling water has high chloride concentration, or halogen demand is high, it is common to find acidic material that contains high concentrations of chloride

FIGURE 27.21 Carbon steel structural components in caster spray chamber covered with thick, highly stratified layers of iron oxide, some of which peel away from the surfaces.

beneath and between the iron oxide layers. The acidic conditions that were produced beneath the oxide layers accelerate the corrosion dramatically.

Corrosion inhibitors that are added to the spray water to control corrosion on internal surfaces of spray water supply piping cannot be relied upon to control atmospheric corrosion on carbon and low-alloy steel components within the spray chamber. The inhibitors in the spray water deposit onto the strand and are largely lost when the spray water contacts it. In addition, water does not flow across all components in all locations within the spray chamber, and condensed steam that collects on surfaces will not contain corrosion inhibitors. Due to the susceptibility to atmospheric oxygen corrosion, original and reconditioned carbon steel components are typically coated with suitable resistant organic coatings such as epoxy (Fig. 27.22). The use of resistant coatings or materials is the

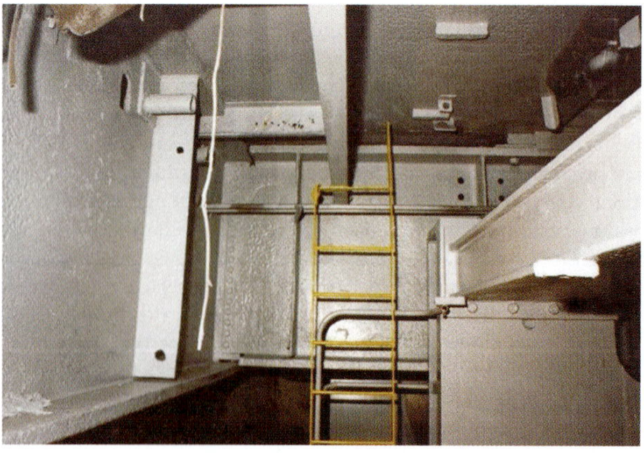

FIGURE 27.22 Reconditioned carbon steel components coated with resistant epoxy coating.

FIGURE 27.23 Organic coating on caster spray chamber wall that degraded over time, causing it to become brittle, followed by cracking, bubbling, and peeling.

most effective method for controlling atmospheric oxygen corrosion. However, due to exposure to high temperatures and aggressive conditions, organic coatings tend to degrade over time, may become brittle, and crack, bubble, and peel (Fig. 27.23). Consequently, repair or replacement of the coating is required.

The type, extent, and severity of corrosion damage within continuous caster spray chambers may vary significantly between different casters. The design, operating conditions, water chemistry and treatment, and mold powder composition are some of the major factors that influence corrosion in a spray chamber. Corrosion damage may vary within a single caster due to different conditions within distinct zones. In casters that produce strands with small cross sections such as billets, it is common practice to feed suitable oil lubricants between the solidifying steel and the mold to prevent sticking and casting defects. For strands with larger cross sections, such as slabs and large blooms, oils do not provide proper lubrication. Therefore, a solid powdery or granular material called mold powder is fed between the steel and the mold. The mold powder melts when heated by the steel to produce a thick, glassy fluid that provides proper lubrication and heat transfer between the mold and the strand. In certain casters that use mold powders, very corrosive acidic water may be produced beneath the mold. The contribution of the mold powder to corrosion of spray chamber components depends upon the composition of the mold powder, the water chemistry, and water volume.

Perhaps the most significant contributor to acid formation is the use of fluoride-containing mold powder. When fluoride-containing mold powder is melted during casting, gaseous and solid fluorine compounds are produced which dissolve into cascading spray water to produce hydrofluoric acid. This acidic water may produce very deep metal loss due to acid corrosion that penetrates at high corrosion rates within localized areas only where it flows or drips (Fig. 27.24). In certain casters that use aggressive mold powders and have low-alkalinity spray water, substantial acid corrosion may occur on components located directly beneath the mold. The amount of corrosion may vary significantly, even within a few inches. In some casters, active corrosion occurs in isolated locations along

FIGURE 27.24 Localized, very deep metal loss due to acid corrosion where acidic water flowed.

the water drippage path where the water is acidic. It is recommended practice to ensure that the spray water has a minimum alkalinity of about 150 ppm. This will help to neutralize acids that may be produced by mold powder by-products. In certain casters that use low-alkalinity spray water, it may be necessary to add a suitable alkalinity supplement. In some casters, stainless steel shields are installed to deflect aggressive waters away from key structural components.

It is also common to find thick, stratified layers of calcium fluoride–rich deposits on portions of structures beneath the mold (Fig. 27.25). These layers form due

FIGURE 27.25 Thick, stratified layers of calcium fluoride–rich deposits on portions of structures beneath the mold.

to the reaction of dissolved calcium in the water with fluoride ions provided by mold powder by-products. Such layers tend to protect components that they cover by producing a watertight seal. In some cases, however, calcium fluoride deposits become excessively thick, and must be jackhammered in order to remove segments and other components.

Out of necessity, certain key components are made of, or are covered with, materials that are resistant to corrosion in continuous caster spray chamber environments. For instance, spray water supply piping is commonly made of 304 or 316 grade austenitic stainless steels. These stainless steels are essentially immune to atmospheric oxygen corrosion that would reduce piping life. More importantly, the use of stainless steel piping prevents oxygen corrosion and the resultant iron oxide corrosion product formation on the internal surface. This ensures that corrosion products will not form on the internal surfaces of the piping that can spall to block flow through narrow spray nozzle orifices downstream. Blocked spray nozzle orifices will cause uneven cooling of the strand, which can result in the production of casting defects.

Control of Spray Chamber Corrosion

Testing can be done to monitor corrosion rates over time with suitable corrosion coupons. Monitoring should be conducted at locations of concern for maintenance personnel. Measurements of beam and component thickness may be made at common locations during inspections to determine corrosion rates and make remaining life measurements. If it is determined that corrosion rates are excessive, and that acid or chloride was responsible, then appropriate corrosion control measures should be instituted, which may include changes in cooling water chemistry or chemical treatment. Coupon test results will monitor the effects of the water chemistry and chemical treatment program and alert for the need to make adjustments. If chemical treatment and water chemistry changes cannot be made, or are ineffective, then alternate corrosion control methods may be needed, such as the use of alternate materials or coatings, shields, and wash sprays. In some casters the application of organic coatings such as epoxy have been considered as a replacement for a failed original and/or existing coating that was made of an inferior material. In order to choose a proper coating material, it may be desirable to conduct an exposure test using coupons coated with various candidate coating materials. For such a test, the coated coupons should be left in place for at least six months to ensure proper evaluation.

Case History 27.4

Industry:	Steel
Specimen:	Blast furnace hot blast gate valve
Orientation:	Vertical
Environment:	Internal: Open recirculating cooling water External: Refractory
Service time:	6 years
Material:	Carbon steel

The cooling jackets inside the gate inlet slot of the hot blast valve body cracked through in separate locations (Fig. 27.26). Metallographic examination conducted on a section of a jacket removed from one of the crack locations revealed that failure was primarily caused by corrosion fatigue cracking, resulting from the application of cyclic tensile stresses. The primary direction of crack growth was transverse, indicating that the cyclic stresses were applied by longitudinal

FIGURE 27.26 Cracks in cooling jackets inside the gate inlet slot of a hot blast valve. Bottom photo is a close-up of circled area within the slot.

expansion and contraction. The most probable source of stress was temperature variations in service.

Oxygen corrosion produced iron oxide layers on the surfaces inside the cooling water channel (Fig. 27.27). The loose chips of iron oxide that were present inside the cooling water channels may have spalled from surfaces inside the cooling water system for the valve, or they may have been transported from piping and other sources upstream. Though the oxide and deposit layers and loose chips covered the surfaces of the channel, no evidence of thermal oxidation or overheating was found on the channel. However, thick layers or pluggage of cooling water channels will cause metal temperature changes to increase. This will cause the metal to expand and contract more substantially to increase the applied, cyclic stresses, which will promote corrosion fatigue cracking.

FIGURE 27.27 Iron oxide layers on the internal surface of the cracked cooling water channel.

Oxygen corrosion may be controlled most effectively through the use of adequate concentrations of proper corrosion inhibitors. However, inhibitors are much less effective when flow rates are low or stagnant during extended idle periods and when substantial amounts of deposits and corrosion products cover surfaces. During maintenance operations, surfaces should be inspected to determine if thick accumulations of corrosion products and deposits cover surfaces. This is particularly important after long-term service of the valves. If cooling water channel surfaces are covered with thick accumulations, suitable cleaning processes should be employed for their removal.

Nondestructive testing techniques such as ultrasonic testing may also be done during maintenance procedures. Such testing would detect excessively deep cracks and monitor for the progression of crack growth over time. This would prevent reinstallation of a valve containing cracks and components which are deeply cracked and close to failure.

Case History 27.5

Industry:	Steel
Specimen location:	Continuous caster
Specimen:	Rolls
Environment:	Hot cast steel strand, spray cooling water, steam, condensate, and mold powder by-products
Sample specifications:	Clad carbon or low-alloy steel

During continuous casting of steel, the strand is conveyed down the caster by a series of rolls that are affixed to water-cooled bearings. The "footrolls" that are installed at the outlet of the mold are commonly made of carbon or low-alloy steel, and are replaced after short intervals, typically associated with mold maintenance. Drive rolls located further down the caster typically remain in service for much longer periods of time. These rolls are commonly clad with wear- and corrosion-resistant martensitic or precipitation-hardened stainless steel alloys that extend roll service life considerably. Such stainless steel cladding is much more resistant to atmospheric oxygen corrosion that may occur between casts when temperatures are low enough to support liquid water. Stainless steels are also much more resistant to high-temperature thermal oxidation than carbon or low-alloy steels. Thermal oxidation occurs when the metal temperature exceeds a certain threshold, causing the exposed surface metal to transform to metal oxide to a significant degree. Stainless steel cladding used on caster rolls tends to resist thermal oxidation up to about 1450°F (788°C), while carbon steel will thermally oxidize appreciably above about 1025°F (552°C). In addition, stainless steel cladding is susceptible to sensitization when metal temperatures are high for a long period of time. For more information about sensitization, see Chap. 26, "Weld Defects." Sensitization further reduces resistance of the cladding due to both aqueous corrosion and high-temperature thermal oxidation.

Surfaces of continuous caster rolls are also susceptible to thermal fatigue and corrosion fatigue cracking due to substantial and rapid temperature variations. The roll surface temperature increases during contact with the hot strand, causing the metal to expand. When the roll turns and the strand no longer contacts it, the surface metal cools, causing it to contract. Cooling water from the spray cooling system will enhance the cooling rate, which will increase the amount of thermal contraction of the metal at the surface (Fig. 27.28). The rapid, intense cyclic heating and cooling, which causes significant, rapid expansion and contraction, imparts substantial thermal stresses on the surface metal. The stresses are applied in three dimensions, causing cracks to form in the transverse and longitudinal directions as well as at various irregular angles on the surface. Cracks also penetrate the subsurface, travelling laterally (Fig. 27.29). Where such subsurface cracks meet, pieces of surface metal break away, producing rough surface contours. Under severe conditions, cracks may penetrate completely through a roll (Fig. 27.30). Such severe cracking is more likely when full-length rolls are used. Full-length rolls will be more heavily stressed than multiple, split rolls that are separated by bearings.

Attempts are made to reduce thermal oxidation and thermal fatigue cracking on rolls by reducing metal temperatures. Thermal fatigue cracking may be reduced

FIGURE 27.28 Cooling water sprays increased the amount of thermal contraction of the hot roll surface.

FIGURE 27.29 Metallographic cross section of thermal fatigue cracks on clad continuous caster roll surface. Cracks formed in the transverse and longitudinal directions as well as at various, irregular angles on the surface. Cracks also penetrated the subsurface, travelling laterally.

FIGURE 27.30 Crack which penetrated completely through a roll initiated at the surface.

by minimizing temperature variations at the surface. Roll cooling may be provided by cooling water that flows through internal cooling water channels. Such cooling water channels have been produced with many different designs. External cooling is supplied to some degree by the spray system. In some casters, separate water sprays are installed that are intentionally directed onto the rolls to provide cooling. However, as discussed earlier, rapid, severe external cooling provided by any water source water may cause thermal fatigue cracking.

FIGURE 27.31 Thermal fatigue cracks on working surfaces of a cemented carbide roll that was used for hot rolling of steel components.

Similar thermal fatigue cracking may occur on surfaces of rolls used for hot rolling of steel components (Fig. 27.31). Rod rolls may be made of hard and brittle materials such as cemented carbides. Hard cemented carbides are very wear resistant, but they are susceptible to thermal fatigue cracking due to their brittle nature.

Case History 27.6

Industry:	Steel
Specimen location:	Continuous caster
Specimen:	Magnetic stirring system for mold
Environment:	Low-conductivity cooling water
Sample specifications:	Mild steel cooling water supply piping

Magnetic stirrers are devices that are used to create eddy currents to stir molten metal within the mold of a continuous caster during the casting process to improve steel quality. Cooling water quality requirements for magnetic stirrers are the most stringent of any steel system. Distilled water is used exclusively, since the system must operate with strict, low conductivity limits. The use of chemical treatment is also restricted by conductivity limits. This makes for a difficult situation since distilled water is highly corrosive to mild steel system components. Corrosion products add conductivity to the water, which necessitates frequent water replacement, typically at three- to six-month intervals. This, in turn, creates more corrosion. The best way to control corrosion under these conditions is to start with a clean system that is free of corrosion products and to use corrosion inhibitors and biocides that do not increase the conductivity of the cooling water. Water replacement frequency is reduced. However, the water may collect contaminants, usually during mold maintenance. Automatic backwashing filters may be used to remove contaminants and corrosion products. Magnetic stirring systems are susceptible to bacterial growth, so a good non-oxidizing biocide program is necessary that would not increase conductivity to excessive levels. Oxidizing biocides cannot be used due to conductivity restrictions and because they will tend to degrade the corrosion inhibitors.

References

1. More information about once-through and recirculating cooling water systems is available in *The Nalco Water Handbook*, 3d ed., ed. D. J. Flynn, McGraw-Hill, New York, 2009.
2. More information about chemical treatment of cooling systems is available in Flynn, *The Nalco Water Handbook*, 3d ed.
3. A. L. Lewis and K. U. Raju, "Evaluation of Two Electrostatic and Magnetic Antiscaling and Anticorrosion Devices," National Association of Corrosion Engineers, Corrosion '97, Paper #443, Saudi Arabian Oil Company, 1997.

Glossary

Acid A chemical species having the tendency to donate a proton (H^+). pH values of 0 or greater, but less than 7, are considered to be acidic.

Active metal A metal that is susceptible to corrosion. Contrast with noble metal.

Adenosine triphosphate (ATP) A biomolecule that is required for cell activity and is a key indicator of the viability of an organism.

Admiralty brass A copper-zinc alloy containing 1 percent tin and usually a small fraction of a percent of one of three inhibitors, including arsenic, antimony, and phosphorus, to control dezincification. Typical applications for this material are surface condenser, evaporator, and heat exchanger tubing, condenser tubing plates, and ferrules.

Aerobic Having air or uncombined oxygen.

Algae A group of aquatic organisms containing chlorophyll.

Alkaline Synonym for basic. A base is a substance that tends to accept a proton (H^+). pH values above 7 and up to 14 are considered to be alkaline (basic).

Alloy A substance having metallic properties and being composed of two or more chemical elements, of which at least one is a metal.

Alloy steel Steel with additions of specified quantities of alloying elements (typically other than carbon) that produce desired changes in mechanical or physical properties.

Alloy system A family of alloys having in common a single, compositionally predominant metal.

Aluminum brass A copper-zinc alloy containing 2 wt% aluminum and 0.02 to 0.06 wt% arsenic for inhibition of dezincification. Typical applications for this material are condenser, evaporator, and heat exchanger tubing, and ferrules. It is often specified when erosion resistance exceeding that of normal brasses is required.

Aluminum bronze A family of copper alloys with aluminum as the principal alloying element. Iron, nickel, silicon, and various other elements are also used in some aluminum bronzes. Most aluminum bronzes contain at least 5 wt% aluminum. Compared to other copper-based alloys, aluminum bronzes have good resistance to corrosion in seawater, as well as to erosion-corrosion and cavitation.

Amphoteric Capable of reacting chemically either as an acid or a base. In reference to certain metals, signifies their propensity to corrode at both high and low pH.

Anaerobic Without air or uncombined oxygen.

Anneal To heat treat such that the microstructure and, consequently, the properties of a material are altered. Often conducted in order to relieve residual stresses that result from welding, or to soften a metal that has been cold worked.

Annealing A generic term denoting a treatment consisting of heating to and holding at a suitable temperature, followed by cooling at a suitable rate, used primarily to soften metallic materials, but also to produce desired changes in other properties or in microstructure simultaneously. When the term is used by itself, full annealing is implied. When the treatment is applied only for the relief of stress, it is properly called stress relieving or stress-relief annealing.

Anode In a corrosion cell, the area over which corrosion occurs and metal ions enter solution; oxidation is the principal reaction.

Anodizing Forming a conversion coating on a metal surface by anodic oxidation; most frequently applied to aluminum.

Applied stress Stress that results from the application of an external load, e.g., due to internal pressurization, thermal expansion and contraction, or vibrations.

ATP See *Adenosine triphosphate (ATP).*

Austenite A face-centered cubic solid solution of carbon or other elements in gamma iron.

Austenitic stainless steel A typically nonmagnetic stainless steel that has alloying additions of chromium and nickel, in proportions that stabilize the face-centered cubic austenite phase.

Autocatalytic corrosion A corrosion process that is self-sustaining.

Backing In welding, a material placed under or behind a joint to enhance the quality of the weld at the root. The backing may be a metal ring or strip, a pass of weld metal, or a nonmetal such as carbon, granular flux, or a protective gas.

Baffle A component that is used to direct the flow of fluids. It may also be used to provide tube support in shell-and-tube heat exchangers.

Base metal (1) In welding, the metal to be welded. (2) After welding, that part of the metal that was not melted.

Biomass A collection of organisms combined with deposited material that includes microbial growth.

Boroscope An optical device (e.g., an optical fiber) used to inspect an otherwise inaccessible space, such as the internal surface of a heat exchanger tube.

Brass An alloy consisting primarily of copper (greater than 50 wt%), with varying proportions of zinc. Smaller amounts of other elements may also be added.

Brittle fracture Material separation accompanied by little or no macroscopic plastic deformation.

Bronze An alloy consisting primarily of copper, with varying combinations and proportions of elements including tin, zinc, manganese, silicon, nickel, aluminum, or iron.

Butt weld A weld that attaches two mating components of similar geometry at their ends. For a cylindrical tube or pipe it may be called a circumferential weld.

Cast iron An iron-based alloy with greater carbon content than steels. Most commercial cast iron alloys contain 3.0 to 4.5 wt% carbon and 1 to 3 wt% silicon.

Cathode In a corrosion cell, the area over which reduction is the principal reaction. It is usually an area that is not attacked.

Cathodic protection Partial or complete protection of a metal from corrosion by making it a cathode, using either a galvanic or an impressed current method. In the galvanic method, a sacrificial anode is employed; the anodic material corrodes sacrificially to preserve the component that is to be protected.

Caustic A hydroxide of a light metal, such as sodium hydroxide or potassium hydroxide.

Cavitation The formation and instantaneous collapse of vapor or gas bubbles within a liquid subjected to rapid and intense pressure changes.

Cavitation damage The degradation of a solid body resulting from its exposure to cavitation. This may include loss of material, surface deformation, or changes in properties or appearance.

Cementite A compound of iron and carbon, known chemically as iron carbide and having the chemical formula Fe_3C.

Circumferential weld See *Butt weld.*

Cladding A layer of metal of different composition than the substrate metal that it covers. This layer is typically used to control corrosion and/or wear. For tubes it is typically applied by welding or coextrusion.

Closed recirculating A cooling system in which water is recirculated in a closed loop. Such systems experience little, if any, exposure to the atmosphere.

Cold flow Deformation of a material resulting from stresses applied at ambient temperatures.

Cold working Plastic deformation of a metal below its recrystallization temperature.

Composite tubing Tubing comprised of an outer tube metallurgically bonded to an inner tube, usually formed by coextrusion. Also referred to as *coextruded or clad tubing*.

Concentration cell An electrolytic cell, the electromotive force of which is caused by a difference in concentration of some component in the electrolyte. This difference leads to the formation of discrete cathode and anode regions.

Coring Compositional inhomogeneities (differences) throughout a metal resulting from solidification of the molten metal.

Corrosion The chemical or electrochemical reaction between a material, and its environment that produces a deterioration of the material and its properties.

Corrosion fatigue The process in which a metal fractures prematurely under conditions of simultaneous corrosion and repeated cyclic loading at lower stress levels or fewer cycles than would be required in the absence of the corrosive environment.

Corrosion inhibitor Chemical that is added to cooling water to control corrosion.

Corrosion product A substance that results from a corrosion process; a solid material based on the oxidized metal or metals that have been corroded.

Corrosivity The tendency of an environment to cause corrosion in a given system.

Crack A thick-walled, brittle fracture that occurs without appreciable plastic deformation.

Crevice corrosion A type of concentration cell corrosion; corrosion caused by the concentration of dissolved salts and metal ions, or depletion of oxygen in crevices or pockets remote from the principal fluid stream, with a resultant buildup of differential cells that ultimately cause deep pitting.

Crystallographic defects Imperfections in the crystal lattice of a material that increase its stored energy.

Cupronickel Alloys used for surface condenser and heat exchanger tubes that contain copper concentrations ranging from 70 to 90 percent, with the balance consisting of nickel. Typical compositions are 90 wt%

copper/10 wt% nickel (90–10) and 70 wt% copper/30 wt% nickel (70–30).

Cyclic stress A stress whose magnitude fluctuates. Contrast with static stress.

Dealloying (See also *Selective leaching*.) The selective corrosion of one or more components of a solid solution alloy. Also called *Parting* or *Selective leaching*.

Defect An imperfection in a material that contributes significantly to failure or limited serviceability. Contrast with flaw.

Deformation twins A special type of grain boundary produced by mirror-image positions of the atoms in the crystal lattice on both sides of the boundary. Occurs in some metals due to applied mechanical shear forces.

Dendritic microstructure A crystalline structure that has a pattern similar to tree branches that forms during solidification. The branches are called dendrites. The term *interdendritic* refers to locations between the dendrites. Dendritic microstructures can be found in cast components and weld beads.

Denickelification Type of dealloying in which nickel is selectively removed from a nickel-containing alloy, most commonly copper-nickel alloys (cupronickels).

Depassivating agent A substance, usually an ion such as chloride, whose reaction with a metal surface destroys its passive character.

Depassivation (activation) The changing of a metal surface from a chemically nonreactive (passive) condition to a reactive condition.

Depression Term used for metal loss that occurs within a small, localized area where the depth of metal loss is smaller than the width of the area exposed. (Compare to *Pit*.)

Dezincification Type of dealloying in which zinc is preferentially removed from a zinc-containing alloy, most commonly copper-zinc alloys containing less than 85 percent copper.

Diffusion Transport of mass that occurs by atomic motion.

Dissolved solids Compounds (generally inorganic salts) present in solution in water, often including ions of calcium, magnesium, silicon, and others that are typically dissolved in natural waters, as well as metal ions and inorganic treatment chemicals.

Ductile fracture Separation characterized by tearing of metal accompanied by appreciable plastic deformation and expenditure of considerable energy.

Ductility The ability of a material to deform plastically without fracturing.

Duplex alloy An alloy that consists of two distinct phases (see *Phase*).

Dye penetrant A type of nondestructive testing that utilizes a dye to detect defects, such as tight cracks that are not observable by unaided visual examination alone.

Eddy current A type of nondestructive testing that utilizes electromagnetic inductance to detect flaws (e.g., pits and cracks) in an electronically conductive material. May also be used to measure material thickness and distinguish between different metals.

EDS See *Energy dispersive x-ray spectroscopy.*

Electrolyte A nonmetallic electrical conductor in which current is carried by the motion of ions. For corrosion, the electrolyte is generally an aqueous solution.

Element mapping An analytical technique, typically conducted using a scanning electron microscope with an energy dispersive x-ray spectrometer, that shows the distribution and relative amounts of elements across an examined area.

Energy dispersive x-ray spectroscopy (EDS) Analysis of characteristic x-rays emitted from a material due to its interaction with an electron beam. EDS produces information about the concentrations of specific elements in the material.

EPS See *Extracellular polymeric substances (EPS).*

Erosion Destruction of metals or other materials by the abrasive action of moving fluids. Erosion may be accelerated by the presence of solid particles or matter in suspension. When corrosion occurs simultaneously, the term *erosion-corrosion* is often used.

Eutectic A reversible reaction in which a liquid solution is converted into two or more distinct solid phases.

Extrados The exterior curve of a bend or arch.

Extracellular polymeric substances (EPS) Secretions that form a fibrous, slimy matrix.

Failure A general term used to imply that a part in service (1) has become completely inoperable, (2) is still operable but is incapable of satisfactorily performing its intended function, or (3) has deteriorated seriously, to the point that it has become unreliable or unsafe for continued use.

Fatigue Damage caused by cyclic or fluctuating tensile stresses having a maximum value less than the tensile strength of the material. Results in thick-walled, brittle-type fractures.

Fatigue limit Stress below which failure by fatigue apparently will not occur for any number of cycles.

Ferrite A body-centered cubic solid solution of one or more elements in alpha iron.

Ferrule Insert that is used on the tubeside at the inlet end of a tube used in a shell-and-tube heat exchanger.

Ferrous hydroxide A corrosion product of iron, $Fe(OH)_2$.

Fins Thin sheets of metal that are attached perpendicular to the external surface of a tube, in parallel and in close proximity, to provide greater surface area for heat transfer.

Fissure A cracklike penetration that penetrates partially through a component.

Flash rust Iron oxide, typically orange, brown, or red hematite, that forms on carbon, alloy steel, and cast iron surfaces due to exposure to residual water that is oxygenated by air during idle periods.

Flaw As used in this book, an imperfection in a material that does not affect its usefulness or serviceability. Contrast with *Defect*.

Flow-induced vibration Tube vibration resulting from the mechanical effects of fluids impinging on tube surfaces. Such vibrations may induce wear or cracking, especially if the tubes vibrate at their natural frequency.

Flux (1) A substance that is capable of dissolving and removing metal oxides and other undesirable materials and can prevent oxidation. (2) The rate of transfer of fluid, particles, or energy across a given surface.

Forging A process in which metal is heated and shaped by plastic deformation by the application of compressive force.

Fretting A type of wear that occurs between tight-fitting surfaces subjected to cyclic, relative motion of extremely small amplitude. Usually, fretting is accelerated by spalled corrosion products.

Galvanic Associated with the terms *galvanic cell*, *galvanic corrosion*, and *galvanic series*. Galvanic corrosion occurs when a galvanic cell is produced by dissimilar conductors, which may consist of dissimilar metals, or similar metals exposed to dissimilar electrolytes. The galvanic series is a chart that lists dissimilar metals. Contacting metals that are widely separated on the list are susceptible to producing galvanic corrosion.

Galvanic cell A cell in which chemical change is the source of electrical energy. It usually consists of two dissimilar conductors in contact with each other and with an electrolyte or of two similar conductors in contact with each other and with dissimilar electrolytes.

Galvanic corrosion Corrosion associated with the current of a galvanic cell consisting of two dissimilar conductors in an electrolyte or two similar conductors in dissimilar electrolytes.

Galvanic potential The magnitude of the driving force in an electrochemical reaction resulting from the coupling of dissimilar materials exposed to a common, corrosive environment.

Galvanic series A series of metals and alloys arranged according to their relative electrode potentials in a specified environment.

Gas porosity Fine holes or pores within a metal that are caused by entrapped gas or by the evolution of dissolved gas during solidification.

Grain An individual crystal in a polycrystalline metal or alloy.

Grain boundary An interface between grains that corresponds to the transition from one crystallographic orientation to another.

Graphitic corrosion Corrosion of gray or nodular cast iron in which the iron matrix is selectively corroded, leaving a porous mass of graphite and iron oxide behind; it occurs in relatively mild aqueous solutions and on buried pipe.

Hardfacing Application of a harder material to a base metal for the purpose of wear resistance or to replace metal lost during service.

Hardness (material) Resistance of a material to deformation, as measured by surface indentation or abrasion.

Hardness (water) Concentration of multivalent cations in water, most commonly calcium and magnesium.

HAZ See *Heat-affected zone (HAZ).*

Header A reservoir, usually a larger diameter pipe or tube, that collects or supplies fluid to or from several attached parallel tubes.

Heat-affected zone (HAZ) In welding, that portion of the base metal that was not melted during welding, but whose microstructure and mechanical properties were altered by the heat.

Heat flux (See also *Flux.*) Heat transfer rate per unit area.

Heat treatment Controlled heating and cooling of metal to change its physical and mechanical properties while keeping its shape constant.

Hematite A form of iron oxide, Fe_2O_3, that is gray to bright red in color. The presence of significant quantities of hematite in waterside corrosion products is often associated with oxygen corrosion.

Holidays Discontinuities in a coating (such as porosity, cracks, gaps, and similar flaws) that allow areas of base metal to be exposed to any corrosive environment that contacts the coated surface.

Hot working Metal forming, such as rolling or bending, that is conducted above a metal's recrystallization temperature.

Hoop stress The circumferentially oriented stress that develops in a cylindrical component, such as a tube, either due to internal or external pressurization or from residual stresses that often result from fabrication.

Hydrated ferric oxide (ferric hydroxide) A flaky, red to brown corrosion product of iron or steel that forms upon exposure to subterranean, atmospheric, or aqueous environments, $Fe_2O_3 \cdot 3H_2O$.

Hydro-blasting Technique for cleaning that uses a high-velocity stream of pressurized water to remove material from surfaces.

Hydrolysis A double decomposition reaction in which a compound chemically reacts with water.

Impingement To strike, e.g., by a rapidly flowing fluid stream upon a solid surface.

Inclusions Particles of foreign material in a metallic matrix. The particles are usually compounds (such as oxides, sulfides, or silicates), but they may be any substance that is foreign to (and essentially insoluble in) the matrix.

Infrared spectroscopy Chemical analysis method that utilizes the absorption of infrared radiation by interatomic bonds in organic compounds. Used for identification of organic molecules.

Inhibited brass Brass that contains a small concentration of arsenic, phosphorus, or antimony in order to control dealloying.

Interdendritic See *Dendritic microstructure.*

Intergranular Occurring between crystals or grains. Also, *intercrystalline.*

Intergranular corrosion Corrosion occurring preferentially along grain boundaries, usually with slight or negligible attack on the adjacent grains.

Intrados The interior curve of a bend or arch.

Laminar flow Streamlined, smooth flow of a fluid, as opposed to turbulent flow.

Lap A surface imperfection, with the appearance of a seam, caused by hot metal, fins, or sharp corners being folded over then being rolled or forged into the surface but without being welded.

Lattice The regular geometric arrangement of atoms in a crystalline solid, such as a metal.

Layup A practice followed during idle periods to control or prevent corrosion damage that may be caused by exposure to atmospheric conditions.

Low-alloy steel Steel that contains relatively low alloy additions (typically between 2 and 10 percent total alloy content).

Magnetic particle inspection Nondestructive testing method that utilizes magnetic fields and small magnetic particles, such as iron filings, to detect defects in a ferromagnetic component.

Magnetite A magnetic form of iron oxide, Fe_3O_4. Magnetite is dark gray to black, and forms a protective layer on iron surfaces.

Mandrel A tool that is used on the tubeside during tube manufacturing to maintain proper shape.

Martensite For steel alloys, a supersaturated solid solution of carbon in iron characterized by a microstructure having a needlelike morphology. Produced by quenching (rapid cooling) of austenite, preventing diffusion from occurring. Very hard and brittle.

Matrix The principal phase in which constituents are distributed.

Microbiologically influenced corrosion (MIC) Deterioration of metals as a result of the metabolic activities of microorganisms.

Microstructure The appearance and distribution of grains and phases in a metal determined by microscopic examination of a polished specimen. Etching is often used to reveal microstructural details of the metal.

Mild steel Carbon steel having a maximum carbon content of approximately 0.25 wt% that also contains 0.4 to 0.7 wt% manganese and 0.1 to 0.5 wt% silicon.

Morphology The form and structure of something.

NDT See *Nondestructive testing (NDT).*

Neumann bands Deformation twins specifically formed by shock loading in metals with a body-centered cubic crystal structure, such as iron.

Noble metal A metal with marked resistance to chemical reaction, particularly to oxidation and to solution by inorganic acids. Contrast with *Active metal.*

Nondestructive testing (NDT) Inspection methods that can detect defects, flaws, pits, cracks, and metal loss without significant alteration of the component to be inspected.

Once-through Cooling system in which water is received from the plant supply, passed through the process equipment, and returned to a receiving body of water.

Open recirculating A cooling system in which water is taken from a cooling tower or evaporation pond, passes through the process equipment, then returns to the evaporation unit. This system requires fresh makeup water to replenish that lost by evaporation and blowdown.

ORP See *Oxidation/reduction potential (ORP).*

Oxidation/reduction potential (ORP) A value representing the net sum of contributions from reducing agents and oxidizers in a liquid to the overall solution potential. The value depends on temperature and the reference electrode used for measurement.

Oxidizing Condition that allows for the reaction where elements or ions lose electrons, resulting in an increase in positive valence of the species.

Pad weld A weld that covers an area on the external surface of a tube to replace lost wall thickness or to provide erosion resistance.

Parting Synonym for dealloying.

Passivation The changing of a chemically active surface of a metal to a much less reactive state.

Passivity A condition in which a metal, because of an impervious covering of oxide or other compound, has a potential much more positive than that of the metal in its active state.

PCR See *Polymerase chain reaction (PCR).*

Pearlite A microstructural aggregate in steel and cast iron consisting of alternate lamellae (layers) of ferrite and cementite.

Penetration In welding, the distance from the original surface of the base metal to that point at which fusion ceased.

pH The negative logarithm of the hydrogen ion activity; it denotes the degree of acidity or basicity of a solution. At 25°C (77°F), 7.0 is the neutral value. Decreasing values below 7.0 indicate increasing acidity; increasing values above 7.0 indicate increasing basicity.

Phase A physically and chemically homogeneous and distinct portion of a material system.

Phase transformation Reaction that occurs, typically on heating or cooling, that results in a change in phase or phases.

Pit Term used for metal loss that occurs within a small, localized area where the depth of metal loss is greater than the width of the exposed area. (Compare to *Depression.*)

Pitting Forming small, sharp cavities in a metal surface by corrosion.

Plain carbon steel Steel containing carbon up to about 2 wt% and only residual quantities of other elements except those added for deoxidation.

Plankton Minute animal and plant life found in water.

Planktonic bacteria Bacteria that are present in water.

Plastic deformation Deformation that is permanent after the release of an applied load due to crystallographic displacements that are not recovered.

Plate-and-frame heat exchanger An exchanger that consists of a stack of thin plates supported in a frame. The plates are typically corrugated. The two fluids flow along opposite sides of each plate.

Polymerase chain reaction (PCR) Method for targeting sequences of nucleic acid (DNA or RNA) and increasing the copy number of the target sequence to obtain useful quantities of nucleic acid. A genetic technique for identifying species of live or dead microorganisms.

Quench In metallurgical terms, to rapidly cool the metal from elevated temperatures.

Radiographic inspection A nondestructive testing technique that uses penetrating radiation (x-rays, gamma rays, etc.) to record differences in material density, thickness, and composition on film or other forms of radiation detectors.

Ratchet marks Steps on a fatigue or corrosion fatigue crack face caused by multiple cracks that originate from slightly different planes. When the cracks eventually join together, steps form.

Recrystallization temperature Temperature above which a new set of strain-free grains will form from a previously cold-worked material in approximately an hour.

Reducing Condition that allows for the reaction where elements or ions gain electrons, so that their valence is reduced.

Residual stress Stress that remains in a material even when it is free from external forces or thermal gradients. Residual stresses often result from welding, casting, cold forming, and other manufacturing processes.

Rifling Shallow, rectangular grooves that are formed in a spiraling pattern on the internal surface of an enhanced tube used in a heat exchanger, particularly chillers. The grooves are designed to promote swirling flow and increase heat transfer.

Rolling A term used to describe mechanical deformation of tube ends into holes in tubesheets to impart a tight, leak-resistant seal.

Root crack A crack either in a weld or in the heat-affected zone at the root of a weld.

Root of joint In welding, the portion of a weld joint where the base metal components are closest to each other before welding.

Root of weld The points at which the weld metal intersects the base metal surfaces either nearest to or coincident with the root of the joint.

Rust A corrosion product consisting of hydrated oxides of iron.

Sacrificial anode See *Cathodic protection.*

Scanning electron microscopy (SEM) Analytical technique using a microscope that utilizes an electron beam to produce a variety of signals at the surface of a solid material. May generate visual images as well as information about chemical composition and crystal structure. See also *Energy dispersive x-ray spectroscopy.*

Seam weld The location where a sheet or plate of metal is formed into a tube shape and is joined at the longitudinally oriented mating ends by a weld.

Seam welding Making a longitudinal weld in sheet metal or tubing.

SEM See *Scanning electron microscopy (SEM).*

Selective leaching Corrosion in which one element is preferentially removed from an alloy, leaving a residue (often porous) of the elements that are more resistant to the particular environment.

Sensitization In austenitic stainless steels, precipitation of chromium-rich carbides at grain boundaries within the approximate metal temperature range of 900 to 1560°F (480° to 850°C). Metal adjacent to the grain boundaries has a decreased corrosion resistance due to localized chromium depletion. May be prevented through the use of low-carbon grades.

Sessile Permanently attached; not free to move.

Sessile bacteria Bacteria that are attached to surfaces.

Shell and tube A heat exchanger that consists of a series of parallel tubes that are fastened into round holes drilled through disk-shaped tubesheets. The assembly is encased in a vessel that separates the tubeside and shellside environments.

Shellside The external surface of a tube or pipe in a shell-and-tube unit.

Shielding Barriers that are typically used to limit impingement erosion or corrosion.

Siderite Ferrous carbonate, $FeCO_3$.

Siphonic gas exsolution The formation of gas bubbles in a fluid caused by decreasing fluid pressure associated with flow.

Spalling The breaking off of chips of material from a surface.

Spheroidal carbides Iron or alloy carbides that are essentially spherical in shape.

Spray coating A layer of a hard material that is applied to a surface to control erosion using a thermal or plasma spraying process.

Stagnant The condition of being motionless.

Stainless steel Ferrous (iron-based) alloys that contain a minimum of 10.5 wt% chromium. The addition of a sufficient amount of chromium results in the formation of a chromium-based oxide layer that provides for improved corrosion resistance in many environments compared to plain carbon or low-alloy steels.

Static stress A stress whose magnitude remains at a constant value. Contrast with *Cyclic stress.*

Stress Generally, load (measured in units of force) divided by the cross-sectional area. Common units are pounds per square inch (psi) or pascals (Pa). See also *Applied stress* and *Residual stress.*

Stress concentrator or stress raiser A location or feature where applied tensile stress will be amplified. Usually on the surface (e.g., pits, depressions, fissures, weld toes, sharp angles on machined surfaces) or internal (e.g., casting porosity or weld porosity).

Stress corrosion cracking Failure by cracking under the combined action of a specific corrodent stress, either external (applied) stress or internal (residual) stress. Cracking may be either intergranular or transgranular, depending on the metal and the corrosive medium.

Stress frequency The number of times a stress cycle is repeated in a unit of time.

Stress raisers Changes in contour or discontinuities in a structure that cause local increases in stress.

Stress relief Heating treatment to reduce residual stresses arising from processes such as welding, cold forming, and casting.

Striations Small grooves or channels, often parallel to each other.

Stringers Elongated inclusions of impurities or secondary phases in a metal.

Surface condenser A shell-and-tube heat exchanger that is used to condense steam on the shellside surfaces of the tubes. Cooling water flows through the tubes to promote condensation.

Suspended solids Very small particles of solid material that are suspended in the water. These may contain calcium, magnesium, silicon, and other elements that are typically found in natural waters, as well as metals or their corrosion products and possibly components of inorganic treatment chemicals.

Swaging Forming of a metal by forcing it to conform to a die, usually with cold work. Examples include tube flaring or rolling and fabrication of seamless tubing using a mandrel.

Synergism Cooperative action of discrete factors such that the total effect is greater than the sum of the effects taken independently.

TDS See *Total dissolved solids (TDS).*

Tensile strength The maximum load-carrying capability of a material, as determined by a tensile test. It is the stress calculated from the ratio of the maximum load to the original cross-sectional area. Also called *ultimate tensile strength.*

Thermal conductivity A measurement for the ability of a material to conduct heat for a unit distance within a unit of time.

Thermal expansion and contraction Changes in the dimensions of a material due to changes in temperature, since atomic vibrations increase with increasing temperature. Expansion occurs on heating and contraction on cooling.

Thermal expansion coefficient Fractional change in a material's length, divided by the change in temperature.

Thermal fatigue Specific case of corrosion fatigue caused by alternate heating of metal and quenching by liquid water.

Thermal oxidation Conversion of the metal to oxide at elevated temperatures, typically above the oxidation limit.

Thermally formed oxide Layer produced when the metal or alloy is converted to oxide at elevated temperatures.

Total dissolved solids (TDS) A measurement in ppm of the amount of dissolved solids in the water (see *Dissolved solids*). The amount of TDS must be properly controlled.

Transgranular Occurring through or across crystals or grains. Also, *intracrystalline* or *transcrystalline*.

Tube bore (1) A circular opening within a tubesheet or header where a tube is inserted and housed. (2) The cylindrical space within the tubeside.

Tube wall The material between the inside diameter and outside diameter of a tube at any location.

Tubercle Localized, structured iron oxide corrosion products in the form of a knoblike mound.

Tubesheet A thick plate that contains drilled holes that house tubes. Tubes may be rolled, swaged, or welded to maintain a tight mechanical connection and seal within the holes.

Tubeside The internal surface of a tube.

Tunnel pits Elongated areas of localized metal loss due to preferential attack in some areas due to compositional inhomogeneities within the metal. Tunnel pits are often observed along chloride stress corrosion cracks in austenitic stainless steel.

Turbidity A unit of measurement used to describe the relative amount of suspended material in water based on water clarity. A common unit of measurement is the nephelometric turbidity unit (NTU), which is determined by several different methods.

Turbulence In fluids, departure from laminar (smooth) flow.

Twin A particular type of grain boundary in which the lattice of the crystal has mirror symmetry across the boundary. Twins can be thermally or mechanically produced by annealing or shock loading, respectively. Twins often appear as heavy parallel lines in micrographs. See also *Deformation twins* and *Neumann bands*.

Ultrasonic testing A nondestructive test in which an ultrasonic beam is applied to sound-conductive materials having elastic properties. The test is used to determine wall thickness, and it can locate inhomogeneities or structural discontinuities within the material.

Underbead crack A subsurface crack in the base metal near a weld.

Undercut In weldments, a groove melted into the base metal adjacent to the toe of a weld that is left unfilled.

Undercut pit A pit that has significant lateral propagation of the pit beneath the surface. Such pit morphology is often associated with chloride pitting of stainless steel, some kinds of microbiologically influenced corrosion, and attack due to strong acids.

Underdeposit corrosion A type of waterside corrosion that is promoted by the concentration of aggressive agents beneath deposit layers.

Weld Two or more pieces of metal that are joined by heat, pressure, or both, with or without filler metal. The joint is produced by fusion or recrystallization across the interface.

Weld bead A deposit of filler metal from a single welding pass.

Weld metal The portion of a weld that is melted during welding.

Weld overlay A layer produced on a metal surface by the application of multiple weld beads.

Welding current The current flowing through a welding circuit during the making of a weld.

Weldment A component with base metal constituents that are joined by welding.

X-ray examination A nondestructive technique utilizing x-ray radiation to detect flaws and differences in thickness in a component.

Yellow metal A copper-based alloy.

Yield strength The stress required to produce a specified, usually small, amount of plastic strain (elongation).

Bibliography

American Water Works Association Research Foundation, *Internal Corrosion of Water Distribution Systems*, 2d ed., American Water Works Association, Denver, CO, 1996.

ASM Handbook, 10th ed., Volume 11, *Failure Analysis and Prevention*, edited by William T. Becker and Roch J. Shipley, ASM International, Materials Park, OH, 2002.

ASM Handbook, 10th ed., Volume 13B, *Corrosion: Materials*, edited by S. D. Cramer and B. S. Covino, Jr., ASM International, Materials Park, OH, 2005.

ASM Handbook, 9th ed., Volume 17, *Nondestructive Evaluation and Quality Control*, ASM International, Materials Park, OH, 1989.

ASM International, *Characterization and Failure Analysis of Plastics*, ASM International, Materials Park, OH, 2003.

ASM Metals Handbook, 8th ed., Volume 1, *Properties and Selection of Metals, Aluminum Properties and Physical Metallurgy*, American Society for Metals, Metals Park, OH, 1961.

ASM Metals Handbook, 8th ed., Volume 2, *Heat Treating, Cleaning and Finishing*, American Society for Metals, Metals Park, OH, 1964.

ASM Metals Handbook, 9th ed., Volume 13, *Corrosion*, edited by L. J. Korb, ASM International, Materials Park, OH, 1987.

ASTM D2688-11, Standard Test Method for Corrosivity of Water in the Absence of Heat Transfer, ASTM International, West Conshohocken, PA, 2011.

ASTM G1–03 (2011), Standard Practice for Preparing, Cleaning, and Evaluating Corrosion Test Specimens, ASTM International, West Conshohocken, PA, 2011.

ASTM G4–01 (2008), Standard Guide for Conducting Corrosion Tests in Field Applications, ASTM International, West Conshohocken, PA, 2008.

Avery, R. E., S. Lamb, C. A. Powell, and A. H. Tuthill, *Stainless Steel for Potable Water Treatment Plants*, Nickel Development Institute, Technical Literature 10087, undated.

Bachus, L., and A. Custodio, *Centrifugal Pumps and Everything You Need to Know About Pumps*, The Bachus Company Inc., Nashville, TN, 2003.

Boffardi, B. P., "Water Treatments," in *Corrosion Tests and Standards: Application and Interpretation*, edited by R. Baboian, American Society for Testing and Materials, Philadelphia, PA, 1995.

Borenstein, S. W., *Microbiologically Influenced Corrosion Handbook*, Industrial Press Inc., New York, 1994.

Caruso, L., and H. T. Micheals, "The Resistance of Copper-Nickel Alloys to Ammonia," *Corrosion in Simulated Steam Condenser Environments*, Proceedings of the American Power Conference, 1980.

Cohen, Arthur, "Corrosion of Copper and Copper Alloys," in *ASM Handbook*, Volume 13B, *Corrosion: Materials*, ASM International, Materials Park, OH, 2005.

Davies, M., and P. J. B. Scott, *Guide to the Use of Materials in Waters*, NACE International, Houston, TX, 2003.

Davis, J. R. (ed.), *Corrosion of Aluminum and Aluminum Alloys*, ASM International, Materials Park, OH, 1999.

Dexter, S. C., "Microbiologically Influenced Corrosion," in *ASM Handbook*, Volume 13A, *Corrosion: Fundamentals, Testing, and Protection*, ASM International, Metals Park, OH, 2003.

Ezrin, M., *Plastics Failure Guide—Cause and Prevention*, Hanser Publishers, Munich, 1996.

Ferguson, John F., Otto von Franque, and Michael R. Schock, "Corrosion of Copper in Potable Water Systems," in *Internal Corrosion of Water Distribution Systems*, 2d ed., American Water Works Association Research Foundation, Denver, CO, 1996.

Flynn, D. J. (ed.), *The Nalco Water Handbook*, 3d ed., McGraw-Hill, New York, 2009.

Fontana, M. G., and N. D. Greene, *Corrosion Engineering*, 2d ed., McGraw-Hill, New York, 1978.

Francis, R., *Galvanic Corrosion: A Practical Guide for Engineers*, NACE International, Houston, TX, 2001.

Frayne, C., *Cooling Water Treatment Principles and Practice*, Chemical Publishing Co., Inc., New York, 1999.

Hammitt, Frederick G., and Frank J. Heymann, "Liquid-Erosion Failures," in *ASM Handbook*, Volume 11, *Failure Analysis and Prevention*, ASM International, Metals Park, OH, 1986.

Javaherdashti, R., *Microbiologically Influenced Corrosion: An Engineering Insight*, Springer, London, 2008.

Jones, Russel H. (ed.), *Stress-Corrosion Cracking of Stainless Steels*, ASM International, Materials Park, OH, 1992.

Kovach, Curtis H. (ed.), *High Performance Stainless Steels*, Nickel Development Institute, Toronto, 2000.

Larson, T. E., *Corrosion by Domestic Waters*, ISWS-75-BUL59, State of Illinois, Department of Registration and Education.

Lewis, A. L., and K. U. Raju, "Evaluation of Two Electrostatic and Magnetic Antiscaling and Anticorrosion Devices," Corrosion '97, Paper #443, National Association of Corrosion Engineers, Saudi Arabian Oil Company, 1997.

Little, B., P. Wagner, and F. Mansfield, "Microbiologically Influenced Corrosion of Metals and Alloys," *International Materials Review*, 1991.

Little, B. L., and J. S. Lee, *Microbiologically Influenced Corrosion*, John Wiley & Sons, Hoboken, NJ, 2007.

Matsukawa, Y., H. Chuta, M. Miyashita, M. Yoshikawa, Y. Miyata, and S. Asakura, "Galvanic Series of Metals Conventionally Used in Tap Water With and Without Flow and Its Comparison to That in Seawater," *Corrosion*, December 2011.

McCoy, J. W., *The Chemical Treatment of Cooling Water*, 2d ed., Chemical Publishing Company, New York, 1983.

Moalli, J., *Plastics Failure—Analysis and Prevention*, William Andrew Publishing/ Plastics Design Library, Norwich, NY, 2001.

Novak, Charles J., "Structure and Constitution of Wrought Austenitic Stainless Steels," in *Handbook of Stainless Steels*, edited by Donald Peckner and I. M. Bernstein, 1977.

Peckner, D., and I. M. Bernstein, *Handbook of Stainless Steel*, McGraw-Hill, New York, 1977.

Puckorius, P., "Cooling Water System Corrosion Guidelines, Part 1," *Process Cooling*, July 2003.

Revie, R. W. (ed.), *Uhlig's Corrosion Handbook*, 2d ed., John Wiley and Sons, Inc., Hoboken, NJ, 2000.

Revie, R. W. (ed.), *Uhlig's Corrosion Handbook*, 3d ed., John Wiley and Sons, Inc., Hoboken, NJ, 2011.

Sandvik product literature, Sandvik SAF 2507, November 28, 2013.

Scheirs, J., *Compositional and Failure Analysis of Polymers—A Practical Approach*, John Wiley and Sons, New York, 2000.

Schweitzer, P. A. (ed.), *Corrosion Engineering Handbook*, Marcel Dekker, Inc., New York, 1996.

Schweitzer, P. A., *Mechanical and Corrosion-Resistant Properties of Plastics and Elastomers*, CRC Press, New York, 2000.

Sequeira, C. A. C., "Copper and Copper Alloys," in *Uhlig's Corrosion Handbook*, 3d ed., edited by R. Winston Revie, John Wiley and Sons, Inc., Hoboken, NJ, 2011.

Shreir, L. L., R. A. Jarman, and G. T. Burstein (eds.), *Corrosion Metal Environment Reactions*, 3d ed., Butterworth Heinemann, Oxford, 1994.

Slunder, C. J., and W. K. Boyd, *Zinc: Its Corrosion Resistance*, 2d ed., International Lead Zinc Research Organization, New York, 1983.

Thompson, D. H., "Stress Corrosion Cracking of Copper Metals," in *Stress Corrosion Cracking of Metals: A State of the Art*, American Society for Testing and Materials, Philadelphia, PA, 1972, pp. 39–57.

Uhlig, H. H., *The Corrosion Handbook*, 1st ed., John Wiley and Sons, Inc., New York, 1948.

USEPA, *Technologies for Upgrading Existing or Designing New Drinking Water Treatment Facilities*, 1984.

USEPA, *Corrosion in Potable Water Systems*, 1982.

Vargel, Christian, *Corrosion of Aluminum*, Elsevier, New York, 2004.

Warke, W. R., "Stress-Corrosion Cracking," in *ASM Handbook*, Volume 11, *Failure Analysis and Prevention*, ASM International, Materials Park, OH, 2002.

White, Richard A., *Materials Selection for Petroleum Refineries and Gathering Facilities*, NACE International, Houston, TX, 1998.

White Rust: An Industry Update and Guide Paper, Association of Water Technologies (AWT), Rockville, MD, 2002.

Whitman, L. W., R. Russell, and V. Altieri, *Industrial & Engineering Chemistry*, American Chemical Society, 1924.

Zelver, N., W. G. Characklis, and F. L. Roe, CTI Paper No. TP239A, Annual Meeting of the Cooling Tower Institute, Houston, TX, 1981.

Index

A

B

C

L

M

T